THE REGULATION AND POLICY OF LATIN AMERICAN ENERGY TRANSITIONS

T0311935

THE REGULATION AND POLICY OF LATIN AMERICAN ENERGY TRANSITIONS

Edited by

LUCAS NOURA GUIMARÃES
EDP Brasil; Brazilian Institute of Energy Law Studies

ELSEVIER

Elsevier
Radarweg 29, PO Box 211, 1000 AE Amsterdam, Netherlands
The Boulevard, Langford Lane, Kidlington, Oxford OX5 1GB, United Kingdom
50 Hampshire Street, 5th Floor, Cambridge, MA 02139, United States

Notices
Knowledge and best practice in this field are constantly changing. As new research and experience broaden our understanding, changes in research methods, professional practices, or medical treatment may become necessary.

Practitioners and researchers must always rely on their own experience and knowledge in evaluating and using any information, methods, compounds, or experiments described herein. In using such information or methods they should be mindful of their own safety and the safety of others, including parties for whom they have a professional responsibility.

To the fullest extent of the law, neither the Publisher nor the authors, contributors, or editors, assume any liability for any injury and/or damage to persons or property as a matter of products liability, negligence or otherwise, or from any use or operation of any methods, products, instructions, or ideas contained in the material herein.

British Library Cataloguing-in-Publication Data
A catalogue record for this book is available from the British Library

Library of Congress Cataloging-in-Publication Data
A catalog record for this book is available from the Library of Congress

ISBN: 978-0-12-819521-5

For Information on all Elsevier publications
visit our website at https://www.elsevier.com/books-and-journals

Publisher: Brian Romer
Acquisitions Editor: Graham Nisbet
Editorial Project Manager: Michelle Fisher
Production Project Manager: Sruthi Satheesh
Cover Designer: Christian Bilbow

Typeset by MPS Limited, Chennai, India

Dedication

To Carol, for always giving me the energy to write books about energy matters.

Contents

Part 1

CHANGING THE ENERGY LANDSCAPE: POLICY AND ECONOMIC ISSUES

1. Is there a Latin American electricity transition? A snapshot of intraregional differences

LUCAS NOURA GUIMARÃES

2. Using market mechanisms for a reliable energy transition in Latin America

ALEXANDRE VIANA

3. Assessing the macroeconomic effects of sustainable energy transitions in Costa Rica and Chile: a multisectoral balance-of-payments-constrained growth approach

FRANCISCO EBELING

Part 2

ENERGY TRANSITION TAKING SHAPE: REGULATORY ISSUES

List of Contributors

Lourdes Alonso-Serna University of Manchester, Manchester, United Kingdom

Carlo Altamirano-Allende Arizona State University, Tempe, AZ, United States

Matheus Linck Bassani Law, Federal University of Rio Grande do Sul, Porto Alegre, Brazil

Silvina Cecilia Carrizo National Scientific and Technical Research Council (CONICET), Buenos Aires, Argentina; National University of Northwest Buenos Aires (UNNOBA), Buenos Aires, Argentina; Centre of Studies on Territory, Energy and Environment (TEAM), Buenos Aires, Argentina; National University of La Plata (UNLP), La Plata, Argentina; School of Architecture and Urbanism (FAU), La Plata, Argentina; Centre of Urban and Territorial Research (CIUT), La Plata, Argentina

Marco Aurelio Lenzi Castro Electricity Regulatory Specialist, ANEEL (The Brazilian Agency for Electric Energy Regulation), Brasilia, Brazil

Cecilia Dastres Head of Content, Methodology and Partnerships Unit, Participation and Community Engagement Division, Ministry of Energy, Chile

Danielle Mendes Thame Denny University College London Faculty of Laws, London, United Kingdom; Fundação Armando Álvares Penteado, São Paulo, Brazil; Universidade Paulista, São Paulo, Brazil

Luiz Gustavo Silva de Oliveira Copernicus Institute of Sustainable Development, Utrecht University, Utrecht, The Netherlands; International Centre for Integrated assessment and Sustainable development, Maastricht University, Maastricht, The Netherlands

Ana Laura Rodríguez D'Espada Technological University (UTEC), Montevideo, Uruguay

Francisco Ebeling Carl von Ossietzky University Oldenburg, Oldenburg, Germany

Marie Forget Laboratory on Environments, Dynamics and Mountain Territories (EDYTEM), Mixed Research Unit (UMR) 5204, University of Savoie Mont Blanc (USMB), Chambéry, France

Dustin Welch García Seattle University, Seattle, WA, United States

Lucas Noura Guimarães EDP Brasil, São Paulo, Brazil; Brazilian Institute of Energy Law Studies, São Paulo, Brazil

Sirkka Heinonen Finland Futures Research Centre, University of Turku, Helsinki, Finland

Maria Alejandra Ise National Scientific and Technical Research Council (CONICET), Buenos Aires, Argentina; National University of Northwest Buenos Aires (UNNOBA), Buenos Aires, Argentina; Centre of Studies on Territory, Energy and Environment (TEAM), Buenos Aires, Argentina

Joni Karjalainen Finland Futures Research Centre, University of Turku, Helsinki, Finland

Antonella Mazzone King's College London, London, United Kingdom

Adolfo Mejía-Montero University of Edinburgh, Edinburgh, United Kingdom

Francisco Merino Coordinator of Generación Comunitaria Partnership Framework, Participation and Community Engagement Division, Ministry of Energy, Chile

Adolfo Mejía Montero University of Edinburgh, Edinburgh, United Kingdom

Edisson Morales Morales & Asociados Law Firm, Quito, Ecuador

Pablo Morales Morales & Asociados Law Firm, Quito, Ecuador

Hai-Vu Phan Edison International, Rosemead, CA, United States

Ignacio Sabbatella National Scientific and Technical Research Council, University of Buenos Aires (CONICET-UBA), Buenos Aires, Argentina

Miguel Saldivia Olave University of Cambridge, Cambridge, United Kingdom; Andean Geothermal Center of Excellence (CEGA), University of Chile, Santiago, Chile

Thauan Santos Post-Graduate Programme in Maritime Studies, Brazilian Naval War College (PPGEM/EGN), Rio de Janeiro, Brazil

Carlos Tornel Department of Sustainability Studies, Iberoamericana University, Mexico City, Mexico

Noora Vähäkari Finland Futures Research Centre, University of Turku, Helsinki, Finland

Sofía Vargas-Payera Andean Geothermal Center of Excellence (CEGA), University of Chile, Santiago, Chile

Alexandre Viana Thymos Energia, SP, Brazil

Author biographies

Adolfo Mejía Montero is a PhD candidate at the University of Edinburgh, United Kingdom. His research aims to contribute to energy justice scholarship by focusing on the case of wind energy in Oaxaca, Mexico.

Adolfo has been a lecturer for the Interactive Economics Museum and a research assistant for the Mexican Commission of Science and Technology, as well as a consultant on solar and community renewable projects in Mexico and the United Kingdom. Parallel to his PhD studies, he has performed as a teaching assistant in the University of Edinburgh, Scotland, and was a member of the Renewable Energy and Landscape Quality action from the European Cooperation in Science and Technology.

Adolfo obtained a Bachelor's degree in Physics from the *Universidad Nacional Autónoma de Mexico* and an MSc on Sustainable Energy Systems from the University of Edinburgh, Scottland.

Alexandre Guedes Viana is a partner for Energy Markets and Regulation at Thymos Energia and a member of the International Association of Energy Economics. He is an expert in energy economics and international power markets.

Previously, Alexandre was the Energy Trading Director at SPIC Brasil and Auction and Regulated Market Executive Manager at Câmara de Comercialização de Energia Elétrica (CCEE). SPIC is a large hydrogenerator with a trading desk of short-term power contracts. CCEE is the entity responsible for operating the Brazilian power market. He also worked as a volunteer consultant of USAID and USEA, promoting markets and regulations to attract investors for developing countries in Africa and South East Asia.

Alexandre obtained a Bachelor of Economics from the University Center of Santo André, a Master of Economics from the Catholic University of Sao Paulo (PUC-SP), a PhD in Power Systems from the University of Sao Paulo (USP), and currently he is a postdoctoral student at USP.

Ana Laura Rodríguez is a professor in charge of the subject "Legal Regulation of the Renewable Energy" at the Uruguay Technological University (UTEC) and a partner, coresponsible in the charge of Natural Resources Area in a Lexin Legal & Business Consulting firm. She is an expert in the field of Energy Law regulation, particularly in renewable energy and power markets.

Previous to her current jobs, she was the legal chief at the Electric Market Administration and the attorney at the National Administration of Fuels Alcohol and Cement.

Ana obtained a degree from Law and another from Public Notary, both from the University of Montevideo, Uruguay. She has also done several postgraduate courses in Brazil, Chile, and Uruguay focused on energy, and others related to multilateralism, negotiation, and leadership in Switzerland, Belgium, France, Spain, and Chile.

Antonella Mazzone is a multidisciplinary scholar with a background in humanities and social science. She recently completed her PhD in "Energy Transition in the Brazilian Amazon. A Gender Perspective" at King's College London.

She holds two postgraduate degrees in Energy Economics and International Business from the Federal University of Rio de Janeiro (Brazil) and a BA in Language, Literature, and Philosophy from the University of Bari (Italy). Antonella is also an illustrator who uses visual methodologies for research and converting research outputs into visual materials. She is currently developing a digital graphic novel of her research in the Amazon.

Her main research interests focus on energy safety nets, gender, and intersectionality perspectives on energy and indigenous knowledge. Antonella's current research explores the gendered impacts of renewable technologies and indigenous knowledge in ethno-cultural communities in the Brazilian Amazon.

She is currently working on a project exploring the effectiveness of "Energy Safety Nets" in low-income and marginalized groups in Brazil. The project has been commissioned by the Overseas Development Institute, CAFOD, and SE4ALL.

Carlo Altamirano-Allende is a PhD candidate at the School for the Future of Innovation in Society at Arizona State University, USA. His research deals with the social dimensions of rapid energy systems change and the political ecology of Mexico's energy transition.

He is the cofounder of the Grassroots Energy Innovation Lab at Arizona State University and cochair of the Eradicating Poverty through Energy Innovation global network. During his PhD tenure, he has taught courses on energy policy, sustainable development, and public engagement with science and technology.

Carlo is a Fulbright scholar and a US National Academies of Sciences, Engineering, and Medicine's Mirzayan fellow. He obtained a Bachelor's degree and an MSc in Physics from the National Autonomous University of Mexico (UNAM).

Carlos Tornel is an associate professor at the Department of International Studies at Iberoamericana University in México City and a program associate at the Mexican Climate Initiative (ICM).

Previous to his current job, Carlos has worked with other Mexican NGOs, including the Mexican Center for Environmental Law (CEMDA) and the Interamerican Association in Defense of the Environment (AIDA).

Carlos holds an MSc in Environmental Policy and Education from the London School of Economics (LSE) and an MSc in Environmental Policy and Energy Regulation from the Latin American Faculty of Social Sciences (FLACSO), Mexico.

Cecilia Dastres is a chief coordinator of content and methodology in the Chilean Energy Ministry public participation and social dialog division.

Since 2014 Cecilia has been responsible to elaborate the Chilean government participation and social dialog division's methodologies and mechanisms. She has consulting experience in sustainability for mining, energy, and other infrastructure companies in Chile. Cecilia also accounts with experience in consultancy for national and international organizations, in the area of civil participation, local development, and entrepreneurship.

Cecilia obtained a Bachelor's degree in Sociology from the Pontifícia Universidad Católica de Chile and a Master degree in management and public policy from the Universidad de Chile.

Danielle Mendes Thame Denny is a Visiting Academic at University College London, Faculty of Laws, London, United Kingdom and licensed full professor at Fundação Armando Álvares Penteado and Universidade Paulista, São Paulo, Brazil. She is a member of the research group Energy and Environment at Universidade Católica de Santos (UNISANTOS), Brazil.

During 2017 she was a Visiting Associate Researcher at the Whitney and Betty MacMillan Center for International and Area Studies of Yale University, USA. From 2015 to 2017 she was granted a merit fellowship from Coordenação de Aperfeiçoamento de Pessoal de Nível Superior (CAPES).

Danielle obtained a Bachelor of Laws from Pontifícia Universidade Católica de São Paulo, three postgraduate degrees from different universities, a Master degree from Faculdade Cásper Líbero, and a Juris Doctor on International Environmental Law from UNISANTOS.

Dustin Welch García is currently an adjunct professor at Seattle University, USA. He has expertise in off-grid energy infrastructure, clean energy transitions, public—private partnerships, and rural development in Latin America.

Prior to earning his PhD he cofounded an NGO dedicated to the installation of off-grid solar energy systems in rural communities of Peru. In addition, he was a Fulbright scholar and Inter-American Foundation Grassroots Development Fellow.

Dustin earned a BA in Spanish Language from Western Washington University, an MA in Latin American, Caribbean, and Iberian Studies from the University of Wisconsin-Madison, and a PhD in International Studies from the University of Washington, USA.

Edisson Morales Pazmiño is an associate attorney in Morales and Asociados Law Firm, based in Ecuador. He is an expert in Corporative Law.

Edisson obtained a Bachelor of Laws from Universidad de las Américas of Ecuador (UDLA), a Master of International Laws from Universität Heidelberg, Germany, and a Master in International Economics and Investments from Universidad de Chile.

Francisco Ebeling Barros is a PhD candidate in Economics at the Carl-Von-Ossietzky Universität Oldenburg, Germany. He is an expert in the fields of energy and petroleum economics and energy policy.

Before this, Francisco was an economics and energy policy analyst at the Brazilian Petroleum Institute and a research assistant at the Energy Economics Group of the Federal University of Rio de Janeiro.

Francisco obtained his Bachelor (Economics) and Master degrees (Public Policy, Strategy, and Development) from the Federal University of Rio de Janeiro, Brazil.

Francisco Merino is the Chilean Energy Ministry coordinator of the public participation and social dialog division in community generation.

Since 2016 Francisco has been working in the Chilean Energy Ministry. He has been managing and implementing the community generation mechanism framed in the 2018—22 Energy Route, the indigenous energy chapter, and the associative and sustainable local development policy. He accounts for more than 10 years in the academic and professional experience in Chile, the United States, the United Kingdom, The Netherlands, and Belgium, related to energy and community dialog.

Francisco obtained a degree in Law from the Universidad Finis Terrae in Chile and an LL.M in natural resources and policy from the Centre for Energy, Petroleum, Mining, Law and Policy (CEPMLP), University of Dundee, Scottland.

Hai-Vu Phan is a Senior Policy Analyst at Edison International, one of the largest electric utility companies in the United States and an industry leader in renewable energy.

Through the Fulbright Public Policy Fellowship, she worked as a technical assistant at the Supervisory Agency of Investment in Energy and Mining (OSINERGMIN) in Peru. Prior to this, she served as an Ambassador Intern at the Institute of Advanced

Development Studies (INESAD) in Bolivia. She has also been a Capital Fellow, working as a policy consultant for the California State Senate.

Hai-Vu earned a BA in Political Science and International Studies from Yale University and an MA and a PhD in Political Science and International Relations from the University of Southern California, USA.

Ignacio Sabbatella is a researcher at National Scientific and Technical Research Council (CONICET). His focus is on energy policies and regional energy integration, particularly oil and gas in South America.

He is also a graduate level professor at the National University of Rio Negro (UNRN) and the Latin American Faculty of Social Sciences (FLACSO). Previously, he served as a consultant at the Argentinian Lower House, and he was a professor at the Faculty of Social Sciences of the University of Buenos Aires (UBA), Argentina.

He has a BA in Political Science, a Master in Social Sciences Research, and a PhD in Social Sciences at UBA.

Joni Karjalainen is a doctoral researcher at the University of Turku, Finland Futures Research Centre, studying the contextual dynamics of the adoption of renewable energy technologies and the long-term transformation of the energy system.

His research focus is in the role of innovation collaboration, knowledge creation, and new business models. He has worked in research projects, such as *Great Electrification in Peer-to-Peer Society*, *Neo-Carbon Energy—Transformative Energy Futures 2050*, *Access to Sustainable Energy for All*, and *Kenya and Tanzania Beyond 2015*. He has been a visiting scholar at the Institute for Futures Research in South Africa and at the University of Sussex, Science Policy Research Unit (SPRU) in the United Kingdom and holds extensive international experience.

Joni holds a Master of Social Sciences in Political Science from the University of Helsinki, Finland.

Lourdes Alonso Serna is a PhD candidate at the University of Manchester, United Kingdom. Her PhD research deals with the social aspects of energy transitions in Mexico, in particular the case of wind energy in the Isthmus of Tehuantepec, Oaxaca.

Between 2009 and 2016, Lourdes was an associate lecturer at Universidad del Mar, a public university in the state of Oaxaca. At Universidad del Mar, Lourdes taught several courses in the BA of International Relations.

Lourdes obtained a BA in International Relations and a Master in Latin American Studies from the Universidad Nacional Autónoma de México.

Lucas Noura Guimarães is energy legal counsel at EDP Brasil, a member of the Advisory Board of the Brazilian Institute for Energy Law Studies and general secretary of the Energy Commission of the São Paulo Bar Association. He is an expert in the field of Energy Law regulation and renewable energy integration.

Previous to his current job, Lucas was energy associate in law firms, where he advocated in cases related to Brazilian Energy Law regulation. He also worked as an advisor of the Attorney General in the Brazilian Electricity Regulatory Agency.

Lucas obtained a Bachelor of Laws and a Master of Laws from the University Center of Brasília, Brazil, and a PhD in Law from the Free University of Berlin, Germany. He is a Néstor Kirchner Fellow from the New School University, New York, USA.

Luiz Gustavo Silva de Oliveira is a PhD candidate at the Copernicus Institute of Sustainable Development, Utrecht University, The Netherlands. He is concluding his PhD research on designing policies for fostering the Brazilian biogas innovation system.

His previous experiences comprise positions at EPE (Brazilian Energy Research Office), at UNEP (United Nations Environment Programme), MCTI (Brazilian Ministry of Science, Technology and Innovation), and Dalkia Brasil where he worked, respectively, as an Energy Research Analyst, External Consultant, and Design Engineer.

Luiz Gustavo obtained a BSc in Electrical Engineering from Federal Fluminense University (UFF) and an MSc in Energy Planning from Rio de Janeiro Federal University, Brazil. Currently, he is finishing his PhD in Innovation Studies at Utrecht University.

Marco Aurelio Lenzi Castro has more than 15 years of experience in the energy industry, working as an Electricity Regulatory Specialist at the Brazilian Electricity Regulatory Agency, with focus on electricity distribution and generation rules.

Marco Aurelio has also worked for the Inter-American Development Bank as an energy specialist, supervising energy projects in Brazil. Previously, he worked for the Brazilian Independent Electrical System Operator analyzing data from the major hydropower plants in Brazil.

Marco Aurelio got his Bachelor and Master degrees in Electrical Engineering from University of Brasilia, Brazil, received a Certificate in Theory and Operation of a Modern National Economy from The George Washington University, and has a Master degree in Advanced Renewable and Sustainable Energy Engineering from Monash University, Australia.

Maria Alejandra Ise is a scholarship holder at National Scientific and Technical Research Council (CONICET), working on a doctoral thesis on solar energy and territorial transformations.

She is a member of TEAM Centre of Studies on Territory, Energy and Environment at National University of Northwest Buenos Aires Province, Argentina.

Alejandra holds a Bachelor's degree in International Relations, and she is currently enrolled in a Renewable Energy Master at National University of Northwest Buenos Aires Province, and in a PhD program at National University of La Plata, Argentina.

Marie Forget is an associate professor at Savoie Mont Blanc University and researcher in the EDYTEM Laboratory (Environment, Dynamics and Mountain Territories) of the National French Research Center (CNRS). She is an expert in the field of territorial trajectories in relation to energy transitions.

Marie is currently the Director of the Geography Department of the Savoie Mont Blanc University and teaches postgraduated degrees on the energy transition, natural resources extraction, and sustainable development. She is also leading a study group on environmental humanities.

Marie obtained a Master and a PhD in Geography, Environment, and Land Use Planning from the University Lyon 2, France.

Matheus Linck Bassani is a Legal Adviser of the Attorney General's Office of the State Accounting Prosecutor of the State of Rio Grande do Sul, Brazil, Licensed Lawyer of the Brazilian Bar (OAB/RS) and a visiting professor of the postgraduation course "The New International Law" at the Faculty of Law of Federal University of Rio Grande do Sul (UFRGS), Brazil.

Matheus is specialized in the Energy, Consumer, Environmental, International Public and Private Law fields, and a member of the Red Iberomericana de Derecho de la Energía (RIDE).

Matheus holds a PhD (2019) and Master (2014) at Law at the Federal University of Rio Grande do Sul (UFRGS), Brazil. He obtained the Diplôme d'Université en Droit Comparée et Européen (DU), at the University of Savoie

Mont Blanc, France. He was a visiting researcher of the Center for Energy, Petroleum and Mineral Law and Policy, at the University of Dundee, Scotland (2013) and of the Max Planck Institute for Comparative Public Law and International Law, Heidelberg, Germany (2018).

Miguel Saldivia Olave is a PhD researcher at University of Cambridge, United Kingdom. He is an environmental lawyer and expert in Energy Law and development of renewable energy projects. His PhD research is focused on regulations on renewable energies in developing countries.

As part of his work experience, Miguel worked as an attorney in the Environmental Law Department of Carey law firm, in Santiago de Chile. He also has been a researcher in the Environmental Law Centre, Solar Energy Centre, and Geothermal Energy Centre, all based in University of Chile.

Miguel holds a Bachelor of Laws (University of Chile), a Bachelor of Communication (University of Chile) and a Master of Laws (University College London).

Noora Vähäkari is a project manager in international projects in sustainable energy, climate change, and development research. Her duties include the promotion of education in the fields of renewable energy and sustainable development.

Noora has worked on the socioeconomic impacts of energy projects in countries, such as Chile, Laos, and Cambodia. She has also experience in India, Myanmar, Cuba, Peru, and Colombia and has worked in the facilitation of bilingual and multinational participatory workshops related to sustainability issues and renewable energy planning. She has previously worked in civil society development cooperation projects.

Noora holds an MSc in Development Geography from the University of Helsinki, Finland.

Pablo Morales Andrade is a partner and a lawyer in Morales & Asociados Law Firm since 2006 and a professor of Public Law at Simon Bolivar University of Ecuador. He is an expert in economic regulation in strategic industries.

Previous to his current job, Pablo was a legal advisor for the Ecuadorian Council of Foreign Commerce and Investments COMEXI.

Pablo obtained a Bachelor of Laws from San Francisco de Quito University, Ecuador, a Master of Economic Regulation in Strategic Sectors from Bonn University, Germany, and Master of Business Law from Berkeley University, California.

Silvina Cecilia Carrizo is an independent Researcher of National Scientific and Technical Research Council (CONICET), at Centre on Territorial and Urban Research (CIUT), National University of La Plata, Argentina.

She directs TEAM Centre of Studies on Territory, Energy and Environment at National University of Northwest Buenos Aires Province, Argentina, where she also directs a Renewable Energy Master.

Silvina is an architect and obtained a Master and PhD in Geography, planning and urbanism from Sorbonne University, France.

Sirkka Heinonen is a professor in Futures Studies at the University of Turku, Finland. Her aim is to promote futures learning to catalyze innovative solutions to global challenges amidst social and technological change.

She is a member of the Club of Rome, a chair of the Helsinki Node of the Millennium Project, a guest professor for University of Science and Technology (USTC) of China, as well as a member of the editorial board of six academic futures journals. She is an expert in studying the perceptions of time and progress, futures thinking, risks of technology on society and nature, and in developing and applying various methodologies. Heinonen has also contributed to the Finnish Government's climate and energy scenario work.

Sirkka has a PhD from the University of Helsinki, Finland.

Sofía Vargas-Payera is a social researcher at the Andean Geothermal Centre of Excellence (CEGA), University of Chile. Her experience has been focused on the interface between social and cultural aspects of geothermal energy and leading geothermal direct projects in Chile.

She has strong experience on designing communication strategies that improve the coordination between governmental agencies, scientific institutions, and the civil society and engage local communities to increase social acceptance of geothermal energy.

Sofía holds an MA in Communication from Simon Fraser University, Canada, and social communicator from University of Chile (2009).

Thauan Santos is a professor at the Brazilian Naval War College (EGN). He is an expert in the field of energy security, regional integration, sea economics, and defense economics.

Previous to his current job, Thauan was a professor at the Federal University of Rio de Janeiro (UFRJ), Rio de Janeiro State University (UERJ), and Pontifical Catholic University of Rio de Janeiro (PUC-Rio) in International Relations, Economics and Defense areas. Besides, he has worked on different national projects, such as IES-Brazil and third Brazilian Inventory on GHG.

Thauan obtained a Bachelor of Economics from UFRJ, a Master in International Relations from the PUC-Rio, and a PhD in Energy Planning from Alberto Luiz Coimbra Institute for Graduate Studies and Research in Engineering (COPPE/UFRJ), Brazil.

Foreword

The world's energy markets, as everyone already knows, are undergoing transformation at rapid rate. The pace of change is even more pronounced in the electricity sector where in fact two separate transformations are taking place at the same time and at the opposite ends of the power sector's long value chain.

On the upstream side of the industry's infrastructure, generation is rapidly becoming decarbonized in many parts of the world, for example, in places where ambitious targets have been set as in California or the United Kingdom to achieve near total decarbonization by 2045 or 2050, respectively. This means massive investments in renewable resources, primarily solar and wind, that are now competitive with fossil fuel options. Among the many challenges of this transformation is how to integrate ever-increasing amounts of variable renewable generation into the network while maintaining reliability and keeping the price manageable. These phenomena are well known and need little elaboration.

What may be less well known is the equally significant transformation that is taking place at the other end of the industry's infrastructure, namely, at grid's edge—where the distribution network meets the customers, and beyond. As noted in a forthcoming volume,[1] arguably the most exciting changes are likely to take place behind-the-meter (BTM), on the customer side of the meter where massive investments are likely to take place by consumers, prosumers (i.e., consumers who produce some of their power), and prosumagers (i.e., prosumers who have storage) in all sorts of new devices and technologies, including rooftop solar photovoltaic (PV) panels, electric vehicles (EVs) as well as other forms of distributed storage, not just batteries.[2] As the former head of the US Federal Energy Regulatory Commission, Jon Wellinghoff, famously put it, "PVs + EVs + storage = game over"—alluding to the end of the traditional distribution and retailing business model.

What makes this rapid transformation at the two ends of the spectrum particularly noteworthy is that the regulated assets that sit between the two ends—transmission and distribution (T&D) network—are not changing fast enough to keep up with the extra demands that are being placed on them from both ends, and this is likely to be a major issue for the regulators and policy makers in the years ahead.

Of course, the pace of transformation is uneven across the world with some countries moving much faster in adopting the three major drivers of change, or the 3Ds, that is

- decarbonization,
- decentralization, and
- digitalization.

---

[1] Sioshansi, F. (Ed.), forthcoming in 2020. Behind and Beyond the Meter: Digitalization, Aggregation, Optimization, Monetization. Academic Press.

[2] Sioshansi, F. (Ed.), 2019. Consumer, Prosumer, Prosumager. Academic Press.

Broadly speaking, Europe, the United States, and a few other counties, including Australia, are experiencing rapid change in all three areas while a number of developing economies are making progress in one or more of the three. Latin America, Africa, the Middle East, and many parts of Southeast Asia are yet to embrace the transformations that have taken root in other countries.

In the case of Latin America a continent blessed with plentiful and varied renewable resources, the progress toward 3Ds is still lagging, a subject further explored in this volume. While decarbonization appears to be gaining momentum in a number of Latin American counties, the movement toward decentralization and digitalization is still lagging behind most OECD economies.

Yet when and if Latin America pivots more aggressively toward the 3Ds, there are a number of useful lessons, it can learn from others who are grappling with some of the challenges.

The most obvious, of course, is how to integrate large percentage of variable renewable generation into the electricity mix given their inherent variability, especially for wind and solar—and to a much less extent hydro with large storage reservoirs, geothermal, biomass, and biogas.

This challenge, as manifested in the famous "California Duck Curve,"[3] is well known and therefore not dwelt on in this space.

The other perennial question is how deep a decarbonization is feasible, affordable, or indeed desirable. This is no longer an academic question, for example, in the case of California where state law requires virtually 100% renewable generation by 2045.[4]

These issues, while interesting, are also dealt with elsewhere in the rich and growing literature. What is emerging as the new challenge is how to keep the upstream and the downstream segments of the industry's sprawling infrastructure together—preventing a disconnect caused by the more-or-less static middle segment.

As noted in a recent thought provoking policy brief[5] by Nicolò Rossetto, Piero Carlo Dos Reis, and Jean-Michel Glachant of the Florence School of Regulation (FSR), the regulated T&D networks are caught between the radical changes transforming both the upstream and downstream of the electric power sector. Making matters worse is that the regulators and policy makers are increasingly challenged in dealing with the rapid pace of change.

The FSR policy brief focuses on the impact of the upstream and downstream developments on the grid—both T&D—which lies in between the upstream and the downstream. As the authors point out, the grid "...stands in the middle, not just in a physical sense, but also because their business is directly affected by what is happening upstream and downstream." The authors note,

> Digitalization is driving the emergence of new business models in the electricity sector. Unlike the case of green generators, these new models tend to be 'asset light', and the offering of innovative services that target particular groups of customers represents their essence.

The historical ways in which electricity was generated in large central power plants and

---

[3] Details at CAISO website.

[4] EEnergy Informer, 2019. 100% Renewable: Nice but Is It Affordable. EEnergy Informer. Available from: <www.eenergyinformer.com>.

[5] Rossetto, N., Dos Reis, P.C., Glachant, J.-M., 2019. New business models in electricity: the heavy, the light, and the ghost. In: Policy Brief. Florence School of Regulation. Available from: <http://cadmus.eui.eu/bitstream/handle/1814/63464/PB_2019_08_FSR_Energy.pdf?sequence = 4&isAllowed = y>.

delivered to passive customers through a one-way T&D network—as everyone knows—is radically changing to one where consumers can generate, store, and consume a significant portion of their energy needs locally.

This, however, is only the first step of the downstream transformation, soon to be followed by the ability to share or trade with others using the distribution network, generally referred to as peer-to-peer trading. More exciting opportunities are possible with the increased digitalization of BTM assets, which in turn can be aggregated into large portfolios of flexible load and generation and optimized using artificial intelligence and machine learning.

And this may be the biggest challenge facing regulators and policy makers everywhere, and not just in Latin America.

In this context, policy makers must

- explore the latest advances in digitalization of BTM assets, including distributed generation, distributed storage, and EVs, and—more important—how these assets can be aggregated and remotely monitored unleashing tremendous value and a myriad of innovative services and business models;
- examine what lies BTM of typical customers, and why managing these assets increasingly matter;
- assess how smart aggregators with intelligent software are creating value by optimizing how energy may be generated, consumed, stored, or potentially shared or traded between consumers, prosumers, and prosumagers; and
- explore new business models that are likely to disrupt the traditional interface between the incumbents and their customers.

As eloquently stated by Mark Kolesar, Chair of the Alberta Utilities Commission,[6]

> The evolving nature of electric generation, consumption, storage and the distribution system has significant implications for the grid, incumbent utilities, consumers, grid managers and the regulatory framework. These are among the central matters the Alberta Utilities Commission (is) examining in its ongoing distribution inquiry.

The province of Alberta, of course, is not the first or only jurisdiction where the regulators are looking into the future of the distribution business, and it certainly will not be the last. The reasons for such inquiries may be found in the aforementioned FSR policy brief.

> The digitalization of the electricity sector alters this (the historical) situation profoundly. Digital technologies improve the availability, usability, storability and transmissibility of information, thereby allowing a reduction of transaction costs and more efficient use of resources.

Digitalization, the third of the 3Ds, will allow new aggregators and smart intermediaries to enable consumers, prosumers, and prosumagers to engage in trading offering a new option for them to participate in "a decentralized community of buyers and sellers, not necessarily spatially close to each other," as noted in the FSR brief.

In such a future, consumers who are not necessarily "active" will nevertheless be able to engage with the distribution network, or be sufficiently motivated to follow prices at the wholesale level or congestion on the local grid and so on.

Examples of useful applications of BTM aggregation and optimization are already emerging especially in places where there is already a significant concentration of PVs and EVs. Other obvious examples appear in the form of semiindependent grids or energy communities.

---

[6] http://www.auc.ab.ca/regulatory_documents/Consultations/2018-12-06-Announcement.pdf and
http://www.auc.ab.ca/regulatory_documents/Consultations/2018-12-06-Announcement.pdf

As noted in the FSR brief,

> Electricity grids – both transmission and distribution – are caught in between the greening of electricity and the digitalization of retail-size units.

And this goes to the core of the dilemmas facing the regulators, who are the ultimate arbitrators of policies on what can and cannot be delivered by whom and at what price.

This is particularly important since the traditional business model of grid companies is, by and large, driven by the investments in regulated T&D assets, which offers a steady and riskless revenue streams.

Moreover, any regulator worth her/his keep will tell you that their main job is to keep a level playing field for all the stakeholders. This responsibility, which was never easy, will become far more complicated in a future where customers are likely to be *stratified* into different segments—consumer, prosumer, prosumager, and the more exotic varieties.

For what it may be worth, I am convinced that developments at the grid's edge and BTM are likely to alter the electricity business model as much as—if not more than—those taking place on the generation side of the business.

While some of the preceding may be new to some of the readers of this volume, there are signs that the transformation of the electric power sector driven by the 3Ds will engulf Latin America sooner rather than later. This suggests that the time is ripe to examine what is happening elsewhere and draw useful lessons to guide the Latin American power sector along the path to the ultimate Nirvana.

Fereidoon Sioshansi
Menlo Energy Economics, San Francisco,
CA, United States

Foreword

Latin America is one of the most exciting places for the energy industry.

The region has around 8% of the world population and 6% of its GDP and is formed by 21 low-middle- to high-middle-income countries with robust energy demand growth, driven largely by economic growth, urbanization, higher living standards, and the successful expansion of electricity access, which currently reaches close to 95% of the population. This translates into a constant need of new greenfield projects and an appetite for infrastructure investments.

A distinctive feature of Latin America lies on the predominance of hydropower in the power generation mix, which enables a low-carbon region already today. Hydro accounts for about 50% of the region's generation capacity, with plenty of potential yet to develop. The hydro reservoirs serve as water batteries and jointly with the countries' transmission grid have been used to modulate the seasonal and variable production of renewables, which in turn have complementary production and locational patterns among themselves and with hydro and load.

This portfolio effect of sources in different time scales has enabled a cost-effective integration of nonhydropower renewables, and the installed capacity of bioenergy, wind, and solar has more than quadrupled between 2005 and 2018. Fossil fuel generation also plays a role in the region, with gas-fired generation as the leading technology, integrated via cross-border pipelines, LNG, and local production fields.

Transport and industry dominate regional energy consumption. Transport represents a larger share than in other major regions of the world, due mainly to a less efficient vehicle fleet and differing modal composition, which also contributes to air pollution. A higher energy use in industry partly reflects the economic structure of Latin America and the important role of energy-intensive industries. A relatively small residential consumption is partly due to the lower use of space heating appliances due to mild weather, although the use of cooling in some subregions is rapidly increasing.

This background challenges us to define the role and potential of the global energy transition in the region. At first, it is important to recognize that there is not "an" energy transition but energy transitions, differing in terms of motivation and objectives, drivers, and governance. Their main outcome is an energy transformation—mostly on technology developments, behavioral, and governance changes—that is rapidly stimulating the development of regulation and policymaking to a new paradigm. Flexible policies, regulation, processes, and institutional culture to "keep up" in a rapidly changing environment will be needed, along with roles and responsibilities in line with evolving system needs.

This will not be different in Latin America. Looking forward, there is potential for substantial transformations in the region's energy industry. While hydropower remains the main electricity generation source, natural gas and nonhydropower renewables are the fastest growing expansion sources. Natural gas and bioenergy may play an important and cost-effective role as transition fuels to reduce emissions in the industry and transportation segments toward, where it makes sense, the electrification of the economy with competitive renewables. Off-grid renewables can encourage small-scale economic activity and entrepreneurship. Some renewables can play a key role in decoupling the energy sector from water availability and land use, contributing to reduce the well-known interlinkage between the water, energy, and food supply.

Its young population is eager to energy supply empowerment with low-carbon footprint that will be added with similar needs of commercial and industrial consumers and create a marketplace for new technologies, including distributed supply and algorithms in a decentralized way. On the other hand, the increased recurrence of climate events, concerns about security of supply, and inequalities introduce challenges to this transformation. And they all demand a more robust, comprehensive approach to the energy sector management to deal with an ongoing transformation.

It is exactly on the topic of this book—regulation and policy—that the energy transitions in Latin America will be defined. The pace defined by policymakers to foster business opportunities that translate into economic efficiency, emissions reduction and access, will be key to enable a long-term structural change in the region's energy systems.

The Regulation and Policy of Latin American Energy Transitions is a well-worth reading that examines the energy revolution through which the Latin-American energy landscape is/should be going through. It aims to contribute in making energy transition a scholarly field but also fosters a research and policy agenda that can guide the way toward the implementation of an environment-friendly, cost-effective, and socially acceptable Latin-American energy transformation.

Chapters are designed to explore policy, economic, regulatory, and socio-technical issues that are enabling or hampering the energy transitions in Latin-American countries and how future policies may drive a complete transition in a renewable-rich—or carbon neutral—Latin America. Case studies and business models are also addressed in the book.

This book intends to fill an important gap in energy transition literature, since the unique renewable-rich Latin America makes it an interesting case on how—or if—energy transitions can scale up to another level. Moreover, a late Latin-American energy transition has a smaller learning curve, in comparison to vanguard countries, benefiting from the trial-and-error experience from predecessor countries, bringing new regulatory possibilities worth analyzing.

Gathering knowledgeable writers to produce 19 chapters is not an easy task, but the outcome is outstanding and will make a significant contribution to the global debate on the energy transition. The editor is commended for the pioneer and ambitious work on such a timely topic.

Luiz Augusto Barroso
CEO, PSR, Rio de Janeiro, Brazil

Disclaimer

The opinions expressed in the chapters represent the author's opinions solely, and not that of the companies, universities, and institutes that they represent.

Introduction

In general, transitions are long-term processes that may take years or decades to unfold (Köhler et al., 2019). Humankind has already faced energy transitions[1] in the past, including examples such as the transition from animal power to windmills, from windmills to coal and from coal to oil. However, the perspective of energy systems as a flux—and not as a stable and resilient system—has never been so evident as today. More than at any other time in the past century, multiple changes in energy systems are unfolding at the same time (Miller et al., 2013), bringing with it different risks (Hanger-Kopp et al., 2019):

- Concerns about the potential impacts of climate change have emphasized debates, policy initiatives, and financial investments related to the adoption of new technologies to explore renewable energy sources, in contrast with the exploitation of fossil fuels.
- Almost as a counterforce, other technologies, still in line with the fossil fuel exploitation's status quo—"everything must change so that everything can stay the same" (Lampedusa, 1991)—are set in motion, such as hydraulic fracking, extraction of natural gas from shale formations, carbon capture and storage, hydrogen cells, geoengineering, and among other techno-fixes (Fauset, 2010; Scheer, 2010).
- Within this energy transition, other niche-transitions are taking shape, for example, the transitions in energy efficiency (Jenkins and Hopkins, 2019), or from cars equipped with internal combustion engines to electric vehicles (Sovacool and Hirsh, 2009), or from lead batteries to lithium-ion batteries.
- The infrastructure of the power sector is rapidly changing from a centralized design—where electricity is transmitted unidirectionally in high voltage from big power plants, located far away from the load centers, to passive utilities and consumers—to a decentralized design, characterized by a shift in the roles of actors playing in the power sector arena (rise of prosumers,[2] prosumagers, aggregators, microgrids operators[3]), as well as the scale and location of power production.
- Consequently, this panorama gives space to manifold social tensions,[4] related to

---

[1] "Energy transitions" has many definitions, being examples of which the following: "a particularly significant set of changes to the patterns of energy use in a society, potentially affecting resources, carriers, converters and services," (O'connor, 2010); "the switch from an economic system dependent on one or a series of energy sources and technologies to another," (Fouquet and Pearson, 2012); "the time that elapses between the introduction of a new primary energy source, or prime mover, and its rise to claiming a substantial share of the overall market," (Smil, 2010).

[2] See Chapter 7, The prosumer legal protection in Latin America: a requirement for an energy transition, of this volume.

[3] See Chapter 9, Urban microgrids: benefits, challenges and business models, of this volume.

[4] See Part 3 of this volume.

ecosystems preservation, indigenous rights enforceability, worker health, communities empowerment, control over and access to affordable, reliable, and sustainable energy sources.

- Gaining support from the advanced technology era humankind is inserted, this paradigm shift is underpinned by digitalization: behind-the-meter solutions (Sioshansi, forthcoming), Internet of Things and big data, self-healing technologies, blockchain, peer-to-peer solutions, energy digital platforms (Kloppenburg and Boekelo, 2019), among others.

That said, the speed at which a transition can take place is a vital element of consideration (Sovacool, 2016). In terms of climate change, CO_2 emissions lock-in and the need to shift away from fossil fuels certainly prove that point, given that a transformation of the energy system, which fails to keep the global temperature rise below 2°C above preindustrial levels, might be insufficient for humanity.

This scenario has given rise to a myriad of studies, peer-reviewed journals, reports, and so on, focusing not only on energy transitions but also on sustainability transitions. The numbers boomed from 2012 to 2013, and they keep growing (Köhler et al., 2019), with many scholars, universities, think tanks, and research centers interested in the topic.

It is hard to keep pace with the niche-studies and segments being created to deal with the "multi" feature posed by sustainability transitions. Regardless of other divisions and classifications, the themes addressing different aspects of transitions research could be divided as follows (Köhler et al., 2019):

- understanding transitions;
- power and politics in transitions;
- governing transitions;

- civil society, culture, and social movements in transitions;
- organizations and industries in sustainability transitions;
- transitions in practice and everyday life;
- geography of transitions: spaces, scales, places;
- ethical aspects of transitions: distribution, justice, poverty; and
- reflections on methodologies for transitions research.

Within sustainability transitions' studies, the topic of energy and electricity transitions is becoming an important interdisciplinary field of study, responsible to address the many ways under which the energetic status quo of a given country and/or energy system is altered, how to achieve a 100% renewable energy system in line with environmental pressures,[5] as well as to assess the social, economic, and environmental changes that such transition encompasses.

Despite the attention that this topic is receiving worldwide, especially in Europe, Latin American governments and society still lack the sense of urgency promoted by environmental pressures and the dependence of finite oil. The abundance of renewable energy sources in Latin America should be a driver for energy and/or electricity transitions and not a reason to justify its postponement.

There is no doubt that Latin America in general is facing a major change in its energy system. While being true that the digitalization of the energy systems is still a far reality, decentralization and decabornization of their energy matrixes are already occupying the central stage. However, in relation to other countries worldwide, Latin American countries can be regarded as late adopters or periphery (Grubler, 2012), given that the continent still lags behind technological

---

[5] See, for example, Brown et al. (2018), Hansen et al. (2019), Aghahosseini et al. (2017), Barbosa et al. (2017).

development, which increases the import needs. As one of the consequences of this late adoption, there is a risk of increasing carbon lock-in in Latin America, since early adopters might tend to export their outdated fossil-related technology to late adopters, therefore enhancing the inertia (Lund, 2006) of the energy systems of Latin American countries.

Besides that, the processes and phases (Wilson and Grubler, 2011) each country is dealing with in their energy transitions vary greatly. Unpredictable regulatory and legal frameworks, unstable economic and political scenarios complicate further a uniform approach of energy transitions in Latin American countries. It should not be forgotten that Latin America—as opposed to European Union—has not benefitted from a history of regional integration in the energy field—exception being made by SIEPLAC initiative in Central America.

Given this complex, varied, and intricate scenario, this book attempts to reflect it by bringing not only the theoretical aspects of energy transitions in Latin America but also case studies on how each country is designing its energy or electricity transition.

Avoiding focusing only on the policy and regulatory aspects of energy transitions, as well as on the availability of energy sources and the technological choices it comprises, this book also sheds light on the social and political dimensions of energy systems and energy transitions. Intentionally, the addition of the sociotechnical aspects related to energy transitions helps deliver a wider understanding of this complex process.

This book, comprising 19 chapters, is divided into three parts.

Part 1 of the book is focused on the policy and economic aspects surrounding the energy transitions taking place in Latin American and Caribbean countries.

In Chapter 1, Is there a Latin American electricity transition? A snapshot of intraregional differences, Lucas Noura Guimarães analyzes to what extent it is possible to speak of a Latin American electricity transition, given the intraregional and inner-country differences in terms of availability of renewable energy sources, INDCs, electricity transition paths, and existent electricity matrixes.

The chapter demonstrates that a closer look into Latin American and Caribbean countries reveals a heterogeneous configuration, with some countries relying greatly on oil imports, others already with a diversified electricity matrix, and others flirting with techno-fixes, making rather difficult to apply a one-size-fits-all approach. On the other hand, the chapter makes an argument favoring the subcontinent's interconnection of the grid to boast its electricity transition, transforming the power sector in an environment-friendly fashion and with the least cost for the population.

The chapter's main contribution is to shed some light on the paramount importance of analyzing from which diversification governments and stakeholders talk about and on the greatest challenge for Latin American and Caribbean, which is to attend its expected demand growth while maintaining its overall electricity matrix clean, avoiding being victim of a "transition-related renewables curse."

In Chapter 2, Using market mechanisms for a reliable energy transition in Latin America, Alexandre Viana discusses different market mechanisms that have been used in Latin America to foster renewables and promote energy transition. This chapter highlights the relevance of creating a friendly environment for private investments instead of using only subsidies as a way of attracting investments. The author also defends that artificial subsidies are presenting signals of exhaustion, and a well-designed power market is a more reliable solution for energy transition.

Focusing on Brazil as a case study, the author examines market experiences that foster the introduction of renewables in a reliable and long-term sustainable way, arguing that special short-term programs might produce poor results and generate criticism from the broad public.

The chapter's main contribution is to provide a toolkit of options for policy-makers, academics, and experts that are willing to increase the relevance of renewables in their countries and to promote reliable and responsible energy transition, including a broader power access for poorer regions.

In Chapter 3, Assessing the macroeconomic effects of sustainable energy transitions in Costa Rica and Chile: a multisectoral balance-of-payments-constrained growth approach, Francisco Ebeling examines whether the sustainable energy transitions have also been leading to sustainable GDP growth in selected Latin American countries (Chile, Uruguay, Costa Rica). The author analyses if a larger diffusion of renewable energy in the countries' energy mix has permitted them to alleviate balance of payment constraints. In that context, sustainable GDP growth is theoretically possible when countries are alleviated from their external constraint on growth.

The balance of payments constrained growth approach sustains that in the long-term the main drivers of that rate are foreign demand and the income elasticities of exports and imports. The more a country manages to specialize in those sectors with higher export income elasticities of demand, the less will it run short-term balance-of-payments deficits, which will have to be financed by short-term capital flows. In order to examine whether a larger share of renewable energy mix is aiding to lift the external constraint, the econometric method developed by Gouveia and Lima to estimate the multisectoral BPCG rate is used, as it allows to look at the evolution of the income elasticities of energy exports and energy imports of the country.

The chapter's main contribution is to offer a clearer insight about the rather undertheorized

issue about how sustainable energy transitions affect a country's macroeconomic long-run outlook.

In Chapter 4, Energy (and climate) challenges in South America: what planning for what transition?, Thauan Santos and Ignacio Sabbatella analyze the link between energy planning and energy transition in South America, particularly considering climate challenges.

The authors evaluate the existence or not of a real energy transition in South America, trying to identify how it happens, based on what metrics and with what objectives. Also, they examine the relevance or not of regional organizations in promoting this transition, stressing the role of greater mutual regional planning.

The chapter's main contributions are twofold. First, what is actually happening in South America is a mere diversification of the sources of power generation rather than a real and structural regional energy transition. Second, regional bodies dedicated to the energy and climate issues have contributed little either because of its institutional nature or because of the recent events of regional political economy.

In Chapter 5, Prompting an energy transition: how policy entrepreneurs passed Peru's first renewable energy legislation for the grid, Hai-Vu Phan investigates how a renewable energy transition is launched by analyzing the political processes behind passing Peru's first policy for large-scale renewable energy production despite strong opposition. While this policy was a profound first step, the chapter also discusses some of the remaining challenges to accelerating the transition.

The author reconstructs the political context and dynamics surrounding the passage of this pivotal renewable energy policy by interviewing relevant stakeholders, consulting government publications, and analyzing market reports. These primary and secondary resources allow her to process trace how political actors were able to maneuver past strong opposition and pass this controversial policy.

The chapter's main contribution is to dissect the dynamic behind passing Peru's first large-scale renewable energy policy, illuminating strategies that may be useful for other Latin American countries wishing to pass difficult but necessary laws to accelerate an energy transition. The chapter also spotlights continuing challenges that political actors must consider if they want to expand the transition.

In Chapter 6, The challenges and contradictions of Peru's *Proyecto Masivo de Energía Solar*, Dustin Welch García analyzes the first off-grid solar energy auction carried out by the Peruvian government and the subsequent public—private partnership (PPP) formed with the winner of the auction. The aim is to show how unseen, disjointed contract elements may disrupt the visions and estimations made by PPP design committees, which may ultimately lead to this project's failure to meet the expectations and needs of the historically neglected communities it is designed to serve.

While much PPP-focused literature has offered analyses of risk allocation between the public and private partners in broad financial and operational terms, this chapter presents an in-depth analysis of the mechanics at play in the contact zone between contract design and on-the-ground solar energy service, which can serve as a roadmap for other governments as they design and implement rural electrification projects in partnership with private actors.

The chapter's main contribution is revealing how the Peruvian state's poor understanding of the ground-level complexities of off-grid solar energy provision is reflected in the *proyecto masivo* design. The limited technical and financial role allocated for the private firm, Ergon, has not properly accounted for the logistical and social challenges of providing solar energy service in remote communities of the country. This chapter also argues that the *proyecto masivo* contract strains the country's energy public distribution companies with burdensome activities that these companies are unwilling to carry out and will ultimately cause them financial harm. In addition, the chapter reveals how both private and public companies are obligated to collaborate with one another, yet the duties allocated to each sector as well as the state-mandated remuneration levels tied to their responsibilities, ensuring that each sector will operate with indifference toward the other.

Part 2 of the book is focused on how current regulatory issues are hampering the development of energy transitions in the subcontinent, and what needs to be implemented to unlock the full potential of renewables exploitation in Latin America and Caribbean.

In Chapter 7, The prosumer legal protection in Latin America: a requirement for an energy transition, Matheus Linck Bassani evaluates on a comparative basis the legal aspects of the prosumer in the most electricity demanding and populated Latin American countries, analyzing whether the domestic regulatory frameworks provide conditions to adopt decentralized energy generation and identifying the obstacles to reach this new model.

The author examines whether consumer or civil Law can protect the prosumer, depending on the regulatory framework of each country to be analyzed.

The chapter's main contribution is to identify and demonstrate on which Latin American countries the prosumer could be considered legally protected by consumer and civil law, due to the existent contract relations within the utilities.

In Chapter 8, Challenges of South American energy transition: energy efficiency and distributed generation, Maria Alejandra Ise, Silvina Carrizo, and Marie Forget examine the array of distributed generation and efficiency policies in the region and analyze differences in their implementation, with emphasis on national trajectories of energy transitions toward more sustainable systems.

The chapter analyzes, for each South American country, the public policies and incentive mechanisms that seek to promote energy efficiency and distributed generation, as well as the schemes adopted for its implementation and associated territorial transformations. The analysis takes into consideration national differences in terms of resource availability and socioeconomic situations.

The chapter's main contribution is to identify the state of development of energy efficiency and distributed generation systems in South America, verifying which countries achieved a consolidated transition process and those with little experience or in experimental phases. The chapter will allow for deeper insight into regulatory schemes, incentive mechanisms, and national trajectories of these new energy alternatives.

In Chapter 9, Urban microgrids: benefits, challenges, and business models, Marco Aurelio Lenzi Castro describes the main benefits that an urban microgrid can provide to its participants, utilities, and the overall society, highlighting the technical, economic, and regulatory challenges that such project must overcome to become feasible, as well as assessing innovative business models that maximize the economic benefits.

The author assesses the key drivers to develop urban microgrids in developed and developing countries, with focus on Brazil, and discuss the changes that could be done in the regulatory framework to support suitable business models that reduce the uncertainty for investors.

The chapter's main contribution is to shed light to the potential benefits that urban microgrids can provide not only for its participants but also for utilities and the energy system, and how the regulatory framework could be improved to allow innovative business models

that bring economic feasibility for these projects.

In Chapter 10, Regulatory aspects of energy efficiency in Uruguay, Ana Laura Rodríguez D'Espada identifies the public policies and legal instruments used to encourage and involve the productive, residential, and government sectors in the change of consumption behavior toward the optimization of the energy use as well as the mechanisms that generate public awareness on the importance of a cultural change toward energy efficiency.

The chapter's main contribution is to offer a clearer insight about the variety of financial and legal instruments in order to achieve an avoided energy consumption goal drafted in the Uruguayan Energy Efficiency Plan-2024. The national legal regulations based on a long-term energy policy, together with international measures (NDCs and NAMAs within the framework of the Paris Agreement), determined that different actors have contributed to this transformation.

In Chapter 11, The Ecuadorian energy matrix: *from a fossil-fuel dependency to a renewable production*, Pablo Morales and Edisson Morales analyze the energy transition in Ecuador, stating what changes Ecuador made in economics, politics, and legal terms, and which ones Ecuador must do in the following years to reach a clean energy transition.

The authors examine the political, economic, and legal actions taken since 2008 in Ecuador regarding the energy transition, also analyzing the regulatory framework.

The chapter's main contribution is to offer a clearer insight about the current state of energy transition in Ecuador and to state what should be done by governments to consolidate a clean energy transition.

In Chapter 12, Environmental impact assessment and public participation of geothermal energy projects: the cases of Chile, Costa Rica, Colombia, and Mexico, Miguel Saldivia Olave and Sofía Vargas-Payera examine the contribution of geothermal energy to the energy transition in Latin America, focused on those countries with a high potential of this clean energy, such as Chile, Colombia, México, and Costa Rica.

This chapter aims to discuss the compliance of geothermal energy regulations with some international principles of sustainable development such as (1) the sustainable use of natural resources and (2) the principle of public participation.

The chapter's main contribution is to identify some legal gaps on regulations on geothermal energy which critical from the perspective of sustainable development. These pending matters include the lack of public participation and the absence of environmental impact assessment in certain geothermal projects.

In Chapter 13, Competitive renewables as the key to energy transition—RenovaBio: the Brazilian biofuel regulation, Danielle Mendes Thame Denny examines the regulatory framework established by Law 13,576/2017, which created the Brazilian National Policy on Biofuels (RenovaBio), and by Resolution ANP 758/2018, which provides for a certification and a criteria for the emission of tradable notes to producers and importers, which are the basis of the trade mechanism established by RenovaBio.

The chapter's main contribution is to offer a clearer insight about the possibilities, consequences, and influences of a regulatory framework on biofuels in Brazil to Latin America's energy transition. It also envisages a possibility of increasing the use of technology to manage the certification process and notes trade.

In Chapter 14, The role and impacts of policies in hampering the biogas transition in Brazil, Luiz Gustavo Silva de Oliveira discusses the evolution of Biogas field in Brazil and examines the systemic problems that hamper further development and diffusion of biogas technologies in the country. More

specifically, this analysis focuses on conflicts between possible solutions for systemic problems and established policies that influence biogas technologies.

This chapter departs from technological innovation system framework to examine the systemic problems of biogas field and borrows from the policy mix conceptual framework to investigate possible conflicts of solutions and established policies.

The chapter's main contribution is to indicate three main groups of problems that limit further development of biogas energies in Brazil: the low level of knowledge about biogas by players in biogas field, the divergent frames and financial conditions, and the limited spectrum of interactions. Moreover, the findings also show the need for horizontal and vertical coordination in order to avoid inconsistencies with established policy mixes.

Part 3 of the book sheds some light on sociotechnical issues, making a clear statement that social and technology aspects play a paramount role in the success or failure of energy transitions and must not be disregarded.

In Chapter 15, Foresight for Chile's energy transition—unleashing societal transformations, Joni Karjalainen, Noora Vähäkari, and Sirkka Heinonen assume a future-oriented, long-term, and holistic view to examine the prospects of an energy transition into a fully renewable energy system. The book chapter focuses on Chile, which has set ambitious renewable energy targets and adopted growing shares of solar energy in the 2010s.

The authors first explain how foresight can be used to explore a transition into a renewable energy system. Scenarios are a commonly used foresight tool, and as compelling and evidence-based narratives they can be used to explore long-term transitions. In scenario-making, there are growing calls to move beyond extrapolating from past or present trends and instead use transition, backcasting, and participatory approaches to find sustainable energy solutions. The authors use radical and discontinuous sociocultural scenarios that describe changes in the energy landscape, with a semibackcasting approach. The chapter demonstrates how a sociocultural approach to scenario building can account for the broad economic and political context of transitions. The results from a futures workshop, organized in Santiago de Chile with Chilean innovation, foresight and technology experts, are presented.

The scenarios were tested and discussed in Chile to explore a transition into a renewable energy system. Scenario-testing integrates stakeholder views and recognizes sociocultural factors, and the identified opportunities and challenges in Chile are further discussed.

The chapter's main contribution is to conceptualize transitions beyond a simplistic, managerialistic view. Instead, a futures-oriented and transformative approach is promoted for unleashing the potential innovation benefits embedded in energy transitions.

In Chapter 16, Integrating social and justice dimensions to energy transitions: the case of Mexico, Carlos Tornel examines how the energy transition is being framed through a narrow approach, where technological innovation and a managerial governance have perpetuated and increased spatial and social inequalities in the energy sector.

The author examines the regulatory framework of Social Impact Assessments and Indigenous Consultations that not only renders social issues into technical problems (and solutions) but also perpetuates practices of centralization, extractivism, and an uneven distribution of costs and benefits from the energy transition.

The chapter's main contribution is to offer clear insights as to how to increase renewable energy integration by incorporating the notion of socio-energy systems, which places considerable emphasis in the interaction between

technical and social dimensions of the energy transition. The author focuses on two case studies in Mexico that underpin new ways to design, imagine, plan, and prospect the energy sector incorporating the notion of socio-energy systems.

In Chapter 17, The role of social resistance in shaping energy transition policy in Mexico: the case of wind power in Oaxaca, Adolfo Mejia-Montero, Lourdes Alonso-Serna, and Carlo Altamirano-Allende analyze the importance of national policies on attending the social dimension of energy transitions in Mexico by focusing on different key wind energy developments in the southern state of Oaxaca, where controversies have arisen. They illustrate the influence that organized grassroots movements of resistance are having in the shaping and the design of the social aspects of the national energy transition policy in Mexico.

The authors use a narrative of grassroots resistance movements on different wind energy projects in parallel to the generation of policies shaping the energy transition regulatory framework in Mexico in order to analyze their mutual influence.

The chapter's main contribution is to illustrate the complexities of the social dimensions of the governance of energy transitions in the context of indigenous land tenures, grassroots organizing, and policy-making for renewable energy markets.

In Chapter 18, Energy transition in isolated communities of the Brazilian Amazon, Antonella Mazzone examines the impacts of sustainable decentralized solutions on poverty reduction in four isolated villages of the Brazilian Amazon. Specifically, the chapter looks at the opportunities and limitations of PV mini-grid systems in the creation of new jobs and increase local production.

The author examines the impacts of PV mini-grid systems through a mixed-method framework in which qualitative methods explain why certain technologies may not work in some specific context due to cultural differences. Quantitative methods instead help the visualization of the data and provide a tangible tool to measure the change in poverty reduction.

The chapter's main contribution is to offer a new approach to a timely issue: whether energy transition can effectively deliver sustainable development and provides interesting insights on the limitations and opportunities of these technologies in this particular geography.

In Chapter 19, An inclusive and participative model for energy transition in Latin America: the case of Chilean *Generación Comunitaria*, Francisco Merino, Adolfo Mejía Montero, and Cecilia Dastres describe the process in which the Chilean energy ministry develops the community energy mechanism, detailing its objectives and components, along with its conceptual foundations. On a parallel, this narrative will be accompanied by an analysis from an academic perspective on how this mechanism resonates with literature on energy justice, community generation, renewable energy social acceptance, and climate change, as well with other academic work on Latin American low carbon transition.

The authors use an approach of government perspective to narrate how the Chilean energy ministry developed the community energy mechanism, in combination with an academic analysis on how this mechanism resonates with the applied principles of energy justice, as well as other relevant academic literature.

The chapter's main contribution is to merge a government narrative of the community generation mechanism with insights from academia regarding resonance with energy justice scholarship. It concludes reflecting briefly on the biggest challenges presented by the Chilean community energy mechanism, and the potential replicability that this process in Chile could find in other Latin American contexts.

References

Aghahosseini, A., Bogdanov, D., Breyer, C., 2017. A techno-economic study of an entirely renewable energy-based power supply for North America for 2030 conditions. Energies 10, 1171.

Barbosa, L., Bogdanov, D., Vainikka, P., Breyer, C., 2017. Hydro, wind and solar power as a base for a 100% renewable energy supply for South and Central America. PLoS One 12, e0173820.

Brown, T.W., Bischof-Niemz, T., Blok, K., Breyer, C., Mathiesen, B.V., 2018. Response to 'Burden of proof: a comprehensive review of the feasibility of 100% renewable-electricity systems'. Renew. Sustain. Energy Rev. 92, 834–847.

Fauset, C., 2010. The techno-fix approach to climate change and the energy crisis. In: Abramsky, K. (Ed.), Sparking a Worldwide Energy Revolution. AK Press, Oakland, CA, p. 301.

Fouquet, R., Pearson, P.J.G., 2012. Past and prospective energy transitions: insights from history. Energy Policy 50, 1–7.

Grubler, A., 2012. Energy transitions research insights and cautionary tales. Energy Policy 50, 8–18.

Hanger-Kopp, S., Lieu, J., Nikas, A. (Eds.), 2019. Narratives of Low-Carbon Transitions — Understanding Risks and Uncertainties. Routledge.

Hansen, K., Breyer, C., Lund, H., 2019. Status and perspectives on 100% renewable energy systems. Energy 175, 471–480.

Jenkins, K.E.H., Hopkins, D. (Eds.), 2019. Transitions in Energy Efficiency and Demand. Routledge.

Kloppenburg, S., Boekelo, M., 2019. Digital platforms and the future of energy provisioning: promises and perils for the next phase of the energy transition. Energy Res. Soc. Sci. 49, 68–73.

Köhler, J., et al., 2019. An agenda for sustainability transitions research: state of the art and future directions. Environ. Innov. Soc. Transitions 31, 1–32.

Lampedusa, G.T.D., 1991. The Leopard. Everyman's Library, London.

Lund, P., 2006. Market penetration rates of new energy technologies. Energy Policy 34, 3317–3326.

Miller, C., Iles, A., Jones, C., 2013. The social dimensions of energy transitions. Sci. Cult. (Lond.) 22, 135–148.

O'connor, P.A., 2010. Energy transitions. In: The Pardee Papers, no. 12.

Scheer, H., 2010. Der Energetische Imperativ. Kunstmann Verlag, Munique, pp. 85–112.

Sioshansi, F. (Ed.), forthcoming. Behind and Beyond the Meter — Utilization, Optimization and Monetization. Associated Press.

Smil, V., 2010. Energy Myths and Realities: Bringing Science to the Energy Policy Debate. Rowman and Littlefield, Washington, DC.

Sovacool, B.K., 2016. How long will it take? Conceptualizing the temporal dynamics of energy transitions. Energy Res. Soc. Sci. 13, 202–215.

Sovacool, B.K., Hirsh, R.F., 2009. Beyond batteries: an examination of the benefits and barriers to plug-in hybrid electric vehicles (PHEVs) and a vehicle-to-grid (V2G) transition. Energy Policy 37, 1095–1103.

Wilson, C., Grubler, A., 2011. Lessons from the history of technological change for clean energy scenarios and policies. Nat. Resour. Forum 35, 165–184.

PART 1

Changing the energy landscape: policy and economic issues

1

Is there a Latin American electricity transition? A snapshot of intraregional differences

Lucas Noura Guimarães[1,2]

[1]EDP Brasil, São Paulo, Brazil [2]Brazilian Institute of Energy Law Studies, São Paulo, Brazil

1.1 Introduction

As the introduction of this volume has presented, energy transitions happen worldwide and have many different drivers, motivations, issues, and risks, being an analysis of climate change and its impacts on the power sector of Latin American and the Caribbean (LAC) countries only one of the possible approaches. Given the huge variety and complexity of energy transition studies,[1] as well as the lack of literature on LAC energy transitions, this chapter focuses merely on shedding some light on the relations between climate change and the power sector in LAC, in order to verify to what extent it is possible to speak of a Latin American electricity transition.

It is well known that Latin America—and South America specially—is a region of the Earth with plenty of natural resources. Water, wind currents, solar irradiation,[2] and geothermal heat[3] are abundant and widely scattered around this portion of the Earth. Besides that, there is a considerable amount of land for biomass, not to mention the vast reserves of fossil fuels—oil in Venezuela, gas in Argentina, and the presalt layer along the Brazilian coast. Due to that configuration, 56.6% of the power generation is delivered to the region's grid by means of renewable energy sources (Inter-American Development Bank, 2017a,b,c). Studies show that this number can reach 100% in the subcontinent (Barbosa et al., 2017; Inter-American Development Bank, 2017a,b,c;

---

[1] Scholars have studied energy transitions for quite a while and from many different points of view, for example, Smil (2017), Rogge et al. (2017), Cherp et al. (2017), Smil (2016), Sovacool (2016), Araújo (2014), Grubler (2012), and Fouquet and Pearson (2012).

[2] For example, the Atacama Desert that has one of the highest and most continuous solar irradiation levels on Earth.

[3] Especially in the Caribbean (Inter-American Development Bank, 2017a,b,c).

The Regulation and Policy of Latin American Energy Transitions
DOI: https://doi.org/10.1016/B978-0-12-819521-5.00001-2

Gils et al., 2017; Zapata et al., 2018; Blazqueza et al., 2018; Hansen et al., 2019).[4]

Nevertheless, when compared to the United States and the European countries, LAC countries consume less electricity,[5] which has direct impact on the region's development. This feature, detailed analysis of which goes beyond the scope of this chapter, can be attributed to political, economical, and technological reasons. Along the past century, LAC countries often lacked the political will,[6] the economic environment, and/or the technological knowhow to explore its natural resources in their full potential. The untapped potential for wind and solar energy, as well as biomass and geothermal[7] energy, is still huge,[8] even if environmental restrictions are taken into consideration.[9]

Once seen as a development problem in the past, this "delay" in the promotion of adequate investment in renewables is now considered by scholars, stakeholders, and public agents. The so far unexplored renewable potential is the free ticket to the LAC electricity transition, given that it can, at the same time, promote economic development, social welfare and avoid jeopardizing the environment.

Studies show that between 2016 and 2040, power generation is expected to grow in LAC countries an average of 2.73% yearly (from 1429.67 to 2657.50 TW h) (Lee et al., 2018). Coupling this growth with the exploitation of the untapped potential of renewables is of paramount importance to achieve an environmental-friendly electricity transition. The constant reduction in the costs of renewable energies, making them a less expensive source of new electricity than the least cost fossil fuel alternative (International Renewable Energy Agency, 2019), can help guarantee this coupling.

But when analyzing in more detail the climate and geography of LAC countries, considering the lack of regional integration, as well as their electricity matrixes, one might come to ask if there is really "a" Latin American electricity transition. Let alone geography and climate, despite the overall abundance of renewable energy sources, not all countries are "blessed by God" in this matter—specially the highly oil-dependent countries in the

---

[4] Skeptic about the possibility of a 100% renewable electricity sector, see Blazqueza et al. (2018). See also Hansen et al. (2019).

[5] As of 2014, LAC consumed 2128 MW h per capita, while the European Union consumed 6352 MW h per capita and the United States, 1325 MW h per capita (The World Bank, 2019).

[6] A recent assessment with almost 300 stakeholders and public and private agents of the power sector in LAC countries showed that 65% of them consider corruption to be the main obstacle to the sustainable development in LAC countries followed by "lack of energy planning" and "lack of private investment," (Organización Latinoamericana De Energía—Olade, 2018).

[7] For an analysis of geothermal exploitation in different countries, see Chapter 12.

[8] Alone the six tiny islands of the Eastern Caribbean—Antigua and Barbuda, Dominica, Grenada, Saint Kitts and Nevis, Saint Lucia and Saint Vincent and the Grenadines—which are highly dependent on oil imports, have an untapped renewable potential of more than 280 MW, Inter-American Development Bank (2016). Only along the Brazilian Southeastern coast, studies identified an untapped potential for offshore wind of more than 100 GW, Pimenta et al. (2008).

[9] For example, in Brazil, where the untapped potential for hydro is located inside the Amazon. Within the scope of the environmental licensing of São Luiz do Tapajós power plant, the Brazilian Institute for Environment (IBAMA) closed the case in August 2016, arguing that there was no possible technical arrangement that could slightly comply with the minimum environmental requirements, drawing a clear line in terms of what is environmentally feasible. Both the Federal Prosecution and the National Indian Foundation (FUNAI) stated that the construction of the power plant was not feasible.

Caribbean region. Besides that—except the SIEPAC project[10] and apart from the political efforts in regional flora[11]—LAC countries cannot be considered energy integrated, neither in the physical nor in the market-based aspects.

Moreover, taking each country's intended nationally determined contributions (INDCs) as example, proposed within the scope of the Paris Agreement and transformed into nationally determined contributions automatically after the ratification of the agreement,[12] it is easy to see that not all LAC countries depart from the same starting point—nor are walking the same energy transition path. The same result is achieved when considering the different degrees of energy market development for each country of the region.[13]

On the other hand the benefits of a largely integrated energy grid—one that crosses national borders and interconnects different types of energy matrixes and energy systems—are well documented in the literature (Liu, 2016; de Castro et al., 2016; Breyer et al., 2015; Czisch, 2005). Despite adding complexity to the grid operation, the larger the grid, the greater the flexibility and the options to meet power demand at optimal economic and technical conditions. That said, it should also be investigated in what measure an interconnected region allows for a unification of the electricity transition discourses and processes of LAC countries.

The main purpose of this introductory chapter is to use scholars' literature and international organizations' reports and assessments to verify to which degree it is possible to see LAC countries as a whole, when it comes to study, analyze, and interpret the respective countries' pathways to set in motion their electricity transitions.

This chapter is divided into five sections. After Section 1.1, there are four more sections that are as follows:

- Section 1.2 provides an overview of the LAC countries' starting points regarding their state-of-the-art electricity matrixes;
- Section 1.3 covers the LAC countries' different energy paths, taking the INDCs related to the power sector as an example;
- Section 1.4 takes a broader approach and examines the possibilities to regionalize the Latin American electricity transition, despite the solo efforts of each country; and
- Section 1.5 brings the conclusion of the chapter.

1.2 Different starting points: assessing the electricity matrixes

As widely known, the main goal of the Paris Agreement is to strengthen the global response to the threat of climate change by keeping the global temperature rise this century well below 2°C above preindustrial levels and to pursue efforts to limit the temperature increase even further to 1.5°C.

How this will be achieved is left entirely to each country's sovereign decision. It is true

[10] The Project consists of a 1800 km long, 230 kV single circuit, transmission line crossing six Central American countries—Guatemala, El Salvador, Honduras, Nicaragua, Costa Rica, and Panama. Available from: <https://www.eprsiepac.com/contenido/> (accessed 31.05.19); Inter-American Development Bank (2013).

[11] For an analysis on this matter, see Chapter 4, Energy (and climate) challenges in South America: what planning for what transition?.

[12] As of May 2019, 185 Parties—out of 197—have ratified the Paris Agreement. Available from: <https://unfccc.int/process/the-paris-agreement/status-of-ratification> (accessed 29.05.19).

[13] For an analysis on this matter, see Chapter 2, Using market mechanisms for a reliable energy transition in Latin America.

that the Paris Agreement promote a long-term goal for adaptation, that is,

> to increase the ability to adapt to the adverse impacts of climate change and foster climate resilience and low greenhouse gas emissions development, in a manner that does not threaten food production.

The countries also agreed to "work towards making finance flows consistent with a pathway towards low greenhouse gas emissions and climate-resilient development." But the means to achieve that and which areas to tackle—transportation sector, agriculture, forestry and land use (AFOLU), and/or power sector—are entirely open.

While it is clear that each country should contribute to the reduction of greenhouse gas (GHG) emissions, their contributions will range widely and will take into consideration not only the current GHG emissions but also historical data. Moreover, each country has a specific capacity to contribute to the fight against climate change, which is influenced by political, social, economical, technological, and environmental factors. Considering this scenario, it would be hard to speak of "the" Latin American electricity transition.

1.2.1 The role of the energy sector in Latin American and the Caribbean countries' greenhouse gas emissions

Before proceeding to the analysis of each country's electricity matrix, it should be noted

that within the scope of fighting climate change, the GHG emissions in the energy and in the power sectors are not the main targets in some LAC countries, contrarily to what is seen in other regions of the world, such as Europe.

Brazil, for example, is ranking fourth—as of 2014—in the world when it comes to national contributions to global warming (Matthews et al., 2014) mainly because of AFOLU emissions. As of 2010 only 29% of total national emissions came from the energy sector, while AFOLU emissions accounted for 756 Mt CO_2eq.[14] As of 2017 alone the agribusiness industry responded for more than 20% of Brazil's total GHG emissions,[15] with 240 Mt CO_2eq.[16]

Peru and Ecuador are also main AFOLU emitters, totaling 478.7 Mt CO_2eq in 2010. Taken LAC countries all together, AFOLU is the main source of GHG emissions on the continent (Postic et al., 2017).

Notwithstanding the importance of AFOLU in the total GHG emissions in LAC countries, given the existent space to solve social and economic inequalities and to expand and meet energy demand in LAC countries, the role assumed by the power sector in mitigating climate change should not be underestimated. Pressures on the environment related to agriculture—embargos on meat imports against countries that opt to deforest in order to raise cattle or suppression of international financial sponsoring for countries that fail to protect forests and biomes, for example—might lead to a

[14]   More than twice the European AFOLU emissions, see Postic et al. (2017).

[15]   From these 20%, almost 65% come from enteric fermentation, fact of which makes room for the development of biogas in Brazil, see Guimarães (2019a,b), Guimarães and Santos (2019), Palermo et al. (2014), and Ribeiro and Raiher (2013).

[16]   Available from: <http://monitoragropecuario.seeg.eco.br/> (accessed 31.06.19). The System for Estimating Greenhouse Gas Emissions (SEEG) is an initiative to produce annual estimates of GHG emissions in Brazil. SEEG covers 48 years (1970–2017) of GHG estimates, providing a unique dataset with more than 2 million up-to-date records at national and subnational levels for five sectors that are emission sources: agriculture, energy, industrial processes and product use, land use change, and waste. It is the largest public database on GHG emissions of a country in the world. Regarding SEEG, see de Azevedo (2018).

future stabilization or decrease of AFOLU-related emissions, while the economic growth and increase in power demand in LAC countries might add relevance to the participation of the energy sector in the total GHG emissions.

Therefore it is of utmost importance for the climate that this growth in the energy sector is not fulfilled by "new" solutions and technologies—such as natural gas, nuclear, carbon capture and storage, shale gas and fracking, hydrogen, geoengineering, among other techno-fixes (Fauset, 2010; Scheer, 2010)—which implementation—benefiting from a narrative that appeals to the necessity to diversify the energy matrix and suffering from a short-sighted view of the transition—might culminate in absorbing the natural advantage that LAC countries have in renewables and in kidnaping the possibility of a long-term renewable electricity transition.

Studies point that, in LAC countries, the increase in power generation goes hand in hand with a sharp drop in the share of fossils and nuclear between 2010 and 2050 (Postic et al., 2017). However, a closer look might prove otherwise for some countries. In the case of Brazil, for example, not only power generation from hydro is dropping,[17] but also the gap is being filled not only by other renewables but also by natural gas,[18] discourse of which favoring it has been gaining momentum due to the exploitation of the presalt layer. The same applies to the exploitation of unconventional gas in Argentina.

Within the scope of electricity transitions, departing from an electricity matrix that is already clean and renewable poses the risk of oversimplifying the diversification argument. The fact that Brazil, Colombia, or Costa Rica, already in 2019, have a good share of renewables should not be understood as an allowance to explore more fossil fuels, for the sake of diversification. It is paramount to analyze from which diversification governments and stakeholders are talking about. The LAC countries' advantage in renewables should not be turned into a new sort of "renewables curse," hampering the maintenance and growth of the renewables' share in the electricity matrixes of LAC countries.

1.2.2 Different on the surface, but with a common target

Let alone the electricity matrixes of each LAC country, the different starting points to counteract climate change and to transform the power sector could make a solid argument favoring solo actions, as seen in Table 1.1.[19]

---

[17] Worldwide, hydro will remain the largest source of renewable energy at least until 2040, International Energy Agency (2019).

[18] According to the National Energy Balance—published yearly by the Energy Research Office, a public think tank integrated to the Brazilian Ministry of Mines and Energy—in 2016, 68.1% of the power offered to the grid came from hydro, 5.4% from wind energy, and 9.1% from natural gas. The National Energy Balance 2018, containing data from 2017, shows a slight reduction on the participation of hydro in the power supply (65.2%). This reduction was equally absorbed by the expansion of wind energy and natural gas: 1.4% increase for each source. Besides that, in these 2 years, the participation of the other fossil fuels remained the same: 6.6% of the total electricity offered to the grid, Empresa De Pesquisa Energética (2017), Empresa De Pesquisa Energética (2018). This data reveals that natural gas is not replacing the more pollutant and expensive thermal power plants, but the cheap and clean hydropower plants. Data from the Brazilian National Agency for Electric Energy corroborates this argument, considering that, as of 2018, 3.1 GW of natural gas-fired power plants are being implemented, while only 2.5 GW of wind and solar projects combined are on the way, Agência Nacional De Energia Elétrica (ANEEL) (2017).

[19] Not listed are the dependencies and other territories, except French Guiana.

TABLE 1.1 Electricity matrixes of each Latin American and the Caribbean countries.

Country	Electricity matrix		
Antigua and Barbuda (329 GW h)[a]	Fossils—100%		
Argentina (136,600 GW h)[b]	Fossils—66%	Hydro—26.5%	Nuclear—5.6%
	Small hydro—1.3%	Wind—0.4%	Biomass—0.2%
	Solar—0%		
Bahamas (1930 GW h)[c]	Fossils—100%		
Barbados (1024 GW h)[d]	Fossils—93.8%	Biomass—5.6%	Solar—0.6%
Belize (371 GW h)[e]	Hydro—43%	Fossils—43%	Biomass—13%
	Solar—1%		
Bolivia (8763 GW h)[f]	Fossils—79.3%	Hydro—19.6%	Biomass—0.7%
	Wind—0.4%		
Brazil (578,898 GW h)[g]	Hydro—65.2%	Fossils—17.1%	Biomass—8.2%
	Wind—6.8%	Nuclear—2.5%	Solar—0.13%
Chile (73,391 GW h)[h]	Fossils—63.7%	Hydro—23.1%	Solar—4.5%
	Wind—4.2%	Small hydro—2.3%	Biomass—2.2%
Colombia (66,255 GW h)[i]	Hydro—79.8%	Fossils—13.7%	Small hydro—5.5%
	Biomass—0.9%	Wind—0.1%	Solar—0%
Costa Rica (10,782 GW h)[j]	Hydro—74.4%	Geothermal—12.4%	Wind—10.6%
	Fossils—1.8%	Small hydro—1.3%	Biomass—0.7%
Cuba (19,366 GW h)[k]	Fossils—96%	Biomass—3%	Hydro—1%
Dominica[l]	Fossils—71.5%	Hydro—28.5%	
Dominican Republic (17,025 GW h)[m]	Fossils—88%	Hydro—9%	Wind—1.5%
	Biomass—1%		
Ecuador (27,396 GW h)[n]	Hydro—57.8%	Fossils—40%	Biomass—1.8%
	Wind—0.3%	Solar—0.1%	
El Salvador (4877 GW h)[o]	Geothermal—30.3%	Fossils—30.3%	Hydro—29.8%
	Biomass—8.3%	Solar—1.5%	
French Guiana	N/A		
Grenada (196 GW h)[p]	Fossils—100%		
Guyana	N/A		
Guatemala (10,878 GW h)[q]	Fossils—40.9%	Hydro—36.3%	Biomass—16.3%
	Geothermal—2.7%	Wind—2%	Solar—1%
Haiti (1092 GW h)[r]	Fossils—80%	Hydro—20%	
Honduras (8782 GW h)[s]	Fossils—50.2%	Hydro—26.8%	Solar—10%
	Wind—6.5%	Biomass—6.5%	
Jamaica (4112 GW h)[t]	Fossils—96.7%	Hydro—3.3%	

(Continued)

1. Changing the energy landscape: policy and economic issues

TABLE 1.1 (Continued)

Country	Electricity matrix		
Mexico (319,364 GW h)[u]	Fossils—81.3%	Hydro—9.7%	Wind—3.3%
	Nuclear—3.3%	Geothermal—1.9%	Small hydro—1.3%
	Biomass—0.5%	Solar—0.1%	
Nicaragua (4541 GW h)[v]	Fossils—47.1%	Wind—16.1%	Geothermal—15.5%
	Biomass—11.7%	Hydro—9.4%	Small hydro—0.1%
Panama (10,733 GW h)[w]	Hydro—61%	Fossils—32.5%	Wind—5.7%
	Solar—0.7%	Biomass—0.2%	
Paraguay (48,412 GW h)[x]	Hydro—99.7%	Biomass—0.3%	
Peru (48,326 GW h)[y]	Hydro—47.6%	Fossils—47%	Small hydro—2.3%
	Wind—2.2%	Solar—0.5%	Biomass—0.4%
Saint Kitts and Nevis (206 GW h)[z]	Fossils—94.33%	Wind—3.9%	Solar—1.77%
Saint Lucia (385 GW h)[aa]	Fossils—99.9%	Solar—0.1%	
St. Vincent and Grenadines (140 GW h)[bb]	Fossils—88%	Hydro—11%	Solar—1%
Suriname (1170 GW h)[cc]	Fossils—64%	Hydro—35%	Solar—1%
Trinidad and Tobago		Fossils—100%	
Uruguay (12,022 GW h)[dd]	Hydro—63.2%	Wind—24.8%	Biomass—6.5%
	Fossils—4.2%	Solar—1.2%	
Venezuela (127,854 GW h)[ee]	Hydro—64%	Fossils—35.9%	Wind—0.1%

[a] As of 2011, International Renewable Energy Agency (2016a,b).
[b] As of 2016, Inter-American Development Bank (2017a,b,c).
[c] As of 2015, National Renewable Energy Laboratory (2015a,b,c).
[d] As of 2015, Energy Division (2017).
[e] As of 2016, Inter-American Development Bank (2017a,b,c).
[f] As of 2016, Inter-American Development Bank (2017a,b,c).
[g] As of 2017, Empresa De Pesquisa Energética (2018).
[h] As of 2016, Inter-American Development Bank (2017a,b,c).
[i] As of 2016, Inter-American Development Bank (2017a,b,c).
[j] As of 2016, Inter-American Development Bank (2017a,b,c).
[k] As of 2014, Environmental Defense Fund (2017).
[l] As of 2014, INDC of the Commonwealth of Dominica.
[m] As of 2014, International Renewable Energy Agency (2016a,b).
[n] As of 2016, Inter-American Development Bank (2017a,b,c).
[o] As of 2016, Inter-American Development Bank (2017a,b,c).
[p] As of 2011, International Renewable Energy Agency (2012).
[q] As of 2016, Inter-American Development Bank (2017a,b,c).
[r] As of 2017, Available from: <http://global-climatescope.org/results/ht#power-market> (accessed 29.06.19.).
[s] As of 2016, Inter-American Development Bank (2017a,b,c).
[t] As of 2014, Available from: <https://www.mset.gov.jm/overview-jamaicas-electricity-sector> (accessed 5.06.19.).
[u] As of 2016, Inter-American Development Bank (2017a,b,c).
[v] As of 2016, Inter-American Development Bank (2017a,b,c).
[w] As of 2016, Inter-American Development Bank (2017a,b,c).
[x] As of 2016, Inter-American Development Bank (2017a,b,c).
[y] As of 2016, Inter-American Development Bank (2017a,b,c).
[z] As of 2015, National Renewable Energy Laboratory (2015a,b,c).
[aa] As of 2015, National Renewable Energy Laboratory (2015a,b,c).
[bb] As of 2014, Inter-American Development Bank (2017a,b,c).
[cc] As of 2016, Raghoebarsing and Reinders (2019).
[dd] As of 2016, Inter-American Development Bank (2017a,b,c).
[ee] As of 2016, Inter-American Development Bank (2017a,b,c).

1. Changing the energy landscape: policy and economic issues

Indeed, Table 1.1 clearly depicts that while some countries—especially in South America—rely greatly on hydro, other countries—in the Caribbean, mainly, but also Chile, Mexico, and Argentina—are extremely oil dependent to cover their populations' and industrial sectors' power demand. Besides that, other-than-hydro renewables, although still gaining terrain and with a small overall participation, bring the potential to augment this intraregional difference—Uruguay, Nicaragua, Brazil, and Costa Rica being already examples of that.

Considering that, a one-size-fits-all approach is not likely to work in promoting a LAC electricity transition, since the share of different energy sources varies from country to country. While Colombia is looking at how to depend less on hydro, St. Vincent and the Grenadines should try to lower its oil imports, and Peru and Mexico are working on the deployment of their potential in small hydro and wind, respectively.

Nevertheless, it can also be argued that the region in general suffers from a lack of diversification of its electricity matrix. Considering the data gathered in Table 1.1, it is easy to note that fossil energy sources, broadly taken, and hydro are the only two sources of energy with noteworthy importance—except in Nicaragua, which has a diversified electricity matrix.

This can be explained by the huge oil reservoirs under the subcontinent and the need to increase stability and reliability in power systems. On the other side, difficulties related to technological innovation in renewable energies—cost of imports of technologies and lack of know-how—contributed to the formation of electricity matrixes seen in Table 1.1.

In an era of elevated climate pressure, not only dependence on nonrenewable, expensive fossil fuels puts countries in peril but also the dependence on climate-sensitive energy sources, such as hydropower (Lucena et al., 2009, 2018; Queiroz et al., 2019; Turner et al., 2017; Zhang et al., 2018; Berga, 2016) and wind,[20] could draw attention from countries relying heavily on such sources, such as Paraguay, Uruguay, Panama, Colombia, Brazil, and Venezuela.

Diversification of the electricity matrix, therefore, can be stated as a common goal that LAC countries must pursue. As seen in Section 2.1, the open questions are which and how much diversification?

1.3 Different electricity transitions: the case of intended nationally determined contributions

Despite the common goal of diversification of energy sources, when looking at the INDCs and their participation in electricity transitions, the intraregional differences are increased. That is mainly due to the fact that LAC has its renewable energy sources scattered along its countries. Moreover, renewable energy sources are not constantly available throughout the year.

For instance, while Chile can benefit from the solar irradiation in the Atacama Desert, Brazil can explore the onshore and offshore wind currents in its northeastern coast, and the Eastern Caribbean can partially meet its future power demand with geothermal energy.

Once again, not only because LAC countries have different starting points, but also because they have different sources to explore, is that a silver bullet would not work properly in implementing a region-wide renewable electricity transition.

Although a silver bullet might not hit all countries equally, it cannot be disregarded that they mostly need to diversify away from fossil

---

[20] Although the opposite could also be true, that is, climate change could improve wind conditions, see Lucena et al. (2010).

fuels and hydro.[21] This complexity given by the dichotomy between "what unites us"—low degree of sources diversification—and "what separates us"—different starting points, different sources to explore, different transition rhythms—poses the question if electricity transition in LAC is only about diversification, mainly to dilute among other sources the high concentration of fossil fuels and hydro participation in the electricity matrixes of LAC countries. More than that could not be agreed on the regional level.

Considering that they point to the level of renewable energy penetration a country wants to achieve, as well as the time to achieve it,[22] the INDCs depicted in Table 1.2 might shed some light on the efforts LAC countries are making to cope with the Paris Agreement, helping to check if "a" Latin American electricity transition is on its way or not.

Although the INDCs corroborate the idea previously stated—LAC countries are trying to diversify away from hydro and fossil fuels—it is clear that their variety fulminates any attempt to speak of "a" Latin American electricity transition. Not only the time horizons vary—2020, 2025, and 2030—but also the energy sources and the growth of each source in a country's respective electricity matrix can greatly differ from one country to another, fact of which reflects the uneven distribution of resources on the subcontinent.

The abovementioned diversity brings, also impacts, on the renewable source financially incentivized. Despite the countries' different rhythms in the implementation of their electricity transitions, the fact that LAC is formed by developing countries has led, until the middle of the 2010s, to a bigger regional development of the wind industry, when compared to investments in solar.[23] This difference can find its roots in the distributed character of the solar projects, which involves the creation of new business models, highly dependent on political, economic, regulatory, and legal stability—features lacking in some LAC countries. On the other hand, centralized wind farms require less innovation in terms of business models, therefore demanding lesser changing efforts from the energy industry.

From 2016 onward this trend has inverted, and now the investment volume in solar is twice as bigger as the investment in wind energy in developing countries—USD 115 billion against USD 54.8 (Frankfurt School-UNEP Centre and Bloomberg New Energy Finance, 2018).

While being true that most of the investment in renewables from developing economies came from China, it cannot be disregarded that Brazil and Mexico figure among the top 10 countries investing in renewables (Frankfurt School-UNEP Centre and Bloomberg New Energy Finance, 2018). It is no coincidence that those two countries passed specific legislation

---

[21] Nevertheless, a study shows that hydro continues to be an important source of energy in Latin America. In 2018 mainly because of Belo Monte operations, Brazil overtook the United States in the installed capacity rank for hydro, occupying the second position, only behind China. South America was the second fastest growing region, adding 4855 MW in installed hydropower capacity in 2018. Ecuador, Chile, Peru, and Colombia added c. 800 MW of installed capacity in 2018, and Brazil another 3.8 GW, International Hydropower Association (2019).

[22] Factors influencing the rhythm of an energy transition include, among others, technological progress, access to financing possibilities, and design of public policies and of adequate regulatory and legal frameworks, FGV Energia and Konrad Adenauer Stiftung (2019).

[23] Frankfurt School-UNEP Centre and Bloomberg New Energy Finance (2016). In general, since 2015 developing countries are investing more in renewables than developed countries. In this study, China, India, and Brazil were considered developing economies.

TABLE 1.2 Intended nationally determined contribution (INDC) targets of each Latin American and the Caribbean (LAC) countries for the power sector.

Country	INDCs
Antigua and Barbuda	By 2030 100% of the electricity demand in the water sector and other essential services will be met through off-grid renewable sources
	By 2030 achieve an energy matrix with 50 MW of electricity from renewable sources both on and off grid in the public and private sectors
Argentina	In the energy sector, two central policy objectives were defined: the diversification of the energy matrix and the promotion of rational and efficient use of energy. In this sense the country has a regulatory structure with strategic long-term plans that promote, among other measures, a larger participation of nonconventional renewable sources, hydroelectricity, nuclear power, and the replacement of fossil fuels by biofuels. Moreover, a set of programs and actions intended to reduce energy consumption intensity has been established
Bahamas	Promote the use of less carbon intensive fuels. Accordingly, the government has defined the policy framework for a low-carbon development plan through the National Energy Policy that sets a target to achieve a minimum of 30% renewables in the energy mix by 2030 and will allow for a 10% residential energy self-generation program within the year
Barbados	Renewable energy: contributing 65% of total peak electrical demand by 2030. The country has made huge strides in this regard; for example, distributed solar PV installation is growing exponentially and this trend is expected to continue. Other planned measures include waste-to-energy and biomass generation plants, wind, distributed and centralized solar PV, and capture and use of landfill gas for energy generation
	Electrical energy efficiency: a 22% reduction in electricity consumption compared to a BAU scenario in 2029
Belize	Increase the share of renewable energy in Belize's electricity mix by 85% by 2027, with a 62% CO_2 emissions reduction compared to a BAU scenario
Bolivia	Increase participation of renewable energy to 79% by 2030
	Increase participation of alternative energy and other energy (steam combined cycle) from 2% in 2010 to 9% in 2030 in the total electrical system, which implies an increase of 1228 MW by 2030
	Increase power generation to 13,387 MW in the electricity sector by 2030, compared to 1625 MW by 2010
	Change and diversification of the energy matrix with renewable energy growth through the construction of hydropower plants (small and medium, large, and multipurpose), boost alternative energy (wind, biomass, geothermal, and solar), and use other sources of energy (steam combined cycle)

Brazil	1. Increase the share of sustainable biofuels in the energy mix to approximately 18% by 2030, by expanding biofuel consumption, increasing ethanol supply, including by increasing the share of advanced biofuels (second generation), and increasing the share of biodiesel in the diesel mix
	2. Achieve 45% of renewables in the energy mix by 2030, including the following:
	  a. expanding the use of renewable energy sources other than hydropower in the total energy mix to between 28% and 33% by 2030;
	  b. expanding the use of nonfossil fuel energy sources domestically, increasing the share of renewables (other than hydropower) in the power supply to at least 23% by 2030, including by raising the share of wind, biomass, and solar; and
	  c. achieving 10% efficiency gains in the electricity sector by 2030

(Continued)

TABLE 1.2 (Continued)

Country	INDCs
Chile	Thirty percent reduction in the marginal costs of electric energy by 2018, 20% of the energetic matrix should be made up of nonconventional renewable energies by 2025, and a 20% reduction in the energy consumption forecast by 2025
Colombia	No specific targets for the power sector in INDC
Costa Rica	Achieve and maintain a 100% renewable energy matrix by 2030
Cuba	Achieve 24% of the country's electricity from renewable sources by 2030, with an installed capacity of up to 2 GW
Dominica	By 2030 98.6% total emission reductions in the energy sector (mainly from harnessing of geothermal resources, with 2×3.5 MW power plants by 2020 and another one with 3.5 MW by 2025)
Dominican Republic	No specific targets for the power sector in INDC
Ecuador	Ecuador intends to reduce its emissions in the energy sector to 20.4%−25% below the BAU scenario. However, a potential for reducing emissions even further in the energy sector, to a level between 37.5% and 45.8% with respect to the BAU baseline has also been calculated
	Optimization of gas use for power generation from the gas associated to oil exploitation at different capacity levels. With the use of these gases, electricity will be generated and transmitted to the Amazon region for the use in oil industry, water pumping, and camps and communities in the covered areas, replacing the traditional use of diesel for these ends. A second phase of this program focuses on linking this energy to the national interconnected system
	The introduction of an installed capacity of electric generation from hydroelectric plants of 2828 MW additional to the BAU in the first scenario and an extra 4382 MW in the second, depending upon international circumstances
El Salvador	At least 12% growth of the country's electricity from renewable sources by 2025
French Guiana	N/A
Grenada	Grenada plans a 30% reduction in emissions through electricity production by 2025 with 10% from renewables and 20% from energy efficiency measures. To achieve this goal, Grenada needs to produce 20 MW h of electricity from renewable sources at a conservative 45% portfolio capacity factor. This will emerge in the form of 10 MW from solar, 15 MW from geothermal, and 2 MW from wind. Energy efficiency actions to reduce emissions include retrofitting of all buildings (20% reduction), establishment of policies for energy efficiency building codes for all building sectors (30% reduction), and implementation of energy efficiency in hotels (20% reduction)
Guyana	Guyana is committed to eliminating its near complete dependence on fossil fuels. Given its solar, wind, and hydropower potential and relatively small national demand, it is believed that with adequate and timely financial support, Guyana can develop a 100% renewable power supply by 2025. Assessment of the potential of the renewable power sources will be undertaken to determine the most cost-effective and efficient means of developing this potential
Guatemala	At least 10% growth of the country's electricity from renewable sources by 2030 (from 69.72% to 80% of renewable participation in the electricity matrix)
Haiti	By 2030 increase to 47% the participation of renewables in the electricity matrix (hydro 24.5%, wind 9.4%, solar 7.5%, and biomass 5.6%)
	By 2030 installation of four wind parks (50 MW), 60 MW from hydro, 30 MW from solar, and 20 MW from biomass

(*Continued*)

1. Changing the energy landscape: policy and economic issues

TABLE 1.2   (Continued)

Country	INDCs
Honduras	No specific targets for the power sector in INDC
Jamaica	Jamaica realizes its energy resource potential through the development of renewable energy sources by increasing the share of renewable sources of energy in its primary energy mix to 20% by 2030
Mexico	No specific targets for the power sector in INDC
Nicaragua	By 2030 60% share of renewables in the electricity matrix
Panama	By 2050 30% share of nonconventional renewables in the electricity matrix
Paraguay	No specific targets for the power sector in INDC
Peru	No specific targets for the power sector in INDC
Saint Kitts and Nevis	It is intended that the policies and measures would increase the use of renewable energy sources by 50% (by 2030), being 35 MW from geothermal sources, 1.85 MW from solar, 7.6 MW from wind, and 0.5 MW from waste resources
Saint Lucia	The country has adopted a 35% renewable energy target by 2025 and 50% by 2030 based on a mix of geothermal, wind, and solar energy sources.
St. Vincent and Grenadines	The plans for renewable energy generation are focused on the development of the country's proposed geothermal power plant. The facility will generate approximately 50% of the national annual electricity consumption needs. In addition, the national energy utility is renovating existing hydropower facilities to improve efficiency and generation capacity as well as enabling and encouraging the installation of small-scale PV in the private and public sectors
Suriname	Suriname is keen to continue to transition its energy sector to ensure it stays above 25% renewable by 2025. In consideration are a hydropower project with a potential output of 168 MW; a biofuel project that could realize the introduction of ethanol in gasoline with 60% of vehicles utilizing the blend and at the same time produce 25 MW of power and 62 MW from thermal energy
Trinidad and Tobago	Trinidad and Tobago already produces all of its electricity from natural gas and is working toward achieving greater efficiency through combined cycle generation at all its power plants. This sector would therefore be at the edge of low carbon emissions with renewable energy being the next stage for reducing emissions even further. The objective therefore is to achieve the optimal energy mix with the lowest greenhouse gas emissions in order to achieve sustainable development, including the decoupling of emissions and economic growth
Uruguay	Uruguay will continue to develop maintaining the current low emission intensity level in the energy sector, and even achieving an intensity reduction of 25% from 1990 values by 2030, by means of domestic resources, and a potential 40% total reduction with additional means of implementation
	By 2025 Uruguay wants to achieve 32% of installed capacity from wind power, 5% of installed capacity from solar energy, and 4% of installed capacity from biomass
Venezuela	No specific targets for the power sector in INDC

BAU, Business-As-Usual; *PV*, Photovoltaics.

related to the development of distributed generation[24] and energy transition.[25]

Therefore as economic growth, political stability, and clear regulatory frameworks are verified, LAC becomes more homogeneous in terms of which energy sources receive greater amount of investment flows. Investments tend to be more "country-wise" and less "region-neutral" if policy conditions and institutional environments are distinct among LAC countries. The difference in the financial mechanisms' development degree will influence the shape and rhythm according to which the electricity transition in LAC will occur.

1.4 Regionalizing the electricity transition

Despite the intraregional differences seen earlier, it is noteworthy that as more variable renewable energy sources[26] are injected into the grid, the greater the risk to energy security.[27] Given that LAC is renewable-rich and the renewable potential is far from being totally explored, and considering its goal of moving away from fossil fuels and hydro, the region needs to come with solutions to counteract the variability of renewable energy.

There are many ways according to which renewable energy intermittency—wind and solar—or seasonality—biomass—can be managed: sector coupling, national and international transmission grid interconnection, supply- and demand-side management, dispatch intervals' shortening, load shifting, energy storage, and establishment of microgrids (Aghahosseini et al., 2019; Brown et al., 2018). While many of them are conceptualized and designed to work inside one country's own national borders, for the purpose of this chapter's investigation, power exchange between countries and countries' regions can, at the same time, counteract renewable variability and help create a common discourse around "a" Latin American electricity transition.

In general, grid interconnection allows for more transmission options to transfer energy surpluses from a given part of a country to another region of the subcontinent, avoiding and/or postponing new investments in new generation capacity, energy storage technologies, implementation of auxiliary services, and baseload generation, as well as opening and expanding the electricity markets and lowering generation costs (Aghahosseini et al., 2017, 2019).

Especially for LAC, there is a great potential for sources complementarity (Viviescas et al., 2019), given the fact LAC is a region that occupies different parts of the globe (the north–south axis, ranging from Patagonia to the US Southern border with Mexico[28]) and has five different time zones (the west–east axis). Those features allow not only for the exploration of the hydro–wind–solar

---

[24] Guimarães (2019a,b), Guimarães and Santos (2019), Silva et al. (2018). In this book, see also Chapter 7, The prosumer legal protection in Latin America: a requirement for an energy transition and Chapter 8, Challenges of South American energy transition: energy efficiency and distributed generation.

[25] See Chapter 16, Integrating social and justice dimensions to energy transitions: the case of Mexico and Chapter 17, The role of social resistance in shaping energy transition policy in Mexico: the case of wind power in Oaxaca.

[26] Regarding wind and solar energy, more conservative discourses tend to consider the intermittency of wind and solar energy as a factor limiting their expansion in national grids. It is highly debatable if the increase of intermittent renewable energy sources really hampers the ability of a grid to meet demand and if there is a real need to use more stable energy sources as baseload generation, see Sovacool (2009), Lovins (2017), Gross et al. (2006), and Brischke (2005).

[27] A study on the relations between energy security, renewable energy sources, and grid interconnection can be found in Guimarães (2018).

[28] Brazil is the most latitudinal country in the world, covering 39 degrees, and Chile comes third.

complementarity but also for load shifting given the different peak hours.

Considering that, more wind energy farms in Northeast Brazil can meet the demand of Venezuelan citizens, the solar energy in the Atacama Desert can flow into Brazil to attend peak hours, and the geothermal surpluses of the Caribbean countries can be exported into the inland Mexican grid.

The potential to explore south—south complementarities is mentioned in the Brazilian INDC's report, where the country

> will undertake best efforts to enhance cooperation initiatives (...) particularly in the areas of: forest monitoring systems; biofuels capacity-building and technology transfer; low carbon and resilient agriculture; restoration and reforestation activities; management of protected areas; increased resilience through social inclusion and protection programmes (...).[29]

Interestingly enough, no word is mentioned to south—south grid interconnection, despite all the well-documented advantages and potentials of that approach.

Unfortunately, South America has no project to create a synchronous supergrid. A detailed digression on the roots leading to the lack of a regional approach for energy exceeds the scope of this chapter, but faulty politic and economic stability, on one side, and confusing institutional overlapping, on the other, has led countries to turn into their own borders and adopt solo actions regarding the development of their national grids.[30] Besides that, it makes little sense—from a nationalist point of view—for a developing country to share a negotiation table with other developing country to discuss energy, if their respective energy resources should be used to promote their own national developments. With so many unexplored resources within each country's borders, with so much growth potential in terms of economic development, grid interconnection was and is never an urgent need. Therefore regional cooperation in energy has always been seen in LAC as a kind of "last-resort measure." Within a scenario of political complexity and economic instability, grid interconnection becomes a too hard solution, making room for other nationally guided approaches, such as the construction of transmission networks or the development of the liquefied natural gas (LNG) market.

Moreover, other factors hampering an integrated energy scenario in LAC are as follows: long distances between load centers, vast availability of renewable energy sources—which encourages solo actions—and, according to projections, cost-competitiveness of local energy storage (Aghahosseini et al., 2019).

The only example of a regionally integrated approach for energy is offered by SIEPAC in Central America. In South America, countries have always chosen to enter into bilateral negotiations to explore generation capacity—for example, the binational hydropower plants of Itaipu (Brazil—Paraguay), Salto Grande (Argentina—Uruguay), and Yaciretá (Argentina—Paraguay) or the wind park Artilleros (Uruguay—Brazil). In fact, some transmission lines cross some countries[31] but the independent transmission operators do not work jointly nor are the countries seen as one single market. Those interconnections are mostly used to export eventual energy surpluses.

---

[29] Available from: <https://www4.unfccc.int/sites/ndcstaging/PublishedDocuments/BrazilFirst/BRAZILiNDCenglishFINAL.pdf> (accessed 28.05.19).

[30] A detailed analysis is found in Guimarães (2019a,b), Guimarães and Santos (2019).

[31] Brazil—Venezuela, Brazil—Uruguay, Brazil—Argentina, Colombia—Ecuador, Colombia—Venezuela, Ecuador—Peru, Chile—Argentina, Argentina—Uruguay, Argentina—Paraguay. The energy exchange level accounts for only 4% of the total energy consumption in South America (considering the energy produced by binational power plants).

It is highlighted in the literature that a Pan-American network, connecting North and South America, is capable of overcoming the variability of renewables at the least cost, as well as contributing to decrease the required generation and storage capacities.[32] Therefore the knot that needs to be untied, in order to promote the LAC energy transition, is not related to resources' scarcity, but to the lack of political will to adopt an integrated approach. Once the political and economic stability, as well as a clear regulatory framework are achieved, the vast availability of renewable energy sources—especially wind and solar—become part of the solution, contributing to the installed capacities of LAC countries. Nevertheless, complexities related to supergrid operation should not also be underestimated.

With more grid interconnections linking the northern and the southern hemisphere, a decrease in the need for excess generation, energy storage, and fuel is noticeable. Besides that, the levelized cost of electricity drops around 15% in an integrated scenario.[33] Therefore a regional approach toward an integrated Latin American electricity transition is not only possible in terms of resources availability but also cost-effective and technically reliable. Nevertheless, it depends on the construction of a political and economic "common ground" to ease the flow of investments into the region. When that is achieved, investments in renewables will not only guarantee a safe and clean electricity transition but also promote regional economic development.

1.5 Conclusion

Given the environmental pressure on the deployment of more fossil fuels and on the expansion of hydro generation, the greatest challenge for LAC is to attend its expected demand growth while maintaining its overall electricity matrix clean. Being a renewable-rich subcontinent, the region is a perfect case of—possible—economic growth decoupled from growth of GHG emissions.

Although it is clear that LAC needs to diversify away from fossil fuels and hydro, a closer look into LAC countries reveals a rather heterogeneous configuration, with some countries relying greatly on oil imports—Caribbean countries—others already with a diversified electricity matrix—Uruguay, Nicaragua, and Costa Rica—and others flirting with techno-fixes—such as fracking in Argentina and LNG in Brazil. This heterogeneity is further increased by the variety of resources and the wide area in the subcontinent where those resources are located. Those features, along with political and economic instabilities verified in some countries, create an environment less fertile for a regionalized approach toward "a" coherent Latin American electricity transition.

It is widely proven that an interconnected grid in LAC would bolster its electricity transition, transforming the power sector in an environmental-friendly fashion and with the least cost for the population. To make it possible, public and private investments are needed, which depend on clear energy policies and stable regulatory frameworks. With the

[32] Brazil–Venezuela, Brazil–Uruguay, Brazil–Argentina, Colombia–Ecuador, Colombia–Venezuela, Ecuador–Peru, Chile–Argentina, Argentina–Uruguay, Argentina–Paraguay. The energy exchange level accounts for only 4% of the total energy consumption in South America (considering the energy produced by binational power plants), p. 189.

[33] Brazil–Venezuela, Brazil–Uruguay, Brazil–Argentina, Colombia–Ecuador, Colombia–Venezuela, Ecuador–Peru, Chile–Argentina, Argentina–Uruguay, Argentina–Paraguay. The energy exchange level accounts for only 4% of the total energy consumption in South America (considering the energy produced by binational power plants), p. 203.

rapid transformation of the technologies used in the exploration of renewable energy sources, the task becomes more challenging.

The region also needs to avoid being victim of a "transition-related renewables curse," given that departing from an electricity matrix that is already clean and renewable poses the risk of oversimplifying the diversification argument. LAC countries in general have great untapped potential in wind and solar energy, for instance. It is paramount to analyze from which diversification governments and stakeholders are talking about, so the maintenance and growth of the renewables' share in the electricity matrixes of LAC countries are not hampered.

References

Agência Nacional De Energia Elétrica (ANEEL), 2017. Banco de Informações de Geração. ANEEL, Brasília.

Aghahosseini, A., Bogdanov, D., Barbosa, L., Breyer, C., 2019. Analysing the feasibility of powering the Americas with renewable energy and inter-regional grid interconnections by 2030. Renew. Sustain. Energy Rev. 105, 187–188.

Aghahosseini, A., Bogdanov, D., Breyer, C., 2017. A techno-economic study of an entirely renewable energy-based power supply for North America for 2030 conditions. Energies 10, 1171.

Araújo, K., 2014. The emerging field of energy transitions: progress, challenges, and opportunities. Energy Res. Soc. Sci. 1, 112–121.

Barbosa, L., Bogdanov, D., Vainikka, P., Breyer, C., 2017. Hydro, wind and solar power as a base for a 100% renewable energy supply for South and Central America. PLoS One. 12, e0173820.

Berga, L., 2016. The role of hydropower in climate change mitigation and adaptation: a review. Engineering 2, 313–318.

Blazqueza, J., Fuentes-Bracamontesa, R., Bollinob, C.A., Nezamuddin, N., 2018. The renewable energy policy Paradox. Renew. Sustain. Energy Rev. 82, 1–5.

Breyer, C., Bogdanov, D., Komoto, K., Ehara, T., Song, J., Enebish, N., 2015. North-East Asian super grid: renewable energy mix and economics. Jpn. J. Appl. Phys. 54.

Brischke, L.-A., 2005. Series 6, Nr. 530 Modell einer zukünftigen Stromversorgung Deutschlands mit hohen Beiträgen regenerativer Energien auf der Basis eines Mehr-Knoten-Netzes. VDI Fortschritt-Bericht, Düsseldorf, p. 27.

Brown, T.W., Bischof-Niemz, T., Blok, K., Breyer, C., Lund, H., Mathiesen, B.V., 2018. Response to 'Burden of proof: a comprehensive review of the feasibility of 100% renewable-energy systems. Renew. Sustain. Energy Rev. 92, 834–847.

Cherp, A., Vinichenko, V., Jewell, J., Suzuki, M., Antal, M., 2017. Comparing electricity transitions: a historical analysis of nuclear, wind and solar power in Germany and Japan. Energy Policy 101, 612–628.

Czisch, G., 2005. Szenarien zur zukünftigen Stromversorgung – Kostenoptimierte Variationen zur Versorgung Europas und seiner Nachbarn mit Strom aus erneuerbaren Energien. University of Kassel.

de Azevedo, T.R., et al., 2018. SEEG initiative estimates of Brazilian greenhouse gas emissions from 1970 to 2015. Sci. Data 5, 180045.

de Castro, N., Brandão, R., Rosental, R., Dorado, P., 2016. Brazil and the international electrical integration: background, current status and perspectives. Energy Procedia 106, 204–214.

Empresa De Pesquisa Energética, 2017. Balanço Energético Nacional 2017. EPE, Rio de Janeiro.

Empresa De Pesquisa Energética, 2018. Balanço Energético Nacional 2018. EPE, Rio de Janeiro, p. 16.

Energy Division, 2017. Barbados National Energy Policy (2017–2037).

Environmental Defense Fund, 2017. The Cuban Electric Grid. Lessons and Recommendations for Cuba's Electric Sector. Environmental Defense Fund, p. 11.

Fauset, C., 2010. The Techno-fix approach to climate change and the energy crisis. In: Abramsky, Kolia (Ed.), Sparking a Worldwide Energy Revolution. AK Press, Oakland, CA, p. 301.

FGV Energia, Konrad Adenauer Stiftung. A Comparative Analysis of Energy Transition in Latin America and Europe. Available from: <https://fgvenergia.fgv.br/sites/fgvenergia.fgv.br/files/artigos/paper_kas-fgv_ing_web.pdf> (accessed 01.06.19.).

Fouquet, R., Pearson, P., 2012. Past and prospective energy transitions: insights from history. Energy Policy 50, 1–7.

Frankfurt School-UNEP Centre, Bloomberg New Energy Finance, 2016. Global Trends in Renewable Investment 2016. FS-UNEP Centre, Frankfurt am Main.

Frankfurt School-UNEP Centre, Bloomberg New Energy Finance, 2018. Global Trends in Renewable Investment 2018. FS-UNEP Centre, Frankfurt am Main, p. 21.

Gils, H.C., Simon, S., Soria, R., 2017. 100% renewable energy supply for Brazil—the role of sector coupling and regional development. Energies 10.

Gross, R., Heptonstall, P., Anderson, D., Green, T., Leach, M., Skea, J., 2006. The Costs and Impacts of Intermittency: An Assessment of the Evidence on the Costs and Impacts of Intermittent Generation on the British Electricity Network. UK Energy Research Centre, London, p. 4.

Grubler, A., 2012. Energy transitions research: insights and cautionary tales. Energy Policy 50, 8–16.

Guimarães, L.N.M.R., 2018. Energieversorgungssicherheit im Europarecht mittels der Förderung erneuerbarer Energien und der Interkonnektion der Netze. Peter Lang Verlag, Berlin.

Guimarães, L.N.M.R., 2019a. Challenges to the promotion of distributed energy resources in Latin America: a Brazilian case study. In: Sioshansi, Fereidoon (Ed.), Consumer, Prosumer, Prosumager: How Service Innovations Will Disrupt the Utility Business Model. Academic Press, London, pp. 235–258.

Guimarães, L.N.M.R., 2019b. Integração energética. In: Ribeiro, E.S. (Ed.), Direito do Mercosul., second ed. UniCEUB, ICPD, Brasília, p. 1139.

Guimarães, L.N.M.R., Santos, R.M.M., 2019. Agronegócio e o Mercado de energia: o caso do biogás. In: Peres, T.B., Favacho, F. (Eds.), Estudos de Direito do Agronegócio, vol. IV. Chiado Books, São Paulo, p. 607.

Hansen, K., Breyer, C., Lund, H., 2019. Status and perspectives on 100% renewable energy systems. Energy 175, 471–480.

Inter-American Development Bank, 2013. Energy Integration in Central America: Full Steam Ahead. Inter-American Development Bank, Washington, DC.

Inter-American Development Bank, 2016. Challenges and opportunities for the energy sector in the Eastern Caribbean – achieving an unrealized potential. In: Technical Note No. IDB-TN-1139. Inter-American Development Bank, p. 14.

Inter-American Development Bank, 2017a. La red del futuro: desarrollo de una rede eléctrica limpia y sostenible para América Latina. Inter-American Development Bank.

Inter-American Development Bank, 2017b. Unlocking geothermal power – how the Eastern Caribbean could become a geothermal powerhouse. In: Technical Note No. IDB-TN-1256. Inter-American Development Bank.

Inter-American Development Bank, 2017c. Challenges and Opportunities for the Energy Sector in the Eastern Caribbean: St. Vincent and the Grenadines Energy Dossier. Inter-American Development Bank.

International Energy Agency, 2019. World Energy Outlook 2018. IEA, Paris.

International Hydropower Association, 2019. 2019 Hydropower Status Report: Sector Trends and Insights. IHA, London.

International Renewable Energy Agency, 2012. Renewables Readiness Assessment: Grenada. IRENA, Abu Dhabi, pp. 28–29.

International Renewable Energy Agency, 2016a. Renewables Readiness Assessment: Antigua & Barbuba. IRENA, Abu Dhabi, p. 6.

International Renewable Energy Agency, 2016b. REmap 2030 – Renewable Energy Prospects: Dominican Republic. IRENA, Abu Dhabi, p. 2.

International Renewable Energy Agency, 2019. Renewable Power Generation Costs in 2018. IRENA, Abu Dhabi.

Lee, W., Segura, K., Schuschny, A., 2018. Presentación comparativa de algunos resultados provenientes de estudios prospectivos internacionales referidos a América Latina y el Caribe. ENERLAC – Revista de Energía de Latinoamérica y el Caribe 2 (Nr. 2), 156.

Liu, Z., 2016. Global Energy Interconnection. Elsevier, London.

Lovins, A.B., 2017. Reliably integrating variable renewables: moving grid flexibility resources from models to results. Electr. J. 30.

de Lucena, A.F.P., Hejazi, M., Vasquez-Arroyo, E., et al., 2018. Interactions between climate change mitigation and adaptation: the case of hydropower in Brazil. Energy 164, 1161–1177.

de Lucena, A.F.P., Szklo, A.S., Schaeffer, R., et al., 2009. The vulnerability of renewable energy to climate change in Brazil. Energy Policy 37, 879–889.

de Lucena, A.F.P., Szklo, A.S., Schaeffer, R., Dutra, R.M., 2010. The vulnerability of wind power to climate change in Brazil. Renew. Energy 35, 904–912.

Matthews, D.H., Graham, T.L., Keverian, S., Lamontagne, C., Seto, D., Smith, T.J., 2014. National contributions to observed global warming. Environ. Res. Lett. 9, 5.

National Renewable Energy Laboratory, 2015a. Energy Transition Initiative. Energy Snapshot: Bahamas. US Department of Energy.

National Renewable Energy Laboratory, 2015b. Energy Transition Initiative. Energy Snapshot: Saint Lucia. US Department of Energy.

National Renewable Energy Laboratory, 2015c. Energy Transition Initiative. Energy Snapshot: The Federation of Saint Christopher and Nevis. US Department of Energy.

Organización Latinoamericana De Energía – Olade, 2018. Barómetro de la Energía de América Latina y el Caribe 2018 – Las Perspectivas del Desarrollo del Sector Energético en la Región. Organización Latinoamericana De Energía – Olade, p. 10.

Palermo, G.C., de Almeida D'avignon, A.L., Freitas, M.A.V., 2014. Reduction of emissions from Brazilian cattle raising and the generation of energy: intensification and confinement potentials. Energy Policy 68, 28–38.

Pimenta, F., Kempton, W., Garvine, R., 2008. Combining meteorological stations and satellite data to evaluate the offshore wind power resource of Southeastern Brazil. Renew. Energy 33, 2375–2387.

Postic, S., Selosse, S., Maïzi, N., 2017. Energy contribution to Latin American INDCs: analyzing sub-regional trends with a times model. Energy Policy 101, 171–172.

de Queiroz, A.R., Faria, V., Lima, L., Lima, J., 2019. Hydropower revenues under the threat of climate change in Brazil. Renew. Energy 133, 873–882.

Raghoebarsing, A., Reinders, A., 2019. The role of photovoltaics (PV) in the present and future situation of Suriname. Energies 12.

dos Santos Ribeiro, M.F., Raiher, A.P., 2013. Potentialities of energy generation from waste and feedstock produced by agricultural sector in Brazil: the case of the State of Paraná. Energy Policy 60, 208–216.

Rogge, K., Kern, F., Howlett, M., 2017. Conceptual and empirical advances in analyzing policy mixes for energy transitions. Energy Res. Soc. Sci. 33, 1–10.

Scheer, H., 2010. Der Energetische Imperativ. Kunstmann Verlag, Munique, pp. 85–112.

da Silva, S.T., Dutra, C., Guimarães, L.N.M.R., 2018. Solar energy and the dawn of 'solar cities' in Brazil. In: Costa, J.A.F., Ribeiro, M.R.S., Xavier Júnior, E.C., Gabriel, V.D.R. (Eds.), Energy Law and Regulation in Brazil. Springer.

Smil, V., 2016. Examining energy transitions: a dozen insights based on performance. Energy Res. Soc. Sci. 22, 194–197.

Smil, V., 2017. Energy Transitions: Global and National Perspectives, second ed. Praeger, Santa Barbara, CA.

Sovacool, B.K., 2009. The intermittency of wind, solar and renewable electricity generators: technical barrier or rhetorical excuse? Util. Policy, 17. p. 288.

Sovacool, B.K., 2016. How long will it take? Conceptualizing the temporal dynamics of energy transitions. Energy Res. Soc. Sci. 13, 202–215.

The World Bank. World Development Indicators. Available from: <https://databank.worldbank.org/data/source/world-development-indicators/preview/on> (accessed o1.06.19.).

Turner, S., Hejazi, M., Kim, S., Clarke, L., Edmonds, J., 2017. Climate impacts on hydropower and consequences for global electricity supply investment needs. Energy 141, 2081–2090.

Viviescas, C., Lima, L., Diuana, F.A., Vasquez, E., Ludovique, C., Silva, G.N., et al., 2019. Contribution of variable renewable energy to increase energy security in Latin America: complementarity and climate change impacts on wind and solar resources. Renew. Sustain. Energy Rev. 113, 109232.

Zapata, S., Castaneda, M., Jimenez, M., Aristizabal, A.J., Franco, C.J., Dyner, I., 2018. Long-term effects of 100% renewable generation on the Colombian power market, Sustain. Energy Technol. Assess., 30. pp. 183–191.

Zhang, X., Li, H.-Y., Deng, Z.D., et al., 2018. Impacts of climate change, policy and water-energy-food nexus on hydropower development. Renew. Energy 116-A, 827–834.

Using market mechanisms for a reliable energy transition in Latin America

Alexandre Viana

Thymos Energia, SP, Brazil

2.1 Introduction: Latin America's natural tendency for renewables

Latin America has a tendency to utilize renewable energy sources due to the abundance of natural resources in the region and the tradition of some countries to use hydro-technology for power generation and more recently, the use of small renewables such as wind, biomass, and solar photovoltaic (PV). A quarter of the region's primary energy sources comes from renewables, twice the world average,[1] and in the power sector there are countries, such as Brazil and Colombia, where renewable technologies account for more than 60% of the installed capacity. The region is also considered the world's leading renewable energy grid with 53%, which is more than double the world average of 22% (Balza, 2016).

Latin American countries are also attractive for investment in renewables. The Climatescope 2018 study (BLOOMBERG-BNEF, 2018a) profiles business environment in 103 developing countries, with four countries in the Latin America region making the top 10 list: Chile (1), Brazil (4), Mexico (8), and Peru (9).

However, Latin America has very diversified degrees of market development, reflecting the diversity of the region in economic, political, and social terms. This chapter considers market development as the degree of competition and insertion of new capacity of renewables in the grid without a compulsory obligated purchase process. Greater additions of installed capacity without government schemes mean a higher degree of market development.

Therefore there is a group of countries that introduces renewables exclusively through government programs and another group that uses centrally organized power auctions as the first stage of market development and competition. Finally, there is a third group, already negotiating investments from bilateral agreements among players [independent power producers (IPPs), trading companies, and consumers].

The Latin American experience shows that moving toward market mechanisms is more economically reliable than exclusively utilizing

---

[1] IRENA — International Renewable Energy Agency. Latin America and the Caribbean. Abu Dhabi, United Arab Emirates. Available from: <https://www.irena.org/lac> (accessed 09.03.19.).

The Regulation and Policy of Latin American Energy Transitions
DOI: https://doi.org/10.1016/B978-0-12-819521-5.00002-4

and maintaining government programs, which do not promote competition and can result in higher costs for society and greater inequality due to subsidies that favor investors of renewables (Mcgee and Greiner, 2019). Therefore this chapter explores the use of market mechanisms in Latin America with a special focus on Brazil, a country with a tradition in renewables and the largest economy and power market in the region.

This chapter is organized as follows:

- Section 2.2 describes some key concepts of renewables in power markets and establishes the framework for analyzing the market development stage for countries in the region.
- Section 2.3 presents an overview of Latin American countries and classifies them in terms of development under the analytical framework presented in Section 2.2.
- The Brazilian experience is explored in Section 2.4, including a description of the development of renewables in the country and an analysis of the different policies adopted by the policy makers to encourage these technologies, in addition to the trends in this market.
- Section 2.5 concludes the analysis.

2.2 Promoting renewable-power markets

This section discusses the different development levels of renewable-power markets as well as means to promote market frameworks and tools to achieve a higher share of renewables in power systems, observing an economically sustainable path.

2.2.1 Stages of a renewable-power market

International experiences demonstrate that market design (Vries and Verzijlberg, 2018)

has a pivotal point regarding how new renewable-power plants are introduced in the power mix including how system operators keep the stability of the system, who is financing power projects (Mazzucatto and Semieniuk, 2018), and the way consumers pay for that. In addition, it is possible to observe a greater consumer participation in the choice of supplier and this tends to influence market mechanisms to introduce new renewable technologies, even if some government measures are initially adopted.

Using the theoretical framework developed by Hunt (2002) on market design and international experience over the last 20 years, it is possible to classify international power markets for renewable technologies in four stages:

1. early stage;
2. initial market stage;
3. market consolidation stage; and
4. advanced market stage.

The *early stage* is characterized almost exclusively by the conduct of government policies for the introduction of renewable technologies. Decisions to introduce new technologies and volumes to be contracted are defined through policy decisions that may or may not have technical or economic support. In general, in order to technically support decisions, a national energy plan is developed, and this plan is the main driver for the policies that will be implemented. In such a context, competition is not characterized as the main point in the power-contracting process, which can occur through public calls for projects that meet the established requirements. Feed-in-tariff programs are common at this stage, with a compulsory pass-through of costs to final tariffs.

The *initial market stage* observes the presence of the foremost competitive mechanisms for the contracting of renewable technologies, still in a context of strong regulatory and political framework, for the definition of the volumes and funding of these technologies. International practice demonstrates the use of reverse auctions as the main market tool at this stage and Latin America is a leader in reverse power auctions, highlighting auctions undertaken in Brazil, Chile, Peru, and more recently, Mexico. Technical studies, such as the national energy plan, have a remarkable influence over contracting policies, but it is also possible to note the presence of market aspects. Both supply and demand have relevant correlation with market conditions. Supply considers equipment costs and financing, while demand observes economic growth and social development for contracting new installed capacity. However, the regulation on supply and demand is still strong, with a regulatory agency defining pass-through limits of the tariffs and who will be eligible to participate in the auctions.

In the *market consolidation stage* the role of policy makers and the national energy plan is to provide guidelines and incentives for the introduction of technologies, but there is no direct intervention in contracting when it occurs in the free-power market. Central contracting mechanisms, such as regulated reverse power auctions, are still present, but there is already a relevant level of long-term contracting of renewables undertaken directly by generators, traders, and consumers, responding to competitive prices and expectations formed by market fundamentals.

Finally, there is the *advanced market stage* in which the role of policy makers and studies conducted by government entities are only a technical support for society and the decision to consume and invest in renewables is private, either from the political or economic perspective. In the United States, European Union (EU), and Australia, contracting processes are mainly carried out bilaterally or in power exchanges for renewable-technology products with long-term power purchase agreements (PPAs). In EU, the negotiation of rolling contracts at the European Energy Exchange (EEX) is worth noting, which emulate long-term PPAs (EPEX Spot, 2019). In such an environment, regulation does not have a significant role for leading investments or the final price to be paid by consumers.

2.2.2 Using market mechanisms to promote renewables

Throughout the 1980s and 1990s, several measures were introduced in developed countries, as well as in some developing countries, to promote renewable technologies to decarbonize economies, specifically in the power sector. However, the introduction of renewables through market mechanisms began to occur consistently throughout the decade of 2000 and gained momentum by the end of that decade. After 2010 a significant price reduction of wind and solar PV technologies (Blasquez et al., 2018; BLOOMBERG-BNEF, 2018b; IRENA – International Renewable Energy Agency, 2016) and an increasing number of countries using power auctions (IRENA – International Renewable Energy Agency, 2017) can be observed.[2]

By analyzing the market tools used to trade commodities and what have been used to contract new renewable projects, it is possible to identify three mechanisms: (1) centralized auctions; (2) bilateral OTC negotiations; and (3) electronic platforms and power exchanges.

2.2.2.1 Centralized auctions

Centralized auctions with long-term PPAs have consolidated over the last 10 years as the main tool for renewables, especially in

---

[2] Market mechanisms are processes to contract power from renewables with some degree of competition; in this chapter, we consider centralized auctions, bilateral negotiations over-the-counter (OTC), electronic platforms, and power exchanges.

emerging markets. In general, a government entity promotes a centralized reverse auction (lowest bids are the winners) and compulsorily allocates contracts to the distributors or to a single central buyer, with contracting costs passed on to consumers' tariffs (Kruger et al., 2018). It is also possible to conduct centralized auctions with the participation of buyers and sellers and the intersection of supply and demand curves determining the price.

In regulated markets, cost transfer and allocation are easier, although a higher tariff cost is also more likely, given that consumers are passive and do not have the freedom to choose the supplier. In liberalized markets the cost of contracting is passed on to retailers and they compete to attract consumers and must offset these costs by reducing margins or by contracting, at lower costs, the remaining required power to supply their customers.

PPAs resulting from auctions tend to have a minimum duration of 10 years and a maximum of 20 years with such long periods necessary in enabling the auction seller to obtain financing at competitive interest rates. A trade-off that exists is the cost of financing versus the reduction of prices promoted by the technology evolution. In other words the cost of financing may fall with longer contracts; however, given the technological evolution, the auctioneer could recontract the power on a shorter horizon at lower prices.

In general, auctions are the first market tool used to promote renewables as they encourage competition and benefit from economies of scale in centralized contracting processes with larger volumes.

2.2.2.2 Bilateral over-the-counter contracts

Bilateral OTC contracts are characterized by the free purchase and sale of power negotiated between sellers and buyers. Players are free to negotiate price, duration, and volume. Nonetheless, in order to consolidate a dynamic and liquid market, a legal and regulatory framework is required, allowing sellers to offer their projects directly to consumers, and in some cases, market rules are required to encourage consumers to buy power from renewable projects.

The OTC option tends to require a period to create liquidity and needs a market environment in which many consumers have the freedom to choose suppliers. This market environment usually encompasses clear rules managed by a regulatory agency, a strong market operator, and credible price formation, which make the project viable.

This kind of trading involves not only the best price but also a network of business relationships and the reputation of market participants in honoring agreements. As the OTC market evolves, there is also a demand for price references, including forward curves, market quotations promoted by trading desks, or an electronic platform that records the contracts and prices.

2.2.2.3 Electronic platforms or power exchanges

An electronic platform or power exchange consists of an anonymous marketplace regarding bid and ask prices and the selection of best offers to make a contract, publishing contract prices that will become references for traded products (Biggar and Hesamzadeh, 2014; Renewable Exchange, 2019). This marketplace demands common market rules, including financial guarantees and ethical trading and competition procedures, avoiding market manipulation.

However, a relevant characteristic of this type of marketplace is the duration of the contracts, which rarely exceed 12 months, a period insufficient to obtain funding for the construction of a new power plant. This short duration drives investors to sell a part of the project in long-term PPAs (auctions or OTC) and only the remaining energy is sold in such an environment. An alternative, which is being put into practice to make funding possible for new projects, is rolling contracts. For this alternative an auction is held 3–6 months prior to the delivery date for a 6-month PPA, in which is participating a pool of buyers who can resell these contracts over the period. A new auction, aiming to roll another cycle of contracts, is again held at least 3 months before the expiration of the current cycle, allowing a continuous flow of payments. In practice, rolling contracts emulate PPAs from auctions that are held by single buyers or central entities.

2.3 Development stage of renewables in Latin America

This section covers the main Latin America power markets with a special focus on renewable technologies and share of participation in the power mix.

2.3.1 An overview of Latin America

Although Latin America has several common characteristics, there is a diversity of policies and degrees of market development for renewables in the region (Gándara, 2018; Molina et al., 2018). This chapter analyzes the development stage of Latin American markets using the framework presented in Section 2.2 (Fig. 2.1), which results in the map of Fig. 2.2.

The abundance of renewables is a natural advantage in the region, reflecting on higher capacity factors and better technical conditions to build financially sustainable projects. However, none of the observed countries have reached an advanced market stage, with new projects becoming financially viable only by private contracts without any kind of policy or government intervention. From an economic perspective the reason for not having advanced market stage countries in the region is the absence of a robust financial market, allowing funding without the support of development banks or subsidized loans. Generally, Latin American countries try to combine private funding with governmental support and it ends up creating a crowding-out[3] effect.

Fig. 2.2 shows that Latin America is moving toward a higher level of market mechanisms for renewables in power grids and cleaner power consumption. Countries, such as Mexico, Peru, Colombia, and Argentina, are already using auctions as the main tool to contract new installed capacity. Fig. 2.3 shows that 58% of the analyzed countries in this study utilize some type of market mechanism to attract new projects.

2.3.2 Countries at early stage

This subsection presents comments about Bolivia, Costa Rica, Ecuador, Haiti, Honduras, Nicaragua, Panama, Paraguay, and Dominican Republic.

Bolivia has an electricity sector, heavily anchored in natural gas without robust competition, with renewables representing around 18% of the installed capacity. The country approved a national plan for increasing the share of small renewables in the grid, including the publication of Decree 2048, in order to achieve clean energy development (Zegada, 2016). However, in concrete terms, there are no robust measures in place to start using market mechanisms.

Over the last 10 years, *Costa Rica* has sought to attract new renewable projects using public calls and the central role[4] of ICE − Instituto

---

[3] Crowding-out phenomenon or effect is an economic theory that defends a large participation of public sector tends to push-out of the market private players and, in some cases, eliminates private sector spending or funding.

[4] ICE—Instituto Costarricense de Electricidad is responsible for all functions in the power system (planning, operations, generation, transmission, and distribution), including the purchase of some new renewable-power projects from private players as a way of complementing its own generation. In accordance with market design theory, ICE is a single buyer for new private power plants.

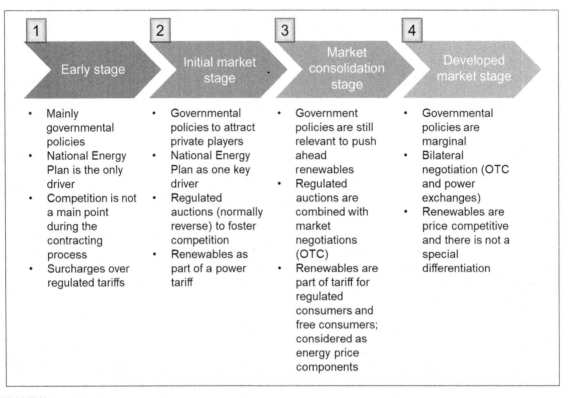

FIGURE 2.1 Stages of a renewable-power market.

Costarricense de Electricidad (2014), seeking to achieve the ambitious goal of 100% renewable generation by 2021. Costa Rica's case is evidence that the nonuse of market mechanisms is sometimes due to political options, given that the country is a stable and consolidated democracy in Central America and would be able to promote auctions, for example.

Small renewables are around 10% of the installed capacity in *Ecuador*, while large hydropower plants represent 50% and fossil fuel the remaining 40%. The country has strong centralized supervision in the electricity sector and has signaled in the last 3 years a desire to improve investment in clean energy, but the goals are vague, lacking numbers, and a pathway forward to achieve a cleaner power mix (Ponce-Jara et al., 2018).

Haiti is a country undergoing reconstruction, facing serious problems in the electricity sector and a precarious supply of electricity even in the country's capital. The total installed capacity is only 284 MW, with 80% coming from fossil fuels and the remaining 20% from the Péligre hydropower plant. The country has a goal of achieving a 47% share in renewables by 2030 and this goal is even more challenging due to the need of attracting private players as well as financial and fiscal conditions. Moreover, there is no provision for market mechanisms in the next 5 years.

Honduras has a significant share of solar PV technology, which totals 17% of the installed capacity, balancing the high fossil fuel share of 37%. To encourage renewables the government ensures priority to dispatch renewables and

FIGURE 2.2   Stages of the renewable-power market in Latin America.

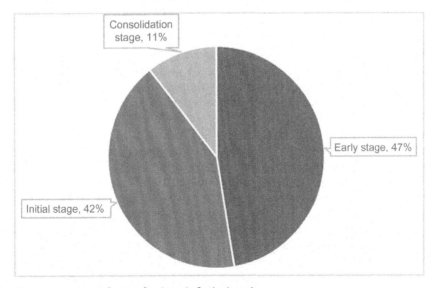

FIGURE 2.3   Countries using market mechanisms in Latin America.

I. Changing the energy landscape: Policy and economic issues

tax incentives, as well (IRENA − International Renewable Energy Agency, 2018a). These actions are in line with targets set in the national energy plan to achieve a 70% share in renewables (including large hydropower) by 2020. There are no upcoming auctions or market mechanisms in the short term and the country has used classic feed-in-tariff mechanisms to pay the current contracted projects.

Nicaragua is characterized by high power tariffs due to the need of importing oil to generate electricity. Renewables already represent a relevant percentage of 56% of installed capacity, but an addition of wind and solar PV might help in reducing tariffs (IDB − Inter-American Development Bank, 2015). There is private participation in generation with IPPs, but they sell power mainly in centrally regulated mechanisms. Renewables have fiscal incentives, but there is still a lack of market mechanisms for more consistent growth.

Panama has no formal restrictions on private participation in the renewable-power market, but there is no market friendly environment. The country also has a goal of interconnecting with Colombia by 2021 and thus increasing the use of renewables, importing and exporting power through the interconnection (IRENA − International Renewable Energy Agency, 2018b). The country already has an interconnection with Costa Rica as part of the Central American market called Sistemas de Interconexión Eléctrica de los Países de América Central (SIEPAC).

The electricity sector in *Paraguay* is characterized by large binational hydro-plants, Itaipu with Brazil and Yacyretá with Argentina. The country is a net exporter of power and although there are no formal restrictions for IPPs, the current context does not provide economic incentives for using market mechanisms given the size of the economy and the proportion of installed capacity. Due to large hydro and the unique country context, there is no foreseeable growth or emergence of a market for wind, solar PV, or small hydro.

Dominican Republic relies heavily on fossil fuel for supply. The country has tried to change this scenario with central programs for attracting small renewables for which there is a target of 25% of clean energy generation by 2020 (IRENA − International Renewable Energy Agency, 2016; NREL − National Renewable Energy Laboratory, 2015a,b), but on analyzing figures about system evolution, this target will not be met (BLOOMBERG-BNEF, 2018a,b). There are no market mechanisms that would be introduced in the next 3−4 years and this is related to the strong participation of public companies in the sector.

2.3.3 Countries at initial market stage

This subsection presents comments about Argentina, Colombia, El Salvador, Guatemala, Jamaica, Mexico, Peru, and Uruguay.

After two decades of crises and problems *Argentina's* electricity sector has been reorganized and one of the government's main goals is to increase the country's installed capacity through renewables. For this goal the country began to use auction mechanisms with successive rounds held once or twice a year, relying on support of the World Bank. Those auctions are raising mixed comments from market analysts and specialists, classifying them into two groups: supporters and critics. Supporters are advocating that the country is entering into a new economic and political cycle and the current results are evidences of that. Critics are showing many auction failures related to auction design and technical qualifications, ending up in a risk of noncompletion for a large share of power plants. It is expected that auctions and other market mechanisms will become more common after the conclusion of the first cycle of power plant construction. Nonetheless, the current share of small renewables in operation is small and it is unlikely to meet the goal of 8% of installed capacity from renewables before 2022 (Yaneva et al., 2018).

The *Colombian* market is dominated by large hydropower plants that represent 65% of the total installed capacity, while fossil technologies account for the remaining 35% (Pérez and Olascuaga, 2017). The country is initiating policies for attracting private investors to small renewables (Cortés and Londoño, 2017; ICEX – España Exportación e Inversiones, 2018) and USAID (US Government Global Development Agency) is helping to put this plan in place with studies to hold centralized reverse auctions. The first auction held by the country was frustrated due to noncompliance with regulatory requirements related to competition, reflecting the Colombian market design, which although friendly to competition, is quite concentrated.

El Salvador is interconnected with Guatemala and Honduras and is also part of SIEPAC (Martinez, 2016), with installed capacity anchored in oil and diesel (41%) and complemented by large hydropower plants (29%). The use of small renewables is increasing, with an emphasis on geothermal and solar PV. Since 2013 the country has held auctions to attract private investment and new auctions are expected in the upcoming years.

Guatemala is a country, which has private and state participation in generation, also being part of SIEPAC. The country is interconnected with Honduras and El Salvador, and to the north there is an interconnection to the Mexican grid. Auctions were held over the past 5 years for wind, solar PV, and hydro, but new tenders are not planned for the next 5 years. However, the country's national energy plan (Gobierno de Guatemala – Ministerio de Energía y Minas, 2019) has set a target of 64% of renewable generation by 2032, which is unlikely to be achieved.

Jamaica is an island where fossil fuel accounts for 90% of generation and renewables have a share of 10% (NREL – National Renewable Energy Laboratory, 2015b). In 2016 the country started holding auctions to contract renewable projects and reduce the dependence of polluting technologies, which is in line with the national energy plan to reach 20% clean energy generation by 2030 (McIntyre et al., 2016).

Mexico[5] was the last large country in Latin America to undertake profound reforms in market design. Fossil fuel represents 81% of the installed capacity and large hydro around 10%, while small renewables are still a small share of the mix. The reforms have created a favorable market environment for renewables, with auctions to contract energy and clean energy certificates. A significant increase in renewable technologies is expected in the following years with several rounds of auctions, observing that bilateral private contracts are still under development at an initial stage.

Peru has used auctions with PPAs in US dollars to attract international investors to renewable projects. Currently, the generation mix is 52% hydro and 37% fossil fuel. New renewable plants are expected to be built in the following years (Yaneva, 2018), either through auctions or bilateral contracts between IPPs and large energy consumers.

Uruguay is a small market managed in a regulated manner by the UTE Company, which is linked to the Ministry of Industry, Mines, and Energy. However, even under this condition, the country held auctions for wind projects aiming to reduce the dependence from fossil fuel technologies. Surplus generation is exported to Brazil and Argentina, noting that Uruguay (given the size of the domestic

---

[5] For a deeper analysis about Mexico, see Chapter 16, Integrating social and justice dimensions to energy transitions: the case of Mexico, and Chapter 17, The role of social resistance in shaping energy transition policy in Mexico: the case of wind power in Oaxaca.

market) will hardly develop more advanced market mechanisms unless a regional market is created with a similar approach used in the EU.

2.3.4 Countries at market consolidation stage

Two Latin American countries are classified in this stage: Chile and Brazil. Despite the difference in size of the consumer market, with Chile having an annual consumption of 78 TW h (Kruger et al., 2018) and Brazil with a consumption of 583 TW h (ONS – Operador Nacional do Sistema, 2019a,b), it is possible to observe several similar market elements, and in terms of renewable technologies, Chile tends to copy solutions that have been adopted by Brazil.

Chile had a total installed capacity of 24 GW in 2018, with approximately 50% fossil fuel and 50% renewable technologies, especially hydro, which totals 7 GW. Solar PV technology tends to show strong growth in the upcoming years, noting that by 2018 it already represented approximately 2.8 GW. The contracting process of small renewables occurs in Chile both through centrally organized auctions and through freely negotiated contracts between IPPs and large consumers, demonstrating a mature use of market tools (ONS – Operador Nacional do Sistema, 2019a,b).

The Chilean grid, called National Electrical System, is divided in four areas: Sistema Interconectado Central (SIC), Sistema Interconectado del Norte Grande (SING), Sistema Aysén de Electricidad (SEA), and Sistema Magellan de Electricidad (SEM). SIC and SING are interconnected and are the two largest grids. This interconnection occurred in late 2017 and tends to be a disruptive milestone to the country's market design. The possibility of exporting surplus power between the two areas, optimizing the system and pushing down the spot price, is the purpose of

the interconnection. This point is paradoxically complex, because as the new renewable projects push the spot price downward, which initially is a good economic signal for the consumer, new projects face difficulty in entering the market, given that the price to pay these new assets is lower. SEA and SEM are the two smallest noninterconnected grids in the south of the country.

Brazil is the first country to promote large-scale auctions for contracting power with a lower tariff approach, which is also known internationally as reverse power auction. Given the country's degree of development, it is possible to observe relevant national and international players constructing new renewable plants with PPAs obtained from regulated auctions as well as contracts negotiated in the free market.

2.4 Using market mechanism: the Brazilian case

In this section, the Brazilian case is addressed in detail, providing insights for other Latin American countries to use market mechanisms for attracting and developing renewable technologies.

2.4.1 Overview

Brazil is the 10th largest world market, in terms of power consumption, with an annual volume of 583 TW h and an installed capacity of 161.7 GW (ONS – Operador Nacional do Sistema, 2019b). Hydro-technology is the most relevant in the country with 67.4% of the total installed capacity, followed by thermal sources with 21.4%. One technology whose relevance is increasing is wind power, which was practically irrelevant until 2006, but had reached 8.9% of the installed capacity by 2018. Solar PV technology tends to follow the same path of wind power, observing that in 2015, it represented practically zero at the utility scale level and held a 1.1% share by the end of 2018 with an installed capacity of 1.8 GW.

The abundance of natural resources is a natural incentive for the development of renewable technologies in Brazil (Viana and Ramos, 2018), something that is reinforced by the scale of the country's market. However, in order to attract investment in renewables, it is necessary to create a market environment that guarantees adequate return for investors with predictable contracts and constant cash flow of receivables.

Brazil has created a robust but complex regulatory framework that allows investors to participate in regulated auctions centrally organized with a lower tariff approach associated to long-term PPAs. Financial guarantees of these PPAs are the receivables from distributors, in an escrow account model, for the sellers. Meanwhile, in the free market, there is a dynamic trading environment between IPPs, trading companies, and large consumers to enable new projects, noting that consumers demand not only lower prices but also a reduction in the "carbon footprint." Using the framework proposed in Section 2.2, it is possible to classify Brazil as a consolidated market, which is moving toward the advanced stage, given that although relevant, regulated auctions are diminishing their relevance for an increasingly independent free market for new plants.

The following subsections describe mechanisms that Brazil has used over time to reach the current stage where no more new feed-in-tariff programs are used, but only competitive regulated auctions or OTC contracts in the free market.

2.4.2 Feed-in-tariff as a first step for renewables

In order to diversify the technology mix of the country, in 2002 the government created the Program for the Incentive of Alternative Energy Sources (PROINFA), a classic feed-in-tariff program run by the federal state-owned company Eletrobrás (BRASIL, 2002). Nonetheless, it is necessary to highlight the context of the program's creation, as the country was facing severe power rationing between 2001 and 2002, which policy makers addressed using expensive thermal plants.

The objective of PROINFA was to contract 3.3 GW of renewable technologies, with equal division between wind, biomass, and small hydropower plants. Developers and investors would comply with technical and regulatory approval requirements for the projects, which would be contracted at regulated rates for a period of 20 years.

It is possible to identify that the program was successful in initiating the diversification of the installed capacity of renewables, especially for biomass and wind technologies. Moreover, biomass technology became competitive and some power plants were awarded contracts in regulated auctions undertaken in 2005, while wind power plants had exclusive procurements in 2009.

However, the high cost of contracted power plants under the program (above US\$ 120/MW h) and technological evolution led the Brazilian government to favor the use of regulated auctions. The main reason for this replacement is the lower price trend due to competition promoted by auction mechanism.

2.4.3 Regulated auctions

In 2004 Brazil promoted a second market reform in the power sector with the goal of increasing the security of supply by contracting new plants without disregarding prices and consumer affordability (Araujo et al., 2008). This reform created two trading environments, one dedicated to regulated consumers with the contracting process managed by distributors and undertaken, especially by auctions, and a second environment where medium- and large-sized consumers can freely contract power from IPPs or trading companies (Viana, 2018).

Nevertheless, regulated power auctions have been consolidated as the main tool for the promotion of renewable technologies between 2005 and 2015. Although they serve a role in the Brazilian market, auctions are losing their relevance since 2016 due to economic crisis that ended up in a lower growth of the regulated market managed by the distributors, resulting in lower demand in the auctions.

It is necessary to point out that in the case of Brazilian auctions, they are more important for the introduction of small renewables (biomass, wind, solar PV, and small hydroelectric power plants) than to large hydro, because they are traditionally consolidated in the country and generally more competitive due to economies of scale.

Biomass technology had the first winning projects in auctions promoted in 2005. In 2007 and 2008 additional efforts were made by policy makers to conduct dedicated auctions for this technology. The Brazilian case for biomass is intriguing, because despite the wide availability of resources, the technology represents only 6.0% of the contracted power in auctions for new plants (CCEE – Câmara de Comercialização de Energia Elétrica, 2018a,b). The discussion for this modest outcome would be related to the financing structure for this type of project and the options that Brazilian farmers have in the market for sugar or alcohol production, both presenting higher return rates.

Wind technology is a success story when considering regulated auctions. The first technology-oriented procurement was organized in 2009, and since this milestone, the technology has won many projects at auctions, with decreasing prices and a robust share of 19.3% of the total contracted power in auctions

for new plants (CCEE – Câmara de Comercialização de Energia Elétrica, 2018a,b). A relevant issue in the context of the auctions was the role of the Brazilian National Social Development Bank (BNDES) with subsidized rates for projects that met minimum local content requirements, which ultimately stimulated the creation of a national industrial supply chain and, consequently, larger scale and lower costs.

Solar PV technology started participating in auctions in 2014 with a technology-oriented process with the goal of contracting the first utility scale projects. Just like the success of wind, the Brazilian policy makers are expecting a repeat for solar PV, with successive auctions that allow gains of scale and reduction of prices (Viana and Ramos, 2018). One difference that arises is that the BNDES is reducing resources and loans with subsidized rates and even if these loans were maintained, it is not clear that there will be a further development of local industry given the strength of Chinese manufacturers. Something also relevant is the use of technology in distributed generation, with the net metering approach, and credits that might be compensated in up to 60 months.

The last relevant happening about renewables in the Brazilian market was the auction held in May 2019 for the isolated system of Roraima, a state in the middle of the Amazon forest and closer to Venezuela. Normally this state, which is not interconnected with the rest of Brazil, imported part of the generation from Venezuela's system and the supply condition deteriorated very quickly after the political turmoil in Venezuela in 2017. Then, Brazilian authorities organized an auction to contract projects for supplying the region and this tender attracted a diversity of private players. Outcomes were a combination of diesel, gas, renewable projects from biomass and hybrid power plants using biomass, biomass fuel, and solar PV. It is important to note that renewables represented 125,324 MW of contracted capacity, or 42.5% of the total (CCEE — Câmara de Comercialização de Energia Elétrica, 2018a,b), a relevant outcome for a region that used to rely, almost exclusively, on fossil fuel power plants.

The success of Brazilian auctions in attracting new investors to renewable technologies is due to the following aspects of design:

- Successive auctions creating a certain level of demand, which allows predictability for manufacturers, developers, and agents in the market.
- Size of the Brazilian market.
- Contracts with clauses that mitigate sellers' risk of delivery
- Financial guarantees and a regulatory framework, which significantly mitigate the risk of sellers.
- Financing promoted by the BNDES, observing that since 2017, private banks and the regional development bank for the northeast region of Brazil are key players, as well.

Other countries in the region (Argentina, Uruguay, Mexico, and Colombia) are using the Brazilian experience as a reference for their local markets. However, it is necessary to exercise some caution, especially considering macroeconomic aspects of each country such as exchange rate and domestic inflation.

2.4.4 Free market initiatives

The Brazilian free market for renewable technologies is booming and it is possible to watch three types of initiatives: (1) private auctions for renewable technologies, (2) OTC operations, and (3) BBCE (Balcão Brasileiro de Comercialização de Energia[6]) electronic platform.

- *Private auctions for renewable technologies*: Large consumers, with good credit rating, and developers are conducting private auctions for contracting solar PV or wind projects. The common practice is a 10-year PPA with delivery starting in the third year, and those PPAs are consumer price index (CPI)-indexed on a yearly basis.
- *OTC operations*: Large companies in the market, both buyers and sellers, directly seeking commercial partners or using trading companies, offering customized contracts in terms of quantity, length, and prices. Contracts are used as financial guarantees to obtain financing.
- *BBCE electronic platform*: Most IPPs, large consumers, and trading companies use the BBCE electronic platform as a mechanism for buying and selling small blocks of energy for up to 1 year. Thus some renewable projects use the platform to sell a small remaining block of an already feasible project aiming for an upside, or to create a price reference for future negotiations in other auctions or OTC operations.

The Brazilian free market is very dynamic, representing 30% of the total consumption and with a churn rate of 4.0, meaning the contract levels are fourfold the power consumption. This is the result of a 20-year journey that started with the first market reforms in the late

---

[6] BBCE is a private marketplace created in 2012, the ownership of which is divided into quotas that were acquired by market players, mainly trading companies and generators. This marketplace had a small relevance for local market in the period 2012—15, but it gained a spot in 2016 as long as free market increased its share, and after 2017, it became a price reference for contracts in free market with a duration and delivery date up to 1 year. However, there is no clearing house associated to this platform, increasing the financial risks. In order to mitigate credit risk, there is a tool that allows each company to establish the limits for its selected counterparties.

1990s. Therefore the creation of a robust market environment takes time and needs to observe relevant details such as financial guarantees, rule of law, monitoring, and the establishment of a reliable market operator.

2.4.5 Trends for a near future

The Brazilian market is in a transformation process and observing the framework of Section 2.2 it is possible to observe that the country is moving toward the "advanced stage." There is pressure among market players over policy makers for full market deregulation and measures to encourage new renewable-power plants through competitive market mechanisms.

There are also discussions at the government level and among private agents for the creation of a power exchange, highlighting the debate during the public consultation MME 33/2017 and more recently in the context of power-sector-modernization task force. Assuming those initiatives and the European example, renewables could be financially viable with specific products copying the rolling-contract approach used by the EEX.

Another trend in Brazil is the termination of the grid tariff discounts for new renewable-power plants. The current legal and regulatory system grants discounts of 50% on grid tariffs for the IPPs and consumers of these projects, which tends to be unsustainable given the significant share growth of renewable technologies. This topic was also mentioned during the process of public consultation MME 33/2017 and promotes heated debates.

Finally, the Brazilian market will adopt hourly pricing by 2021, replacing the current weekly model with three load levels. The hourly price allows better economic signals for the dispatch, allocating resources more efficiently and promoting stimulus to use the resources during the system's peak hours.

With this approach the expectation is that solar PV, battery storage, and reversible hydro will be the winners, while wind technology will depend on the generation profile of the plant site.

2.4.6 Lessons from the Brazilian case

The Brazilian case, in using market mechanisms for renewables, presents the following lessons:

- New technologies may initially require feed-in-tariff programs, but a reliable transition to market mechanisms is recommended in order to generate competition and consequently drive prices downward.
- Centralized, regulated auctions are efficient in creating a certain level of demand and to provide financial guarantees.
- A large free market diminishes the pressure for governmental actions and over public budget; then, it is important to stimulate a free market in which large consumers, IPPs, and trading companies develop projects and enter long-term contracts.
- An electronic platform for the free market can provide price benchmarks.
- Market mechanisms tend to be more efficient and reliable in attracting investment for renewables than noncompetitive government programs.
- The scale of the Brazilian market facilitates the adoption of market mechanisms and the development of a local industry. Smaller countries may have difficulties in replicating this practice; so, regional unions with cooperation would be recommended.

2.5 Conclusion

Latin America is a privileged continent with its availability of resources for renewable-power generation. The diversity of ecosystems allows a complementary use of different technologies, especially if more interconnection projects are promoted such as SIEPAC in Central America.

From an energy research perspective, it should be noted that Latin America has huge potential to be the leading region in energy transition, especially in power generation. However, there are many economic disparities among the countries, hindering greater cooperation, and even countries with larger consumer markets, such as Brazil and Mexico, still cannot be considered developed nations able to sponsor a broad regional program.

Despite the political, economic, and social problems of the region, it is possible to stimulate the formation of regional markets with the goal of allowing economies of scale and promoting the development of market instruments for new investments in renewable-power plants. The single buyer approach, despite its creation in 1978 in the United States, is a first step in countries that have not yet developed any kind of market instrument. In countries that already have private players and some competition, the entry of new investors and players should be encouraged together with a market-oriented energy policy.

Obviously, the challenges are significant, but facing them is necessary for a region with substantial resources to be utilized around an economic growth agenda with robust power generation from renewable technologies.

Then, summarizing the discussion, it is observable that countries in the continent are at different stages of market development for renewable technologies. This chapter presented an overview of 19 Latin American countries, classifying them into the following three groups:

- early stage,
- initial market stage, and
- market consolidation stage.

There are no countries on the continent that qualify for the advanced market stage so far, but perspectives point to Chile and Brazil reaching this stage in the next 5 years, given growing private participation for new power plants and contracts are freely negotiated in the free market.

Brazilian market analysis provides insights and lessons, which can be used for reliable energy transition using market mechanisms and developing local markets:

- Feed-in-tariff programs should be used only initially, but a reliable transition to market mechanisms is recommended in order to generate competition and consequently drive prices downward.
- An electronic platform for the free market can provide price benchmarks for renewables.
- Market mechanisms tend to be more efficient and reliable in attracting investment for renewables than noncompetitive government programs.
- Scale helps the adoption of market mechanisms; therefore smaller countries should consider regional cooperation for creating a larger market.

Nonetheless, the energy transition process is complex and normally, it relies on initial government measures to drive the use of renewable technologies. Therefore after this first hurdle, an economic reliable energy transition to competitive market mechanisms is recommended, resulting in more efficient resource allocation in the long term.

References

Araujo, J.L.R.H., et al., 2008. Reform of the reforms in Brazil: problems and solutions. In: Sioshansi, F. (Ed.), Competitive Electricity Markets. Elsevier.

Balza, L.H., 2016. Lights On? Energy Needs in Latin America and the Caribbean to 2040. IDB, Inter-American Development Bank.

Biggar, D.R., Hesamzadeh, M.R., 2014. The Economics of Electricity Markets. IEEE, West Sussex, United Kingdom.

Blasquez, J., et al., 2018. The renewable energy policy paradox. Renew. Sustain. Energy Rev. 82, 1–5.

BLOOMBERG-BNEF, 2018a. Climatescope 2018. New York. Available from: <http://global-climatescope.org/results> (accessed 09.03.19.).

BLOOMBERG-BNEF, 2018b. New Energy Outlook 2018. New York. Available from: <https://about.bnef.com/new-energy-outlook/#toc-download> (accessed 11.03.19.).

BRASIL, 2002. Lei No. 10.438, April 26, 2002. Creation of PROINFA Program. Available from: <http://www.planalto.gov.br/ccivil_03/leis/2002/L10438.htm> (accessed 23.03.19.).

CCEE – Câmara de Comercialização de Energia Elétrica, 2018a. InfoLeilão Dinâmico – 023 – março/2019. São Paulo. Available from: <https://www.ccee.org.br/portal/faces/pages_publico/o-que-fazemos/infomercado/info_leilao_dinamico?contentId = CCEE_646386&_afrLoop = 2831513565278&_adf.ctrl-state = 7vs838cm1_211#!%40%40%3F_afrLoop%3D2831513565278%26contentId%3DCCEE_646386%26_adf.ctrl-state%3D7vs838cm1_215> (accessed 23.03.19.).

CCEE – Câmara de Comercialização de Energia Elétrica, 2018b. 1 Leilão do Sistema Isolado – Boa Vista. São Paulo. Available from: <http://www.ccee.org.br/portal/faces/oquefazemos_menu_lateral/leiloes?_afrLoop = 1084339949485503&_adf.ctrl-state = 9xm28tcu0_118#!%40%40%3F_afrLoop%3D1084339949485503%26_adf.ctrl-state%3D9xm28tcu0_122> (accessed 20.07.19.).

Cortés, S., Londoño, A.A., 2017. Energías renovables en Colombia: una aproximación desde la economia. Rev. Cienc. Estratégicas 25 (38).

EPEX Spot, 2019. Renewable Energy: Increasingly important role in Europe. Paris. Available from: <https://www.epexspot.com/en/renewables> (accessed 24.03.19.).

Gándara, C.A.G., 2018. Renewable Energy Auctions in Mexico: The Gap Between Design and Implementation. Harvard Kennedy School. Available from: <https://www.hks.harvard.edu/sites/default/files/degree%20programs/MPAID/files/Guadarrama%2C%20Carlos%20SYPA%20final.pdf> (accessed 27.03.19.).

Gobierno de Guatemala – Ministerio de Energía y Minas, 2019. Política Energética 2013-2027. Available from: <http://www.mem.gob.gt/wp-content/uploads/2013/02/PE2013-2027.pdf> (accessed 28.03.19.).

Hunt, S., 2002. Making Competition Work in Electricity. Wiley.

ICE – Instituto Costarricense de Electricidad, 2014. Plan de expansion de la generacion electrica: período 2014-2035.

San Jose, Costa Rica. Available from: <https://www.grupoice.com/wps/wcm/connect/57562039-a158-45d6-8a5b-c4570c5533dd/PEG2014 + rev + jun2014.pdf?MOD = AJPERES&CVID = l0-8jKH> (accessed 27.03.19.).

ICEX – España Exportación e Inversiones, 2018. Energías Renovables em Colombia. Bogotá, Colombia. Available from: <https://www.icex.es/icex/GetDocumento?dDocName = DAX2018783697&site = icexES> (accessed 28.03.19.).

IDB – Inter-American Development Bank, 2015. Renewable Energy Experiences in Nicaragua to Generate Electricity. Washington, DC. Available from: <https://publications.iadb.org/publications/english/document/Renewable-Energy-Experiences-in-Nicaragua-to-Generate-Electricity.pdf> (accessed 27.03.19.).

IRENA – International Renewable Energy Agency, 2016. Renewables Energy Prospects: Dominican Republic. Panama, Abu Dhabi. Available from: <https://www.irena.org/publications/2016/Jul/Renewable-Energy-Prospects-Dominican-Republic> (accessed 27.03.19.).

IRENA – International Renewable Energy Agency, 2017. Renewable Energy Auctions: Analysing 2016. Abu Dhabi, United Arab Emirates. Available from: <https://www.irena.org/-/media/Files/IRENA/Agency/Publication/2017/Jun/IRENA_Renewable_Energy_Auctions_2017.pdf> (accessed 11.03.19.).

IRENA – International Renewable Energy Agency, 2018a. Regional Meeting – Clean Energy Corridor of Central America (CECCA). Section IV: Accelerating Renewables Deployment in Regional Electricity Market. Available from: <https://www.irena.org/-/media/Files/IRENA/Agency/Events/2018/May/Honduras-presentation---accelerating-renewables-in-the-regional-market---23-May.pdf?la = en&hash = 466436443454BE64C66FFB164D09AFEF55605192> (accessed 27.03.19.).

IRENA – International Renewable Energy Agency, 2018b. Renewables Readiness Assessment: Panama. Abu Dhabi. Available from: <https://www.irena.org/-/media/Files/IRENA/Agency/Publication/2018/May/IRENA_RRA_Panama_2018_En.pdf> (accessed 27.03.19.).

Kruger, W., Eberhard, A., Swartz, K., 2018. Renewable Energy Auctions: A Global Overview. University of Cape Town. Available from: <http://www.gsb.uct.ac.za/files/EEG_GlobalAuctionsReport.pdf> (accessed 24.03.19.).

Martinez, L.A., 2016. Sistema de Energía Renovable para El Salvador. In: Congreso de Ingeniería y Arquitectura (CONIA 2015), vol. 1. San Salvador, August 2016.

Mazzucatto, M., Semieniuk, G., 2018. Financing renewable energy: who is financing what and why it matters. Technol. Forecast. Soc. Change 127, 8–22.

Mcgee, J.A., Greiner, P.T., 2019. Renewable energy injustice: the socio-environmental implications of renewable energy consumption. Energy Res. Soc. Sci. 56.

McIntyre, A., et al., 2016. Caribbean energy: macro-related challenges. In: IMF Working Paper WP/16/53. Washington, DC. Available from: <https://www.imf.org/external/pubs/ft/wp/2016/wp1653.pdf> (accessed 28.03.19.).

Molina, J., Scharen-Guivel, N., Hyman, E., 2018. Analysis of renewable energy auctions in El Salvador, Mexico and Peru. In: USAID for Asia Clean Development Forum. Manila. Available from: <https://d2oc0ihd6a5bt.cloudfront.net/wp-content/uploads/sites/837/2018/06/Mikell-O%E2%80%99Mealy-Analysis-of-Renewable-Energy-Auctions-in-El-Salvador-Mexico-and-Peru.pdf> (accessed 27.03.19.).

NREL — National Renewable Energy Laboratory, 2015a. Energy Transition Initiative: Islands. Washington, DC. Available from: <https://www.nrel.gov/docs/fy15osti/64125.pdf> (accessed 27.03.19.).

NREL — National Renewable Energy Laboratory, 2015b. Energy Transition Initiative: Jamaica. Washington, DC. Available from: <https://www.nrel.gov/docs/fy15osti/63945.pdf> (accessed 27.03.19.).

ONS — Operador Nacional do Sistema, 2019a. Resultados da Operação: Carga de Energia 2018. Rio de Janeiro. Available from: <http://www.ons.org.br/Paginas/resultados-da-operacao/historico-da-operacao/carga_energia.aspx> (accessed 28.03.19.).

ONS — Operador Nacional do Sistema, 2019b. Resultados da Operação: Capacidade Instalada, Dezembro 2018. Rio de Janeiro. Available from: <http://www.ons.org.br/Paginas/resultados-da-operacao/historico-da-operacao/capacidade_instalada.aspx> (accessed 23.03.19.).

Pérez, E.J.L., Olascuaga, L.M.H., 2017. Redefinition of the scarcity price: lessons and recommendations from the Colombian electricity market. Trans. Environ. Electr. Eng. 2 (1), 19—26. Available at: <http://teee.eu/index.php/TEEE/article/view/77/54>. http://dex.doi.org/10.22149/teee.v2i1.77.

Ponce-Jara, M.A., et al., 2018. Electricity sector in Ecuador: an overview of the 2007—2017 decade. Energy Policy 113, 513—522.

Renewable Exchange, 2019. Platform. Bristol, England. Available from: <https://www.renewableexchange.co.uk/platform> (accessed 24.03.19.).

Viana, A.G., 2018. Auction as Allocative Mechanism for a New Market Design in Brazil. Polytechnic School of University of Sao Paulo (USP), São Paulo. Available from: <http://www.teses.usp.br/teses/disponiveis/3/3143/tde-06042018-082743/pt-br.php> (accessed 23.03.19.).

Viana, A.G., Ramos, D.S., 2018. Outcomes from the first large-scale solar PV auction in Brazil. Renew. Sustain. Energy Rev. 91, 219—228 (C).

Vries, L.J., Verzijlberg, R.A., 2018. How renewable energy is reshaping Europe's electricity market design. Econ. Energy Environ. Policy 7 (2), 31—49.

Yaneva, M., 2018. The Power Market & Renewables in Peru. Lima. Available from: <https://www.filepicker.io/api/file/ctieyGDvTCq4riOgKrFF> (accessed 29.03.19.).

Yaneva, M., Tisheva, P., Tsanova, T., 2018. Argentina Renewable Energy Report. Buenos Aires. Available from: <http://minaaysp.cba.gov.ar/wp-content/uploads/2018/06/AIRECweek-2018-The-Argentina-Report.pdf> (accessed 27.03.19.).

Zegada, M.E., 2016. The integration of renewables into the Bolivian energy mix — legal framework and policies, Renew. Energy Law Policy Rev., 7. pp. 71—80.

Further reading

IEA — International Energy Agency, 2018. Energy Policies Beyond IEA Countries: Chile 2018. Paris. Available from: <https://www.iea.org/publications/freepublications/publication/EnergyPoliciesBeyondIEACountriesChile2018Review.pdf> (accessed 23.03.19.).

IRENA — International Renewable Energy Agency, 2019. Abu Dhabi, United Arab Emirates. Available from: <https://www.irena.org/lac> (accessed 09.03.19.).

MME — Ministry of Mines and Energy or Ministério de Minas e Energia, 2017. Aprimoramento do marco legal do setor elétrico. Brasília. Available from: <http://www.mme.gov.br/web/guest/consultas-publicas?p_p_id=consultapublicaexterna_WAR_consultapublicaportlet&p_p_lifecycle=0&p_p_state=normal&p_p_mode=view&p_p_col_id=column-1&p_p_col_count=1&_consultapublicaexterna_WAR_consultapublicaportlet_consultaId=33&_consultapublicaexterna_WAR_consultapublicaportlet_mvcPath=%2Fhtml%2Fpublico%2FdadosConsultaPublica.js> (accessed 28.03.19.).

Assessing the macroeconomic effects of sustainable energy transitions in Costa Rica and Chile: a multisectoral balance-of-payments-constrained growth approach

Francisco Ebeling

Carl von Ossietzky University Oldenburg, Oldenburg, Germany

3.1 Introduction

In the last 10 years, when the positive effects of the ongoing sustainable energy transition have become more visible, it has been frequently underscored that deploying an increased share of renewable energy carriers into a country's energy mix may have positive macroeconomic consequences, for instance, on job creation, industrial innovation, and on the balance of payments (Lewis and Wiser, 2007; Heshmati et al., 2015). In this chapter, those macroeconomic impacts are assessed, in particular what concerns industrial innovation and the balance of payments.

In fact, attempts to carry out a sustainable energy transition at the country level have been made in Latin American countries in the past, with both positive and negative macroeconomic results. For instance, Brazil, from the 1950s onward, has successfully deployed—albeit with time lags—a strategy to make its energy mix more sustainable. With that policy, it also envisaged to lift external constraints on its growth rate, having obtained some success. While lifting its external constraint on growth, Brazil had also managed to steer its industrial innovation capacity. However, Brazil's approach to this issue made its foreign-debt problem much worse and costed it, in the 1980s, a lost decade.

Costa Rica and Chile have been regarded as two of the countries in Latin America, which have made the greatest effort to step up their renewable energy-producing capacities. In that context, in this chapter, it is assessed whether these sustainable energy transitions have aided or will aid Costa Rica and Chile to grow and develop themselves economically because they lift a macroeconomic constraint that otherwise acts as a drag on the development of other growth engines, such as autonomous government investments or technological progress.[1] Can Costa Rica and Chile replicate what Brazil did right from the 1970s onward? Can Costa Rica replicate its own success from the postwar years in deploying renewable energy carriers into its energy mix? What can these two countries do to avoid Brazil's mistakes?

Most Latin American countries, however, do not actually import or export renewable energy, at least not in its directly commodified form.[2] There are a few exceptions, for instance, Brazil (which imports hydroelectric power from Paraguay and exports ethanol), Paraguay (which exports hydroelectric power to Brazil), and Colombia (which exports some ethanol). Costa Rica and Chile, by their turn, practically do not directly import or export renewable energy. Thus in order to assess the impacts of renewable energy on those economies, it is necessary to appraise their import and export of fossil fuels over time and to evaluate what a substitution by renewable energy could mean for their macroeconomic performance, for example, due to an increase in the R&D level that is made possible by a lower fossil fuels import bill.

In order to do so, the method developed by Gouvêa and Lima (2013) to estimate the multisectoral balance-of-payments-constrained growth (BPCG) rate is used. Thirlwall (1979, 2002) posits that in the long-term, the main drivers of that rate are foreign demand and the income elasticities of exports and imports

(Thirlwall's law). The more a country manages to specialize itself in those sectors with higher income elasticities for exports, the less will it run short-term balance-of-payments deficits, which will have to be financed by short-term capital flows. Those, by their turn, must be rewarded by higher interest rates. If it is successful in the task of specializing in sectors with higher income elasticities for exports, it will be able to enjoy a higher BPCG rate (McCombie, 1989). In that context, by looking at the sectoral income elasticities, Gouvêa and Lima's (2013) method permits to appraise if fossil fuels are indeed a drag on growth and whether an increase in renewable energy could alleviate the external constraint on growth.

This chapter is organized as follows:

- The first section briefly analyses Brazil's successful policy experience of simultaneously transitioning from a dirtier to a more sustainable energy mix and of lifting its external constraint on growth and draws some lessons.
- The next section succinctly describes Thirlwall's (1979, 2002) approach.
- In the third section, Gouvêa and Lima's (2013) method is reproduced for the Costa Rican and the Chilean cases.
- The fourth section discusses the cases.

3.2 Brazil's sustainable energy transition experience from the 1950s onward

The purpose of this section is to draw some lessons of the Brazilian experience with an increased renewable energy deployment from

---

[1]   There are other countries in the region, which have been excelling in their sustainable energy transitions, such as Uruguay, or who could benefit from accelerating its pace—such as the Dominican Republic—and which could have had been included in this analysis. Nevertheless, the statistical data for those two countries is scarce.

[2]   Energy embodied is another matter, of course. Brazil exports a lot of embodied energy in the ore that is removed, for instance, at a very high energetic cost, from the mines of Minas Gerais state. Such a form of energy, however, seems difficult to consider in a BPCG modeling approach.

the 1950s onward. To what extent is the Brazilian trajectory a model for Costa Rica and Chile and which policy decisions by Brazil should these countries better avoid?

Brazil started to plan and build its big hydroelectric plants in the 1940s and 1950s, respectively. Former president Juscelino Kubitschek's (who was in office from 1955 to 1959) "Target Plan"[3] allocated more than 40% of its resources to the expansion of the energy sector's production capacity as "the lag between demand and installed energy capacity was the most serious sectoral bottleneck in the Brazilian economy" (Jaguaribe, 1968). Thus, although the main reason to build those plants was to solve the bottleneck issue that was strangling the country's industrialization, if the country had not increased its power capacity at that point, it would also have faced very serious balance-of-payments problems. In that context, expanding the country's supply of hydroelectric energy was a cornerstone of the country's import substitution industrialization (ISI) strategy. If the ISI strategy—which required the country to increase its energy supply—had not been to a certain extent successful, Brazil would have continued to import the very goods which that strategy aimed at substituting.

In the 1970s Brazil stepped up its policy to increase its hydroelectric energy supply; in 1973, after the Itaipu Binational Company was founded as part of an undertaking to construct a series of mega hydro plants—more than 500,000 kW each—that had been announced by the first PND (National Development Plan) in 1971. The Itaipu hydroelectric power plant, with an installed capacity of 14,000 MW, represented, when completed, around one-third of the integrated South–Southeast system's capacity, and was, at that time, the largest hydroelectric enterprise in the world. The II PND (of 1973) foresaw the then ongoing construction of Itaipu—the final cost of which topped the US$ 15 billion mark—and of other hydropower plants, such as Tucuruí (which went online in 1984 with an installed capacity of 8000 MW, at a total cost of US$ 6 billion), in order to sustain the massive growth rates of the early 1970s. For this purpose the government growingly relied on foreign loans that the "recycling of petrodollars" had made cheap in the aftermath of the oil nationalizations in the Middle East and of the 1973 oil shock, which substantially increased oil exporting countries' revenues. Thus although in the 1980s, Brazil could successfully complete its hydro park, it then had to face mounting foreign-debt problems. By the end of former president Joao Figueiredo's government, in 1985, the external debt of the electric sector had reached the US$ 20 billion mark (Boa Nova, 1985).

In 1973, still, due to the oil shock, Brazil had to address another very serious problem, as its "petroleum bill" increased substantially after the quadrupling of oil prices. At that point the country imported around two-thirds of the oil it consumed domestically (Calabi, 1983). The country attempted to resolve this another problem by stepping up its oil and gas exploration and production efforts in the offshore Campos basin (an endeavor in which Petrobras eventually achieved success), and by starting, in 1975, the "Proalcool Program," in which massive incentives—which took the form of financing, credit incentives, tax incentives, and subsidized prices—(Magalhães et al., 1991) were given to accelerate the production of ethanol, with the aim of substituting a parcel of the oil Brazil had to import.[4] In a first phase of the

---

[3] Plano de Metas.

[4] It has also been argued that with the implementation of the Proalcool Program, policymakers did not only aim solely at substituting fuels but also at bringing relief to the crisis plagued sugar cane industry, whose political influence was then considerable (Boa Nova, 1985, p. 140).

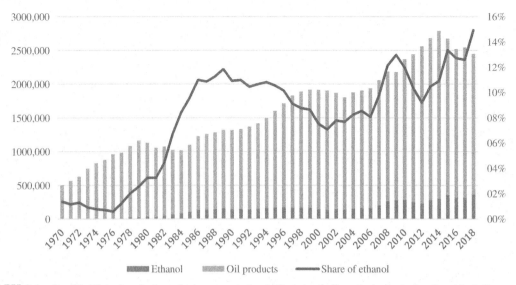

GRAPH 3.1 Brazil's ethanol and oil products consumption in barrels of oil equivalent per day (boe/day). Source: *Data from EPE, National Energy Balance (BEN), 2018. Brazilian Energy Balance 2018 Year 2017/Empresa de Pesquisa Energética. EPE, Rio de Janeiro (EPE, 2018).*

program, ethanol was added to a gasoline blend of 5%−10% ethanol (Grubler et al., 2012).

In 1979 the automobile industry committed itself to Proalcool by a protocol signed by Anfavea, and as of January 1, 1980, it started to produce vehicles moved entirely by alcohol. For 1980 a goal had been set to produce 3 billion liters, which would represent a fivefold increase of the existing production within 5 years (Boa Nova, 1985), and that goal was successfully exceeded. For 1985 again the government set the goal of producing 10.7 billion liters, but this goal was not reached, as by 1985 around 8 billion liters had been produced since the program's inception. In terms of barrels of oil equivalent per day (boe/day), 1985's production was equivalent to 116,000 boe/day: around 11% of Brazil's petroleum consumption in that year (see Graph 3.1). Proalcool was thus regarded as a classic—relatively well-succeeded—import substitution policy (Grubler et al., 2012).

However, in the second half of the 1980s, the Proalcool Program entered a deep crisis—

due to the government's decision to phase out subsidies—of which it would only recover in the early 2000s. In a scenario of increased oil prices, Brazil's supply of oil and natural gas received a new impetus, which also aided the revitalization of the demand for ethanol. For that purpose the fact that flex-fuel cars were introduced into the market in 2003 helped reorganize the hydrated ethanol market, "whose consumption had declined with the scrapping of the fleet of vehicles to alcohol" (Furtado, 2015). Since there was idle capacity and because prices were well above costs of production, the national supply of bioethanol could grow accordingly (Furtado, 2015).

Nevertheless, in the late 2000s Brazil's ethanol producers once again had to face a seemingly endless crisis. One of the main culprits was the government's petroleum fuels' pricing policy, which was designed to fight inflation. Since the second half of the 2000s, Petrobras was forced by the government to keep its prices of fuels below oil prices, a policy decision that not only affected the state company's performance but also had a negative influence on the demand for ethanol: "the pricing to counter inflation resulted in artificially low gasoline prices that competed against ethanol prices which are not regulated at the pump" (Kaup). Likewise, the government's and the private sector's plans to step up the production of "second-generation ethanol," requiring a substantial increase of the R&D effort, have not taken off yet. Nevertheless, despite these problems, in 2018 Brazil's ethanol consumption corresponded to around 15% of the sum of ethanol and oil products the country consumed, a record mark (see Graph 3.1). This suggests that Brazil's ethanol industry indeed has much space to grow.

The fact that Brazil's policy to step up its hydro park resulted in a massive debt crisis and the ups and downs of Brazil's efforts to substitute imported fuels with ethanol suggests that increasing percentage of renewable energy in a country's energy mix is no panacea from the macroeconomic point of view, particularly, when considering its external constraint on growth. Policymakers who want to transform that increase in a long-term (external) competitive advantage must shield the renewable energy policy from other policy considerations (for instance, the desire to control inflation). Another relevant conclusion from the Brazilian case is that the pace of expansion of renewable energy in the country's energy mix has to be carefully dosed, in order to avoid, for instance, an excessive external indebtedness. It should be observed at this point that many developing countries still depend on loans they obtain from multilateral banks for financing their renewable energy expansion. The World Bank, for instance, allocates the bulk of its funding on energy to fossil fuels—fired or hydroelectric power plants (Sovacool and Dworkin, 2014).

3.3 The balance-of-payments-constrained growth approach

Thirlwall (1987) builds his approach on the three Kaldor's growth laws, which state the following: (1) manufacturing is the engine of growth, because it commonly has dynamics of scale and draws on labor resources from sectors with low productivity; (2) productivity growth of manufacturing has a positive effect on output growth as a whole, what has come to be known as "Verdoorn's law"; and (3) the faster manufacturing grows, the faster will be the rate of labor transfer from agriculture to the more dynamic sectors, increasing overall productivity.

As Fujii (2012) explains, Thirlwall's law also emanates from core ideas of Latin American Structuralism. First, peripheral economies tend to suffer balance-of-payments constraints which limit their economic growth. The other idea is that those economies are structurally heterogeneous (Prebisch, 1950, 1959). In a similar vein, Vernengo (2015) argues that Thirlwall's law is a particular case of Prebisch's unequal development theory. One of Prebisch's (1950, 1959) main findings is that, because the import elasticities of the goods produced in the periphery is smaller, the growth rate of the center is larger. Finally, Dávila-Fernández and Amado (2015) suggest that Thirlwall's law is a mathematical derivation of the Prebisch—Singer hypothesis. For them the BPCG rate can be decomposed in an income effect—Thirlwall's law—and the Prebisch—Singer hypothesis. Their distinctive contribution is that they propose to capture the effect of prices on growth parting from the Prebisch—Singer hypothesis of deterioration of the terms of trade of primary products in relation to manufactured ones. That deterioration, they show, does not happen continuously, but through structural breaks, especially when a major innovation reaches mature market stage, which calls upon comprehensive strategies to develop a country's productive structure.

Thus Thirlwall (1979, 2002) develops an export-led growth model in which he also models imports. The author posits that in the long-term, the main drivers of that rate are foreign demand and the income elasticities of exports and imports (Thirlwall, 1979, 2002), which arguably reflect nonprice competitiveness (Hein, 2014). McCombie (1989) summarizes the core idea of the BPCG model in the following way:

> the central tenet of the balance-of-payments-constrained growth model is that a country cannot run a balance-of-payments deficit for any length of time that has to be financed by short-term capital flows and which results in an increasing net foreign-debt-to-GDP ratio. If a country attempts to do this, the operation of the international financial markets will lead to increasing downward pressure on the currency, with the danger of a collapse in the exchange rate and the risk of a resulting depreciation/inflation spiral. (...) In the long run, the basic balance (current account plus long-term capital flows) has to be in equilibrium.

In that context, Thirlwall (2002) finds, "currency depreciation cannot raise a country's growth rate on a permanent basis unless it is continuous. The exchange rate (...) is not an efficient instrument for structural change because it simply makes countries more competitive (temporarily) in the goods that cause the balance of payments problems in the first place." The author argues that changes in price competitiveness will only have short-run temporary effects on the BPCG rate (Thirlwall, 2011). Alternatively called "Thirlwall's law," the BPCG rate can be expressed as

$$y_B = \frac{\varepsilon z}{\pi} = \frac{x}{\pi} \qquad (3.1)$$

where ε stands for the world's income elasticity of demand for a country's exports; and π expresses the domestic income elasticity of demand for its imports.

Thus in order to improve long-run growth perspectives, countries have to use both export promoting and import substitution strategies. The more a country manages to specialize in those sectors with highest export income elasticities of demand, the less will it run short-term balance-of-payments deficits, which have to be financed by short-term capital flows, and the higher will be its BPCG rate (McCombie, 1989). In the BPCG approach, what matters most is a solid, updated, and modern industrial policy.

By their turn, Araujo and Lima (2007) have decomposed the BPCG rate into a "multisectoral Thirlwall's law." The formula asserts,

> a country's growth rate of per capita income is directly proportional to the growth rate of its exports, with such a proportionality being inversely (directly) related to sectoral income elasticities of demand for imports (exports). These income elasticities are weighted by coefficients that measure the share of each sector in total imports and exports, respectively (Hein, 2014).

Besides this contribution, the authors have made BPCG more flexible in order to incorporate the possibility of structural change by applying Pasinetti's multisectoral macrodynamic framework. It is on Araujo and Lima's "multisectoral Thirlwall's law" that Gouvêa and Lima's (2013) methodology, which will be used later, is based.

3.4 Estimation

The objective of this section is to estimate, using Gouvêa and Lima's methodology, the sectoral income and export demand functions, in order to obtain the respective sectoral export and income elasticities for Costa Rica and Chile. With these results, it is possible to assess whether an increase in renewable energy deployment can contribute to the lifting of these two countries' external macroeconomic constraints. Because renewable energy can be domestically produced and thus require much less hard currency to be financed, there is a

good possibility that these two countries—
which are still structurally dependent on the
import of fossil fuels—can improve their external macroeconomic sustainability through an
increase in the former's deployment.

3.4.1 Data

While for Costa Rica, trade data was obtained
from the country's national statistics institute
(INEC),[5] Chilean data was gathered at the country's Central Bank's website.[6] Data regarding
annual gross domestic product (GDP), GDP
growth per capita, and exchange rates, for Costa
Rica, Chile, and the World, was obtained in the
World Banks' databank.[7] For the Costa Rican
case the period chosen was 2002–15, whereas
for Chile it was from 2003 to 2015. This choice
of period was based on data availability and
estimation accuracy.

3.4.2 Step 1: Estimation of demand functions

Following Gouvêa and Lima's methodology,
the first step was to estimate the import and
export demand functions, as to obtain estimates for the sectoral income elasticities. The
equations estimated for Costa Rica's and
Chile's economy were as follows:

$$\ln M_{it} = \pi_i \ln y_t + \psi_i \ln e_t$$
$$\ln X_{it} = \varepsilon_i \ln z_t + \eta_i \ln e_t,$$

where t is the time index; $i = 1, 2, \ldots$; n repre-
sent the sectors; π_i, ψ_i, ε_i, and η_i are, respectively, the income and price elasticities of
sectoral imports and exports; y_t and z_t are,
respectively, expressions of the countries' and
the world's GDPs; and e_t gives the country's
exchange rate with respect to the dollar.

After the application of some preliminary statistical procedures, the equations were estimated
using Johansen's cointegrating methodology.[8]
Tables 3.1–3.3 present the results that were
obtained for the estimations for the income and
price elasticities of imports and exports of the
different sectors for both countries.

The asterisks indicate the number of cointegrating vectors that have been found, at a 5%
significance level, for each sectoral demand
function: one asterisk (*) indicated one cointegrating vector, two asterisks (**) indicated two
cointegrating vectors, and three asterisks (***)
indicated three cointegrating vectors. The relatively large number of cointegrating vectors
that have been found for some of these
demand functions can probably be explained
with reference to the relative small sample that
has been used in the estimation.[9] Further, the
elasticities that appear underlined indicate
whether the results were as expected: positive
signs for the income and price elasticities of
exports, negative for the price elasticities of
imports, and positive for the income elasticities
of imports. Finally, in light gray, the results
that correspond to the countries' respective

---

[5]   Available from: <http://www.inec.go.cr/economia/comercio-exterior-0> (accessed 15.02.19.).

[6]   Available from: <https://www.bcentral.cl/es/web/guest/comercio-exterior> (accessed 15.02.19.).

[7]   Available from: <https://data.worldbank.org/> (accessed 15.02.19.).

[8]   The unit root tests Augmented Dickey–Fuller (ADF) and Kwiatkowski–Phillips–Schmidt–Shin (KPSS) were applied and
indicated that the series used in the equations are nonstationary. Since all variables used are integrated of order 1, the Johansen
methodology was applied in order to assess whether the series could be cointegrated. In the cases, where the null hypothesis of
the existence of at least one cointegration vector could not be rejected, Johansen's methodology was used. The number of lags
chosen was one, since the time series was relatively small. For an explanation of Johansen's methodology (see Johansen, 1995).

[9]   In many cases, however, the choice was between a smaller and a larger value for the elasticities, as the values found
went in the "right" direction.

TABLE 3.1 Costa Rica.

	X P	X Y	M P	M Y
Live animals	3.32	−1.03*	− 1.96	1.12*
Products of the vegetable kingdom	−1.99	1.23**	− 0.23	0.85***
Vegetable fats and oils	−2.32	0.86***	− 1.48	0.78***
Products of the manufacturing industries	−2.95	0.86***	− 1.83	0.91*
Mineral products	−3.60	0.88*	− 0.81	0.47***
Products of the chemical industry	2.06	2.43***	− 0.15	1.49***
Plastic and manufacturing materials	−3.31	0.88*	− 1.77	0.94***
Skins, leather, and fur	−1.27	0.05**	2.93	0.10**
Wood and vegetable board	3.11	−0.90*	− 1.65	0.82***
Wood pulp and other fibrous materials	−2.08	0.78*	− 3.13	−0.33***
Silk, cotton, and textile fibers	−2.14	0.27*	1.67	−2.26*
Footwear, hats, umbrellas, etc.	−3.16	1.08**	− 1.08	1.01***
Manufacture of stone, plaster, cement, glass, and ceramics	−2.92	1.16**	2.42	−0.32***
Fine or cultured pearls	−2.27	1.70*	− 0.04	−0.04***
Common and manufactured metals	−1.97	0.65***	− 0.94	0.23***
Machines and appliances, electrical equipment, etc.	−3.92	0.73*	2.17	−0.43*
Transport material	3.33	−0.84*	1.91	−0.95*
Instruments and equipment for optics, topography, etc.	−2.67	0.70*	− 3.02	1.53*
Miscellaneous goods and products	3.59	−0.79*	1.72	−1.07*
Objects of art and collection	−5.03	1.01***	− 2.96	0.86***

*indicates one cointegrating vector.
**indicate two cointegrating vectors.
***indicate three cointegrating vectors.

TABLE 3.2 Chile: imports.

	M P	M Y
Durable consumer goods	2.13	−1.98**
Semidurable consumer goods	− 0.010	−1.70*
Other consumer goods	− 0.48	0.34***
Energy products	− 5.42	4.55***
Other intermediate goods	− 2.33	2.31***
Capital goods	− 2.06	1.69***

*indicates one cointegrating vector.
**indicate two cointegrating vectors.
***indicate three cointegrating vectors.

TABLE 3.3 Chile: exports.

	X P	X Y
Mining	2.08	−0.90**
Agriculture, forestry, and fishing	−2.27	−0.41***
Foods	−5.71	4.13**
Drinks and tobacco	−0.82	0.70**
Forestry and wooden furniture	−2.01	2.77**
Cellulose, paper, and others	−1.34	1.08**
Chemical products	−0.89	0.83**
Basic metallic industry	−0.84	0.47*
Metal products, machinery, and equipment	−9.62	5.38***
Other industrial products	0.81	0.89***

*indicates one cointegrating vector.
**indicate two cointegrating vectors.
***indicate three cointegrating vectors.

energy sectors are highlighted. For Costa Rica the price and income elasticities of imports of the "mineral products" item—which comprise the countries' oil imports—were as expected. For Chile the item "energy products" also was obtained as expected.

3.4.3 Step 2: Validity tests of Thirlwall's law (aggregated)

In this section the method used by Gouvêa and Lima (2013) is only partially reproduced:

Thirlwall's law is only tested for the aggregate case, and not for the multisectoral one. Since the errors tend to impair the accuracy of a hypothetical aggregation of the sectoral elasticities into national aggregate demand functions, it is contended here that, in order to carry out the rest of the exercise, it suffices to determine whether the countries' growth rates are balance-of-payments constrained or not.

Like Gouvêa and Lima (2013), this section uses McCombie's (1989) methodology as to verify whether the growth of the Costa Rican and the Chilean economy were conditioned by restrictions of foreign exchange. This methodology entails defining a hypothetical income elasticity of demand for imports that generates an equality between the observed growth rate and the estimated growth rate, that is compatible with the external equilibrium. The null hypothesis that this estimated income elasticity of demand for imports is equal to this hypothetical income elasticity of demand is tested, where the nonrejection permits to conclude that the hypothesis that the growth of the country was restricted by the balance of payments cannot be rejected. As Carvalho and Lima (2009) note, "an analogous way of applying this methodology is to compare the expected growth rate of the model and the real growth rate." As was seen earlier, Thirlwall's law model is given by the following expression:

$$y_B = \frac{\varepsilon z}{\pi} = \frac{x}{\pi},$$

where the growth rate of GNP per capita growth is given by the world's income elasticity of demand for a country's exports times the world's GNP per capita growth divided the domestic income elasticity of demand for imports (Thirlwall, 1979, 2002).

Table 3.4 synthetizes the results.

TABLE 3.4 Per capita growth rates—Thirlwall's law and effective.

	Thirlwall's law	Observed
Costa Rica's average growth (2002–15)	2.85[a]	2.8
Chile's average growth (2003–15)	3.2[b]	3.25

[a]*Three and one cointegrating vectors where found for, respectively, the income elasticity of demand for exports and imports.*
[b]*Three cointegrating vectors where found for, respectively, the income elasticity of demand for exports and imports.*

As can be observed in Table 3.4, for the Costa Rican case the model overestimates the average growth very modestly, by around 2%. For the Chilean case the model also fits very well the observed data; the model's average growth rate underestimates the average observed growth by a meager 1.5%. Hence, there is evidence that both the Costa Rican and the Chilean growth rates were BPCG in the period indicated.

3.4.4 Step 3: Assessing structural change in Costa Rica's and Chile's economy

In this section, multisectoral Thirlwall's law is used in order to assess the evolution of the elasticities of exports and imports and, thus, the external restriction's intensity. For that purpose, Gouvêa and Lima's method is used, which consists of pondering, for each year, the sectoral income elasticities that have been estimated previously both for imports and exports.[10] This technique permits to capture whether there has been a structural change in the composition of a country's trade sector. Graphs 3.2–3.5 present some of the results,

---

[10] As Gouvêa and Lima point out, the empirical literature on Thirlwall's law has also used the "rolling regressions technique," consisting of estimating ordinary least squares (OLS) at sequential intervals (Gouvêa and Lima, 2013).

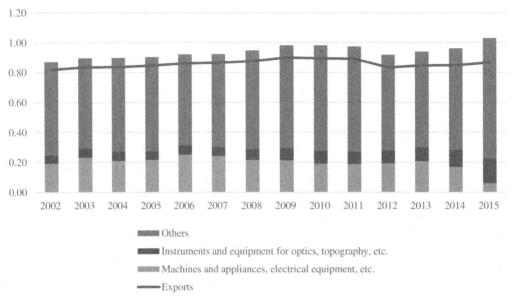

GRAPH 3.2   Exports: Costa Rica.

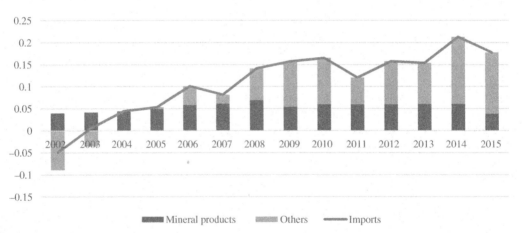

GRAPH 3.3   Imports: Costa Rica.

respectively, for the cases of exports and imports in Costa Rica and Chile.

Graph 3.2 suggests that Costa Rica has been able to maintain a relatively stable pondered income elasticity of demand for exports. The item "machines and appliances, electrical equipment, etc."—which belongs to the realm

of manufacturing—has suffered a steep decline from 2014 onward but has been compensated by "instruments and equipment for optics, topography, etc.," also from that realm. In 2015 that item has given the largest contribution to Costa Rica's pondered export income elasticity of demand.

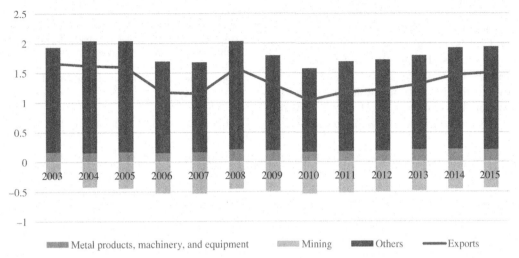

GRAPH 3.4 Exports: Chile.

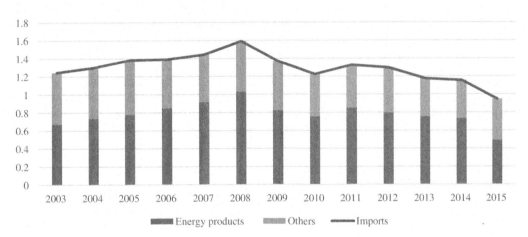

GRAPH 3.5 Imports: Chile.

Graph 3.3 suggests that one of the crucial reasons why Costa Rica's BPCG rate is constrained is its relatively large import bill of "mineral products," which comprises the country's oil bill. Hence, if the country manages to decrease the amount of energy, it imports in the form of fossil fuels—by increasing its production of biofuels or by shifting its car fleet to electric cars powered by renewable energy sources such as solar—Costa Rica should be able to increase its BPCG rate, and

thus its external macroeconomic sustainability will improve.

Graph 3.4 presents the results for Chile's export income elasticity of demand. They suggest that in the period of the analysis, the item "mining" have not been aiding that indicator's health. Indeed, from a peak of US$ 49,083 billion, Chile's copper exports have fallen significantly, having reached the U$30 billion mark in 2016, to recover again in 2018 (when Chile exported around US$ 40 billion in copper).

On the other hand the indicator "metal products, machinery, and equipment," an expression of Chile's export-oriented manufacturing effort, has increased. If Chile's BPCG rate is to rise, that sector should receive more inputs, possibly by redirecting resources that are currently being devoted to the imports of energy toward capital accumulation and to an increased R&D effort.

As can be seen in Graph 3.5, during the period of this analysis, the item "energy products" of Chile's import agenda has given a significant contribution for the country's pondered import income elasticity of demand. The indicator has fallen in 2015, but this was not a structural break, as will be seen in the next section, in that year Chile started to import less natural gas from neighboring Argentina. Hence, it allows one to conclude that reducing Chile's energy imports through an increase of domestic renewable energy production would be an invaluable contribution for the country's overall balance-of-payments health, particularly, in a context where the exports of copper have been oscillating.

3.5 Assessing the cases

In this section, considering the results of the estimation, a brief assessment of the cases is provided. Costa Rica is widely regarded as a success case in the region in what concerns economic and social development. According to OECD (2018),

> Costa Rica has achieved strong well-being and robust economic growth. Almost universal access to education, health care and pensions have contributed to high levels of life satisfaction. (...) Costa Rica has established a world-renowned green trademark and eco-tourism industry by protecting its abundant biodiversity and developing renewable energy sources. Open trade and foreign direct investment are an integral part of Costa Rica's successful growth model. This has underpinned Costa

Rica's structural transformation from an agricultural-based economy to one with a more diversified structure that is integrated into global-value chains. Building on these achievements, Costa Rica has the opportunity to increase its specialization in medium- and high technological intensive sectors. (...).

This success has been attributed to the fact that Costa Rica has a relatively old and robust democracy, and also to an early decision—inscribed in the 1949 constitution—to abolish the country's armed forces, with corresponding savings being invested, among others, to health and education. Costa Rica thus managed to accumulate human capital and became an attractive harbor for FDI in skill and knowledge-intensive sectors. Thus in the late 1990s the company Intel established itself in the country, and it has been hailed "a success story on many levels (...) [and] an example of a well-focused FDI targeting process" (Ferraz et al., 2011), into more technology-intensive technologies. The establishment of Intel—which led to the formation of an electronics cluster—allowed the country to increase its exports and to upgrade its composition. Thus Costa Rica's comparative advantage has pointed to an increased level of sophistication of its exports (OECD, 2018). Costa Rica's dynamism, arguably a "strongly converging commodity importer," owes to the fact that its growth has exceeded its "export pull because of a very dynamic domestic response" (De la Torre and Ize, 2019). The way external deficits have been financed, with "a preponderance of FDI inflows (which facilitate learning and technology transfer) is consistent with Costa Rica's superior performance" (Ferraz et al., 2011). Those FDIs into tradable services tend to be related with favorable domestic spillover effects, due to positive effects on local employment.

However, Costa Rica's experience is still limited to the fact that the hardware cluster still revolves around Intel and that it "did not expand or consolidate to the extent expected considering other, especially Asian experiences" (Ferraz et al., 2011). Part of that bottleneck, as observed by Ferraz et al. (2011), can be found "in the development of local suppliers and in the weakness of linkages between suppliers and foreign investors." Thus the country's export basket is strongly dependent on less sophisticated products and could "benefit from upscaling opportunities in a number of medium- and high-technological intensive industrial sectors" (OECD, 2018). For this to happen, however, one important policy measure would be to improve its education system, which struggles "to keep pace with the growing demand for skilled workers" (OECD, 2018).

The country's 1949 decision to abolish the armed forces—which increased disbursements in other items of public expenditure—also enabled the country to build its hydroelectric park,[11] and thus Costa Rica has to a great extent eliminated its energy generation bottleneck. Costa Rica consistently manages to produce more than 90% of its electricity from renewable energy sources (mainly hydro) (MINAE, 2015). However, the country's state oil refining company[12] has to import 100% of its domestic oil supply (55,000 boe/day), at an annual cost that oscillates between US$1.2 and 1.4 billion dollars (MINAE, 2015). As of 2015, 66% of the country's total energy consumption was supplied by oil and petroleum derivatives (MINAE, 2015). In that vein, Costa Rica's 2015−30 National Energy Plan[13] considers one of its main challenges to develop the country's energy mix in a sustainable fashion both in what concerns environmental sustainability and the macroeconomic equilibrium (related to the country's oil bill). It is considered that a growing demand for fuels could significantly impair midterm macroeconomic performance if unmatched by structural transformations in the energy mix. In that vein, some attempts have been made to increase the country's biofuels production. One of the initiatives was an attempt to pass a biofuels law[14] to regulate the domestic production of biofuels. Furthermore, the 2014−18 national development plan[15] established a goal of mixing 5% of ethanol to the gasoline blend.[15] Although in the 2006−15 period the country managed to mix 2%−4% of ethanol to gasoline, in 2015 that policy was discontinued as even that amount of ethanol had to be imported, at a relatively high cost.[16] By 2018 Costa Rica still struggled to effectively implement a biofuels plan, with production being negligible at that point.

In that context, in 2018 Costa Rica has disclosed the 2018−50 National Decarbonization Plan to reach zero net emissions economy by 2050, in line with the 2015 Paris Climate Change Agreement objectives. Aiming at reducing greenhouse gas emissions, the plan also contains measures in basic infrastructure as well as in the transport, energy, industry, agriculture, waste management, and soil and forest management sectors. With the plan, policymakers have established a roadmap for the modernization of the economy, with measures to generate jobs and boost sustainable growth. The measures will be incorporated by the country with its new national climate action plan, which it will present to UN Climate

---

[11] However, other sources as finance, such as external loans, were also deployed (see Evans, 2010).

[12] Refinadora Costarricense de Petroleo—RECOPE.

[13] Plan Nacional de Energía 2015−30.

[14] Ley de Biocombustibles Expediente Número 18789.

[15] Plan Nacional de Desarrollo (2014−18).

[16] Available from: <https://www.larepublica.net/noticia/nuevo-intento-de-introducir-los-biocombustibles-en-el-pais> (accessed 06.03.19.).

Change in 2020, when signatories of the Paris Agreement agreed to present an update of their nationally determined contributions. According to the country's plan, concrete goals will be pursued in key sectors such as transport, currently one of its major sources of greenhouse gas emissions. It is expected that by 2035, 70% of all buses and taxis will be electric, and a goal of full electrification has been set for 2050.[17] The plan also contains incentives for the gradual abandonment of fossil-fueled cars and their substitution by zero-emission vehicles or by car-sharing schemes. In the realm of electricity production, where currently around 95% of generation comes from renewable energy carriers, a 100% goal was set by 2030.[17] The country also wants to ensure that by 2050, electricity becomes also the primary energy source also in the transportation sector, as well as for residential, commercial, and industrial usages.[17] If Costa Rica indeed should manage to decrease its oil bill due to an increase in domestic renewable energy production for the transportation sector, this could perhaps lead, in the midterm, to an increase in R&D expenditures, a vital step for the country to increase the technological sophistication of its export base. In fact, Costa Rica's 2018–50 National Decarbonization Plan explicitly states that the policies in the key areas are simultaneously devised as to "reverse the increase of greenhouse gas emissions, (. . .) [and] to encourage the modernization and revitalization of the economy through a vision of green growth." (Costa Rica Bicentennial Government). In that context, increased R&D expenditures are both means and ends toward a longer term goal of revitalizing Costa Rica's on a more sustainable basis, both what concerns the economic and the environmental dimensions.

By its turn, Chile's macroeconomic dynamics is still highly dependent on the copper sector's exports. Until the 1973 coup, Chile attempted, without success, an "industrialization behind high protective barriers." The ultraliberal experiment of the Pinochet period did not fare better. Finally, under the concertation (who took office in 1990), the country experienced a huge success. From 1990 onward, growth accelerated, and inflation and interest rates fell. According to Bacha and Fishlow (2011),

> domestic industry (. . .) dramatically increased its savings rates and began to invest. New activities eschewed dependence on tariff protection. (. . .). Foreign investment returned. Copper production that had been 80% public reversed to 70% private by 2000, as facilities came online. As income continuously rose, domestic poverty rapidly fell, although income distribution inequality persisted at preceding high levels. (. . .) Public expenditure contributed to better health, housing, and education. (. . .) R&D outlays increased.

Being highly liberal in its approach to trade, today 40% of Chile's GDP consists of exports, of which copper represents around half.[18] Having signed deals such as the NAFTA, Chilean policymakers believe that they are better off exporting raw materials such as copper "and importing manufactured products of higher quality" than producing them domestically, for a higher price (Bacha and Fishlow, 2011). This set of policies has been hailed as a successful one:

> (. . .) as the price of copper soared upwards after 2003, Chile has demonstrated the effectiveness of compensatory fiscal policy. The gain of an increased export surplus (. . .) has been encapsulated within a sovereign wealth fund, and invested in domestic (. . .) instruments. These resources – now involving more than $15 billion dollars – were accessible to

---

[17] Available from: <https://unfccc.int/news/costa-rica-commits-to-fully-decarbonize-by-2050> (accessed 18.05.19.).

[18] It also exports fruit and vegetables, forest products, fish and wine.

compensate for the dramatic fall in price back to $1.4 in 2009. (...) As a result, Chilean income fell only modestly in 2009. (...) More fundamentally, Chile has become the poster example for the very absence of a natural resource curse. (...) (Bacha and Fishlow, 2011).

It has also been argued that Chile has been able to reduce its income gap with the United States because it has translated economic growth into its increased global export share (export dynamism), which increased as the country's copper industry gained market share in the 1990s and 2000s. Chile arguably managed to combine both due to its "sound macroeconomic management, strong institutions, and high-quality public services" (De la Torre and Ize, 2019). In that vein, Henstridge and Roe (2018) argue, "today Chile is usually characterized as a country that has been unusually successful in managing the macroeconomic consequences of its large extractives wealth."

There are, however, problems with that strategy. The country's export complexity, which has never been very high, has been found wanting lately (De la Torre and Ize, 2019), also because its "share of value-added manufacturing in its copper exports" has decreased, "with the proportion of refined copper in total exports being drastically reduced in favor of the far more primitive copper concentrates" (Palma, 2011). Considering this, there is evidence that the slowdown of growth observed since the late 1990s can be in part attributed to low investment in "upward productive diversification" (Palma, 2011). Another problem is that an "amount equivalent to 10 percent of GDP (plus a similar amount in central bank reserves) is likely to be an excessively conservative amount to hold to insure against the risks of copper price volatility" (Henstridge and Roe, 2018). This is particularly troublesome given that Chile has relatively low levels of public spending on education, health, public pensions, and other social sectors as a share of GDP. Thus although Chile has managed to implement a relatively solid and stable fiscal policy in which its sovereign wealth fund (SWF) compensates the oscillations of the quantities and values of copper sold, the fact that its expenditures in human and social capital are relatively low throw a shadow on the country's future.

These concerns receive another, potentially more troublesome dimension, when considering the energy sector. In recent years, electricity demand has been set to grow fast, and the government estimates that it will be more than double by 2050. Generation capacity has managed to anticipate that growth as it has more than tripled over the last two decades (from around 6500 MW in 1997 to around 23,000 MW by the end of 2017) (OECD/IEA). However, the problem lies in the fact that this expansion has been carried out by the construction of thermoelectric plants, which are fueled by natural gas and coal. Thus Chile has managed to increase its generation capacity at the cost of a higher import bill of those fuels.

Chile started to import natural gas from Argentina in 1998, and those imports peaked in 2004. After Argentina's 2004 energy crisis—when its supplies ran low—Chile started to build two liquified natural gas (LNG) plants to reduce its dependence on Argentina's natural gas. In 2016 Chile consumed 5.2 billion cubic meters of natural gas, which accounted for 12% of the country's total primary energy supply (TPES) (OECD/IEA). The country managed to produce less than one-quarter of its domestic supply (1.2 bcm) (OECD/IEA). After a 14-year interlude, in 2018 Argentina once more begun to export natural gas to Chile.[19] As of 2016, Chile produced 2.5 million tons of

---

[19] Available from: <https://www.reuters.com/article/chile-argentina-gas/argentina-restarts-natural-gas-exports-to-chile-idUSL2N1XA209> (accessed 20.03.19.).

coal, which accounted for 19% of its total coal supply (OECD/IEA). The rest Chile must import. In that year, coal's share in TPES reached the 20% mark, a growth that is related to the sharp fall in gas imports (OECD/IEA). Chile is also highly dependent on oil, which in 2016 accounted for 40% of its TPES and more than 50% of its final consumption (OECD/IEA). In that year the country imported 97% of its internal supply of that fuel (OECD/IEA).

The country has however the means to resolve most of its domestic energy supply problems, as it counts on an enormous potential to upscale its renewable energy production capacity:

> Chile has a vast untapped potential for renewable electricity, which can help (...) reduce import dependency. The government has set a target for a 60% share of renewable power by 2035 and 70% by 2050. To attain a 60% share of renewable power by 2035 at the least cost, the share of solar photovoltaics (PV) and wind power must increase considerably (OECD/IEA).

Further, according to OECD/IEA,

> Chile has enormous potential for various renewable-energy sources. Solar resources are spectacular: the Atacama Desert in the north boasts a direct normal irradiance of more than 9 kilowatt hours (kWh) per square meter (m^2) per day, the highest in the world. In its extreme south, together with Argentina, it has the best onshore wind resources in the world. (...) The development of renewable-energy resources is expected to bring economic and social benefits to the country. (...) Chile recently began to develop its solar and wind capacity. Given that it is still at an initial stage, wind and solar together account for only 1.2% of TPES and for 6.2% of power generation. However, growth has been rapid, and it is expected to continue, thanks to the country's vast potential for wind and solar energy and increasingly competitive costs of renewable-energy generation.

Indeed, in 2016 Chile was the first Latin American country to reach the 1 GW solar generation mark, and by 2018 it was close to the 2 GW mark.[20] In a context in which an "energy revolution" is arguably under way (OECD/IEA), the Chilean government has launched the "Energy Program 2016—25," aimed at developing "an export-oriented national solar-power industry," which will require the "government to increase public funding for energy-related RDI" (OECD/IEA). From this, it follows that there is an alternative virtuous policy constellation for Chile in which it can simultaneously harness its enormous renewable energy potential and reduce its external dependence, both through a decrease of the fossil fuels it has to import and through the creation of a R&D-intensive blooming PV industry. This policy constellation should be able to benefit from linkages with the lithium economy, a resource with which Chile is also richly endowed. The country's huge lithium endowment not only gives it the means to store the renewable energy it produces—which is subject to intermittency—but also puts the country in a relatively good position to also decrease its foreign dependence on oil, as lithium is the most critical resource for building electric vehicles. In that context, it is highlighted that Chile is the world's largest lithium producer, as can be seen in Graph 3.6. In that vein, as Vara (2015) aptly puts it:

> on the national layer, regarding electric cars, Bolivia, Chile and Argentina are expected to provide as usual the natural resource, lithium; while Japan, Germany, or South Korea are expected to industrialize it, and to provide the technology, the batteries and/or the cars – and, in turn, to buy the cars. (...) Bolivia, Chile, and Argentina are currently working on and negotiating from the position of suddenly empowered nation-states in a new geopolitical world. (...).

[20] Available from: <https://energytransition.org/2018/07/is-an-energy-revolution-underway-in-chile/> (accessed 20.03.19.).

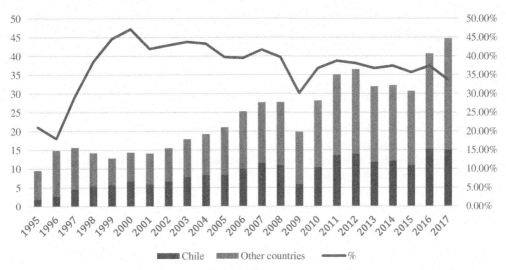

GRAPH 3.6 Chile's lithium production in the world (in thousand tons of Lithium content). *Source: Data from BP, 2018. BP Statistical Review of World Energy 2018 (BP, 2018).*

3.6 Conclusion

The objective of this chapter was to analyze the impact of sustainable energy transitions on the macroeconomic growth trajectories of two Latin American countries, Costa Rica and Chile, as it is possible that these transitions will aid those countries to remove a macroeconomic constraint that otherwise acts as a drag on growth.

In Section 3.2 the example of Brazil, which started to carry out its sustainable energy transition as early as the 1950s, was cited as to show that a sustainable energy transition can bring along many developmental benefits, but that it is no panacea, particularly, from the macroeconomic point of view. A policy lesson from Brazil for the Costa Rican and Chilean cases is that the pace of the transition should be carefully managed as to avoid an excessive indebtment, both foreign and in domestic currency. In that sense the transition should be coordinated with the country's fiscal and monetary policy, as to ensure that it is also sustainable from an economic point of view. In a

word the sustainable energy transition has to be coordinated with the country's overall developmental strategy.

Another reason why the sustainable energy transition is no panacea is that there is no guarantee that the economic surplus that an eventually reduced energy import bill will be redirected toward those more productive expenditures that can lead to an overall larger BPCG rate. That economic surplus could also stem from overall lower domestic energy prices, which could indirectly stimulate investment in productive capital or in R&D. But even lower prices are not guaranteed, as the owners of the carbon infrastructure (both thermoelectric plants or fossil fuels stations) might demand a compensation for their stranded assets. Hence, the sustainable energy transition might get blocked or be slowed down by a series of vested interests, which, by their turn, can slowdown the process through which that transitions lead to an overall larger BPCG rate.

Another open question—which concerns especially Costa Rica—is whether the biofuels technological route is the more viable and economic one, considering that the electric vehicle is gaining terrain very rapidly in the world. Thus choosing the biofuels route may mean that Costa Rica could get "locked in" in a more inefficient pathway, a technological *cul de sac*. There are, however, signs that the country will give preference to the technological route of the electric cars. In that context, when the country should start encouraging the widespread acquisition of electric vehicles to solve the oil import bill problem—which would have to be purchased abroad—there will be negative impacts, at least short- and midterm, on the country's balance of payments. Another problem that affects the Costa Rican case is that energy prices are already high. In order to fully electrify its transportation sector, Costa Rica will have to harness renewable energy sources that probably will have to be subsidized by the government, at least in the short- and midterm. The problem is that when energy subsidies are removed, there is no guarantee that producers will have passed through a learning curve, and prices may remain high. Although there are no guarantees, the signs that come from the international energy industry suggest that renewable energy carriers are passing through a significant learning curve, and that Costa Rica could benefit from this development in a substantial manner if its regulatory policies for this new round of investments are well designed.

The Chilean case is quite another where the stakes are very high, and the country does seem to have been once more "blessed" by its natural resource endowment of lithium. On the other hand, in Costa Rica the financial and natural resources to carry out the sustainable energy transition are relatively scarce, which means that the country has a relatively small margin of maneuver. Chile counts on a vast natural resource base which, with some ability, it can largely convert into human-made capital. It is expected that Chile, which has a well-succeeded experience with its copper SWF to

build on, is prepared to succeed in that endeavor. However, Chile does face some problems. First, its investments in R&D and education have been less than ideal in the last decades. Second, as it has signed free trade agreements, the country faces the complicated task of having to make some of its human-made capital more competitive than, for instance, some of its American competitors. In that context, what plays to Chile's advantage is that only succeeding in the task of reducing its substantial foreign-energy dependence is already a remarkable achievement that very certainly will add up to its population economic and social well-being.

On a final note, it should be underscored that the BPCG approach has a critical heterodox nature what concerns its policy prescriptions, due to its Keynesian and structuralist origins. From the former, it takes its demand-side stance. From the latter, it has absorbed its tendency to look at the constraining and enabling character of structures, not solely at the role of the individual. In that context the BPCG approach tends to recommend industrial policies and policies in the realm of technology and innovation that are bold, demand-side oriented, and that seek to profoundly alter a country's economic structure. From the point of view of the BPCG approach, thus, to be successful sustainable energy transitions should be carried out by daring, risk-loving, and revolutionary minds that are willing to go off the beaten tracks.

References

Araujo, A., Lima, G.T., 2007. A structural economic dynamics approach to balance-of-payment-constrained growth. Cambridge J. Econ. 31 (5), 755–774.

Bacha, E.L., Fishlow, A., 2011. The recent commodity price boom and Latin American growth: more than new bottles for an old wine? The Oxford Handbook of Latin American Economics. Oxford University Press, Oxford.

Boa Nova, A.C., 1985. Energia e classes sociais no Brasil. Edições Loyola, São Paulo.

BP, 2018. BP Statistical Review of World Energy 2018.

Calabi, A.S., 1983. A energia e a economia brasileira: interações econômicas e institucionais no desenvolvimento do setor energético no Brasil. FIPE, Fundação Inst. de Pesquisas Econômicas, São Paulo.

Carvalho, V.R., Lima, G.T., 2009. Estrutura produtiva, restrição externa e crescimento econômico: a experiência brasileira. Econ. Soc., Campinas 18 (1), 31–60.

Costa Rica Bicentennial Government, Decarbonization Plan Commitment of the Bicentennial Government, 4. Available from: https://www.2050pathways.org/wp-content/uploads/2019/02/Decarbonization-Plan-Costa-Rica.pdf.

Dávila-Fernández, M., Amado, A., 2015. Entre a lei de Thirlwall e a hipótese Prebisch-Singer: uma avaliação da dinâmica dos termos de troca em um modelo de crescimento com restrição no Balanço de Pagamentos. Econ. Soc. 24 (1), 87–119.

De la Torre, A., Ize, A., 2019. Latin American economic growth: hopes, disappointments, and prospects. In: Unfulfilled Promises—Latin America Today.

EPE, 2018. Brazilian Energy Balance 2018 Year 2017/ Empresa de Pesquisa Energética. EPE, Rio de Janeiro.

Evans, S., 2010. The Green Republic: A Conservation History of Costa Rica. University of Texas Press.

Ferraz, J.C., Mortimore, M., Tavares, M., 2011. Foreign direct investment in Latin America. The Oxford Handbook of Latin American Economics. Oxford University Press, Oxford.

Fujii, G., 2012. Latin American structuralism. In: King, J.E. (Ed.), The Elgar Companion to Post Keynesian Economics. Edward Elgar, Cheltenham.

Furtado, A.T., 2015. Energia e Desenvolvimento: a necessária transição para um novo modelo de desenvolvimento. In: Pedro de Souza. (Org.). Brasil, Sociedade em Movimento, vol. 1. Paz eTerra, São Paulo/Rio de Janeiro.

Gouvêa, R.R., Lima, G.T., 2013. Mudança estrutural e crescimento sob restrição externa na economia brasileira: uma análise empírica do período 1962-2006 com considerações sobre o II PND. Econ. Soc. 22 (1), 107–139.

Grubler, A., Aguayo, F., Gallagher, K.S., Hekkert, M., Jiang, K., Mytelka, L., et al., 2012. Policies for the Energy Technology Innovation System (ETIS).

Hein, E., 2014. Distribution and Growth After Keynes: A Post Keynesian Guide. Edward Elgar, Cheltenham.

Henstridge, M., Roe, A., 2018. The Macroeconomic Management of Natural Resources, Extractive Industries: The Management of Resources as a Driver of Sustainable Development. Oxford Univeristy Press, Oxford.

Heshmati, A., Abolhosseini, S., Altmann, J., 2015. The Development of Renewable Energy Sources and Its Significance for the Environment. Springer, Singapore.

INEC. Costa Rica's national statistics institute. Available from: <http://www.inec.go.cr/economia/comercio-exterior-0> (accessed 15.02.19.).

Jaguaribe, H., 1968. Economic & Political Development; A Theoretical Approach & A Brazilian Case Study.

Johansen, S., 1995. Identifying restrictions of linear equations with applications to simultaneous equations and cointegration. J. Econ. 69 (1), 111–132.

Kaup, F., The Sugarcane Complex in Brazil. The Role of Innovation in a Dynamic Sector on Its Path Towards Sustainability. Springer, Cham, ZG.

Lewis, J.I., Wiser, R.H., 2007. Fostering a renewable energy technology industry: an international comparison of wind industry policy support mechanisms. Energy Policy 35 (3), 1844–1857.

Magalhães, J.D.A., Kuperman, N., Machado, R.C., 1991. Proálcool: uma avaliação global. Astel, Rio de Janeiro.

McCombie, J.S.L., 1989. Thirlwall's law and balance-of-payments-constrained growth: a comment on the debate. Appl. Econ. 21, 9–20.

MINAE, 2015. Programa de las Naciones Unidas para el Desarrollo PNUD. VII Plan Nacional de Energía 2015–2030. PNUD, San José, CR.

OECD, 2018. OECD Economic Surveys: Costa Rica 2018. p. 26.

OECD/IEA. Energy Policies Beyond IEA Countries: Chile. OECD/IEA, Paris.

Palma, J.G., 2011. Why has productivity growth stagnated in most Latin American countries since the neo-liberal reforms? The Oxford Handbook of Latin American Economics. Oxford University Press, Oxford.

Prebisch, R., 1959. Commercial policy in the underdeveloped countries. Am. Econ. Rev. 49, 251–273.

Prebisch, R., 1950. The Economic Development of Latin America and Its Principal Problems. United Nations, New York.

Sovacool, B.K., Dworkin, M.H., 2014. Global Energy Justice. Cambridge University Press, Cambridge.

Thirlwall, A.P., 2011. Balance of payments constrained growth models: history and overview. PSL Q. Rev. 64, 307–351.

Thirlwall, A.P., 1987. Nicholas Kaldor. New York University Press, New York.

Thirlwall, A.P., 1979. The balance of payments constraint as an explanation of international growth rate differences. Banca Naz. Lav. Q. Rev. 128 (791), 45–53.

Thirlwall, A.P., 2002. The Nature of Economic Growth. Edward Elgar Publishing, Cheltenham.

Vara, A.M., 2015. A South American approach to metamorphosis as a horizon of equality: focusing on controversies over lithium. Curr. Sociol. 63 (1).

Vernengo, M., 2015. Una lectura crítica de la crítica al modelo de Thirlwall. Invest. Econ. 74 (292), 67–80.

Further reading

Chilean Central Bank. Available from: <https://www.bcentral.cl/es/web/guest/comercio-exterior> (accessed 15.02.19.).

Cimoli, M., Porcile, G., Rovira, S., 2010. Structural change and the BOP-constraint: why did Latin America fail to converge? Cambridge J. Econ. 34 (2), 389–411.

McCombie, J.S.L., 2012. Balance-of-payments-constrained Economic Growth. In: King, J.E. (Ed.), The Elgar Companion to Post Keynesian Economics. Edward Elgar, Cheltenham.

Pacheco-López, P., Thirlwall, A.P., 2006. Trade liberalization, the income elasticity of imports and economic growth in Latin America. J. Post Keynesian Econ. 29 (1), 41–66.

World Bank Open Data. Available from: <https://data.worldbank.org/> (accessed 15.02.19.).

4

Energy (and climate) challenges in South America: what planning for what transition?

Thauan Santos[1] and Ignacio Sabbatella[2]

[1]Post-Graduate Programme in Maritime Studies, Brazilian Naval War College (PPGEM/EGN), Rio de Janeiro, Brazil [2]National Scientific and Technical Research Council, University of Buenos Aires (CONICET-UBA), Buenos Aires, Argentina

4.1 Introduction

The objective of this chapter is to analyze the link between planning and energy transition in South America, particularly considering climate challenges. In this sense, the first research question that arises is whether the South American countries are carrying out a real energy transition: "if so, how?", "by what measures?", "what are the goals set?", "what actors are involved?", and "are the efforts driven only at the national level or are there joint actions at the regional level?". These questions are the ones that together guide the main purpose of this chapter.

In order to address these issues a methodological framework based on documentary research was applied through the collection, systematization, and analysis of primary and secondary data. On the one hand, official documents were extracted from the website of the United Nations Framework Convention on Climate Change (UNFCCC): Nationally Determined Contributions (NDCs) of each South American country and Biennial Update Report (BUR) submitted by some countries. Documents from regional integration organizations, such as The Southern Common Market (Mercosur), the Andean Community (CAN), and the Union of South American Nations (UNASUR), were also collected. On the other hand, statistics published on the websites of Latin American Energy Organization (OLADE), International Energy Agency, Climatescope, and the company British Petroleum (BP) were also used.

The chapter is then divided into four sections as follows:

- First, the relationship between planning and energy transition will be addressed at the theoretical level.
- Second, the NDCs of each country will be analyzed according to the research

questions, providing a comparative analysis between them.

- Third, regional energy policies will be examined to determine whether or not regional planning actually exists.
- Finally, the chapter provides some final considerations, followed by the references.

4.2 Energy transition and energy planning

At the same pace as international concern about climate change is growing, the energy sector is changing in shape. Given that around two-thirds of global greenhouse gas (GHG) emissions come from the production and use of energy, the energy sector is the core of each individual effort to address both problems (IEA-IRENA, 2017). Therefore it is crucial for climate change and emissions reduction to have energy as a central sector, and the energy transition must play a strategic role. From the point of view of global commitments, it can be argued that the Paris Agreement, as a result of the Conference of Parties (COP) 21,[1] and the Sustainable Development Goals (SDGs) (Arent et al., 2017) have already put the global mitigation agenda in a new era (Arent et al., 2017)—making 2015 a milestone for the issue.

Energy transition can be understood as a structural change in the system of energy supply and utilization. Sometimes, it is the consequence of technological and economic transformations; sometimes, it is the product of political decisions (Carrizo et al., 2016). The ongoing concept of "energy transition" is used to mention the shift from fossil fuels to renewable energy sources (IRENA, 2019). Although it is not the first energy transition humanity that has ever been witnessed, the current energy transition from fossil fuels is already ongoing and should not be analyzed from a single lens—since it is not an exclusively technological or resource endowment issue (Grayson, 2017).

There are three main aspects that underpin the current transition (IRENA, 2019), all of them directly affected by the technology development (Bueb et al., 2017): (1) energy efficiency, (2) renewables energies, and (3) electrification. Energy efficiency enables guaranteeing economic growth demanding lower energy inputs. By doing so, it is also possible to reduce emissions, which contributes to climate change.[2] Regarding the growth of renewable energies, solar and wind power[3] stand out while other renewables have been growing more slowly. Related to electrification, heat pumps and electric vehicles are pushing for the increase of renewable energies in the transport, industry, and construction.

Even though the three aspects dialog and have close relations, the literature emphasizes energy efficiency and renewable energy as the main drivers of the energy transition, due to its majority drop emission, they still need to expand in all sectors (IRENA, 2018a). In some

---

[1] The Paris Agreement was agreed by virtually every country in the world to reduce GHG emissions in order to limit the average global temperature rise well below 2°C and as close as possible to 1.5°C, thus avoiding the most severe impacts of climate change, such as increased droughts, floods, and severe storms. Given the impossibility of reaching consensus on a single universal emission mitigation formula, it was agreed that each country decided its own goals for the period 2020–30.

[2] BP (2018) (accessed on April 23, 2019). To understand the role of energy efficiency to the South American energy transition, see Chapter 8, Challenges of South American energy transition: energy efficiency and distributed generation.

[3] As an example, solar power added more new capacity in 2017 than did coal, gas, and nuclear plants combined. See IRENA (2018b). "Wind and solar now provide 6% of electricity generation worldwide, up from 0.2% in 2000." (IRENA, 2019: 16).

developed countries, for example, the transformation of their energy mixes with the aim of reducing CO_2 emissions began many years ago, while a few developing countries incorporated renewable energies in order to increase energy supply and improve access to energy electricity in isolated areas (Bersalli et al., 2018). However, since the Paris Agreement and the SDGs, the promotion of renewable energies has accelerated in both cases.

In this context, research and technological innovation play a key role in the energy transition, being central to the long-term transition to a sustainable energy sector (IEA-IRENA, 2017). Smart technologies as well as distributed (decentralized) energy link people and products to power systems (Kuzemko et al., 2018).

Due to the very complex and dynamic nature of the energy transition, it is difficult to make precise predictions (IRENA, 2019). What is known is that energy transition will require significant additional policy interventions (IEA-IRENA, 2017), and the financial system will need to consider broader sustainability and energy transition requirements.[4] Nevertheless, since government agencies and private markets do not like to spur a transition on their own (Fri and Savitz, 2014), the necessary changes in energy planning will require distinguished changes not only in technology but also in political regulations, tariffs, pricing regimes, and the behavior of users.[5]

Therefore energy planning becomes paramount to guarantee not only energy security but the energy transition itself. The planning of a system, whatever it may be, consists on the analysis of its current characteristics, its dynamics of operation, and its possible future evolution, with the intention of acting on it to control its trajectory. In the case of economic systems, three well-differentiated conceptions could be distinguished[6]:

1. Liberalism, which advocates self-regulation or self-control of economic systems, through the free play of market mechanisms, which leads to planning being considered unnecessary
2. The total centralization of power, which requires planning to define the levels of activity of different economic sectors, giving rise to regulatory planning
3. The recognition of shared power in mixed economies, which requires a process of agreement between the different social actors to agree on desirable trajectories of the system and leads to strategic planning (Bouille, 2004)

Energy planning is a tool of energy policy. It can be defined as a systematic and analytical methodology that conveniently processes information on the demand, transformation, and supply of energy, in order to generate strategies to achieve the defined long-term objectives (OLADE, 2017). According to the OLADE manual (OLADE, 2017), the conceptual modalities of energy planning are divided into three:

1. Normative planning: economic aspects are privileged, following technocratic criteria and do not pay attention to feasibility

---

[4] See IRENA (2018a). Global investment in renewable energy has grown steadily for more than a decade, rising from almost USD 50 billion in 2004 to USD 348 billion in 2015 (IEA-IRENA, 2017). The additional cost of long-term energy transition would amount to USD 1.7 trillion annually in 2050—although there would be outweigh and cost-savings related to reduced air pollution, better health and lower environmental damage (IRENA, 2018a).

[5] See Sovacool (2017). Since strong government commitment might reduce risk and lower the cost of financing (IEA-IRENA, 2017).

[6] Section 4.3 will analyze the NDCs of the South American countries, the influence of these more liberal, more centralized, or mixed economic systems can be clearly perceived.

issues. In essence, attention is focused on the coherence between the objectives proposed in the plan and the instruments recommended for it. He fully relies on the power and capacity of the State (conceived as an internally homogeneous actor) for the realization of the plan.

2. Indicative planning: this is the formulation of a plan that represents the desired evolution of the energy system from the perspective of the entity in charge of planning, but the execution is left to the decentralized actors of the system of the system, whether they are public or private. Therefore it has an orientation character. The main difficulty of this planning approach is that it trusts that the economic signals of the markets (prices and profit expectations and other economic incentives) are sufficient to induce the actors and execute the production and/or investment actions proposed in the plan. However, if these actions do not materialize, or are not carried out in a timely manner, it can seriously compromise the security of energy supply, one of the most important objectives of energy planning.

3. Strategic planning: this approach includes mechanisms to build the political viability of the plan. Its strategies and actions are binding in the sense that their implementation and execution are verified in an effective manner. To this end, the subsidiary activity of the planning State is established, which implies concrete mechanisms for its implementation either directly or through public actors existing or created for this purpose.

During the 1950s and 1960s, United Nations Economic Commission for Latin America and the Caribbean (ECLAC) proposed normative planning, although later criticisms of that approach led to disenchantment about the possibilities of planning in the 1970s. During the period of neoliberal reforms, from the 1980s to the end of the 1990s, indicative planning dominated, which was reduced to a simple prospective of the energy system. On the other hand, at the beginning of this century, the necessity and relevance of the intervention resurged through public policies and planning (OLADE, 2017).

However, two issues arise that need to be reflected upon. The first one already mentioned is that energy planning need not be based solely and exclusively on a country's own resources. In this way, it is possible to take place at regional level, being guaranteed through arrangements such as energy integration and/or cooperation, or even through international trade. The second issue is related to the concept of energy security, which, again, is intrinsic and often associated with national sovereignty and self-sufficiency.

Consequently, energy is seen from the point of view of planning and security, as issues intrinsic to sovereignty, autonomy, independence, and self-sufficiency of States. In this sense, it is common to see the concepts of energy security as domestic policies and priorities in the national agenda (Santos, 2014). In the classic reference texts on energy planning, such as Chevalier (1973), Helm (2002, 2011, 2014), Martin (1974, 1990, 2000), and Percebois (1986, 2008), the idea of national sovereignty over such resources remains evident.

Considering regional integration, it can be emphasized that it faces two main challenges. First, the incorporation of renewable energies is a goal that countries are nationally assuming within the framework of the fight against climate change. For example, Kyu and Yeong (2016) studied the case of Northeast Asian Energy Market and they arrived to the conclusion that a country will face great limitations in development potential and costs if it attempts to increase the share of renewable energy. Instead, they consider that building an energy grid between neighboring countries enables eco-friendly energy use. Joint, large-scale development, and sharing of abundant renewable sources would allow a substantial increase of that share.

In the same line, Maldonado et al. (2017) understand that regional integration in Latin America benefits renewable projects since large projects can be developed to meet high demands, take advantage of shared energy resources to interconnect the countries of the region, and reduce costs. In another order of ideas, Hurtado and Souza (2018) warn about the interest of core economies and global governance organizations to promote a massive transfer of green technology to noncentral regions, especially after the 2008 financial crisis, with the objective of relaunching capitalist accumulation. Conversely, the development of renewable energy sources may be an opportunity for peripheral countries to develop endogenous technological capacities from the initial stage of a new leading sector. Undoubtedly, the technological aspect is key to the extension of margins of autonomy of a region such as South America.

According to OLADE, in 2017, primary energy supply in South America comprised 34% oil, 28% natural gas, 10% hydro energy, 6% coal, 1% nuclear, and 21% others. Unlike the global energy matrix, natural gas occupies the second place, and coal has a smaller share even than the hydro energy. Regarding power generation, in the same year, the region accounted 58% hydro, 30.3% nonrenewable thermo, 5% renewable thermo, 1.8% nuclear, 4.4% wind, and 0.5% solar. Due to the existence of a power mix strongly based on hydroelectricity, the emissions of GHG are lower than the world average (Bersalli et al., 2018).

From the point of view of equity, the per capita emissions of developing countries are generally much lower than those of developed countries. South American emissions as a whole represent only 3.2% of the total in 2017, according to BP data, mainly due to the strong

preponderance of hydroelectric power as shown earlier. Taking this into account, the historical responsibility for South America (as well as for Latin America) in the climate change should therefore be very limited—especially if we consider the principle of "common but differentiated responsibility."

Considering the energy planning in South America, there is a relevant possibility of exploring synergies derived from hydrological complementarity, as well as different sources (Paredes et al., 2017; Moura, 2017; Ramos, 2016; Castro et al., 2012). Thus there is evidence of a strong complementarity between the different pluviometric regimes in the region, which suggests the joint planning of the dispatch of hydroelectric dams, construction of new ventures, and joint management of decision-making. In this way, and considering the changes resulting from the energy transition, the regional approach seems more necessary than ever to face the challenges ahead (about this topic, see Chapter 1: Is there a Latin American electricity transition? A snapshot of intraregional differences).

Particularly considering South America region, Santos (2018) creates and analyzes four modeling scenarios for the integration of the power sector in the region:

- Reference integration scenario
- Weak integration scenario
- Moderate integration scenario
- Strong integration scenario

They consider the expansion and new international interconnection lines, new binational hydroelectric plants, new contractual arrangements (swaps), as well as regulatory harmonization. "Undoubtedly, it will require political will and 'diplomatic engineering' to carry out the measures of each scenario in the face of such adverse political-economic context (...), but the modeling exercise ratified the argument that greater electricity integration in Mercosur (and in South America as a whole) leads to a reduction in the need to increase installed capacity, as well as to lower geographic and socio-environmental impacts." (Santos, 2018: 222).

4.3 Nationally Determined Contributions of South American countries

The Paris Agreement, signed in December 2015, requires each Party to prepare and communicate its NDC, which consist of a climate plan that aims at reducing its GHG emissions, taking into account its domestic circumstances and capacities. The 21st session of the COP 21 had already invited all Parties to communicate to the secretariat their intended NDC (iNDCs) by the first quarter of 2015 for those Parties ready to do so. According to the official website, by April 18, 2016, a total of 190 Parties had reported an iNDC (97% of all Parties to the UNFCCC) with a total CO_2 coverage of 94.6%.

As the NDCs are expected to be sent every 5 years to the secretariat of the UNFCCC, the next round of NDCs (new or updated) should be submitted in 2020. Successive NDCs should represent a progression compared to the previous NDC in order to increase the ambition of the goals.

In general, and as stressed in Section 4.1, the energy sector plays an important role in each NDC due to its high contribution in GHG emissions at national level. Therefore the contributions of the energy sector in each South American NDC will be analyzed later in alphabetical order.

Argentina submitted the first revision of its NDC in 2016, but it did not present specific measures for the energy sector. At the COP 23 (2017) meeting in Bonn (Germany), the government presented the National Energy and Climate Change Plan. The mitigation measures consist of four axes:

1. Energy efficiency (efficiency in household appliances, efficient water heaters, solar water heaters, heat pumps, water economizers, street lighting, residential lighting, and thermal enclosures in buildings)
2. Renewable energy (electricity generation from renewable sources connected to the grid, distributed electricity generation, and isolated power generation from the grid)
3. Fuels (blended with biofuels)
4. Large-scale generation (nuclear, hydroelectric, substitution of fossil fuels with natural gas in electricity generation, improvement in the efficiency of thermal power plants)

In the second BUR of 2017 the Argentinian government presented the progress of the mitigation measures. When it comes to the promotion of renewable energies, it highlighted the launch of the RenovAr program as a first step to comply with Law 27,191/2015. The law establishes that renewable energy sources must reach 20% of the power demand in the year 2025. RenovAr program consists of a series of renewable power auctions to spur the development of clean energy projects and includes a series of tax benefits, financing mechanisms and guarantees.

For *Bolivia* the power sector will soon become one of the most wealthy ones through investments in hydropower and alternative energies, increasing the export capacity of energy from renewable sources to neighbor countries. The government foresees State participation in energy generation, trough state-owned *Empresa Nacional de Electricidad*, creating income and implementing policies and redistribution of wealth. In its iNDC the Plurinational State of Bolivia estimates reaching the following goals:

• Increased participation of renewable energy to 79% by 2030 from 39% in 2010. With international cooperation, the share would reach 81%.

- Increased participation of alternative energy and other energy (steam combined cycle) from 2% in 2010 to 9% in 2030 in the total electrical system. While national efforts imply an additional increase of 1228 MW, international cooperation would imply an additional increase of 1378 MW by 2030.

In its NDC, *Brazil* highlights the biofuel program, including cogeneration of electricity using biomass and the current energy mix, which consists of 40% of renewables (75% of renewables in its electricity supply), since it has a strong share of hydropower. In this way, energy mix amounts to three times the world average in renewables, fact of which allows Brazil to qualify "as a low carbon economy."

Likewise, Brazil proposes some energy goals by 2030: increasing the share of sustainable biofuels in the Brazilian energy mix to approximately 18%; achieving 45% of renewables in the energy mix by 2030, increasing the share of renewables (other than hydropower) in the power supply to at least 23%; achieving 10% efficiency gains in the electricity sector. Brazil has implemented auctions to contract new clean energy generation capacity since 2009, despite a 2-year tender hiatus between May 2016 and December 2017. Brazilian Development Bank (BNDES) is the top lender for renewable energy projects, and it applies a strict local content requirement in granting wind and photovoltaic project loans, spurring equipment manufacturers to build factories in the country.[7]

Chile declares in its NDC that energy is one of the priority sectors for mitigation. Two objectives sought in National Energy Agenda led by the Ministry of Energy are the promotion of nonconventional renewable energy (NCRE) and the reduction of energy consumption. The targets presented are 20% of electric power in 2025 made up of NCRE, in accordance with Law 20,698/2013, and 20% reduction in the energy consumption forecast by 2025.

The goal is for 45% of all the power generation capacity installed in the country between 2014 and 2025 to be generated from NCRE. The chosen route to expand this type of sources is the investment of the private sector through public tenders.

In its iNDC, *Colombia* affirms that the energy sector makes up one of the eight Sector Mitigation Action Plans but does not present specific goals to it. It points out that electricity generation matrix is very clean due to the high share of hydroelectricity (68% of the electricity generation in 2010) and an energy consumption well below the international electricity consumption averages.

It should be noted that the Parliament passed Law 1931/2018 on Climate Change, which defines the planning and management instruments of climate change. According to the second BUR the strategic mitigation lines implemented by the Ministry of Mines and Energy are energy efficiency, energy generation, active demand management, and fugitive emissions.

In 2010 the country set a modest target of 6.5% of generation from nonlarge hydro renewables for end users with access to the grid by 2020. In 2015 the share of renewable energies in total power generation exceeded 5%, mainly from small hydro. Recently, Colombia held its first auction to award 23 renewable generation and commercialization contracts. However, no project was awarded because a bidder represented an excess of concentration in the auction. A new auction will be held before the end of the first half of 2019.[8]

---

[7] See http://global-climatescope.org/.

[8] Reuters, "Colombia no logra adjudicar proyectos de energías renovables pese a interés de empresas." https://lta.reuters.com/articulo/energia-colombia-subasta-idLTAKCN1QF30D.

The new government of *Ecuador* informed the Secretariat on September 8, 2017 mentioning that the iNDC presented in October 2015 should not be registered as its first NDC, since it is under review process. In its BUR presented in 2016, Ecuador stated that its electricity matrix in 2013 was made up of 50% of thermoelectric plants, 46% of hydropower and only 1% of NCRE, mainly firewood and cane products.

Guyana highlights its Low-Carbon Development Strategy (LCDS) that was first formulated in 2008. Implementation of the LCDS has been financed primarily with Guyana Reducing Emissions from Deforestation and forest Degradation (REDD) + investment fund (GRIF) resources earned under the Guyana Norway Agreement.

Regarding energy unconditional contributions, Guyana aims at its goal to develop a power mix of wind, solar, biomass, and hydro. The government is promoting several projects, and for that legislation has been enacted to remove import duty and tax barriers for the import of renewable energy equipment. Also, it is developing energy efficiency and public education policies with the expectation that will reduce Guyana's energy consumption. Regarding conditional contributions, the country considers it possible to eliminate its fossil dependence and to develop a 100% renewable power supply by 2025.

Paraguay covers practically all of its electricity demand with three large (binational) hydroelectric plants and even generates surpluses that it exports to its neighboring countries. Electricity market is controlled by state-owned vertically integrated utility *Administración Nacional de Electricidad*. The main activity related to the generation of emissions in the energy sector is due to land transport.

However, Paraguay's National Development Plan 2014–30, mentioned in its NDC, foresees as a goal to increase the consumption of renewable energy by 60% and reduce fossil fuel consumption by 20%. For that, it plans to incorporate technologies for the exploitation of new sources of sustainable energy (solar, wind, and biomass).

The *Peruvian* iNDC mentions the implementation of energy transformation initiatives such as switching fuels to natural gas and promoting renewable energy sources. According to Climatescope, the share of renewable energies in the electricity matrix exceeded 10% in 2015, mainly small hydro. Anyway, the iNDC does not set goals regarding the energy sector.

Suriname presents unconditional and conditional contributions regarding renewable energies. The unconditional contributions are included in the National Energy Plan 2013–33, and several initiatives are already at an advanced stage: solar energy for communities in the hinterland, a study on waste-to-energy at the national landfill, and microhydropower projects in the Interior. The iNDC states that other forms of renewable energy to be explored are wind energy as well as biomass-to-energy. A national energy efficiency program has also been initiated and tariffs on renewable energy products have been eliminated. Regarding conditional contributions, the government expects international funding to explore the potential of biofuels with rice husk, several species of grasses, and microalgae as a biological source. Also, in consideration are a large hydropower project; a biofuel project to mix gasoline and at the same time to produce electricity.

Nearly half of Suriname's power generation comes from hydroelectricity. Within the conditional goals the iNDC proposes renewable participation more than 25% by 2025, but it does not clarify if it refers to others than hydro.

Without any doubt the *Uruguayan* NDC is the most detailed of all South American countries. It makes an accurate diagnosis and lists mitigation and adaptation goals sector by sector. Regarding energy sector, it states that the matrix transformation was possible through a public–private investment accumulated for several years. The renewable energy projects were supported under the investment promotion system. Nontraditional renewable sources (wind, solar, biomass residues) have reached 33% of the power generation in 2015, according to OLADE.

Together with traditional hydropower, they have reached 57% of the primary energy mix the same year, according to NDC. For that reason, total emissions from the energy sector in relation to the country's GDP are very low compared to the world average. The NDC insures that the structural transformation of the power generation matrix would allow the country to halve absolute emissions in 2017 compared to 1990, although power consumption has almost tripled.

The goals that will be implemented unconditionally in the electricity sector by 2025 are the following: 32% of installed capacity from wind power, 5% of installed capacity from solar energy, 4% of installed capacity from biomass. There are other mitigation measures, such as increase power generation from biomass energy for self-consumption by the private—industrial sector; start of the ring closure of the high-voltage power supply network throughout the country to support decentralized electrical power generation from renewable sources; use of solar collectors for domestic hot water in large users, industrial, and residential users; implementation of a residential area pilot plan of smart grids; implementation of the 2024 Energy Efficiency Plan.

Regarding transport, NDC foreknows adoption of biofuels (5% bioethanol blended with gasoline and 5% biodiesel blended in diesel fuel); implementation of mandatory labeling of energy efficiency in light combustion engine vehicles; introduction of electric vehicles in public transport; introduction of utility electric vehicles; installation of the first electrical route of Latin America, installing power systems for electric vehicles in the national routes that connect Colonia—Montevideo—Chuy.

Finally, *Venezuela* underlines that it has a clean energy matrix as the share of hydropower is close to 70%, but at the same time, it makes the country vulnerable to droughts. Therefore the government has built thermoelectric plants to reduce energy vulnerability. Besides, it has initiated the substitution of thermoelectric generation technology based on liquid fuels (diesel and fuel oil) by natural gas and combined cycle technologies. In addition, the government is implementing energy efficiency and rational use of energy policies.

Within the framework of its Plan for Economic and Social Development of the Nation, which "includes a strong social-environmental ecosocialist component whose implementation attacks the main causes of climate change," the Venezuelan government introduces some strategic objectives in energy matters. One of them is to increase clean energy generation, particularly wind energy projects. At this point, the NDC highlights the promotion of technological sovereignty. Another objective is to increase solar energy generation through the installation of solar panel factories, which primarily address the energy demand of isolated populations. By last, the Plan suggests studies for the development of marine energy sources in order to take advantage of the potential of our extensive coasts.

In general, the main focus of NDCs is on power generation, although it generally accounts for only 18% of the final energy consumption in the region, according to OLADE data from 2015. Likewise, it is notable that practically all countries have the goal of incorporating or expanding renewable energy sources in their respective energy matrices, but only one country (Venezuela) makes explicit its willingness to manufacture these technologies on its own and thus safeguard its technological sovereignty. As for the actors, it has been noted that Chile specifies the predominance of private investment in the incorporation of renewable energy, while Bolivia reserves a preponderant place for its state-owned company. For its part, Uruguay highlights the joint action of the public and private sectors and its planning assumes a strategic nature, since it designs from the State a future trajectory for the energy system in conditions of shared power with other actors. The planning of Venezuela and Bolivia seems to be more normative, since it is directly linked with economic objectives through the action of the State. On the other hand, the rest of the NDC analyzed are reduced to an orientative planning that privileges the market mechanisms.

In order to close and summarize this section, Table 4.1 shows the difficulty to compare the incorporation of renewable energies in the electricity generation of each country. The year of reference and the target year are different in each case, and there is no homogeneous way to define the renewable concept: some countries refer exclusively to nonconventional energies (wind, solar, and waste, for example), while others include large hydropower or steam combined cycle.

Finally, it is observed that the participation of renewable energies will not exceed an average of 20%−25% in the year 2025, with the exception of Uruguay. Large hydropower will continue to have a preponderant weight in the regional energy mix. This leads to the question whether there is a real energy transition in progress or, barely, a diversification of the sources of power generation.

TABLE 4.1 Percentage of renewables in power generation by country.

Country	Reference year[a]	NDC goal	Renewable energy sources taken into account
Argentina	1.9% (year 2015)	20% (year 2025)	Small hydro, biomass, wind, and solar
Bolivia	2% (year 2010)	9% (year 2030)	Alternative energy and other energy (steam combined cycle)
Brazil	14.5% (year 2015)	23% (year 2030)	Other than hydro
Chile	9.8% (year 2015)	20% (year 2025)	NCRE
Colombia	5.1% (year 2015)	−	Small hydro, biomass, wind and solar
Ecuador	1% (year 2013)	−	NCRE
Guyana	6.2% (year 2015)	100% (year 2025, with international cooperation)	Wind, solar, biomass, and hydro
Paraguay	100% large hydro (2015)	Increase the consumption of renewable by 60% (year 2030)	Solar, wind, and biomass
Peru	10% (2015)	−	Small hydro, biomass, wind, and solar
Suriname	49% hydro (2015)	25% (year 2025)	No data
Uruguay	33% (year 2015)	Installed power capacity (year 2025): 32% of wind energy, 5% of solar energy, 4% of biomass	
Venezuela	Near 70% large hydro (year 2017)	Promote national development of wind and solar technologies	

[a]When a NDC does not specify a reference year, data for 2015 was used.
NCRE, Nonconventional renewable energies; *NDC*, Nationally Determined Contributions.
Own elaboration based on Climatescope, OLADE, NDCs, and BURs.

4.4 Regional organizations

Latin America in general and South America in particular are characterized by an overlap of regional integration processes, which often deal with similar issues (Santos, 2018). In terms of energy, this occurs between the Southern Common Market (Mercosur), the Andean Community (CAN), the Initiative for the Integration of the Regional Infrastructure of South America (IIRSA), and the Union of South American Nations (UNASUR).[9]

The Southern Common Market (*Mercosur*) is an initiative that dates back to the 1990s and initially counts on Argentina, Brazil, Paraguay, and Uruguay. Regarding market integration modalities, the current energy integration in Mercosur is based on the spot market and (bilateral) contracts, thus there is no effective joint regional planning, nor optimization between systems, but only opportunities exchanges. Even though regional market splitting (full market integration)[10] is desired, it is important to consider that as long as there is not something really regional, it is necessary that the countries at least take into account the plans of their neighbors when preparing their energy expansion and operation plans. Undoubtedly, this will be a first step toward a path of regional energy planning.

Santos (2018) analyzes the evolution of the energy institutional design within Mercosur by using the official Mercosur database, then considering 25 official regulations and documents dealing with energy, electricity, oil, gas, biofuels, and international interconnections between 1993 and 2012. Based on such analysis, five points will be highlighted below. In the first Mercosur document that discusses the energy issue (Resolution 57/1993), Energy policies in Mercosur are drawn. The document highlights

1. favoring integration between the energy markets of the States Parties, with freedom of purchase and sale of energy companies and free transit of energy, respecting the laws in force in each country;
2. promotion of rational use of energy and conservation;
3. admission of the possibility of binational or multilateral energy agreements within or outside the region; and
4. the preparation of regional integrated energy planning studies in accordance with national macroeconomic planning.

It is clear that already in the first document the binational character that will be present in the different agreements stands out.

The second point to be highlighted is the Decision 10/1998, comprised a memorandum of understanding regarding electrical exchanges and electrical integration in Mercosur. Then, it is understood that the document, while pioneering the integration of electricity in Mercosur, aims to remove legal and political barriers to energy exchanges between members, not assuming the role of infrastructure expansion for the transport of energy, which is still precarious among Mercosur countries.

The third point is that Decision 10/1999 bears many similarities to Decision 10/1998, since the text itself appears to have been built upon it. The difference between them is that Decision 10/1998 addresses integration based on electricity while 10/1999 deals with gas exchanges and

[9] Not to mention other regional organizations that deal with the same issue, but have different institutional arrangements and a more technical focus, such as Latin American Energy Organization (OLADE) and Regional Energy Integration Commission (CIER).

[10] "Although it seems unlikely to happen in the South American subcontinent, the most controversial point would be the definition of a single operator, which implies a loss of countries' autonomy and the fear that this operator acts in a discriminatory manner benefiting countries with the largest consumer market." (Santos, 2018: 55).

gas integration between Mercosur States Parties (Pergher, 2016).

The fourth point highlighted can be found in the Framework Agreement on Energy Cooperation (2005). In this document, there is an interest in "advancing the integration of production and transportation," which would necessarily promote the expansion of energy transport infrastructure, as well as joint production among Mercosur countries. However, the rest of Mercosur legislation does not appear to be a normative basis for such interest, and there are only regulations on reducing national barriers to energy exchanges. In addition, the agreement points out great asymmetries in the sector between the Member States of the Agreement. The last point emphasized is that since 2012 there are no recommendations, directives, resolutions, and/or decisions on the official Mercosur website.

It is also interesting to note the existence of a Sub-Working Group within Mercosur that specializes in energy issues (SGT-9). Some of the most relevant documents signed in Mercosur were consequence of the SGT-9, such as those already discussed (Decisions 10/1998 and 10/1999). The existence of an Ad Hoc Group on Biofuels (GAHB) was established by Decision 49/2007. In this sense, energy integration issues are a major topic at the Meeting of Ministers of Mines and Energy (RMME), part of the Common Market Council, as well as of the SGT-9 and Mercosur GAHB, both subordinate to the Common Market Group. On the other hand, environmental issues are undertaken by SGT-6, which works close to a preparatory technical commission for the Meeting of the Environment Ministers (RMMA), having a more political tone.

Thus Meneghini and Voigt (2011) conclude that SGT-9 did not act to structure and coordinate concrete policies or projects on regional energy integration. In the analyzed period, it only worked on (1) the harmonization of energy regulation in order to facilitate exchanges; (2) the elaboration of inventories on the electric sector; (3) feeding databases to foster decisions in other instances; and (4) the analysis of financial, legal, and tax aspects of the sector. Although SGT-9 has daring objectives, they ended up not being reached by several reasons (Santos, 2018).

Still regarding Mercosur, the SGT-9 agenda can be considered broader in some way, including some aspects of sustainable energy. This is reflected in its latest meetings, which refers, for example, to actions related to the rational and efficient use of energy, such as the generation of a base of energy efficiency indicators for Mercosur. However, since SGT-9 discontinued its work and has not met since 2011—without any formal explanation—it is not possible to establish its relationship with the United Nations (UN) 2030 Agenda, which was launched in 2015.

In general terms, it can be argued that the different SGTs do not have the relevance of other Mercosur issue areas, whose nature is born strongly influenced by trade issues (and still remains as such). Besides, one can also connect the loss of relevance of SGT-9 with the creation of UNASUR Energy Council, the corresponding technical forum, and the negotiations of the UNASUR energy integration treaty. This is due to the already mentioned nature of Latin American regional integration processes characterized by the overlap of initiatives, which often deal with the same theme.

The *Andean Community* (CAN) is the continuation of the Andean Pact (1969) and is currently composed of Bolivia, Colombia, Ecuador, and Peru. In terms of energy, CAN has focused mainly on the creation of a regional energy market and the electric interconnection of its countries. For this purpose, it has an Andean Committee of Regulatory Organisms and Regulators of Electricity Services (CANREL) and two working groups: the Group of Electric Service Regulatory Agencies (GTOR), of the regulatory bodies and the Working Group of Electric Service Policy Setting Agencies (GOPLAN), for the planning of the infrastructure. Unlike UNASUR, CAN does not consider the political dimension of energy and is restricted to an eminently technical dimension of electrical interconnection (Silva Pareja, 2018). Currently, CAN (as well as Chile as a country associated to that entity) has approved the general framework for the subregional interconnection of power systems and electricity exchange, under the umbrella of Andean interconnection initiative called SINEA. Basically, "the aim is to create a general community framework that allows integration in the energy market between the countries mentioned. However, in the first stage, Bolivia will participate only as an observer country. SINEA aims to develop an Andean Electricity Corridor, creating an interconnected electrical system between 2014 and 2024" (Santos, 2018).

For its part, *UNASUR* was created in 2008 as result of a series of discussions and meetings. It is noteworthy that from the very beginning energy was a priority in the UNASUR agenda. Among the general documents, declarations, agreements, communiqués, and protocols, UNASUR signed 46 documents related to energy. The following important energy integration frameworks will be analyzed, since they stress the regional role to the regional energy planning: (1) Declaration of Caracas (2005); (2) Declaration of Cochabamba (2006); (3) Declaration of Margarita (2007); and (4) The Guidelines of the South American Energy Strategy (approved in 2010).

The *Declaration of Caracas* (2005), signed in Venezuela, gives priority to energy integration. Therefore the importance of regional infrastructure for energy transport is emphasized, as is the strategic role that energy has in the economic and social development of South America. The *Declaration of Cochabamba* (2006), signed in Bolivia, is often remembered as the "cornerstone of the South American integration process." Its main objective was to stimulate energy integration in order to reduce the region's socioeconomic asymmetries. The "Presidential Declaration on South American Energy Integration" is accomplished, which is the first of its kind, at the South American level, within the framework of developing the institutionalization of this sector. It also represents a more far-reaching view, identifying the principles that will guide the process of political and normative convergence in terms of regional energy integration (CAF, 2013).

Signed in Isla Margarita, the *Declaration of Margarita* (2007) created UNASUR, as well as the South American Energy Council (CES). In addition, it is suggested to prepare a proposal for alignments of the South American Energy Strategy, the Action Plan, and the Energy Treaty of South America documents that originally should have been discussed during the course of the III South American Summit of Nations.

Finally, the *Guidelines of the South American Energy Strategy* were approved in Los Cardales (2010). They promoted the development of renewable and alternative energies, the promotion of energy efficiency, and move toward regional planning, among other points. They also stressed the need to pay attention to climate change and its consequences in the framework of sustainable development. In addition, they favored the technological independence.

Although the UNASUR Energy Council is responsible for formulating general policies that should lead to energy integration of South America, the South American Council of Infrastructure and Planning (COSIPLAN) has been directly playing a crucial role as a forum for political and strategic discussion through consultation, evaluation, cooperation, planning and coordination of efforts, and articulation of programs and projects in order to foster regional infrastructure integration among UNASUR member countries. Since 2011, the

IIRSA was incorporated into the COSIPLAN. IIRSA includes coordination mechanisms among governments, multilateral financial institutions, and the private sectors and aims to coordinate investment plans and programs, as well as to prioritize Integration and Development Hubs (EID).

UNASUR could have a closer approximation to the energy objectives of the global agenda because it is a mechanism of a more political nature than Mercosur (Silva Pareja, 2018). However, in recent years, there has been a regional change of political orientation since the arrival of promarket governments, especially in Brazil and Argentina, and the economic and political crisis that Venezuela is going through. In this context, there is a process of dismantling or disuse of the Mercosur's and UNASUR's mechanisms. Since 2016, for example, the activity of the Energy Council was paralyzed and in 2018 Argentina, Brazil, Chile, Colombia, Paraguay, and Peru decided to suspend their participation in UNASUR. Later, Colombia and Ecuador decided to leave the organization.

Recently, new regional instances, beyond South America, have emerged in which the relationship between energy and climate policies is addressed. The *Forum of the Countries of Latin America and the Caribbean on Sustainable Development* was born at the 36th session of ECLAC in 2016. There, a resolution was passed for the establishment of this annual meeting of the countries of the region to follow up on the SDGs, including N. 7. The current structure foresees the concerted relationship between the UN agencies and other entities, such as the regional integration mechanisms of Latin America and the Caribbean (Silva Pareja, 2018).

Besides, the *Energy and Climate Partnership of the Americas* (ECPA) had actually been created in 2009 at the behest of the United States in order to face the challenges of energy and sustainable development, but it was reactivated with the Second Ministerial Meeting that took

place in Mexico in 2015. ECPA is a hemispheric cooperation mechanism with the aim to share information, knowledge, and best practices. Participation is voluntary, and commitments among its members are not binding. It is proposed to complement existing energy initiatives in the region and contribute to global objectives to move toward cleaner, more efficient, more reliable, and more affordable energy. In addition, it takes into account the peculiarities and specific diagnostic assessments of the four subregions of the Americas.

As could be seen, in fact there are different institutions that deal with energy (and climate) issues at the regional level. Majorly considering South America, they often constitute spaces for debate and alignment of national interests, looking for promoting regional policies and trends. However, given the intergovernmental nature of most of these institutions, there are no binding policies allowing countries to have too much autonomy to formulate their (national) policies; so in practice, there is little regional energy planning, especially if taken into account the current regional political economy.

4.5 Conclusion

Taking into account the questions raised in the Introduction, this chapter sought to answer them from the analysis of data and official documents from the website of the UNFCCC, such as the NDC and BUR submitted by each South American country. With regard to the first question (which has other associated questions), "are the South American countries carrying out a real energy transition? If so, how, by what measures and goals?", all the NDCs' analysis leads us to believe that, in fact, what is actually happening in South America is a mere diversification of the sources of power generation rather than a real and structural regional energy transition. Considering the theoretical discussion related to more liberal, more centralized, or mixed economic systems, Bolivia and Venezuela show the most statesman approach, Uruguay is case of a more mixed system, and the other cases are more promarkets when it comes to the incorporation of renewable energies.

As regards the second question, whether the efforts are driven only at the national level or at the regional level, highlighting therefore the actors involved in the process, official documents from Mercosur, CAN, and UNASUR were used. What has been noticed is that the regional bodies dedicated to the energy and climate issues have contributed little either because of its institutional nature or because of the recent events of regional political economy. It is indeed possible to question the trajectory of the events that promoted (or not) regional integration in South America, which also applies to the energy case. As seen in this chapter, there has been some normative and institutional development in the field of energy integration, which has ended up being incapable of translating into practical results; the most successful cases have binational nature, such as the Itaipu Binacional and the Bolivia—Brazil gas pipeline (GASBOL).[11]

Therefore in practice the normative effort of Mercosur, CAN, and UNASUR was not able to overcome political, technical, economic, and regulatory barriers that prevent the advance of energy integration in Mercosur countries. Due to the intergovernmental nature of them, both institutions end up presenting limitations to their performance. Besides, SGT-9 did not act to structure and coordinate concrete policies or projects on regional energy integration. Trying to map out other regional movements beyond South America, there were also considered the

Forum of the Countries of Latin America and the Caribbean on Sustainable Development and the ECPA—which have recently narrowed the Latin American dialogue with the sustainable development and climate change issues.

The energy transition and energy planning in (the countries of) South America will only be effectively addressed as far as both issues are faced from a regional perspective. In this way, it will be possible to use regional integration, international cooperation, trade agreements, and even the already existing regional institutions to align mutual expectations and suggest common policies. Without any doubt, political will is required for these advances to happen, however, the South American regional context suggests otherwise. Given this current regional scenario of such political and economic uncertainty, it is actually hard to suggest any practical proposals that will reach the commitments undertaken and the established goals.

References

Arent, D., Arndt, C., Miller, M., Tarp, F., Zinaman, O. (Eds.), 2017. The Political Economy of Clean Energy Transitions. Oxford University Press, Oxford, UNU-Wider Studies in Development Economics.

Bersalli, G., Hallack, M., Guzowski, C., Losekann, L., Zabaloy, M.F., 2018. La efectividad de las políticas de promoción de las fuentes renovables de energía: experiencias en América del Sud. Enerlac II (1), 158—174.

Bouille, D., 2004. Economía de la Energía. IDEE/FB, San Carlos de Bariloche.

BP, 2018. Advancing the Energy Transition. BP, London. Available from: <https://www.bp.com/energytransition> (accessed 15.05.19.).

Bueb, J., Richieri Hanania, L., Le Clézio, A., 2017. Border adjustment mechanisms: elements for economic, legal, and political analysis. In: Arent, D., Arndt, C., Miller, M., Tarp, F., Zinaman, O. (Eds.), The Political Economy of Clean Energy Transitions. Oxford

---

[11] Unlike the South American case, Central America has a successful case of regional energy integration known as Central American Electrical Interconnection System (SIEPAC). However, because it is a very particular case and far from the reality and complexity of the one analyzed in this chapter, we avoid treating them together (in the context of Latin America) or even suggest some comparative analysis between both cases.

University Press, Oxford, pp. 60–79. UNU-Wider Studies in Development Economics.

CAF, 2013. Energía: una visión sobre los retos y oportunidades en ALyC: Integración energética. Corporación Andina de Fomento.

Carrizo, S., Núñez Cortés, M., Gil, S., 2016. Transiciones energéticas en la Argentina. Ciencia Hoy 147, 24–29.

Castro, N.J., Leite, A.L.S., Rosental, R., 2012. Integração energética – uma análise entre União Europeia e América do Sul. TDSE 48, GESEL/IE/UFRJ, Rio de Janeiro.

Chevalier, J.-M., 1973. Le Nouvel Enjeu pétrolier. Calmann-Lévy, Paris.

Fri, R.W., Savitz, M.L., 2014. Rethinking energy innovation and social sciences. Energy Res. Soc. Sci. 1, 183–187.

Grayson, M., 2017. Energy transitions. Nature 551 (7682), S133.

Helm, D., 2002. Energy policy: security of supply, sustainability and competition. Energy Policy 30 (3), 173–184.

Helm, D., 2011. Peak oil and energy policy: a critique. Oxford Rev. Econ. Policy 27 (1), 68–91.

Helm, D., 2014. The European framework for energy and climate policies. Energy Policy 64, 29–35.

Hurtado, D., Souza, P., 2018. Geoeconomic uses of global warming: the "green" technological revolution and the role of the semi-periphery. J. World Syst. Res. 24 (1), 123–150.

IEA-IRENA, 2017. Perspectives for the Energy Transition: Investment Needs for a Low-Carbon Energy System. OECD/IEA and IRENA.

IRENA, 2018a. Global Energy Transformation: A Roadmap to 2050. International Renewable Energy Agency, Abu Dhabi.

IRENA, 2018b. Renewable Energy Statistics 2018. International Renewable Energy Agency.

IRENA, 2019. A New World: The Geopolitics of the Energy Transformation. International Renewable Energy Agency, Abu Dhabi.

Kuzemko, C., Keating, M.F., Goldthau, A., 2018. Nexus-thinking in international political economy: what energy and natural resource scholarship can offer international political economy. In: Goldthau, A., Keating, M.F., Kuzemko, C. (Eds.), Handbook of the International Political Economy of Energy and Natural Resources. Edward Elgar Publishing, pp. 1–20.

Kyu, L.S., Yeong, J.H., 2016. The formation of the EU's Energy Union and its implications for the Northeast Asian Energy Market. Northeast Asian Econ. Rev. 4 (1), 11–22.

Maldonado, Y.A.M., Pineda, R.G., Blanco, S.S., 2017. Análisis de integración regional con fuentes de energía renovable en América Latina y el Caribe. Enerlac I (1), 106–125.

Martin, J.-M., 1974. L'energie et le desarroi post-industriel – essai sur la croissance de l'energie. Energy Policy 2 (3), 256–257.

Martin, J.-M., 1990. L'économie Mondiale de L'énergie. a Découverte, Collection Repères, Paris.

Martin, J.-M., 2000. Le changement des technologies de l'énergie: genèse, modalités et hipothèses explicatives. In: Bourgeois, B., et al., (Eds.), Énergie et Changement Technologique. Éditions Économica, Paris, pp. 17–61.

Meneghini, E.M., Voigt, M.R., 2011. O Mercosul e atuação do SGT9 energia: uma avaliação dos processos de integração energética no âmbito do mercado comum. PMUC-FAPESC.

Moura, G.N.P., 2017. Long-term Power Systems Integration Using OSeMOSYS SAMBA – South America Model Base – and the Bargaining Power of Countries: A Cooperative Games Approach (Ph.D. dissertation). PPE/COPPE/UFRJ, Rio de Janeiro, Brazil.

OLADE, 2017. Manual de Planificación Energética. Available from: <http://www.olade.org/wp-content/uploads/2017/06/Manual_Planificacion_Energetica_Espa%C3%B1ol_Final22-05-2017.pdf> (accessed 18.04.19.).

Paredes, J.R., Schaeffer, R., Szklo, A., 2017. Contribución de las energías renovables a la seguridad energética en América Latina. Banco Interamericano de Desarrollo, BID.

Percebois, J., 1986. Gas market prospects and relationship with oil prices. Energy Policy 14 (4), 329–330. 333-346.

Percebois, J., 2008. Electricity liberalization in the European Union: balancing benefits and risks. Energy J. 29 (1), 1–20.

Pergher, H., 2016. A integração energética na América do Sul: uma análise das políticas de integração energética promovidas no Mercosul e na Unasul. RICIC 3 (6), 58–82.

Ramos, D.S., 2016. Electricity markets regional integration: conceptual basis, potential benefits and opportunities for the southern cone. In: Castro, N.J., Rosental, R. (Eds.), Integration and Electric Energy Security in Latin America. Oficina de Livros, Rio de Janeiro, pp. 73–98.

Santos, T., 2014. A Critical Deconstruction of Myths and Misunderstandings About Energy Integration in South America. V SimpoRI, UERJ, Rio de Janeiro.

Santos, T., 2018. Regional Energy Security: Re-Evaluating Concepts and Policies to Promote Energy Integration in Mercosur (Ph.D. thesis). PPE/COPPE/UFRJ.

Silva Pareja, M.C., 2018. Relaciones entre la institucionalidad global y regional en materia de energía: la Agenda 2030 de Naciones Unidas y la Agenda Energética de América del Sur. ENERLAC II (1), 72–90.

Sovacool, B.K., 2017. The history and politics of energy transitions: comparing contested views and finding common ground. In: Arent, D., Arndt, C., Miller, M., Tarp, F., Zinaman, O. (Eds.), The Political Economy of Clean Energy Transitions. Oxford University Press, Oxford, pp. 16–38. UNU-Wider Studies in Development Economics.

Prompting an energy transition: how policy entrepreneurs passed Peru's first renewable energy legislation for the grid

Hai-Vu Phan

Edison International, Rosemead, CA, United States

5.1 Introduction

The theme of this book analyzes the energy transition in Latin America. This chapter contributes to this compilation by analyzing how an energy transition may begin, opening the door for future renewable energy efforts.[1] Through the lens of the policy window framework, this chapter analyzes how Peru was able to pass its first major policy so that renewable energy generators can sell their electricity on the national electric grid. Distributed renewable energy efforts (those technologies not connected to the electric grid) in Peru are discussed in the next chapter by Welch in this volume.

At first glance, Peru looks like a dream country for renewable energy enthusiasts. Aside from great natural conditions that are necessary for renewable energy production, Peru appears to be one of the most supportive countries of the environment and renewable energy. According to Ernst & Young, Peru is the fifth most attractive country in Latin America for renewable energy investment because of its abundant natural resources: solar, wind, hydro, and geothermal, among others (Gestión, 2018). It is part of the United Nations (UN) Framework Convention for Climate Change (UNFCCC), it was an early signatory of the Kyoto Protocol, and it even hosted the 20th Conference of the Parties

---

[1] Hydroelectricity is a renewable resource that has long been an important part of Peru's energy mix. Exploitation of other renewable resources, such as solar, wind, and geothermal, has been more challenging. Because this chapter talks about the advancement of newer renewable energy technologies, the term "renewable energy" as used in this chapter refers to energy from renewable sources excluding hydroelectric dams above 20 MW.

(COP). All of these international efforts aimed to mitigate climate change, in large part through promoting renewable energy. In Lima, there is even a national theme park dedicated to educating citizens about climate change, the only one of its kind in Latin America.

Despite these many supportive gestures, actual policy that supported large-scale renewable energy consumption was lacking. For many years, entrenched political opposition and lack of a strong need successfully thwarted most efforts to pass renewable energy policies. In the 2000s, Peru was in a position of an energy surplus, so there was no energy security problem. A total of 80% of its electricity came from hydroelectricity, so it had an energy mix that was cleaner than most other countries. Renewable energy technology was many times costlier than existing sources. Renewable energy was also less reliable, possibly causing problems for the electric grid that was not built to withstand the intermittent surge of electric load. Existing players in the electric market opposed renewable energy because it was a new competitor, could have caused malfunctions in the electric system, and raised overall prices leading to higher prices of electricity.

Yet in 2008, Peru surprisingly passed its first policy that allowed electricity from renewable energy sources to connect to and sell on the electric grid. This chapter analyzes the political dynamics that allowed this groundbreaking policy, which many hoped would prompt a renewable energy transition in Peru, to pass.

This chapter consists of six sections, in addition to the introduction. Section 5.2 discusses the evolution of the Peruvian energy landscape from the 1990s to the 2000s to provide some context of the Peruvian energy sector. Section 5.3 explains the passage of the country's first grid-connected, renewable energy policy, delving into the Peruvian Trade Promotion Agreement (PTPA), the legislative process in Peru, and how the two combined opened a policy window that allowed for this

policy to pass. Section 5.4 analyzes the concept of the policy window in depth to understand whether what happened in Peru is unique or can be transferable to other contexts. Section 5.5 analyzes the impact of this policy on Peru's market. Section 5.6 touches upon remaining issues that present obstacles to further proliferation of renewable energy. Finally, Section 5.7 concludes that the renewable energy political and economic landscape is once again at a standstill in Peru. There needs to be another big push by taking advantage of another policy window to spur further renewable energy development. The government, in the meantime, can help with this effort by encouraging more distributed renewable energy resources and investing in research and development.

5.2 Evolution of Peru's energy landscape: 1990s—2000s

This section provides some context of Peru's energy sector to show that opposition to renewable energy was powerful and entrenched. In the 1970s and 1980s, petroleum was Peru's primary energy source, meeting 70% of its commercial requirements in 1981 and accounting for 24% of its total exports (UNDP/World Bank Energy Sector Assessment Program, 1984, p. iv). After several unsuccessful explorations in the 1980s and 1990s, the government switched its focus to natural gas, which Shell had discovered in the fields of Camisea, near Cusco. The government persisted despite social, environmental, and technological challenges, and the Camisea pipeline finally came online in 2004 (Wise, 2007). Although the discovery of oil and natural gas brought Peru benefits, it tended to divert the government's attention and finance away from investment in other markets. As a result, hydroelectricity investments stalled, and development in the market and technology of other resources remained nascent.

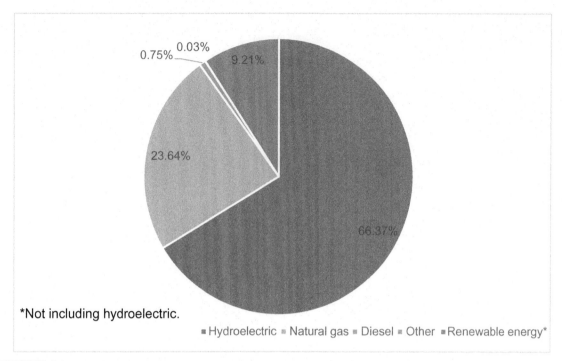

FIGURE 5.1 Electric generation by source April 2019, percentage of total. Source: *COES, 2019. Boletines Mensuales — Abril. Coes.Org.Pe. Available from: <http://www.coes.org.pe/Portal/Publicaciones/Boletines/> (accessed 18.06.19.).*

The electricity sector, though always a priority, developed in the shadows of oil and natural gas exploration. In 1995 hydroelectricity produced 88.1% of Peru's electricity, while 11.9% came from thermal electricity (a category that includes oil, natural gas, and coal sources, and for Peru this was mainly oil at that time) (Tamayo et al., 2016, p. 239). In 2001 hydroelectricity produced 84.7% of Peru's electricity, while hydrocarbons produced 14.5% (World Bank, 2017). By 2015, 50.6% of Peru's electricity came from hydroelectricity while 47.6% came from thermal electricity (at the time coming mainly from natural gas) (Tamayo et al., 2016). Fig. 5.1 shows Peru's electric generation by source in 2019.

The modern Peruvian electricity sector took shape starting in 1992, when the Peruvian government decided to privatize the formerly national sector. The period leading up to the 1990s was especially rough for Peru, marked by hyperinflation, a stagnant economy, high unemployment, growing poverty, natural disasters, terrorist attacks, and rampant nepotism within the government (Abusada Salah et al., 2000). The new law unbundled electricity services—generation, transmission, and distribution—into separate companies and began to privatize them (Campodónico Sánchez, 1999, p. 32). This process of privatization continues into present day, and the government, even today, has an interest in the economic viability of electric companies.[2]

[2] Out of 56 electricity generators in Peru, five are public companies and together produce about 20% of the market's electricity. All 17 transmission companies in Peru are private, and 10 out of 23 distribution companies are public (Corporación FONAFE, 2018; COES, 2019; MINEM, 2017).

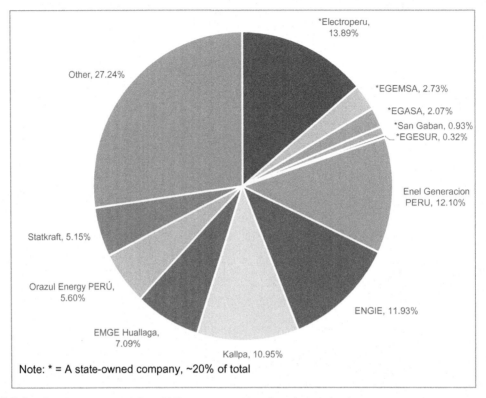

FIGURE 5.2   Electric generators in Peru 2019, percentage of total market production.

Fig. 5.2 shows the generation companies in Peru, with those highlighted in blue being state-owned companies.

Peru in the decades of the 1990s and 2000s was also reestablishing itself in the international community. One way it did so was to be an advocate of climate change initiatives. Peru participated in the UN Conference on Environment and Development, also known as the Earth Summit, in 1992 and signed the UNFCCC that same year. Peru's Congress ratified it in 1993, and the UNFCCC entered into force in Peru in 1994 (UN Climate Change, 2014a). Peru began participating in every COP to the UNFCCC. Peru was also an early signatory of the Kyoto Protocol, which developed out of COP 3 in 1997, signing it in 1998 and ratifying it in 2002 (UN Climate Change, 2014b). It served in the Subsidiary Body for Implementation within the conferences in the role of rapporteur from 1995 to 1998. Not only was Peru aware of the dialog surrounding climate change and its mitigation, its leadership position gave it the capacity to even shape this dialog.[3]

Despite active international participation the picture at home was very different. Renewable

[3] Peru's international commitments have not stopped. At COP 19 in Poland, Mr. Jorge Voto-Bernales of Peru was one of the vice presidents of the conference. Its leadership culminated in 2014, when Peru hosted COP 20 in Lima, where the then—Prime Minister of the Environment, Manuel Pulgar-Vidal, was president of the conference. Peruvians have also continued to serve in other roles; between 2013 and 2016, Tania Zamora Ramos served as an expert in the Consultative Group for the Latin America and Caribbean region (UNFCCC Historical Membership Charts, 2017).

energy had close to no presence in Peru's market before the 2000s. There was zero renewable energy electricity generator selling to the grid (COES, 2006–2018). Renewable energy players were mostly small nongovernmental organizations or social enterprises that tried to promote these technologies to rural residents (ACCIONA, 2018; Ayuda en Acción, 2018; SNV, 2018).

While the discrepancy between international and domestic efforts surprised many external observers, the lack of renewable energy in Peruvian society came as little surprise to its residents. As mentioned earlier in the chapter, in the early 2000s, Peru was in a position of electricity surplus (BNAmericas, 2018). Even though there was an energy crisis in 2004,[4] once the Camisea[5] gas fields started producing in 2004, Peruvians were assured that they would have enough reserve for domestic consumption for 30 or 40 years without fear of shortage (Merco Press, 2008). In other words, Peru did not have an energy security problem. Beyond the capacity surplus, Peru's electricity was already "green" in the sense that hydroelectricity constituted 80% of Peru's electricity production.[6] In addition, renewable energy was several times more expensive than the energy sources that Peru was already using, and introducing it would force consumers to pay higher electricity bills. In 2008 solar energy cost more than USD 200/MW h, compared to market price for fossil fuel energy, which was around USD 50/MW h (Quintanilla, 2016). Renewable energy technology was still incipient and unreliable, possibly harming the electric grid. Finally, the major players in Peru's energy market—generators, transmitters, and distributors—already had an established natural monopoly, and they opposed renewable energy for fear that it would disrupt the current functioning system and raise prices for customers (Ernst & Young, 2016).

Peru did not have any immediate need to introduce a renewable energy policy at this time. In fact, it had strong reasons to oppose it, which many people did. Economically, politically, and environmentally, Peru appeared to have no urgency to push for renewable energy. Despite the opposition and a calm year for the status quo of energy operations, Peru passed its first law to allow renewable energy to feed into its national electric grid in 2008. What allowed Peru to make this policy decision despite a disagreeable political context? Learning about Peru's difficult case and how policy entrepreneurs were able to maneuver through the challenges can shed some light on some of the factors that can make this happen, which may be applicable to other developing countries where renewable energy policy is also laggard because of domestic opposition, such as Brazil and Chile (see Bonilla, 2012; Forero, 2013).

5.3 Passing Peru's first renewable energy legislation for the grid

This section dives into how Peru was able to pass the first renewable energy policy for grid-connected generation. It sees that this

---

[4] This energy crisis was caused by a major drought, which revealed the many shortcomings of relying heavily on hydroelectricity, even though the 2004 drought did not shut down the electric system. Recall that hydroelectricity produced about 85% of Peru's electricity at this time. This prompted the government to search for a different electricity source, which they believed to be Camisea (Comisión MEM-OSINERG, 2005).

[5] In the 1980s, Shell discovered in Camisea, near Cusco, fields of natural gas. However, the consortium that Shell led and the Peruvian government could not come to an agreement on how to develop this resource. The issue was further complicated by social protests from environmentalists and people living on the land. The government persisted despite challenges, and the Camisea pipeline, a publicly owned project, finally came online in 2004 (Wise, 2007).

[6] (COES, 2006–2018, p. 10).

policy was able to pass due to the opening of a policy window. Kingdon (2011) introduced the ideas of the "policy window" in his groundbreaking book, *Agendas, Alternatives, and Public Policies*. In this model a *policy window* is "an opportunity for advocates of proposals to push their pet solutions, or to push attention to their special problems" and is the point where independent streams—problem, policy, and politics—converge (Kingdon, 2011, p. 165). Institutions render some of these windows predictable and others unpredictable. Predictable windows occur because of structures in the legislative process, such as the beginning and ending of a legislative session, the budget period, or policy renewal—all of which are points where debate surrounding an issue can arise. However, some windows are unpredictable due to unforeseen circumstances such as a natural disaster. These types of windows work most outside of institutional constraints because it is when a random event such as a crisis opens an opportunity. Although these events are generally outside of both institutional and actors' control, the actions that actors can take when these events occur are still subject to those allowed by the existing institutions. Exogenous shocks, then, can tap into spaces of opportunity that exist within institutions but are not accessible during routine politics. In extraordinary instances, spaces emerge within which actors are able to produce great change. Institutions are generally designed with regularly occurring and predictable policy windows. For windows that occur outside of routine politics, institutional design renders some actions acceptable and others inacceptable (Kingdon, 2011, p. 165).

Section 5.3.1 explains a free trade agreement occurring at this time with the United States. It argues that a free-trade agreement with the United States prompted Peru to use a legislative power, the subject of Section 5.3.2, that would allow it to pass policies quickly. Section 5.3.3 of this section brings both ideas together to open a policy window, brought on by an exogenous factor of a bilateral trade deal, allowing legislative decree (LD) 1002's passage.

5.3.1 Peru trade promotion agreement with the United States

Peru's opportunity to pass its first renewable energy policy came from an unsuspecting source: a bilateral trade deal. While global developments were happening rapidly in the renewable energy sector, Peru's more immediate concern was negotiating its first permanent trade deal with the United States. The trade agreement would not enter into force until both parties "have completed their legal requirements or on such other date as the Parties may agree" [USTR (Office of the U.S. Trade Representative), 2006]. Once both parties have notified the other that they have completed their obligations, the PTPA would enter into force 60 days from the receipt of notice.

Although the United States and Peru did not have a formal trade agreement prior to 1991, the United States has long been an important trade partner. In the 1980s the United States was Peru's most valuable market. A total of 33.5% of Peru's exports went to the United States, while 39.7% of Peru's imports came from the United States (Lengyel and Ventura-Dias, 2004). In 1991 the United States signed the Andean Trade Preference Act (ATPA) with Peru, Bolivia, Ecuador, and Colombia, which gave these countries tariff privileges for certain Andean products. In 2002 this preferential treatment was expanded under the Andean Trade Promotion and Drug Eradication Act (ATPDEA), enlarging the list of duty-free products sold in the United States from Andean countries to about 5600 [USTR, 2005; USCBP (US Customs and Border Protection), 2017]. The ATPDEA was set to expire on December 31, 2006. Both the ATPA and the ATPDEA were available for renewal, but this would have required both the US Congress' and the president's approval each time, which was inconvenient [OAS (Organization of American States), 2017].

In November of 2003, rather than seeking to renew the ATPDEA, the United States announced its intention to negotiate permanent trade deals with each of the Andean countries (Ministerio de Comercio Exterior y Turismo, 2011). Of the issues that were discussed during the numerous rounds of negotiations, one of the enduring topics was that of environmental protection. One of the declarations in the PTPA's Preamble is to "IMPLEMENT [sic] this Agreement in a manner consistent with environmental protection and conservation, promote sustainable development, and strengthen their cooperation on environmental matters." (USTR, 2006). Gina Cady, Deputy Director for Bilateral Environment Program of the United States Agency for International Development, explains, "This was the first time the U.S. put stringent environmental requirements in a trade agreement, so it is a litmus test to see if trade agreements actually work in promoting environmental terms." (Cady, 2017). In this respect the United States requested that Peru enact laws to protect its environmental endowments, especially its forests and biodiversity, although exact laws and enforcement procedures would be up to Peru's discretion.

Negotiations began in 2004. After 13 rounds of negotiations the PTPA was signed in Washington, DC, in 2006 (Ministerio de Comercio Exterior y Turismo, 2011). The period after its signing, Peru went into a frenzy to implement the many requirements that the PTPA demanded. While there was no deadline to complete these requirements, the parties had an incentive to act quickly since the agreement would not go into effect until both the United States and Peru completed their legal obligations

as stated in the agreement. Once both parties have notified the other that they have completed their obligations, the PTPA would enter into force 60 days from the receipt of notice.

5.3.2 Legislative process in Peru

Immediately after signing the PTPA, the Peruvian government entered into a crucial period during which it was required to make many institutional and policy changes to meet the PTPA's terms.[7]

However, passing a national law in Peru is a multistep and time-consuming procedure. The various check points—policy committee, budget committee, Congress floor, and the president's desk—allow different political elites to review and voice their concerns about each policy. Debates on the Congress floor become public record, with all 130 Congress members able to participate at once. At each stage of deliberation, the bill may be subjected to amendments before it can move forward in the process. The reason for this meticulous vetting is so that the bill can be well-crafted. However, with so many cooks in the kitchen, the original supporters are many times left unsatisfied because the final policy is unrecognizable from its first introduction (Congreso Perú, 2018).

This practice serves Peru well enough for its everyday, nonurgent issues, but it was not going to be adequate in meeting the PTPA's hundreds of demands quickly. When voluminous, fast policy-making is needed, Congress can delegate legislative power to the executive branch, whose hierarchical structure can make decisions more quickly compared to Congress' egalitarian and deliberative nature.[8]

---

[7] The PTPA's institutional and policy requirements included strengthening Peru's financial institutions and protecting intellectual property rights. One of the enduring topics was that of environmental protection (USTR, 2006).

[8] Congress entrusts its powers to the Executive Branch by passing a law to authorize this. Congress must specify in this law the specific subject matter of the laws that the executive branch can pass and the length of time that the executive branch has this power. Congress cannot delegate legislative power to the executive branch related to constitutional reforms, international treaties, organic laws, budgetary laws, and laws affecting the National Account (Congreso Perú, 2018).

This transfer of power can come in three forms. *Urgent decrees* are powerful, but they are narrowed to only economic or finance-related issues and are subjected to revisions by Congress. *Supreme decrees* are also narrow because they are only for changing how ministries and public-sector institutions function. LDs are political vehicles that put the most power into the President's hands as they allow the Executive Branch to pass policies that fall under Congress's authorization, and this decree becomes law. Although the LD is also narrow because the allowed policies must fall under those that have Congress' authorization, sometimes these can be broadly interpreted. In the late 2000s, amidst negotiations of the PTPA, renewable energy supporters found the LD to be the exact policy instrument they needed (Congreso Perú, 2018).

5.3.3 A legislative decree presents a policy window

Because of time pressure, Congress, calling upon Article 104 of the Constitution related to the LD, delegated some of its power to the Executive Branch so that policies could be passed quickly. Law 29157 (2007) gave the Executive Branch the authority to pass policies related to commerce; regulatory and institutional improvements; judicial administration; private investment; technological innovation; the promotion of employment and small- and medium-sized enterprises; strengthening of institutions related to the environment; and improving the competitiveness of agriculture. This law was to be in effect for 180 days starting January 1, 2008—a window of 6 months.

The opportunity window may have opened, but what allowed policy opportunists to successfully seize the moment? To understand policy direction within the Ministry of Energy

and Mining (MINEM), one must look past the minister and at the vice ministers. In Peru the minister is a politician. It is the vice ministers who are the most knowledgeable about their sector. The vice minister at the time of LD 1002 was Pedro Gamio, who served as vice minister from 2006 to 2008. Immediately prior to his appointment as vice minister, Gamio was a policy analyst in the Committee on Energy and Mines within the Peruvian Congress (El Comercio, 2008). In 2007 Gamio was given authority to sign an interinstitutional cooperation agreement with the National Council of the Environment in order to plan how to best manage the Global Environment Fund that Peru received from the UN under the UNFCCC.[9] With this authority, Gamio created a task force to study how renewable energy could help Peru. The task force studied the experiences of other countries, including those of Germany (Gamio, 2017).

The circumstances surrounding what happened is best explained by Gamio himself: "At first we tried to propose a bill through Congress, but we did not receive an adequate response. We ran the risk that the bill would be changed without technical support in the various instances of deliberation." Supposedly, there was a lobby that tried to stop the advancement of renewable energy. Gamio goes on to comment, "The country had just signed a treaty of free trade agreement with the United States, and the Executive Branch had requested delegated powers from Congress in order to modernize our polices and improve the competitiveness of the country. [Using this delegated power], the second LD that we passed is this promotional framework for renewable energy." (Gamio, 2017).

Basically, with his task force, Gamio et al. learned about the benefits that renewable energy brought to other countries. It also made sense for Peru because of its abundance of

---

[9]   Ministerial Resolution 195-2007-MEM/DM.

natural resources that are appropriate for renewable energy.[10]

"A consideration to pass this LD was the enormous potential that Peru had for renewable energy," says Novoa (2017), President of the Peruvian Association of Renewable Energies and one of the people who helped write the language of the LD. The task force initially tried to go through the formal policy channel, but doing so met with too much opposition. This method also ran the risk that the policy would be changed to something completely different or not substantiated by technical studies. However, like surfers in wait of a suitable wave, the players were prepared with the policy already drafted. They were just waiting for the right wave to come that could push them all the way to shore.

It is important to note that the United States did not require Peru to pass such a law. The United States would have signed the PTPA with Peru even if Peru did not pass a renewable energy policy. Renewable energy was not a required, hard ask in the negotiations with the United States. "There was no demand, like the Paris Agreement, to reduce emission. You have to remember that at that time, the United States was not a signatory of the Kyoto Protocol, or Paris, or anything, so in itself it was not demanding of itself, so it did not demand this for the Free Trade Agreement either," says Cesar Butron, President of the Committee of the Economic Operation of the National Interconnected System (COES).[11] Renewable energy proponents, however, used this trade deal and the legislative power that it received as a window of opportunity to pass a policy that some people supported.

Using these two authorities—legislative power and environmental protection—the executive branch passed LD 1002, Legislative Decree to Promote Investment in Electricity Generation with the Use of Renewable Energies, in 2008.

LD 1002 does several things. LD 1002 establishes that the law will promote renewable energy for electricity production, not considering large hydroelectric plants. Qualified renewable energy sources include biomass, wind, solar, geothermal, tidal energy, and hydroelectric plants that are less than 20 MW. It also establishes that the state will have as its goal to produce 5% of its electricity from these renewable energy sources. It charges the MINEM with the task of updating this percentage every 5 years as it sees fit. MINEM is the entity in charge of promoting renewable energy sources, while regional governments are also allowed to promote its use within their own territories. Renewable energy producers also have priority of dispatch. This means that if they feed into the grid, COES is required to use the electricity that they produced before resorting to the electricity that is produced from diesel, natural gas, or other nonrenewable sources. COES must also consider that the variable cost for renewable energy sources is USD 0. Generators are guaranteed a premium: they can sell their electricity at the current market price, but if the market price is below the contract price, then the Supervisory Agency for Investment in Energy and Mining, Peru's energy regulator, can readjust the cost for users so that these generators can receive their contracted price.[12] Last, renewable energy generators only have to pay the incremental incurred cost to use

---

[10] Peru is considered to be a country with high potential for wind, solar, hydro, and geothermal, and a high-medium potential for biomass (Currie, 2016).

[11] COES operates and ensures the proper functioning of Peru's national electric grid, balancing supply and demand to minimize blackouts (Butron, 2017).

[12] This policy essentially creates a subsidy for renewable energy generators, because when market price drops below their guaranteed contract price, electricity consumers are charged more to pay for this difference.

distribution lines (while other electricity producers pay a maintenance fee to use these lines).

LD 1002 was groundbreaking in several ways. First, it was the first law in Peru to create a mechanism for electricity generators who use a renewable energy source to sell on the national electricity grid. Second, it does so by creating an auction system in which generators using renewable energy can all enter, bid, and sell their electricity. This auction also guarantees an annual income to generators if they produce the amount of electricity that they promised to the state. Third, it codifies a goal that Peru would produce 5% of its electricity from nonconventional renewable energy sources, a rate that should be updated every 5 years.[13] Fourth, it gives renewable energy companies priority to dispatch their electricity before other producers can dispatch theirs, guaranteeing that they can always sell their electricity. Overall, this is the most supportive policy of renewable energy on a commercial scale that Peru had passed thus far.

5.4 Prompting an energy transition

Is a policy window, which can help other highly opposed renewable energy policies to pass, a phenomenon unique to Peru? This chapter argues that while a case study of Peru is highly particular, some generalizable lessons can be drawn that may be applicable in other contexts. In this case study, the event that opened a policy window was an exogenous factor that drew great national attention. This allowed for attention to be diverted elsewhere and for renewable energy supporters to act without drawing much notice. It also shifted legislative power from its traditional seat in Congress to the Executive Branch. Although it

was only momentary, the Executive Branch was able to push many of its policy priorities through during this time. This event provides support that a policy window is any event that changes the regular distribution of power in a system.

Another possibly generalizable observation is that institutions may be designed to provide for policy windows because there are structural ways that the balance of power can shift. The balance of power can change if a political actor or group of actors suddenly gains power that it did not previously have. One way that this could occur is if there is a change in actors' functions. Another way that the balance of power can shift is if a certain party that was traditionally in power loses power, that is, an election. Elections can also give voice to new actors that did not have a voice in the past. One can also imagine that the *threat* of shifting the balance of power may be enough to open a policy window. Crises tend to shift the balance of power because when a crisis occurs and the public is unhappy with the response, people tend to blame the leaders who are in place (Achen and Bartels, 2016). There can also be a change in the weight of power, which can be actual or perceived.

Finally, this shift can take the form of a reduction in the number of veto players. Veto players are people who can block the adoption of a policy (Tsebelis, 1995). When the Peruvian Legislature deliberates on a policy, the number of veto players can be as many as 130 members of Congress, plus all the committee analysts and lobbyists who are involved with a policy. When the Executive Branch received legislative power, the veto players were reduced to essentially three. Given that the vice minister wanted the policy, he only needed the approval of the minister and the President. Decision-making was highly concentrated and

---

[13]   Although this goal should be updated every 5 years, so in 2013 and 2018, the 5% goal has not changed since this LD was implemented as of June 2019.

more radical change was possible. To be clear, this is not to argue that this is good or bad, just that this situation is more conducive to radical policy change than when there are more veto players.

Thus a policy window permits major policy change because it shifts or threatens to shift the balance of power as compared to the status quo. It is significant because it allows many policies that would not have passed during a status quo period to pass. It is also when a greater volume of policies pass. Finally, policy windows are important because they permit major policy changes from the current political trajectory. For strategic actors who are trying to push a controversial or highly opposed policy through, policy windows are their opportunity to finally pass their coveted policies. Understanding a policy window as a shift in the balance of power allows policy makers and researchers to identify when a random policy window is opening, ex ante, so that they can prepare their strategy if they want to pass a highly controversial policy, including many renewable energy policy ideas that have not found a political opportunity to become law.

Another lesson from this case is the strong influence that the international system can have on domestic politics. Peru's institutions changed drastically because of the trade deal with the United States "If you look at the basic level, without the trade agreement, Peru would probably be more behind environmentally than it is now. And in the past 10 years, it has done a lot of work. We need to give Peru the time to strengthen its institution," says Cady (2017). The fact that a highly significant environmental and energy policy resulted from a trade deal with the United States is quite incredible. The desire of developing countries to conform to international norms is a strong force, so it is worthwhile to include these normatively desirable terms in international deals, such as when a country is entering into a bilateral trade deal, regional trade deal, or the Organization for Economic Cooperation and Development (OECD). Other terms that are worthwhile to include are environmental protection, human rights, civil liberties, and rural development terms.

This case demonstrates the linkages between the international, state, and individual levels. The three levels are in constant interaction with each other. The developments in the international level are observed by the state and individual levels, and their effects also reverberate on the state and individual level. The closer the levels are to each other, the more likely they are to be able to impact the other, in both directions (Sikkink and Keck, 1998).

5.5 Impact on the market

Ten years after LD 1002's approval, now is a good time to analyze what impact, if any, it has had on the renewable energy market in Peru. LD 1002 has opened the door for renewable energy generators to sell electricity on the national grid for the first time, thereby changing Peru's energy matrix. The first auction occurred in 2009 and is supposed to take place every 2 years. Abiding by their contracts, renewable energy producers started selling their electricity to the grid in 2011. Fig. 5.3 shows the growth of the participation of renewable energy in Peru's electricity grid.[14] It has been growing every year due to the auction system and as companies complete their plants. Renewable energy (not including large hydroelectric plants) currently constitutes slightly less than 5% of Peru's total electricity output. There was supposed to be an auction in 2017, but it did not occur. There is no official reason for the delay, although there is speculation that the government was trying to

---

[14] (COES, 2006−2018).

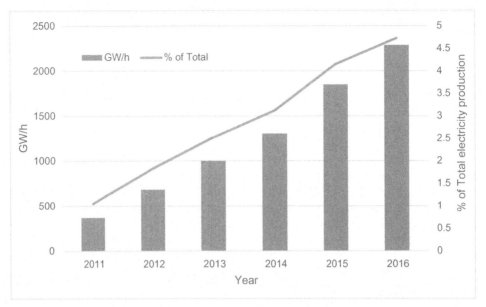

FIGURE 5.3   Electricity production from renewable energy. Source: *COES, 2006–2018.*

make the auctions better or considering increasing the 5% requirement that LD 1002 stipulates to a higher percentage.

As of early 2019, there had been four renewable energy auctions, which have added 6141 GW h/year to the energy mix (OSINERGMIN, 2019). Because of the way that the law is written now—which states that companies that win these auctions must start producing electricity for the grid after 2 years of signing the contract—the only sources that qualify are solar, urban biomass, wind, small hydroelectric systems, and

agricultural biomass.[15] Geothermal plants take about 7 years from development to production, so they still cannot enter these auctions (Rough and Espinoza, 2017).

Fig. 5.4 shows the results of the four auctions that have taken place and the average prices of the bids that were entered. The light green horizontal line shows the average market price, which comes in at about USD 52/ MW h. As can be seen, renewable energy prices have decreased over time.[16] Wind and solar were cheaper than market price for the first time in 2015.

---

[15]   Thus far, companies have been good at meeting the deadline to start selling electricity on the electric grid within 2 years of receiving the bid. (COES, 2006–2018).

[16]   The first auction took place in 2009. In this year the average price for solar was USD 221.1/MW h; urban biomass was USD 110/MW h; wind was USD 80.4/MW h; small hydroelectric was USD 60.0/MW h; and agricultural biomass was USD 52.0/MW h. In the second auction in 2011, solar cost USD 119.9/MW h; urban biomass cost USD 100.0/MW h; wind cost USD 69.0/MW h; small hydroelectric plants cost USD 52.6/MW h; and agricultural biomass did not participate. In the third auction in 2013, only small hydroelectric plants won contracts and their average price was USD 56.5/MW h. The fourth auction took place in 2015, with solar coming in at USD 48.1; urban biomass at USD 77/MW h; wind at USD 37.7/ MW h; and small hydroelectric plants at USD 43.8/MW h. Again, agricultural biomass did not participate (Quintanilla, 2016, p. 14).

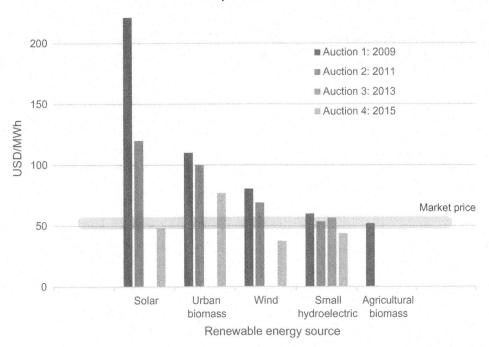

FIGURE 5.4 Peru's renewable energy auction results: average prices of bids. *Source: Quintanilla, E., 2016. Perú: Soluciones Para Un Mercado Eléctrico De Alto Crecimiento - Promoción De Energías Renovables... Y Competitivas. Cuadernos De Energía 48, 14. Available from: <https://www2.deloitte.com/content/dam/Deloitte/ec/Documents/energy-resources/Deloitte-ES-Energia-cuadernos-de-energia-48.pdf> (accessed 16.08.18.).*

Solar electricity was injected into the grid for the first time in 2012, and then wind entered in 2014 (Ramos, 2017). In the first auction, agricultural biomass was the lowest priced, then small hydroelectric, then wind, then urban biomass, and finally solar. Solar energy was twice as expensive as the next most expensive energy source, urban biomass. By the fourth auction the highest priced energy sources were urban biomass, then solar energy, and then small hydroelectric plants; wind energy was the cheapest source. The prices of solar, wind, and small hydroelectric plants are reaching parity with traditional energy sources. As prices drop, consumers will no longer have to pay the difference in these prices from market price to subsidize renewable energy because their prices will be market prices.

LD 1002 has also had some psychological impacts on Peruvians as well. The first long-held distrust was related to Peru's infrastructure. Prior to renewable energy's entrance, many skeptics believed that renewable energy would be disruptive to power lines. Some people were uncertain whether the surge in electricity production caused by certain types of renewable energy production would disrupt the existing grid and cause a breakage. The second distrust was of renewable energy producers. The question was whether renewable energy companies would be able to deliver the amount of electricity written in their contracts. Now that LD 1002 has been in place for about 10 years and renewable energy producers have been feeding electricity into the grid for about 7 years without any major disaster, the concerns about renewable energy have abated.

This has led more people to see the feasibility of incorporating more renewable energy production in the future, especially given the abundance of renewable resources in Peru.[17]

5.6 Remaining issues

Even today, arguments against renewable energy remain strong. Renewable energy's price decrease has been used as argument that the privileges given to renewable energy producers (guaranteed rate of return, priority of dispatch, $0 variable cost, no grid maintenance fee) are no longer necessary and should be removed so that renewable energy producers can compete in the market with other producers, on the same level playing field as everyone else (Quintanilla, 2016, p. 14). The issue with this argument is that according to Peru's current law, generators are not allowed to sign contracts with unregulated users or distributors above the "stable power" that they can produce.[18]

This is to make sure that generators do not sign contracts based on an amount of electricity that they cannot realistically deliver, such as at the level of their installed capacity.[19]

Biomass and hydroelectric plants usually do produce a relatively stable supply of electricity over a month, but solar and wind do not produce at the same level all the time. Therefore, as Peru's law is currently written, eliminating LD 1002 would prevent solar and wind producers from signing any power purchase agreements (PPAs) with users or distributors to sell their electricity.

But Butron does not see this as a major problem. He sees stable power as an administrative issue rather than a physical problem. Being an administrative issue, stable power is a concept that the government can decide to abolish, so that renewable energy companies can sign PPAs and finance their projects (Butron, 2017). He suggests that there is a simple solution to this issue, which is to change regulations to state that when a generator cannot generate stable power, this part of the law will mean that their stable power be calculated as 80% of their installed capacity (Butron, 2017). But Butron says that the MINEM does not consult with COES when drafting any new regulations, and the communication surrounding this issue breaks down.

MINEM, for its part, is also highly aware of the issue for renewable energy producers because of this "stable power" requirement. Peru originally required that energy generators sign PPAs based on their stable power to avoid energy companies signing contracts for a greater amount of energy than they can safely produce. This law is now interpreted as a PPA can only be signed based on the stable power amount that a generator can produced. Because solar and wind companies do not have stable power and produce intermittently based on availability of the sun and wind, they cannot sign PPAs on their own outside of the

---

[17]   Total power potential in Peru of hydroelectricity is 70,000 MW, versus its installed capacity at 3118 MW. Wind energy's total potential is 22,000 MW, and installed capacity is 142 MW. Biomass' potential is 450 MW versus 27.4 MW of installed capacity. Geothermal's potential is 3000 MW, but there is 0 installed capacity. There is no total potential capacity calculated for solar energy, but its installed capacity is 80 MW (Currie, 2016).

[18]   Law 28,832, 2006, Ch. 2, Art. 3. "Stable power" is the maximum power that a plant can produce while maintaining a high level of security over a span of time, usually measured over a month (Sectorelectricidad, 2014).

[19]   Producing at the absolute maximum that their machines can produce for a long period of time may cause the machines to break down and prove to be dangerous. Butron explains why this provision exists: "It was created when Peru had a scarcity and it wanted to avoid this problem of a company lying about how much it can produce... Now that there is no scarcity, now that there is time to respond, it is no longer needed" (Butron, 2017).

auction system. "It is very difficult to change a law, so this office is trying to do whatever we can do by ourselves. We try to look at what is the most efficient way to promote renewable energy," say representatives from the Energy Efficiency Office within MINEM (Rough and Espinoza, 2017). Currently, there is a learning curve to understand and keep up with new technologies, and MINEM is trying to monitor the progress of the current renewable energy installations. Aside from possible ways to change regulations, they are creating an analytical tool that can incorporate flexibility in policies and technology, devising an information technology program that can allow instant input of information about production to show Peru's energy balance in real time, drafting a National Energy Plan, and analyzing ways to improve future renewable energy auctions (Rough and Espinoza, 2017). However, even within MINEM, there still exist many people who hold the older distrust of renewable energy, because it is new and possibly disruptive to the energy market. Within MINEM, there are supporters and opponents, making progress slow but carrying the benefit of being a deliberative process.

Another issue that needs some contemplation is that of geothermal energy. Peru has much geothermal energy potential. There is about 3000 MW of geothermal energy in Peru, mostly concentrated in the southern region (Claros Pacheco, 2015). Geothermal energy source currently cannot enter the auctions through the mechanism that LD 1002 created, since a geothermal plant cannot be installed and start delivering energy into the grid within 2 years of winning the bid. Geothermal plants usually take about 7 years to construct. MINEM is also thinking about ways to amend laws or regulations to permit geothermal plants to compete as an energy source as well.[20]

Of course, skeptics remain, arguing that Peru does not need renewable energy, it is still expensive, incorporating it increases electricity prices for everybody, and the minimal environmental impact does not justify the high cost increase for all users.[21] Butron explains that because everyone is conscious that adopting more renewable energy would increase the rate of electricity, the government tries to fulfill the 5% goal slowly. For example, renewable energy generators were frustrated that in the 2013 auction, no wind, solar, urban biomass, or agricultural biomass company won a bid. The only technology type that won was small hydroelectric plants (Butron, 2017). The ongoing skepticism creates an environment of "reluctant implementation," where the bare minimum of the law is reached, and progress is not made as fast as the available capability allows.

These sentiments show that it is premature to be too optimistic. Renewable energy production currently makes up less than 5% of total electricity production in Peru (excluding large hydroelectric dams). This means that its market presence is still very small versus its potential (Quintanilla, 2016, p. 14). It is also a new presence. Therefore it is still difficult for domestic actors to conjure up effective lobbying to change policies. However, the Paris Agreement passed and requires countries to make changes to reduce their greenhouse gas emissions. Peru, as part of its Paris Agreement, agreed to reduce 30% of its greenhouse gas emissions by 2030 and strengthen its climate change policies (Andina, 2017). If the global norm continues in favor of ratcheting up efforts to mitigate climate change, Peru will likely, as it has been in the past, feel compelled

[20] Rough, Daniella. 2017. "Namas De Energia Renovable (Conectado Y No Conectado) En El Perú". Presentation, Lima, 2017.

[21] Butron (2019). I would like to thank especially Mr. Butron for the careful consideration and feedback on the arguments in this paper.

to increase its efforts, too. If the momentum continues, this will give renewable energy producers more market presence.

According to the terms of the LD, MINEM is supposed to conduct an auction every 2 years. There should have been an auction in 2017, but this was canceled. The president of the Peruvian Society of Renewable Energy, Juan Coronado, believed that a renewable energy auction would take place at the end of 2018, which did not happen. As of the writing of this chapter in June 2019, the fifth auction has still not occurred. The auctions are especially necessary now because Peru's energy consumption pattern grew more than predicted and the oversupply of energy is predicted to end by 2020 or 2021 (Energía Estratégica, 2018). LD 1002 gave renewable energy producers the ability to have a market presence for the first time. The next step is to enhance market presence of renewable energy stakeholders, increasing competition.

5.7 Conclusion

This chapter analyzes in depth how domestic actors tried to kick-start an energy transition in Peru. Their effort occurred during a shift in the balance of power, which opened a policy window. Policy entrepreneurs were prepared with their ideas and took advantage of this opportunity to pass Peru's first large-scale renewable energy policy.

However, the transition is now at a standstill. As of June 2019, the Peruvian government is 2 years behind on holding another auction that is required by LD 1002. There needs to be another big push for renewable energy. One policy area is to increase the mandate for procuring from renewable energy sources. Currently, the recommended maximum is 5%. A more robust policy is to require electricity providers to procure a certain percentage of their electricity from eligible renewable resources, such as solar, wind, and small hydroelectric systems. This way, generators, transmitters, and distributors would all be interested in having more renewable energy plants to provide electricity.

To expand the renewable energy market even more the government can encourage distributed generation by traditional energy consumers. Luckily, more residents are aware of these options in other countries and want this capability in Peru as well. Some residents have installed solar rooftop systems using their own money, even though there is no financial incentive from the government to do so. With rooftop solar panels, each energy consumer can be a producer as well, with unused electricity sold back to the grid. This is being done on a growing scale in many countries, including the United States, Germany, and Japan. But policies must be put in place so that distributed generation can be more secure and profitable for investors to encourage more people to invest, such as net energy metering or a feed-in tariff. One current drawback is lack of information. The government and private sector can increase information for users so that they know what policies are in place and have more information about renewable energy sources (Töfflinger, 2016b).

The government could provide for research and development. At the Pontifical Catholic University of Peru, researchers conduct research on new layers of materials that could be used to improve solar panels. The government can help researchers by providing more funding for research projects and then providing infrastructure and support for commercialization if it reaches that stage.[22]

---

[22]   But the researchers cannot transfer these layers to actual solar cells to test these layers when they are interacting with solar cells, so they must transfer their hypotheses and materials to the Helmholtz-Zentrum Institute in Berlin or another institution that has a solar cell production facility to test this. While this process generates communication between researchers from all over the world, it also introduces many inefficiencies because institutes have to get institutional clearance to collaborate and then share information (Töfflinger, 2016a).

As many Peruvian policy decision makers have noted, the world is moving evermore toward renewable energy use. This means that the number of people with the desire to expand renewable energy consumption will increase, and some of them will be in a decision-making position. With more advocates for renewable energy in the government, there is an increased likelihood that they will be able to take advantage of a policy window—whether opened by an exogenous shock such as the PTPA, an election cycle, or other political events that shake up the composition of the government—and pass policies in support of renewable energy. The study of why highly opposed policies are able to pass is interesting because it presents moments of great opportunity for policy change. Radical policy change is rare but momentous due to its lasting effects. In Peru's case, this radical change was catalyzed by a trade deal with the United States— an unlikely source to create environmental and renewable energy policies.

References

Abusada Salah, R., du Bois, F., Pastor, E.A.M., 2000. La Reforma Incompleta: Rescatando Los Noventa. Centro de Investigación de la Universidad del Pacífico, Lima.

ACCIONA, 2018. ACCIONA En Perú. Acciona.Com. Available from: <https://www.acciona.com/es/en-el-mundo/latinoamerica/peru/> (accessed 12.01.18.).

Achen, C.H., Bartels, L.M., 2016. Democracy for Realists: Why Elections Do Not Produce Responsive Government. Princeton University Press, Princeton, NJ.

Andina, 2017. Acuerdo De París: Los Compromisos De Perú Frente Al Cambio Climático. Andina.Pe. Available from: <https://andina.pe/agencia/noticia-acuerdo-paris-los-compromisos-peru-frente-al-cambio-climatico-669371.aspx> (accessed 06.08.19.).

Ayuda en Acción, 2018. Perú – Memoria Anual De Ayuda En Acción. Memoria.Ayudaenaccion.Org. Available from: <https://memoria.ayudaenaccion.org/memoria-institucional/donde-trabajamos/paises/peru/>. (accessed 12.01.18.).

BNAmericas, 2018. Reportes De Inteligencia: Los Riesgos De La Abundancia En El Sector Eléctrico De Perú. Bnamericas.Com. Available from: <http://www.bnamericas.com/es/intelligence-series/energiaelectrica/los-riesgos-de-la-abundancia-en-el-sector-electrico-de-peru> (accessed 12.01.19.).

Bonilla, N., 2012. A battle over the quest to tap Patagonia's rivers for energy. National Geographic: The Great Energy Challenge Blog. Available from: <http://energyblog.nationalgeographic.com/2012/02/10/battle-over-chiles-attempt-to-tap-patagonias-rivers-for-energy/> (accessed 21.06.17.).

Butron, C., May 16, 2017. "Renewable Energy in Peru." Interview by the Author. In person. Lima.

Butron, C. July 15 2019. Email to the author. "LD 1002".

Cady, G. 2017. Renewable Energy in Peru. Interview by the author. In person. Lima. 6th March.

Campodónico Sánchez, H., 1999. Las Reformas Estructurales Del Sector Eléctrico Peruano Y Las Características De La Inversión 1992-2000. CEPAL, Santiago.

Claros Pacheco, A., 2015. Geothermal Energy in Perú. Presentation, Nairobi.

COES, 2006–2018. "Estadísticas De Operaciones". Estadísticas De Operaciones. COES, Lima. Available from: <http://www.coes.org.pe/Portal/Publicaciones/Estadisticas/> (accessed 12.07.19.).

COES. 2019. "Boletines Mensuales – Abril". Coes.Org.Pe. Available from: <http://www.coes.org.pe/Portal/Publicaciones/Boletines/> (accessed 18.06.19.).

Comisión MEM-OSINERG, 2005. Libro Blanco: Proyecto de ley para asegurar el desarrollo eficiente de la generación eléctrica. Osinerg, Lima.

Congreso Perú, 2018. Proceso Legislativo. Www4. Congreso.Gob.Pe. Available from: <http://www4.congreso.gob.pe/tutor/ciclo/multimedia/proceso.htm> (accessed 06.08.19.).

Corporación FONAFE, 2018. Corporación FONAFE. Fonafe.Gob.Pe. Available from: <https://www.fonafe.gob.pe/pw_content/organizacion/1/Doc/Estructura%20de%20la%20Corporación%20FONAFE.pdf> (accessed 13.03.19.).

Currie, S., 2016. Renewable energy in Latin America: Peru. Available from: <http://www.nortonrosefulbright.com/knowledge/publications/134777/renewable-energy-in-latin-america-peru> (accessed 06.08.19.).

El Comercio, 2008. Renunció Viceministro De Energía Pedro Gamio Aita. Elcomercio.Pe. Available from: <https://elcomercio.pe/ediciononline/HTML/2008-12-06/renuncio-viceministro-energia-pedro-gamio-aita.html> (accessed 06.08.19.).

Energía Estratégica, 2018. Perú Lanza Subasta De Energías Renovables En La Segunda Mitad Del 2018. Energía Estratégica. Available from: <http://www.energiaestrategica.com/peru-lanza-subasta-energias-renovables-la-segunda-mitad-del-2018/> (accessed 06.08.19.).

Ernst & Young, 2016. Renewable Energy Country Attractiveness Index: What Happens When Grid Parity Hits? EYGM Limited, London.

Forero, J., 2013. Brazil's hydroelectric dam boom is bringing tensions as well as energy. Guardian. Available from: <https://www.theguardian.com/world/2013/feb/12/brazil-hydroelectric-jirau-dam> (accessed 05.04.17.).

Gamio, P., December 11, 2017. Email to the Author. "Renewable Energy in Peru".

Gestión, 2018. EY: Perú Es El Quinto País Más Atractivo De Latinoamérica Para Invertir En Energía Renovable. Gestión. Available from: <https://gestion.pe/economia/ey-peru-quinto-pais-atractivo-latinoamerica-invertir-energia-renovable-236521> (accessed 18.03.19.).

Kingdon, J., 2011. Agendas, Alternatives, and Public Policies. Longman, Boston, MA.

Lengyel, M.F., Ventura-Dias, V., 2004. Trade Policy Reforms in Latin America. Palgrave Macmillan, New York, p. 170.

Merco Press, 2008. Peru En Route to Become Regional Natural Gas Power. Merco Press: South Atlantic News Agency. Available from: <http://en.mercopress.com/2008/01/14/peru-en-route-to-become-regional-natural-gas-power> (accessed 16.07.18.).

MINEM, 2017. Anuario Estadístico De Electricidad 2017. Anuario Estadístico De Electricidad. MINEM, Lima. Available from: <http://www.minem.gob.pe/_estadistica.php?idSector=6&idEstadistica=13034> (accessed 25.05.19.).

Ministerio de Comercio Exterior y Turismo, 2011. TLC Perú - EEUU: Sobre El Acuerdo. Acuerdoscomerciales. Gob.Pe. Available from: <http://www.acuerdoscomerciales.gob.pe/index.php?option=com_content&view=category&layout=blog&id=56&Itemid=79> (accessed 06.08.19.).

Novoa, A., August 4, 2017. Renewable Energy in Peru. Interview by the Author. In Person. Lima.

OAS (Organization of American States), 2017. SICE: Trade Policy Developments: ATPA. Sice.Oas.Org. Available from: <http://www.sice.oas.org/TPD/USA_ATPA/USA_ATPA_e.ASP> (accessed 06.08.19.).

OSINERGMIN, 2019. Subastas. Osinergmin.Gob.Pe. Available from: <http://www.osinergmin.gob.pe/empresas/energias-renovables/subastas> (accessed 06.08.19.).

Quintanilla, E., 2016. Perú: Soluciones Para Un Mercado Eléctrico De Alto Crecimiento - Promoción De Energías Renovables... Y Competitivas. Cuadernos De Energía 48, 7–16. Available from: <https://www2.deloitte.com/content/dam/Deloitte/ec/Documents/energy-resources/Deloitte-ES-Energia-cuadernos-de-energia-48.pdf> (accessed 16.08.18.).

Ramos, J., August 10, 2017. LD 1002 and Renewable Energy Policy in Peru. Interview by the Author. In person. Lima.

Rough, D., Espinoza, C., March 21, 2017. Energy Efficiency Policies in Peru. Interview by the Author. In Person. Lima.

Sectorelectricidad. 2014. Perú: Diferencias Entre Potencia Firme, Instalada Y Efectiva. Sectorelectricidad.Com. Available from: <http://www.sectorelectricidad.com/10965/potencia-firme-instalada-efectiva/> (accessed 06.08.19.).

Sikkink, K., Keck, M.E., 1998. Activists Beyond Borders: Advocacy Networks In International Politics. Cornell University Press.

SNV, 2018. "Peru". Snv.Org. Available from <http://www.snv.org/country/peru> (accessed 12.01.18.).

Tamayo, J., Salvador, J., Vásquez, A., Vilches, C.M., 2016. La Industria De La Electricidad En El Perú: 25 Años De Aportes Al Crecimiento Económico Del País, first ed. OSINERGMIN, Lima.

Töfflinger, A., November 22, 2016a. Renewable Energy Policies in Peru. Interview by the Author. In Person. Lima.

Töfflinger, A., December 8, 2016b. Renewable Energy Policies in Peru. Interview by the Author. In Person. Lima.

Tsebelis, G., 1995. Decision Making In Political Systems: Veto Players In Presidentialism, Parliamentarism, Multicameralism And Multipartyism. Br. J. Polit. Sci. 25 (03), 289–325. Available from: https://doi.org/10.1017/s0007123400007225.

UN Climate Change, 2014a. Status of Ratification of the Convention. Unfccc.Int. Available from <http://unfccc.int/essential_background/convention/status_of_ratification/items/2631.php>. (accessed 09.01.18.).

UN Climate Change, 2014b. "Status of Ratification of the Convention". Unfccc.Int. Available from: <http://unfccc.int/essential_background/convention/status_of_ratification/items/2631.php> (accessed 09.01.18.).

UNDP/World Bank Energy Sector Assessment Program, 1984. Peru: Issues and Options in the Energy Sector. UNDP/World Bank, Washington, DC, Report No. 4677-PE.

UNFCCC Historical Membership Charts, 2017. UNFCCC Historical Membership Charts. Unfccc.Int. Available from <http://unfccc.int/files/bodies/election_and_membership/application/pdf/consolidated_historical_membership_charts_of_convention,_kp_and_cma_bodies.pdf> (accessed 25.05.19.).

USCBP (US Customs and Border Protection), 2017. Andean Trade Preference Act (ATPA) – Expiration of Duty-Free Treatment. Help.Cbp.Gov. Available from: <https://help.cbp.gov/app/answers/detail/a_id/325/~/andean-trade-preference-act-%28atpa%29---expiration-of-duty-free-treatment> (accessed 06.08.19.).

USTR, 2005. Interim Environmental Review: U.S.-Andean Free Trade Agreement. Ustr.Gov. Available from: <https://ustr.gov/sites/default/files/Interim-Environmental-Review.pdf> (accessed 06.08.19.).

USTR (Office of the U.S. Trade Representative), 2006. Peru TPA Final Text. Ustr.Gov. Article 23.4. Available from: <https://ustr.gov/trade-agreements/free-trade-agreements/peru-tpa/final-text> (accessed 06.08.19.).

Wise, C., 2007. Peru. In: Weintraub, S. (Ed.), Energy Cooperation in the Western Hemisphere: Benefits and Impediments, first ed. Center for Strategic and International Studies (CSIS), Washington, DC, pp. 302–335.

World Bank, 2017. Access to Electricity (% Of Population) | Data. World Bank Data. Available from: <https://data.worldbank.org/indicator/EG.ELC.ACCS.ZS?locations = PE> (accessed 10.10.17.).

6

The challenges and contradictions of Peru's *Proyecto Masivo de Energía Solar*

Dustin Welch García

Seattle University, Seattle, WA, United States

6.1 Introduction

This chapter explores how the Peruvian state's lack of understanding of the many ground-level challenges of off-grid solar energy provision are manifested in the design of the *proyecto masivo de energía solar*,[1] a public—private partnership (PPP) aimed at installing 150,000 solar home systems (SHS) in unelectrified rural households throughout the country. The following pages contribute to the growing body of energy humanities scholarship that examines the interplay between bureaucratic institutions and energy policy and how they intersect with citizens' access to and use of energy—or lack thereof (Boyer and Szeman, 2014; Kale, 2014; Gupta, 2015; Mitchell, 2011).

What is more, it has become abundantly clear that the application of technologies, such as SHS, alone cannot catalyze widespread clean energy transitions. Rather, in order to enable and sustain such energy infrastructure transitions, it is crucial for countries such as Peru to design

individualized energy programs that effectively deploy public policy and enact regulation, while appropriately incorporating the private sector (Arent et al., 2017) and also meaningfully engaging the actual community members who will be receiving energy service (Fernández-Baldor et al., 2014; Feron and Cordero, 2018).

This chapter contributes to conversations about clean energy transitions, by revealing the internal elements that are most often hidden from sight and tend to only become visible or be discussed when energy infrastructures seem at risk of failing (Lampland and Star, 2009). The research aims to orient the reader to the internal "invisible trouble" of infrastructural systems, by exposing the design of PPP contracts that "are argued through in hardworking committees that rarely see the light of history … manifested in written documents that are rarely perused by any but their most immediate users." (Lampland and Star, 2009, pp. 22, 13–14). Throughout much of Latin America, public auctions[2] have become

---

[1] This translates to English as the "massive solar energy project."

[2] This type of auction is defined by the public announcement or offer of a renewable energy project of a determined size where competition among energy companies is the driving force and the company that submits the lowest bid wins the auction (IRENA, 2015).

The Regulation and Policy of Latin American Energy Transitions
DOI: https://doi.org/10.1016/B978-0-12-819521-5.00006-1

the most common mechanism used to build countries' renewable energy infrastructure (OSINERGMIN, 2016). However, auctions generally have trade-offs between issues such as low prices, pace of delivery of service, and social and economic impact. As a result, a country's priorities toward a range of issues including energy technology preference, socioeconomic development, and policy benefits can be seen in the design of their energy auction (IRENA, 2017). While research on PPPs for infrastructure provision in developing countries has explored many areas, including economic considerations, project finance and management, as well as the more social and political concerns regarding design and renegotiation,[3] there has been little academic work focused specifically on PPPs for the delivery of off-grid solar energy. PPP contracts merit further study because they are incredibly complex to design and involve long-term operational commitments from all parties involved. As such, the public entities responsible for PPP design must be able to both foresee risky scenarios and properly allocate program risks in order to avoid conflicts once the infrastructure is installed and service is underway (Yescombe, 2013). Incidences of renegotiations of PPP contracts in Latin America have shown to be particularly high and have tended to favor the private operator, indicating serious financial losses on part of the government sector (Trebilock and Rosentstock, 2015). The pages that follow demonstrate how unforeseen discord may unsettle the well-intentioned calculations of PPP designers and contribute to energy systems' inability to resolve long

histories of neglect and deficient infrastructure service in rural regions of Peru. Moreover, it is especially crucial to analyze the policies and procedures underlying infrastructure programs such as Peru's *proyecto masivo*, as this initiative bears a double burden in the global energy transition. It represents both a "planet-preserving" effort to fight climate change, while also striving to offer social and economic inclusion benefits to the country's marginalized rural populations (Howe, 2015).

The Peruvian government's *proyecto masivo* was inspired by a small-scale program operated by the NGO, *Luz Solar Andina—LSA*[4] in the Andean region of Cajamarca. This energy initiative, called *Casa Solar*, is widely held by energy industry and development experts to be among the most exemplary energy initiatives in the country. Yet, problematically, the *proyecto masivo* attempts to replicate LSA's success nationally while utilizing a program model that is fundamentally different from that of LSA. While LSA alone carries out the entire social, technical, and financial components of *Casa Solar*, in contrast, the *proyecto masivo* divides these responsibilities between public and private entities. Ergon Peru,[5] the private energy company involved in the *proyecto masivo*, is responsible for installation of the SHS and all technical service related to them. The different public energy distribution companies throughout the country are responsible for carrying out all customer invoicing and bill collection aspects of the program.

This chapter argues that the narrow technical and financial role that Peru has outlined for the private company Ergon has overlooked

---

[3] For more on these topics, see Boardman and Vining (2010), Hellowell and Vecchi (2012), Wettenhall (2010), and Guasch (2004).

[4] Unless otherwise noted, the names of all interviewees, towns, and programs names have been changed to pseudonyms; I have made these changes to protect the identities of my research participants.

[5] This is the company's real name, not a pseudonym.

the interrelated logistical and social challenges of providing ongoing solar energy service in isolated rural communities. The social challenge for Ergon will be explaining SHSs function and payment rules to residents who have never before had access to electricity. The logistical challenge is that the company must engage in costly and laborious trips to reach isolated homes located across vast, rugged terrain to conduct maintenance and service of SHS. That is to say, the social component, the company's ability to teach residents the uses and limits of SHS, as well as payment rules, will affect the rate of SHS's misuse and also residents' rate of payment/delinquency/default for energy service, ultimately influencing Ergon's logistical operations. Relatedly, the rural *campesinos*[6] who were interviewed described how their sense of ownership of the program increased their commitment to making their recurring payments for solar energy service and ultimately led them to feel that their financial cooperation with the program, by consistently paying their bills on time, was a crucial component to *Casa Solar's* success, equally as important as the performance of their SHS and the technical service provided by LSA itself.

This chapter also asserts that the energy service model designated by the *proyecto masivo* burdens public distribution companies with new responsibilities that they are reluctant to engage in and will cause them financial deficits. Successful PPP contract design necessitates that program risk be adequately allocated between the public and private sectors, allowing the public sector to fulfill its needs while giving the private sector flexibility to innovate. Instead, what is often seen is limited communication or collaboration between sectors, with each taking an adversarial position toward the other (Domingues and Zlatkovic, 2015). In the case of the *masivo*, the program's success hinges upon

these private and public actors' ability to carry out energy service in coordination with one another. Rather than adversarial relations, this auction design simply allows Ergon and public actors to operate independently and with indifference toward how the other performs their partnership responsibilities.

This chapter demonstrates that there is in fact much to be learned from Peru's approach to off-grid solar energy provision and PPPs. While much PPP-focused literature has offered analyses of risk allocation between the public and private partners in broad financial and operational terms, this chapter presents an in-depth analysis of the mechanics at play in the contact zone between contract design and on-the-ground solar energy service, which can serve as a roadmap for other governments as they design and implement rural electrification projects in partnership with private actors. This chapter is organized into the following four sections:

- The first section orients the reader to the main arguments and research methods, provides a vignette describing the ground-level challenges of rural solar energy provision, and describes the background and current state of the *proyecto masivo*;
- The second section outlines the *proyecto masivo* auction contract, Ergon's approach to rural energy provision, as well as the flaws in the company's approach to the interrelated social and technical aspects of program operation and maintenance;
- The third section examines the public distribution companies' role in the *proyecto masivo* and the problems that will arise due its treatment as an afterthought in this PPP; and
- The final section offers concluding thoughts and recommendations for governments embarking on similar initiatives.

[6] The roughly translates to English as "peasant farmer."

6.1.1 Methods

The arguments made in this chapter regarding the logistical obstacles and social nuances of energy service and operation have been informed by participant observation of LSA as it carried out operation and maintenance of its *Casa Solar* energy program, as well as 4 months of ethnographic work carried out in a highland community of Cajamarca, called Lahuaymarca, served by this same solar energy program. In addition, data presented here were drawn from over 50 in-depth interviews with technocrats in the energy bureaucracies in Lima, such as Peru's Supervisory Agency for Investment in Energy and Mining (OSINERGMIN) and Ministry of Energy and Mines (MEM), as well as NGO officials, various financial institution officers, private and public energy distribution company representatives, and members of domestic and international academic institutions.

6.1.2 The ground-level challenges of rural solar energy provision

Riding along with the technicians from a small Peruvian NGO, I had the chance to experience the grueling work[7] that is off-grid solar energy service. After a 2.5 hour drive out of Cajamarca, mainly on bumpy, unpaved dirt roads, we arrived at a small school where we parked and got out of the 4×4 truck. The locals told us we could follow a trail to reach the house of an energy customer whose faulty SHS battery the technicians were scheduled to replace. During the 30-minute hike, the technicians, the driver, and I alternated carrying the 80 lb battery on our backs, down a narrow footpath that ascended and descended with

the rugged Andean terrain, snaking our way around and down into a deep valley. Even though both the driver (who was not required to assist the technicians) and I helped transport the battery, it was still a physically exhausting endeavor. As we approached the house, we called out for the owner but she was nowhere to be found. After yelling for her for a few minutes, she appeared and hurriedly made her way down the hillside. We then warily snuck past the family's dog, who barked aggressively, guarding its newborn puppies, only relenting when its owner shushed it. The homeowner then led the technicians inside and they switched out the faulty battery for the new one. Once again facing the burden of carrying the old battery to the truck on our shoulders and backs, this time up valley, we kindly inquired if the homeowner might lend us a donkey to carry the battery. She agreed and quickly made her way up the hill surrounding her house and brought down a donkey that had been grazing, hidden out of sight by trees and shrubs. After loading the battery on the donkey's back and securing it with ropes, the homeowner accompanied us as we made the 45-minute hike back up to where we had parked the truck. This small maintenance operation, involving a 4×4 truck, diesel fuel, a driver, two technicians, replacement equipment, and a borrowed donkey, took nearly 2 hours to complete and is representative of the daily challenges of maintaining and operating SHS in rural Peru. While some maintenance operations are simpler because a customer's home may be located close to the main road or not require replacement of heavy equipment, other trips are even harder than the one described here. In the case that

---

[7] Various energy officials that I interviewed recognized the arduous nature of providing off-grid energy service in isolated rural communities, describing it with dehumanizing terms such as "ant-like work" (*trabajo de hormiga*). Their use of the term in this context harkens back to Peruvian scholars such as MarioVargas Llosa, who has described Peru's indigenous communities as "ant-like societies" (Llosa, V. Questions of Conquest. p. 51) and views "contemporary highland peasants as outside the flow of modern history" (Starn, O. Missing the revolution. p. 64).

residents are not at home when technicians reach their house, the same arduous trip must be carried out again on another day.

6.2 Background and current status of the Proyecto Masivo

In 2013 Peru's MEM announced its first public international auction for off-grid solar energy, which was awarded to Ergon in 2014 with the goal of providing SHS[8] to 150,000 rural households in the north, central, and south regions of the country where the national electric grid has not reached (OSINERGMIN, 2016). The solar energy program that resulted from this auction award, often called the *proyecto masivo*,[9] requires that Ergon design, procure, and install, and provides operation and maintenance of these SHS for a period of 15 years.[10] This business model also involves the participation of public energy distribution companies. The public distributors will carry out the customer invoicing and billing for the energy service that the private operator, Ergon, provides. Essentially, Ergon will install solar energy at the individual household level as well as provide maintenance and

operation for these SHS, as if it were a contractor working for the local public distribution company (Orosco, 2014).

In this business model, 80% of customers' monthly solar energy costs are covered by a national cross-subsidy called FOSE (Electrical Social Compensation Fund).[11] Customers make a fixed monthly payment representing nearly 20% of the established solar tariff, while the difference is covered by another subsidy called FISE (Fund for Energy Social Inclusion).[12] This PPP business model has been intentionally designed to isolate the private company, Ergon, from the financial risks involved in billing rural energy customers. Thus, Ergon receives its monthly FOSE subsidy payment regardless of whether the end user pays their bill for the corresponding month (Orosco, 2014). The FOSE subsidy will cover 80% of the total annual financial remuneration amount Ergon submitted for its energy auction bid; this subsidy will also compensate the public electricity distribution companies for their activities involved in issuing invoices and collecting payments.

It is well known among those working in the energy industry and other related sectors in Peru that the *proyecto masivo* has been beset with complications and is well behind the expected

---

[8] According to the technical specification of the auction, Ergon must equip households with pole-mounted, (12V/DC) solar home systems (SHSs), which consist of (1) solar panel no smaller than 85 W, (2) sealed battery no smaller than 90 A h, (3) charge controller no smaller than 10 A, (4) LED lights not to exceed 10 W, (5) light switch, and (6) electrical socket. Rough estimates of daily uses of SHS of these specifications include (1) 10 W LED lights for 8 hours, (2) radio for 6.5 hours, or (3) black and white TV for 3.5 hours, as well as cell phone charging.

[9] *Proyecto masivo* is the term commonly used to describe the off-grid solar electrification program that resulted from the "International auction for the supply of electricity with renewable energy resources in areas not connected to the grid," first announced in September, 2013. (*Subasta internacional para el suministro de electricidad con recursos energéticos renovables en áreas no conectadas a red.*)

[10] This is known as a BOOT model whereby the contractor builds, owns, operates, and then transfers the infrastructure equipment back to the state upon contract fulfillment (Delmon, 2011).

[11] FOSE is a cross-subsidy where energy customers (generally located in poor rural areas) consuming less than 100 kW h of energy per month receive a discount that is financed by those consumers (generally located in urban areas) using more than this amount of energy per month (OSINERGMIN, 2017).

[12] FISE is subsidy intended to benefit the most vulnerable sectors of Peruvian society that is supported by taxes on the sale and transportation of various forms liquid hydrocarbons (FISE, 2017).

project timeline. While the *proyecto* has garnered praise from some energy industry officials, it has also been critiqued by experts and officials across nearly every sector linked to energy and development. Some of the general points of contention bear mentioning, as they help illustrate the critical landscape surrounding the *proyecto masivo* more broadly. Most prominently, the *proyecto* has received criticism for what experts believe were the political and economic motives for its creation. Many respondents that were interviewed[13] asserted that the *masivo* was born of the former President Ollanta Humala's desire to complete the project by the end of his political term in mid-2016. It was reportedly[14] rushed forward by bureaucrats at the MEM as a politically expedient and innovative business initiative, rather than on the merits of its technical design. Critics of the origins of *proyecto* were quick to point out the stark mismatch between the presidential political cycle (2011–16) and program implementation of this magnitude. The political and business incentives of accelerating this project to completion by the end of Humala's term simply did not match with the protracted technical exigencies of carrying out the *masivo*. For example, officials experienced in planning rural electrification programs claimed that locating and registering half a million potential energy customers across isolated rural areas of the country would alone take 1–2 years to carry out. The timeline established in the *proyecto masivo* framework allowed Ergon just 4.5 months.

Moreover, the MEM has been critiqued for designing the *proyecto masivo* auction without coordinating with other state ministries, nor consulting with domestic universities. It has been observed that coordination between various ministries and agencies, as well as political leadership from multiple sectors rather than a single champion of an initiative, is required to create a healthy climate for PPPs to emerge (Delmon, 2011). Relatedly, many countries have established PPP units dedicated specifically to helping "ensure that the necessary capacity to create, support, and evaluate multiple PPP agreements exists." (OECD, 2010: 11). In contrast, suggesting a very fragmented approach to development initiatives in Peru, many interviewees[15] bemoaned the complete lack of coordination with other ministries, such as the Ministry of Education and Ministry of Health, even though schools and medical posts will be provided SHS as part of *proyecto masivo*. Professors from both engineering universities and technical institutes remarked that no one in their field had been consulted, despite their expertise and abilities to provide input on the technical specifications of the public auction. Relatedly, MEM personnel were broadly perceived by many officials interviewed as unfamiliar with cutting-edge renewable energy technology, and that the technical requirements outlined in the auction framework were too ambiguous and incorporated out-of-date SHS technology.

[13]   Octavio Díaz, interviewed by author, October 17, 2016; Santiago Salgado, interviewed by author, July 27, 2017; María Álvarez, interviewed by author, December 7, 2016; Gerardo Porres, interviewed by author, October 30, 2016; Carla Saenz, interviewed by author, August 18, 2015; Eduardo Jiménez, interviewed by author, July 10, 2015; Horacio Magdelena, interviewed by author, November 2, 2016.

[14]   María Álvarez, interviewed by author, December 7, 2016; Gerardo Porres, interviewed by author, October 30, 2016.

[15]   María Álvarez, interviewed by author, December 7, 2016; Carla Saenz, interviewed by author, August 18, 2015; Horacio Magdelena, interviewed by author, November 2, 2016; Gerardo Porres, interviewed by author, October 30, 2016; Humberto Moreno, interviewed by author, January 31, 2017; Esteban Ramos, interviewed by author, December 7, 2016.

6.3 Ergon and rural solar energy provision

The *proyecto masivo* auction was designed, not only with the nuances of rural energy service in mind, but also as a technical and financial initiative to attract private investors. Studies on energy auctions have shown that setting very rigid or very lax requirements for auction participation by investors can create trade-offs (IRENA, 2017). Setting a high bar for applicants can minimize the number of bidders, which can harm competition and possibly lead to more elevated prices. On the contrary, setting a low bar for requirements may lead to applicants who may be less qualified and eventually lead to project delays (IRENA, 2017). The *masivo* has shown to be an example of the latter. An official from the state energy regulator, OSINERGMIN, and president of committee that designed the *proyecto* auction contract, Manuel Rodríguez, described how the contract outlined low technical requirements and allowed the market to define the appropriate technological characteristics to be deployed in the *proyecto*, saying that the "MEM (...) defined a very traditional technology, in terms of batteries requirements, and then [the contract] evolved, so that the [technical] solution would be provided by the market." This statement not only lends strength to the criticism, noted earlier in this chapter, that MEM is unfamiliar with cutting-edge solar energy technology (or is indifferent to it at best), but also that it allows market actors to largely dictate the specific technical details of the batteries, the most expensive and crucial component to SHS performance and sustainability.

Rodríguez went on to elaborate about the effect of setting high requirements on auction contracts and how he approaches risks and guarantees, saying, "from the experience that I have, as soon as you put up more barriers, you end up with few bidders. It's as simple as that. I prefer to pass the risk on to whoever is going to assume [the *proyecto*] and I put in (...) guarantees for those who submit bids." This means that the state sets low barriers for bidders, allowing them, for example, to present bids utilizing SHS technology of their choice, taking the risk that their chosen technology will perform robustly over the course of the 15-year contract commitment. Meanwhile, the state offered the bid winner a guaranteed subsidy (FOSE) for each of the 150,000 households that receive solar energy, regardless if the homeowner pays their monthly bill.

Yet, as it can be seen, Rodríguez's statements indicate that on the one hand, the general lax requirements of the *masivo* auction were intended to attract many bidders, and on the other, there was an expectation that this approach would also garner bids from serious energy companies. He said,

> We were looking for people who had the experience of having done this. Not anyone can get involved in this business. So if for example, they already had done [projects] in India, they know how to deal with communities, know how to deal with rural areas. In...Peru, what we... do not want is a company that comes to gain experience in this, they should already come with experience.[16]

Instead of a veteran of off-grid energy provision, the *proyecto masivo* was ultimately awarded to Ergon Peru, a subsidiary of the Italian-based holding company Tozzi Green,[17] a company with no prior experience working in Peru, nor any prior experience carrying out rural, off-grid solar energy programs.[18]

---

[16] Unless otherwise noted the preceding quotes are from Manuel Rodríguez, interviewed by author, December 12, 2016.

[17] This is the company's real name, not a pseudonym.

[18] Octavio Díaz, interviewed by author, October 17, 2016.

Tozzi Green formed Ergon expressly for the purpose of submitting a bid for Peru's *proyecto masivo* auction;[19] however, Tozzi Green began focusing on renewable energy in the 1990s, and since then it has conducted *on-grid* renewable energy projects in Italy, Madagascar, and South Africa. It bears mentioning at this point that the former MEM official and one of the two main architects of the *masivo* auction contract, Octavio Díaz, began working for Ergon after it won the auction as its head representative in Peru. While it is not the purpose of this chapter to question the motives or the ethics of this career move, this helps understand how the MEM views off-grid energy provision, as reflected in the *masivo* auction drafted by Díaz, and how Ergon sees these same challenges, are in many respects, one and the same.

All energy industry experts across the NGO, public, and private sectors have voiced criticisms that Ergon, a subsidiary of a company with no experience in this type of solar program, headed by a former MEM bureaucrat, does not have a true grasp on the challenging realities of off-grid solar energy service in rural Peru. Doubts about Ergon's comprehension of the exhausting on-the-ground work that household solar energy service entails also came from the private sector. Ulíses Lopez, of Norte Renovables, illustrated a rural solar energy scenario outlining the challenges of the *proyecto* that Ergon seems unaware of, saying

> To do an installation in a user's house, for example, you may have transported a battery that weighs 30 or 40 kilos on a mule…a panel weighing 15 kilos more, and you arrive after a walk of several hours, and [the homeowner] isn't there, because he is visiting his family…or he has a second home a few

kilometers away and he has gone there for two months. And then you're stressed; when you went [there] two months ago this person said yes, but when you go there…you have no way of notifying [homeowners] beforehand. There are a ton of things that we could analyze — risks, logistical complications — which make this project a unattractive, both for their size and for all these logistical difficulties, for large companies.[20]

Comments such as these demonstrated that López's understanding of energy service in isolated communities of the Peruvian highlands is not without nuance. What makes this statement all the more interesting is that his company, which has in fact participated in the installation and operation of SHS in rural Peru, lost the *masivo* auction to Ergon. What is more, Ergon's bid was half the amount of Norte Renovables', which means that Ergon has bet that it can profitably procure, install, operate, and maintain 150,000 SHS across Peru, for half the annual cost that was proposed by Norte Renovables, despite Ergon's complete lack of prior experience with off-grid solar provision in rural settings. Alternately, Ergon's low bid may in fact prove to be an example of what is known as the "winner's curse," a situation where the auction winner ends up making negative profits. This can result from the company's perception of an extremely competitive auction environment, leading to the submission of an overly aggressive bid proposal.[21] While speculating on Ergon's business strategies is outside the scope of this chapter, the company's drastically low bid, in combination with its inexperience with off-grid solar provision, suggests that the company may have misjudged its competitive environment, be using less costly SHS's components, believes it can operate with very slim profit margins, or simply does not

---

[19] Ergon is a "special-purpose vehicle," which is created as a "legal entity that only operates and owns one specific project/concession during the contract period" (Sarmento and Renneboog, 2016, 102).

[20] Ulises López, interviewed by author, November 10, 2016.

[21] IRENA (2015) (accessed 02.12.17.).

fully grasp the challenging nature of its role in the *proyecto's* service model.

6.3.1 Fees for disconnecting and reconnecting solar home systems

Another element of concern is Díaz's simplistic view regarding Ergon's role in providing maintenance and operation of SHS. He paid no attention to the difficulty of sending technicians to cut power of a single SHS of a homeowner who has fallen behind on payment. Even more troublingly, his "indifference" toward whether or not a customer pays revealed that he sees Ergon as an independent, technical actor, with no recognition of its partnership with the public sector, nor the public distributors' efforts at bill collection. He said,

> Ergon, what it does is installs the (solar home) system, it makes sure that the system is operational, whether the user pays or doesn't pay, Ergon is indifferent. Complying with its obligation, Ergon has to keep [the SHS] operative. If at any time the (public energy) distributor says 'Hey, Ergon, cut their power', at that moment Ergon cuts off power, but the amount charged to cut power is not included in the operation, it is a separate charge.[22]

Perhaps the greatest consequence of Ergon's underestimation of rural energy service is related to how the *masivo* has been designed to financially compensate Ergon for its activities related to cutting power, reconnecting power, and removing SHS from households. This important detail is touched upon only briefly in the auction contract, which states "The Investor has the right to the corresponding remuneration for the disconnection and reconnection of the Autonomous RER [Renewable Energy Resource] Installations. The charge for disconnection or reconnection will be set by OSINERGMIN."

(OSINERGMIN, 2014). While Ergon was allowed to set its annual required remuneration amount through its bid proposal for the *proyecto masivo*, the state's energy regulator, OSINERGMIN, establishes the fixed amounts that Ergon will be allowed to charge rural energy customers for certain required energy services. These include how much Ergon can charge customers for cutting power and reconnecting power to their SHS (usually only billed upon reconnection), as well as reimbursement fees for having to reinstall SHS in another community if the national electric grid expands into Ergon's zones of operation. A former MEM official with sound knowledge of the *proyecto masivo* contract highlighted the problem of some prices being set by the bidder and others by state regulators, saying "The big defect that I see in the auction ... in which all of [the private companies] competed, is that it doesn't take into consideration the cutting (power) and reconnection... as you know, it's paid at the price set by OSINERGMIN."[23] To emphasize the dramatic financial and logistic implications of cutting/reconnecting SHS power for Ergon, he went on to reference the experiences of LSA, saying

> [LSA] is important, because...this year it's made an average of 1,000 cuts (to SHS power) and reconnections...we are talking about 25% of its (total number of) customers. That it is no small thing. I think that Ergon believes it's going to make a few (cuts to power)...but imagine if it has to cut and reconnect 20,000, 30,000 (customers)? It will not be an insignificant (financial) amount, because, [Ergon] thought it was going to have a very small number of incidences...So, the big problem of the auction-...model, is that it combines seemingly firm revenue streams (from the FOSE subsidy of 150,000 customers), but unfortunately [Ergon has to] keep carrying out operations that are [financially] regulated by [OSINERGMIN], but as a private (company) will not necessarily accept them.[24]

---

[22] Octavio Díaz, interviewed by author, October 17, 2016.

[23] Santiago Salagado, interviewed by author, July 27, 2017.

[24] Santiago Salagado, interviewed by author, July 27, 2017.

Cutting and reconnecting SHS power is an arduous, two-step intervention. It involves sending a technician to first cut power to a SHS whose owner has fallen behind in payment, then once the customer pays their outstanding debt, sending a technician back to the same household to reconnect power to the SHS. Regardless of the difficulty of this multipart task, LSA, for example, can only charge a customer the OSINERGMIN-regulated fee of 12 Soles (roughly $3.75). Given this low remuneration amount, the detailed accounts of the physical logistics of rural energy service offered, which is mentioned earlier, and the potentially high number of cuts and reconnections that Ergon may have to carry out, the same MEM official went on to say, "I bet you that there will be problems when [Ergon] begins to operate, because [Ergon] will begin to complain, to fight because they do not like the [remuneration] values, and it's going to happen."[25]

This issue of cutting power for late payment and reconnection once a customer becomes caught up on their bills shows not only a weakness of the *proyecto* auction but also the degree to which Ergon does not understand the true costs involved in maintaining 150,000 SHS in operation. This is a weakness because the *masivo* auction required bidders to submit proposals without them knowing the exact remuneration levels for cutting and reconnecting power to SHS, an activity that, based on LSA's experience, will likely take up a considerable portion of their time and resources. As for Ergon, it appears that they have an underappreciation for the financial and human resources necessary in cutting/reconnecting power, and most importantly, little acknowledgment of how the corollary role of the public sector will greatly affect its customers' rate of payment and default, which in turn, will affect the number of cuts and reconnections it will be required to carry out.

This discord between the potential operational commitments of solar energy service and state-mandated levels of remuneration for these activities is illustrative of the sort of contract design flaws that can eventually lead to contract renegotiation. While it is not the aim of this chapter to thoroughly explore all the causes of and remedies for potential contact renegotiation of the *proyecto masivo*, it is worth placing these conflicts in context, since "contractual renegotiation has typically been seen as undesirable and reflecting the inefficiencies of contracts since it imposes high transaction and social costs and may induce opportunistic behaviour of both private and public parties." (Domingues and Zlatkovic, 2015). More broadly, when renegotiations occur they contribute to the general perception that private sector involvement in infrastructure services leads to poor outcomes. This perception can cause the general public to lose support of private infrastructure provision and, given governments' budget restrictions and growing reluctance to provide public services, can ultimately lead to fewer viable alternative forms of infrastructure provision (Sarmento and Renneboog, 2016).

6.3.2 Technology, community engagement, and subsidies

True to the intention of the *masivo* auction design, Ergon's representative, Octavio Díaz, demonstrated a heavy reliance on technology performance and receipt of the FOSE subsidy to shield Ergon from financial risks of the *proyecto*. In addition, the company is further protected from risky program activities, which have instead been allocated to the public sector. Confident of its limited technical role and buttressed by the FOSE subsidy, Díaz seemed incapable of connecting the dots between how

---

[25] Santiago Salagado, interviewed by author, July 27, 2017.

the customers' grasp on SHS function may drastically influence Ergon's ability to carry out its own responsibilities and ultimately affect its bottom line, saying

> In the case of ERGON, the...responsibility is very focused on the technical part, in fact, I would say exclusively on the technical side. So, to the extent that [Ergon] has sufficient trust in the (performance of the) equipment it's installing... in the sense that you are installing equipment that you trust and will not generate (problems)...Based on that you can say with confidence, "I don't care how far [away customers] are," because if [regulations] give me 5 days, 6 days to get there, I know that when I go, it will not be a problem that I have to replace the panel, or replace the battery, because those cases will be the minority. Most (problems) will be...generated by misuse by the [customer], and a problem of misuse, is no longer the fault of the equipment, you understand? In that sense, the remuneration it is not affected...you have [FOSE subsidy] guaranteed.[26]

This narrow view of rural energy provision expressed by Díaz is reflective of the *masivo* auction itself, which offers only vague guidelines about the private company's responsibilities for explaining energy service to customers. Out of the 108-page contract, only two paragraphs dictate how the private company (Investor) must engage with rural residents (Users) about energy service and bill collection by local public distribution companies (Distributor).

> The Investor taking into consideration the profile of the User and the characteristics of the zone of influence (such as, for example, language, customs, among others), will have the necessary training information material to be handed over to the Users, for the use, maintenance and preservation of the Autonomous RER [renewable energy resource] Installations, as well as the operation of the technical service and the function of the Distributor...
>
> ...The Investor will give training and information to the Users regarding the installations, will give them the installation certificates, the

instructions for use, and the handbook of maintenance and will update the users' information during the Time Term of Enforcement. (OSINERGMIN, 2014)

These ill-defined stipulations not only underscore how the Peruvian state sees the *masivo* through a purely technical and financial lens, but also how various components of the program are treated as if they were unrelated and have no effect on one another. There seemed to be no recognition by Díaz that if community residents do not fully understand the uses and limits of the SHS and the corresponding financial obligations of the program, a vast number of technical issues may arise for Ergon. What is more, Ergon treats this social component of service in cavalier fashion, more as preparation for a one-time technical intervention and perfunctory training, than a series of ongoing activities on which the success of the *proyecto* largely depends.

Díaz described the community engagement process that Ergon will conduct, which, shockingly, entirely bypasses direct engagement with community members, primarily emphasizing coordination with local authorities so that these authorities successfully relay the message to townspeople to be in their homes during the scheduled SHS installation date. He said,

> Once registered...in the list of potential users, Ergon carries out an...awareness-raising effort weeks before the installation. This work is done by our installation contractor... in coordination with our community relations area. The activity consists in coordinating with the authorities and local representatives to make them aware of the program and the installation process that will take place. This work is very important operationally for the installation process, because this depends on ensuring the maximum presence of people in their homes for the dates of installation itself.[27]

---

[26] Octavio Díaz, interviewed by author, July 24, 2017.

[27] Octavio Díaz, personal communication, July 10, 2017.

Equally as troubling were Díaz' comments about how Ergon will conduct training to energy customers about *proyecto masivo* energy service. Again, contracted installers will be tasked with explaining to household individuals all technical and financial aspects of the program at the time of installation.

> Additionally…during the installation process in the home, Ergon's installation contractor is responsible for training the user in the use of the system, and providing all the necessary information to the user so that they can…call the corresponding distribution company's call center.[28]

The importance of social aspects of energy service, such as in-depth training workshops to familiarize new customers with program details, has not only been stressed by development officials in Peru and identified in the literature more broadly,[29] but also it has been emphasized by rural energy customers themselves. The ethnographic fieldwork in a small village served by LSA's *Casa Solar* program in the highlands surrounding Cajamarca, called Lahuaymarca, confirmed that residents placed great value on the informational meetings that LSA held in their community prior to the start of solar energy provision. Residents' testimonies, gathered over the course of 4 months while living in the village, described how LSA's repeated community workshops provided them opportunities to understand both the technical and financial details of *Casa Solar* and gain an overall sense of ownership over the program.

Residents of Lahuaymarca make it clear that, if given a choice, multiple informational meetings facilitated by energy company representatives with the whole community in attendance, would be preferable to one (or even more) meetings run by local community authorities. Their reasons behind this preference begin to reveal why the *masivo* auction contract's loose community engagement guidelines and Ergon's minimalist approach to conveying program information to residents may well prove problematic as the *proyecto* takes effect in the future.

Explaining the technical and financial parameters of the *proyecto* to rural residents is crucial for program success. This is because the national electric grid offers urban residents electricity for 24 hours a day, in contrast, the SHS (like those used in the *masivo*) generally only provide rural residents with light and energy for about 5–7 hours a day and only allow the use of DC tools and appliances (Peru's national electric grid uses AC power). Speaking about the crucial importance of community workshops and communicating this information with rural residents, the manager of LSA, Carla Saenz, emphasized countless times throughout our interviews, that if residents fail to truly grasp the uses and limitations of SHS, they will feel disappointed and let down by the SHS' energy output and eventually stop making their payments or simply reject the project outright and ask Ergon to remove their SHS.

Voices from this community and NGO officials stand in stark contrast with Ergon's stated plans to carry out household trainings as superficially and quickly as possible. Díaz did not show concern for how Ergon's cursory explanations given by contracted installers may lead to residents' poor understanding of SHS's performance or misunderstanding of payment rules. Nor did he seem to grasp how residents' misapprehension could lead to misuse of the SHS or cause delinquent payment, all of which would increase Ergon's maintenance and operation workload. Instead, given

---

[28] Octavio Díaz, personal communication, July 10, 2017.

[29] For example, see Fernández-Baldor et al. (2014), Hancock (2015), Ikejemba et al. (2017), Urmee and Md (2016).

Ergon's limited technical role and guaranteed subsidies, Díaz shrugged off worry about how social elements may affect its operation:

> In our case, there is a guaranteed remuneration... if a user tells us they no longer want (their SHS), the ministry continues paying us. When a user gives up their SHS...we continue to get paid, but we're obligated to take that SHS to another place. It's as if I was being paid for making a SHS available, which is not the same as being paid depending on whether the client wants or does not want the SHS...the project was designed to isolate the 'risk of quantity,' the risk of quantity is isolated in the contract of Ergon, because you have a minimum amount (of SHS customers).[30]

That is to say, the *masivo* has been intentionally designed so that Ergon's base number of 150,000 customers and the corresponding FOSE subsidy is guaranteed. So unlike LSA that loses the FOSE subsidy for a customer who gives up their SHS or stops paying, in a similar situation, Ergon retains its monthly FOSE subsidy payment, as long as it relocates the SHS to another user. Further highlighting Ergon's restricted, technical role, the state's technology-driven understanding of rural energy provision, and the preference it has given to private actors in the *masivo*, Daniel Montenegro, of OSINERGMIN, stated

> The function of the investor is to simply install and charge, practically, those are the conditions that were in the contract, but [Ergon] is not charging, who charges is the (public energy) distributor... I think it is a sweet deal for the investor, because they are only installing, nothing else.[31]

Despite the overtly technically and financially minded approach outlined by Ergon, Díaz still expressed some recognition of the social demands of this project. Seeming to pay lip service to the social concerns surrounding the *proyecto's* service model, he went so far as to recognize them as a general stumbling block that needs to be addressed as part of the *proyecto*. More importantly, for him, this was an issue that should be confronted by the state, and not by the private sector:

> You realize the social problem, or rather the problem of community development, has both components, a technical component and social component. The issue is that you have to cover both. So, it is bad that it's...seen only as a technical solution, because in the end it is not a solution, it is a good business for a private (company) but it is not the solution.[32]

Addressing criticisms voiced by actors in the NGO sector and advocates of renewable energy in Peru, Díaz finally admitted that the design of the *masivo* essentially punted the billing responsibilities of energy service off to the public distributors and remains a largely ill-defined endeavor, stating bluntly "So... when [LSA] tells you that this social component is not (in the auction contract), it's true, it's not within the auction, but (it's) within the project, only that it's on the side of the state."[33] That is to say, at the time of the auction award, the process by which distributors were to expand their rural bill collection efforts for the *masivo* was yet to be specifically outlined by the MEM.

Again, bringing into sharp relief the effects of the auction design, which has allowed Ergon to view itself as an independent actor whose profits are essentially guaranteed, and not part of a cohesive PPP, Díaz said straightforwardly, "I know...the people of [LSA] said

---

[30] Octavio Díaz, interviewed by author, July 24, 2017.

[31] Daniel Montenegro, interviewed by author, November 22, 2016.

[32] Octavio Díaz, interviewed by author, October 17, 2016.

[33] Octavio Díaz, interviewed by author, July 24, 2017.

this will not work. If in reality this does not work out well, this dispute will generate problems, but it will not generate problems for the private [company]."[34] On another occasion, he distanced Ergon even from the state's flawed social development efforts, striking an even more critical tone, saying "Social awareness, regarding...bill collection and (energy) service is a very critical issue. The social sustainability...does not fall on Ergon but on the state apparatus. The state...is not diligent, does not do its studies, it does not do its work in relation to [social issues], so there may be problems."[35] And toward the various potential problems that *masivo* may generate for the public distribution companies is what this chapter turns to next.

6.4 Public distributors' role in the Proyecto Masivo

The many in-person interviews that support the ideas presented in this chapter brought into sharp relief that there was no clear understanding of the public distributors' role in the impending *proyecto masivo*. Various interviews conducted with Ángel Ochoa, a director of the public energy distributor, Caxa Eléctrica, demonstrated that he was unaware of the specific role that his company would play in the imminent *proyecto masivo*, because MEM was yet to deliver mandates to public distributors regarding their bill collection activities in the *masivo*, despite two years having passed since the award of the auction.

In the absence of concrete plans for the distributors' role in the program, Ochoa offered sketches of how he expected the distributors would fulfill their responsibilities, based on

their current operations. Ochoa made it clear he would rather have seen Ergon assume all responsibilities of the *proyecto*. Yet, he expressed optimism that distributors such as Caxa Eléctrica will figure out a way to successfully carry out their soon-to-be imposed duties. Although, he lamented that Caxa Eléctrica was even required to be a part of the initiative because, as he said, "it's going to generate problems for us." This section shows how the role of the public distributors has not been explicitly spelled out in the *masivo* design and demonstrates that public distributors are apprehensive that these new activities will bring financial and logistic challenges, as well as increased debt upon their operations. These interviews also suggest that the auction was, in some ways, based on the false premise that both the private *and* the public distributors would have equal motivation to fulfill their energy service duties in the *masivo*.

Speaking with both current and former officials from Peru's MEM, it was clear that the *masivo* auction had left the details of the public energy distributors' roles in the program largely undefined, and that the state was still lagging behind in determining how they would carry out the duties stipulated in the auction. A MEM official named Héctor Alarcón, who works on productive uses of energy programs in conjunction with a World Bank–funded rural electrification initiative, offered a frank assessment of the state's preparedness to tackle the *proyecto*, stating

> If you ask me if the electricity distribution companies are prepared for the task they're being given, I would categorically say no, absolutely not. They could be prepared in the future, if a period of transition is given and they handle it well. They may be (ready), but it's... a tricky task right now.[36]

---

[34] Octavio Díaz, interviewed by author, October 17, 2016.

[35] Octavio Díaz, interviewed by author, July 24, 2017.

[36] Héctor Alarcón, interviewed by author, January 31, 2017.

It is little wonder then, that this MEM official offered such a critical assessment of the state distributors' current capabilities to address the *masivo* duties. Because how these vexing social activities are to be carried out— interfacing with isolated rural communities regarding bill collection, whose residents often have little experience with making recurring monthly payments—was never explicitly outlined in the *masivo* auction. In fact, they were intentionally left vague in the auction contract, with the expectation that the state would find an adequate solution to these socially burdensome responsibilities sometime down the road. Octavio Díaz, was plainspoken in his description of how its design offered technical, but not social solutions, while unloading these unresolved problems on an unresponsive state apparatus. He described how the social part:

> is not defined...the solution for this was left out of the auction. It is not that the auction gives the solution for the whole thing. Actually the auction and Ergon offer only a partial solution. Because the problem...the technical problem has been solved, not the social problem. The social problem has been postponed...the idea is that it gets solved along the way...the state has to see, that the problem is what we are seeing now— that the state is a little slow at this. There is a big challenge here.[37]

Perhaps most disconcerting were comments made by Roberto Miyagi, a MEM official overseeing the implementation of the *proyecto masivo*. At the time of the interview the *masivo* was still mired in arbitration and Ergon had not installed a single SHS, even though the contracted date for installation of all 150,000 SHSs had passed six months beforehand. Yet, the MEM was still months away from reaching an agreement with public distributors on issues related to bill collection, which states

Surely in two, three months we're going to have it clarified. For example, this issue of billing is on the table being discussed, some ideas are being exchanged. The idea, like we've said at the roundtables, is that this project has to have a way forward... to be both profitable for [public distributors] and that it can be financed with the FOSE (subsidy) resources.[38]

In addition to the state distributors' general lack of preparedness and the unresolved details regarding their duties in the *proyecto*, there was an undeniable sense that these new billing activities in Peru's most remote areas, where public distributors do not have established bill collection infrastructure, will surely create challenges for them. Ángel Ochoa, of Caxa Eléctrica, shared his thoughts on the *proyecto masivo* from the standpoint of a public distributor, by saying

> it's going to generate problems for us because it means increasing our...process of collecting [rural residents'] bills...I don't know how [many SHS] there will be in the area of Cajamarca... Because we don't have collection centers in all areas...Even for the residents this can sometimes create a problem— How to get there? Where to go to pay? In certain areas we have collection centers, but not everywhere. I imagine that those parts that do not already have conventional (grid) electrification, we would have to implement some kind of collection center, that will generate costs.[39]

His comments not only made clear that public distributors will obviously incur new operational costs involved in setting up and operating new bill collection centers, but they also alluded to social elements of rural energy service, the importance of which Ergon has seemed reluctant to recognize. His mention of the new logistical demands placed on residents and their potential confusion about

---

[37] Octavio Díaz, interviewed by author, October 17, 2016.

[38] Roberto Miyagi, interviewed by author, November 21, 2016.

[39] Ángel Ochoa, interviewed by author, April 4, 2017.

where and who to pay for their energy service, speaks to his comprehension of the minute social challenges that have been overlooked both by the *masivo* auction design and Ergon's casual approach to community engagement.

Addressing the dreaded social aspect of energy service that the public distributors must confront, he said,

> People in the area…they'll have to participate. A [public] worker going there? No. It has to be (someone) from there, because the area is very dispersed, it's very expansive…there are places where it isn't accessible, there are no roads, there are trails, suddenly [trucks] can't get through.[40]

While he spoke in a positive tone about the increase in public distributors' workload, his description of these rural community collaborations belied his general optimism. As it was precisely these undertakings, dealing with "the idiosyncrasies of the people," as Ochoa called it, which posed a challenge for public distribution companies that the *masivo* auction dealt to the state and Ergon planned to avoid.

6.4.1 Remuneration and the auction's false premise

This final section brings into focus how the *masivo* contract mandates that public distribution companies be reimbursed at rates set not by the companies themselves, but at levels established by OSINERGMIN. While state officials claimed these remuneration levels are well calculated and sufficient, doubt remains as to whether or not this will hold true in practice. Ultimately, the potentially low level of compensation for public distributors suggests that much of the *proyecto masivo* auction was predicated on a false premise. This was the

premise that OSINERGMIN-mandated reimbursement would serve as sufficient motivation for public distributors to execute their new partnership role as purposefully as Ergon. During one interview, OSINERGMIN's Daniel Montenegro walked me through tables and statistics, confidently assuring me that state regulator's calculations pertaining to *proyecto masivo* subsidies and remuneration were thorough and indisputable, saying

> Here is the cost of billing for each province, then we do a weighted analysis to try to get the total cost in each region, north, central, and south. Here is the cost of billing, how much it will cost, how much you should pay to each (distribution) company…everything is supported by…analysis, which…says that the distribution company must be paid for extra activities it performs… Everything is well detailed here.[41]

Alluding to public distributors' likely discontent with OSINERGMIN's calculations, Díaz assessed, "We, Ergon, have a guarantee that we will get paid what we asked for in the auction, but the distributor will not, the distributor has a guarantee that they will get paid what OSINERGMIN says."[42] He went on to rebut Montenegro's assertion, calling into question whether OSINERGMIN's numbers would support the public distribution companies' activities in practice. Not only did Díaz claim that the reimbursement rates are insufficient, but alluded to how MEM will likely override norms protecting distribution companies, using political sway to compel public distributors into assuming tasks that will put them in the red, saying

> The problem that…happens a lot in Peru—it's one thing that's written on paper and another what happens in real life. On paper, the public

---

[40] Ángel Ochoa, interviewed by author, April 4, 2017.

[41] Daniel Montenegro, interviewed by author, November 22, 2016.

[42] Octavio Díaz, interviewed by author, July 24, 2017.

(distribution) company is protected, in the sense that it can reject an assignment. 'Look…you are giving me a commission but…according to what you tell me, my accounts don't add up. So, I reject the order.' But in this case, the [Ministry of Energy and Mines] has exercised its power that it has with public companies and said 'You accept the orders even if your accounts don't add up,'…without respecting the economic balance of these duties, and…has left the responsibility to define how much to pay, to OSINERGMIN. Currently, OSINERGMIN…defines how much the distributor is paid…So…there is a serious problem because the (public distribution) companies…complain that [OSINERGMIN is] recognizing them at such a low rate for an activity that actually has a higher cost. That problem right now is not solved. But happily, this problem does not affect Ergon.[43]

As previously outlined, one of the principal controversies surrounding the operation of the *proyecto* is the public distribution companies' dissatisfaction with reimbursement for their program involvement. The issue of insufficient compensation levels is revealing in various ways. In one respect, it shows how the *masivo* auction was built on the false premise that public distributors would carry out their program functions with similar efficacy and motivation as the private sector. In another respect, it again highlights the program's isolated and fragmented design, which has created an operational dynamic where public and private actors have little incentive (or even need) to operate in cohesion with one another. Santiago Salgado, a former MEM official, offered critical insight into the contradictions of the auction's premise, saying

> The big problem…the error of our PPP, is that…the public company really has no motivation in doing…rural electrification projects…there is no interest by the distribution companies, in

intervening in the rural electricity sector, be it through the extension of (grid) networks or even of photovoltaics. There is no interest, because…it costs so much and it's believed that the distributors have reached their limit and they are not going to do more. But nevertheless the [PPP] mobilizes them, and the [PPP] is supported by the public (distribution companies)…they are not given a small role…the [public distributors] are being asked…to distribute invoices and do bill charging…The model contradicts its premise in that the public (distributors) are not interested in doing the rural electrification activities…given that these OSINERGMIN (remuneration price) signals are not enough… everyone has the hope it will turn out, but there will be problems.[44]

He also alluded to what Díaz mentioned previously, about the public distribution companies relenting to pressure from MEM to carry out responsibilities in the *proyecto*. His comments also supported the notion that many of the auction details regarding how public distributors would be compensated were ill defined or misleading:

> the big problem there that is not solved, because in the end the public will accept it, always the public ends up accepting everything, it is that when the agreement with the ministry was signed, it was believed, and therefore the public (distributors) accepted…that they would be recognized based on their incurred costs. And this is not the case, they are recognized by the costs that OSINERGMIN regulates…So, but the public (distributors)…assume…these additional activities that are not part of their routine business activities, but additional one that the state is giving them to perform.[45]

Perhaps most consequential of all his criticisms were his observations, which lend strength to the argument made throughout this chapter, that the *masivo* design lacks articulation between involved actors. Salgado said,

[43] Octavio Díaz, interviewed by author, July 24, 2017.

[44] Santiago Salgado, interviewed by author, July 27, 2017.

[45] Santiago Salgado, interviewed by author, July 27, 2017.

The public (distributors) they have no interest in doing [bill collection], if they distribute (bills) and [customers] don't pay, they don't not pay. They'll present their numbers to OSINERGMIN and ask to be paid; 'It's is not my problem,'…I believe that under these conditions [public distributors] will not have any interest in carrying out effective (bill) collection, but they are things that will come out little by little.[46]

The indifference that he foresees expressed by public distributors mirrors the same lack of concern voiced by Ergon. This is because the design of the *masivo* allows both Ergon and public distributors to disregard the collective mechanisms that will sustain the *proyecto's* operation, with each actor compelled only to conduct their contractual responsibilities; Ergon incentivized to make a profit, the public sector constrained to minimize deficit.

6.5 Conclusion

Drawing on the testimonies of officials from across Peruvian energy and development sectors, as well as participant observation of rural energy programs, this chapter has shed light on the challenges and contradictions of the Peru' *proyecto masivo* auction design. From its inception, the *proyecto* was meant to adhere to an unrealistic political cycle timeline, which necessarily precluded a better auction contract. What resulted was a contract that relies heavily on a private company to deploy a robust technology to solve longstanding energy access deficiencies in rural areas. This technological and market-based approach saw the omission of explicit guidelines designating responsibility for raising community awareness of SHS function or program details. At the same time, public distribution companies were delegated the unwanted task of establishing new billing operations in some of the most

remote corners of the country. The state-regulated remuneration levels for program activities, which will adversely affect both Ergon and public companies, draw further attention to the state's trivialized view of the logistical challenges inherent to off-grid energy service. And it has been shown that the *masivo* auction has been propped up under a faulty proposition, as remuneration mechanisms are isolated across sectors, allowing Ergon and public distributors to operate with indifference to how the other fulfills their duties in this "partnership."

Although the *proyecto masivo* is only in its initial stages, the arguments forwarded in this chapter aim to orient the reader to infrastructural systems' unforeseen internal problems, to highlight the uncertainty and potential volatility in the way that the actors and relational mechanisms in the *proyecto masivo* cohere with one another. In regards to the future design of public auctions and PPPs, Peru's MEM would do well to envision them as more than narrow, apolitical tools, and rather, as dynamic infrastructure services embedded in larger political and historical currents. To begin to accomplish this, a robust and transparent dialog is needed between Peruvian ministries so that public auctions incorporate cross-institutional input that is attendant to the social and geographical realities of the country's remotest populations located even beyond the reach of the *proyecto masivo*. Going further, the Peruvian government should form a PPP-dedicated department that plans, coordinates, and oversees PPPs over their entire life cycle, which will help ensure their long-term viability and success. This will necessitate that PPP design and implementation periods span multiple political cycles and will require improved state institutional capacity and flexibility. Under these conditions the *proyecto masivo* and other PPPs can surely accomplish more than ribbon cutting

[46] Santiago Salgado, interviewed by author, July 27, 2017.

and inauguration photos. Through the responsive administration of cohesive collaborations between public and private actors, Peru and other developing countries can begin to transition to a less carbon-dependent future while also addressing longstanding legacies of exclusion and infrastructural deficiency in their most remote, marginalized communities.

References

Arent, D.J., Arndt, C., Miller, M., Tarp, F., Zinaman, O., 2017. The political economy of clean energy transitions: a study prepared by the united nations university world institute for development economics research (UNU-WIDER), Studies in Development Economics, first ed. Oxford University Press, Oxford.

Boardman, A., Vining, A., 2010. Assessing the economic worth of public–private partnerships. In: Hodge, G.A., Greve, C., Boardman, A.E. (Eds.), International Handbook in Public-Private Partnerships. Edward Elgar, UK.

Boyer, D., Szeman, I., 2014. The Rise of Energy Humanities. University Affairs. Available from <http://www.universityaffairs.ca/opinion/in-my-opinion/the-rise-of-energy-humanities/>.

Delmon, J., 2011. Public-Private Partnership Projects in Infrastructure: An Essential Guide for Policy Makers. Cambridge University Press, New York.

Domingues, S., Zlatkovic, D., 2015. Renegotiating PPP contracts: reinforcing the 'P' in partnership. Transport Rev. 35 (2), 204–225.

Fernández-Baldor, Á., et al., 2014. Are technological projects reducing social inequalities and improving people's well-being? A capability approach analysis of renewable energy-based electrification projects in Cajamarca, Peru. J. Hum. Dev. Capabilities 15 (1), 13–27.

Feron, S., Cordero, R., 2018. Is Peru prepared for large-scale sustainable rural electrification? Sustainability 10 (1683), 1–20.

FISE (Fondo de Inclusión Social Energético), 2017. Available from <http://www.fise.gob.pe>.

Guasch, J.L., 2004. Granting and renegotiating infrastructure concessions. Washington, DC: World Bank.

Gupta, A., 2015. An anthropology of electricity from the global south. Cult. Anthropol. 30 (4), 555–568.

Hancock, K., 2015. The expanding horizon of renewable energy in sub-Saharan Africa: leading research in the social sciences. Energy Res. Soc. Sci. 5, 1–8.

Hellowell, M., Vecchi, V., 2012. An evaluation of the projected returns to investors on 10 PFI projects commissioned by the national health service. Financ. Account. Manage. 28 (1), 77–100.

Howe, C., 2015. Latin America in the Anthropocene: energy transitions and climate change mitigations. J. Latin Am. Caribb. Anthropol. 20 (2), 231–241.

Ikejemba, E., et al., 2017. The empirical reality & sustainable management failures of renewable projects in Sub-Saharan Africa (part 1 of 2). Renew. Energy 102, 234–240.

IRENA (International Renewable Energy Agency), 2015. Renewable Energy Auctions: A Guide to Design. Available from <http://irena.org/DocumentDownloads/Publications/Renewable_Energy_Auctions_A_Guide_to_Design.pdf>.

IRENA (International Renewable Energy Agency), 2017. Renewable Energy Auctions: Analyzing 2016. IRENA, Abu Dhabi.

Kale, S., 2014. Electrifying India: Regional Political Economies of Development. Stanford University Press, Stanford.

Lampland, M., Star, S.L., 2009. Standards and Their Stories: How Quantifying, Classifying, and Formalizing Practices Shape Everyday Life. Cornell University Press, Ithaca.

Mitchell, T., 2011. Carbon Democracy: Political Power in the Age of Oil. Verso, London.

OECD (Organisation for Economic Co-operation and Development), 2010. Dedicated Public-Private Partnership Units: A Survey of Institutional and Governance Structures. OECD, Paris.

Orosco, D.Z., 2014. Lineamientos de la Política Pública de Inclusión Eléctrica con Sistemas Fotovoltaicos Autónomos. Ministerio de Energía y Minas, Lima. Available from <http://www.minem.gob.pe/index2.php>.

OSINERGMIN (Organismo Supervisor de la Inversión en Energía y Minería), 2014. Tender Documents for Auction of Electricity Supply With Renewable Energy Resources in Areas Not Connected to Network. OSINERGMIN. Available from <http://www2.osinerg.gob.pe/EnergiasRenovables/contenido/1eraSubastaOffGrid.html>.

OSINERGMIN (Organismo Supervisor de la Inversión en Energía y Minería), 2016. La Industria de la Electricidad en el Peru: 25 Años de Aportamiento al Crecimiento Económico del País. OSINERGMIN, Lima.

OSINERGMIN (Organismo Supervisor de la Inversión en Energía y Minería), 2017. FOSE (Fondo de Compensación Social Eléctrica). OSINERGMIN. Available from <http://www.osinerg.gob.pe/newweb/pages/Publico/49.htm>.

Sarmento, J.M., Renneboog, L., 2016. Anatomy of public-private partnerships: their creation, financing and renegotiations. Int. J. Manag. Projects Bus. 9 (1), 94–122.

Trebilock, M., Rosentstock, M., 2015. Infrastructure public-private partnerships in the developing world: lessons from recent experience. J. Dev. Stud. 51 (4), 335–354.

Urmee, T., Md, A., 2016. Social, cultural and political dimensions of off-grid renewable energy programs in developing countries. Renew. Energy 93 (2016), 159–167.

Wettenhall, R., 2010. Mixes and partnerships through time. In: Hodge, G.A., Greve, C., Boardman, A.E. (Eds.), International Handbook in Public-Private Partnerships. Edward Elgar, Cheltenham.

Yescombe, E.R., 2013. PPPs and project finance. In: de Vries, P., Yehoue, E.B. (Eds.), The Routledge Companion to Public-Private Partnerships. Routledge Press, New York, pp. 227–246.

Further reading

Hart, O., 1995. Firms, Contracts, and Financial Structure. Oxford University Press, New York.

Hughes, T., 1983. Networks of Power: Electrification in Western Society. The Johns Hopkins University Press, London & Baltimore, MD.

Llosa, M.V., December 1990. Questions of Conquest: What Columbus Wrought and What He Did Not. Harper's Magazine, pp. 45–53.

Nye, D., 1992. Electrifying America: Social Meaning of New Technology, 1880-1940. MIT Press, Cambridge.

Starn, O., 1991. Missing the revolution: anthropologists and the war in Peru. Cult. Anthropol. 6 (1), 63–91.

Willems, T., van Dooren, W., 2016. (De) Politicisation dynamics in public-private partnerships (PPPs): lessons from a Comparison between UK and Flemish PPP policy. Public Manage. Rev. 18 (2), 199–220.

PART 2

Energy transition taking shape: Regulatory issues

The prosumer legal protection in Latin America: a requirement for an energy transition

Matheus Linck Bassani

Law, Federal University of Rio Grande do Sul, Porto Alegre, Brazil

7.1 Introduction

This chapter addresses the prosumer of electricity as a very important issue to promote an energy transition. The transition (Labussière and Nadaï, 2018) from conventional to renewable energy sources requires substantial changes in the energy matrix (primary source) and in consumption. Replacing primary sources, such as gas, oil, and coal, with renewable sources, such as solar, wind, and geothermal is the first step to avoiding the depletion of natural resources and the emission of greenhouse gases. Consequently, consumption should be more sustainable, either by its reduction (quantitative) or by its efficiency (qualitative).

The transition is not instantaneous, and it requires a long-term perspective [temporal dimension (Labussière and Nadaï, 2018, p. 323)], will and norms to facilitate the adoption of environmental friendly practices, protecting the right of future generations[1] to enjoy terrestrial space, without being pruned the opportunities (Sen, 2010) of a dignified and healthy life.

Decentralized generation has been considered as a pulverized mean of achieving the generation and consumption of electricity through a sustainable (Schrijver, 2007) manner. This means that, encompassing both functions of production and consumption, the now prosumer (agglutination of the words "producer" and "consumer") still needs to be protected and avoids risks when adopting this modality. The problem to be analyzed is whether the prosumer could still be protected under the Consumer Protection Law.

---

[1] BRASIL, Brazilian Federal Constitution. Art. 225. Everyone has the right to an ecologically balanced environment, a good of common use for the people and essential to a healthy quality of life, imposing on the public authorities and the community the duty to defend and preserve it for present and future generations.

The Regulation and Policy of Latin American Energy Transitions
DOI: https://doi.org/10.1016/B978-0-12-819521-5.00007-3

This chapter consists of three sections other than Introduction section:

- Section 7.2 approaches the activity of the consumer, considered the main criteria to frame the prosumer as a consumer, considering the application of consumer law. Then, this concept will be applied into the power sector;
- Section 7.3 analyzes whether the consumer laws of Brazil, Argentina, and Colombia, and the regulations on distributed generation could approach this new role of the consumer, called prosumer, without losing legal protection.
- Finally, the conclusion (Section 7.4) is drawn.

7.2 The prosumer as an active consumer: a perspective from the power sector

The notion of consumer activity[2] is based on the premise that consumers do not only act by purchasing and enjoying products and services and withdrawing them from the market, but that they act during the contractual relationship through various conducts that may include collaborative measures, critical actions (positive and negative), control measures, and others that may emerge over time.

A sound perspective has been identified within the European Union through a more pragmatic approach. The "Council and European Parliament Directive on common rules for the internal electricity market COM/2016/0864 (Winter Package)" establishes a definition of an active customer in Chapter I, Article 2, item 6:

> Chapter I. Article 2. Definitions
>
> 6. '**active customer**' means a customer or a group of jointly acting customers who **consume, store or sell electricity generated** on their premises, including through aggregators, or participate in demand response or energy efficiency schemes provided that these activities do not constitute their primary commercial or professional activity; (...) (European Union, 2016).

It is pointed out that it is considered an active customer, individually and collectively, who consumes, stores, and even sells energy, as long as it is not its main activity of income, commercial, or professional, precisely so that it does not fit as a supplier.

The main point that should be highlighted is the rate of activity that can be included in the concept of an active customer. The term customer may be broader than consumer and comes from a long-term relationship usually arising from regulated market services, such as in the telecommunication, gas, and electricity sectors (Micklitz, 2013, p. 300 e following). Customer and consumer exercise the same function as withdrawing the product and service from the market and may be considered for academic purposes as synonyms. With respect to small and medium sized companies, their weaker position in the contract relationship must be verified, considering the value and supply chains (Micklitz, 2016, p. 33). The theory of

---

[2] Hannah Arendt, in performing deep analysis on discourse and action, describes the meaning of the word act in which men are driven since their birth: "in its most general sense, it means to take initiative, initiate (as indicated by the Greek word archein, "begin", "lead" and finally, "govern"), impress movement to something (which is the original meaning of the Latin term "agree"). In acting and speaking, men "actively reveal their unique personal identities, and thus make their appearance in the human world; differently it occurs in complete silence and total passivity" that one can hide" (Arendt, 2016, pp. 218–222).

depth finalism[3] applied in Brazilian Law raises the vulnerability (Marques and Miragem, 2014) criterion for the classification of these subjects as consumers according to the concrete case.[4]

In this context the active consumer could generate, consume, store, and sell electricity as long as it is not for professional or commercial purposes. In addition to being a factual and economic recipient, it is explicitly demonstrated that the consumer performs other parallel functions without, however, mischaracterizing their main function while remaining under the protection of consumer legislation.

This understanding is proven when the consumer also acts in the planning and organizing of the demand for electricity (demand response), that is, due to the increased control of the network and automation, new products and services will emerge and will be contracted as needed. Two characteristics of the active consumer can be identified from this movement that will still be sprayed in the future: (1) use of technological capacity and tools capable of generating electricity or using electricity storage; (2) implementation of management consumption. A change in the pattern of traditional consumption, which is still rooted in centralized power generation and consumption, is observable (Volkova et al., 2018). The prosumers will be closer to the grid operation and will represent a massive source of influence on the constant and instantaneous generation and consumption, whose electricity prices will be fixed in real time, emerging the need to adopt a management system (Petit, 2019).

Consumer activity,[5] therefore, does not mischaracterize its conception already defined in art. 2 of the Brazilian Consumer Protection Code (CDC),[6] interpreted by the application of the theory of the depth finalism (Marques, 2016, p. 308). Consuming, producing, generating, storing, selling, that is, parallel functions to the main function—which is to consume factually and economically, excluding the professional or commercial purpose—are increasingly present in consumers within the consumer society (see Beaudrillard, 2001).

From this active characteristic, it emerges a concept identified in different areas but mostly in the energy field, which seeks to agglutinate the functions of consuming and producing, called prosumer. However, scholars do not provide a legal concept on this active function of the consumer. To tackle this blurred area, the prosumer focused on the decentralized generation of electricity is analyzed below in a more specific way.

As will be seen in this section, the power sector has been the one in which the prosumer has been frequently cited (Jacobs, 2016; Cseres, 2018; Leal-Arcas et al., 2018; Lavrijssen, and Carrillo Parra, 2017; DA SILVA et al., 2018). The liberalization of the electricity sector in the European Union was extremely relevant to trigger new functions for the consumer. Lucia Reisch and Hans-W. Micklitz approached the regulation of electricity from the consumer's perspective, identifying it as a market actor, a citizen, and also a producer in the informal sense:

> In consumer policy discourse, it has been proposed (Reisch, 2004) that consumers should not be conceptualised exclusively as market players, but also as citizens and producers in households performing so-called "informal work." This is in

---

[3] For further explanation, see Section 7.3.1.

[4] See Brazilian Superior Court of Justice (STJ) decisions: Recurso Ordinário em Mandado de Segurança (RMS); REsp 1195642/RJ; AgRg no AREsp 735.249/SC; STJ. REsp 1599535/RS.

[5] Sobre uma abordagem focada ao comportamento do consumidor: Bianchi (1998).

[6] Brasil. CDC. Art. 2: A consumer is any natural or legal person who purchases or uses a product or service as a final recipient.

accordance with household behavior theory (Cécora, 1991) that recognises three basic collective entities that frame consumer behavior: enterprises, the state, and households. Empirically, consumers do not behave as fully informed, economically rational, isolated market actors — the way demand is conceptualised in neoclassical consumption theory (Reisch and Micklitz, 2006, p. 401).

As an actor in the market, the consumer purchases products and services and is located on the opposite side of the supplier, being able to articulate through collective organizations, whose acts and interests are influenced and guaranteed by the rules of consumption. The authors cite the term prosumer as one of the forms of action in the market, in the sense of proarticulation (the other sense is the counter articulation, which has the sense of balancing the power over the markets): "In prototyping and buycotts, consumers are actively involved. As so-called 'prosumers', they co-produce in one-way or another the desired product or service." In the power sector, the consumer might act in the market only after the liberalization policies (Reisch and Micklitz, 2006, pp. 402–403).

In this context, although there are few approaches in the literature on the subject, the concept of prosumer has been applied to the decentralized generation of electricity. Katarin Cseres states renewable energy sources are catalysts for an energy transition, and recent technological developments, such as smart meters, solar panels, and electricity storage, enable a proactive role for electricity consumers (Cseres, 2018, p. 227). Consumer activity becomes essential to promote the decentralization of generation from renewable sources. In this sense the protection of this new role is also necessary to verify an active consumer

performance in mixed transactions or with dual purpose (Cseres, 2018, pp. 237–238).[7]

The word prosumer should be considered in a broader sense because modern energy consumers, in addition to generating electricity, act in network services such as storage (by batteries), regulation, and demand response (Lavrijssen, and Carrillo Parra, 2017). In this sense, it is possible to deduce that, according to Sharon Jacobs' understanding, the prosumer can accumulate other functions besides consuming and producing by involving other activities, with an expansion of the concept (Jacobs, 2016, pp. 519–580). Examples of electricity prosumers were reported by opinion of the European Parliament, including residential, community/cooperative, commercial, and public:

- residential prosumers — citizens who produce electricity on their property, mainly by installing solar PV panels on their rooftops or through micro combined heat and power (micro-CHP);
- community/cooperative energy — citizen-led renewable energy cooperatives ('Res Coops'), housing associations, foundations, charities, which are not commercial actors, but produce energy meant for self-consumption, mainly by solar PV panels and wind turbines;
- commercial prosumers — SMEs, department stores, office buildings, industry and other business entities whose main business activity is not electricity production, but which self-consume the electricity they produce, mainly with rooftop PV panels and CHP, leading to significant cost savings;
- public prosumers — schools, hospitals and other public institutions that self- generate electricity(Šajn and European Union, 2018).

---

[7] This is the understanding of the French judiciary by maintaining consumer protection in contracts with mixed use. "35. Usages mixte: l'usage mixte par um salarié d'um véhicule loué à des fins privées, mais utilisé également dans l'exercice de sa profession, ne suffit pas à faire perdre à l'intéressé sa qualité de consommateur." Grenoble, 13 juin 1991: JCP 1992. II. 21819, note Paisant.—V. aussi. Aix-em-Provence, 26 mai 2005: prec. note 32 (França, 2017).

Based on the concept of prosumer or prosumption, resulting from the agglutination of the terms producer/consumer, the term prosumage emerged to aggregate another function: "prosumage additionally includes energy storage that can be used to increase self-consumption (producers, consumers and storage)" (Schill et al., 2017).

Storage is directly connected to the variation of the price of electricity, that is, if the price of kW h is above the pre-fixed price (feed-in tariff (FIT)[8] is used to encourage the adoption of decentralized generation), it is more advantageous to stop selling to the utility and to increase its own consumption. The increasing storage electricity need for greater autonomy at peak times can be seen through the following finding:

> Accordingly, it was more profitable to substitute grid consumption, which has an average retail price of approximately 0.29 Euro/kWh, with self-consumed electricity (difference "A" between retail price and levelized costs of PV) than feeding this electricity to the grid under the feed-in tariff of around 0.12 Euro/kWh20 (difference "B" between FIT and LCOE, assuming that LCOE are actually lower than the FIT). Accordingly, there is room for prosumagers to increase self-consumption even if this involves additional (storage) costs. It should be noted that volumetric charging of retail price constituents such as the EEG (the German Renewable Energy Sources Act) surcharge, grid fees, and taxes contributes considerably to this setting in which the levelized costs of PV are cheaper

than after-tax retail prices, which can be referred to as "socket parity" (cf. IEA 2014). Back in 2009, the situation was different: retail prices were lower and feed-in tariffs much higher (left panel) such that the difference "A" between the retail price and levelized costs of PV was negative. Accordingly, this setting provided no incentives for self-consumption.[9]

It is interesting to highlight an example of a regulatory framework created to foster a fast energy transition: Law No. 1096 of 2015 of the State of Washington, United States, which provides the promotion of a more efficient and reliable electricity distribution system. According to item 1 of section 1 (corresponding to article 1), the reasons for issuing the legislation arise from the rapid changes in the scenario of generation for own consumption and energy autonomy, and the reduction in the revenue of utilities due to the reduction in the volume of electricity that passes through the network, that is, the tariff criterion, is based on the volume of electricity in the network (volumetric tariff to pay infrastructure costs); with the increase in distributed generation, the purchase of electricity from the utility and also its revenue has been reduced. Section 1.2 states that the State of Washington needs viable distributors and that the rapid growth of electricity generation owned by consumers "who both produce and consume electricity, known as prosumers, is challenging and rendering obsolete the careful balance of

---

[8] "A feed-in tariff establishes a secure contract for wholesale electricity sale at a set price that results in a rate of return attractive to investors and developers" (Ferrey, 2011, p. 228).

[9] "As in most other European countries (cf. European Commission, 2015), grid fees and other parts of retail prices, including surcharges for renewable energy support, are paid volumetrically in Germany, that is on a per kWh basis. Volumetric pricing generally tends to incentivize prosumage (IEA, 2014). Yet a restructuring of grid tariffs is being discussed, which may eventually result in a shift toward more capacity-based pricing schemes. The feed-in tariff (FIT) for small-scale PV installations has been decreasing strongly through 2016, triggered by falling levelized costs of electricity (LCOE)." By 2016, the FIT was much lower than the retail price (right panel of Figure 2). (. . .) (Schill et al., 2017).

values established by our current net metering law" (The State of Washington Bill Text, USA, 2015).[10] There is a definition of prosumer in the legislation with some variables:

(11) "Prosumer" means:

a. A customer-generator as defined in RCW 80.60.010[11];

b. An electric utility customer with a production meter connected to a utility's distribution system that measures production of electricity generated on the customer's premises intended to offset part or all of the customer's electricity requirements; or

c. A utility customer who enters into a special arrangement with a utility to:

i. Obtain premium services, such as enhanced reliability or voltage control, requiring extraordinary capital investment; or

ii. Provide premium services, such as demand response, energy storage, and load management (The State of Washington Bill Text, USA, 2015).

Section 8 of the referred Law provides on the need to regulate this relationship between operator and prosumer in view of the energy transition (The State of Washington Bill Text, USA, 2015).

---

[10]   The term "net metering" has the meaning of compensating the prosumer who sends energy to the grid, in the figurative sense of the "reverse meter." Power utilities can limit this sending to the grid when they do not need it, as occurred in Maryland, United States, which was limited it to 125% of total monthly unit consumption (Ferrey, 2011, p. 232).

[11]   Rev. Code Wash. (ARCW) § 80.60.010. Statutes current through the 2018 Regular Session. Annotated Revised Code of Washington. Title 80 Public Utilities (Chs. 80.01–80.98). Chapter 80.60. Net Metering of Electricity (§§ 80.60.005–80.60.040). 80.60.010. Definitions.

The definitions in this section apply throughout this chapter unless the context clearly indicates otherwise.

(1) "Commission" means the utilities and transportation commission.

*(2) "**Customer-generator" means a user of a net metering system.***

(3) "Electrical company" means a company owned by investors that meets the definition of RCW 80.04.010.

(4) "Electric cooperative" means a cooperative or association organized under chapter 23.86 or 24.06 RCW.

(5) "Electric utility" means any electrical company, public utility district, irrigation district, port district, electric cooperative, or municipal electric utility that is engaged in the business of distributing electricity to retail electric customers in the state.
   (...)

(9) "Net metering" means measuring the difference between the electricity supplied by an electric utility and the electricity generated by a customer-generator over the applicable billing period.

(10) "Net metering system" means a fuel cell, a facility that produces electricity and used and useful thermal energy from a common fuel source, or a facility for the production of electrical energy that generates renewable energy, and that:

   (a) Has an electrical generating capacity of not more than one hundred kilowatts;

   (b) Is located on the customer-generator's premises;

   (c) Operates in parallel with the electric utility's transmission and distribution facilities; and

   (d) Is intended primarily to offset part or all of the customer-generator's requirements for electricity.
   (...)

(14) "Renewable energy" means energy generated by a facility that uses water, wind, solar energy, or biogas from animal waste as a fuel.

Another US regulation addressed the issue. The New York Public Service Commission considered it necessary to review the energy distribution model due to its analysis of the strong trends in technological, market, and environmental policy evolution. A new regulatory framework was approved in order to reform the distribution business model and align tariff practices with regulatory and public policy objectives. In this review, customers were classified into three types:

Traditional consumers—Those customers who do not choose to actively manage their energy usage, or for whom it is difficult to do so[12];

Active consumers—Those customers who undertake DER measures that allow them to actively modulate their usage in response to rate signals with the purpose of reducing their bills;

Prosumers—Those customers who install or participate in DER including generation or other technologies that allow them to provide services to the grid (New York Public Service Commission, USA, 2016).

From the classification above, traditional consumers can be equated with captive consumers of electricity, passive subjects who do not care about managing the use of energy mainly due to the complexity of the information. The concept of active consumer proposes that the subject be adapted to the best condition of supply and quality of supply, keeping only in the demand position. The prosumer, finally, acts consuming and also generating energy or providing another service to the system operator. Although not expressed in the last concept, the prosumer also has the active function of generating electricity.

The North American system (New York) has a clear focus on the economic efficiency of the system, which also applies to how it deals with consumption policies and how it considers the role of the consumer in the market (see Micklitz, 2016). Without going into this merit the concept of prosumer is attached to the sense that the subject can accumulate other functions. Prossumage (consumer, producer, and storage) indicates that the subjects can add numerous functions without losing the protection that the law grants them. The fact of also storing electric energy, which would be a service, or performing a fourth, fifth, or infinite activities, is not enough to prevent the law incidence.

Three challenges have been identified in seeking a definition of prosumer: (1) ensuring that prosumers maintain their status as consumers and their rights as such; (2) ensuring that the definition of prosumers is sufficiently clear, but sufficiently flexible to encompass different activities and actors, especially as the market develops; and (3) understanding the implications of prosumers as companies for the purposes of competition and, in particular, for state aid. According to Josh Roberts, the prosumer should be legally defined as a special class of active consumer in order to continue to benefit from the already established consumer protection (Roberts, 2016). Therefore it hasn't been agreed with this perspective due to the possibility to weaken the legal protection already established by consumer law.

7.3 The prosumer protection in Latin American countries

Once the concept of prosumer is clarified, it seeks to identify and analyze if the legislation about decentralized generation of electricity approaches the concept of the subject and

---

[12] "Consumers who rent their homes, reside in multi-family or mixed-use facilities, and/or do not have individual metering may lack either an economic incentive or practical access to manage their energy usage by investing in Distributed Energy Resource (DER)."

verify if the consumer law could frame it for the contractual protection. To illustrate, recent bibliography (Guimarães, 2019, p. 243) identified some countries that have already foreseen a decentralized generation system, such as Brazil, Argentina, Chile, Uruguay, Guatemala, Colombia, and Costa Rica.

The consumer's power contract holds relational characteristic due to the captivity of the contractor with respect to the object (electricity) and to the need for sufficient and continuous supply at an affordable tariff, which are the foundations of energy security.[13] The contractual relationship emerging from the decentralized generation of electricity has the goal of promoting the social—economic—environmental purpose—sustainable development—more precisely the sustainable consumption (Miragem, 2014) of electricity. This is another argument to provide legal protection to the prosumer once sustainable development is in the track to be considered a legal basis. Michel Prieur analyzes the possibility of implementing a sustainable development law, approaching "(. . .) sustainable development is not transformed into a new specific right, but lies between: maintaining an objective or a purpose outside the law, serving to guide public policies; it becomes a principle of action, whether informal or legally enshrined; access

the level of standard used specifically by the judge to arbitrate disputes and act as a substitute for the principle of proportionality; it appears as a simple process conditioning neomodern governance and institutional decision-making structures" (Prieur, 2015, p. 21).

The environment safeguarding should not be just focused on the maintenance of the standard of living or individual interests. The human being has a responsibility due to his power asymmetry feasible by his capacity of reasoning regarding other living beings (Sen, 2011, p. 285). The standard of living and the satisfaction of needs must be added to the expansion of freedom, including the freedom to satisfy our needs, which would alter the idea of sustainable development to a notion of sustainable freedom "to include the preservation and, where possible, the expansion of the substantive freedoms and capabilities of today's people "without compromising the ability of future generations" to have similar or greater freedom."[14,15]

Generating and consuming energy in a sustainable manner starts from a greater protagonist premise of the concept of sustainable development within the scope of public policies in order to increase environmental conservation and, at the same time, enable the expansion of production and trade to promote

---

[13] To avoid shortages, it must be ensured: (1) the energy supply (imported or domestic) must be sufficient; if it is below consumption, it may pose a risk to the country; (2) the energy supply (imported or domestic) must be continuous, under penalty of causing interruptions or cuts causing social, political, and economic damage; and (3) the energy (imported or domestic) must be reasonably priced, enabling the payment capacity (Conant and Gold, 1981).

[14] Amartya Sen seeks to merge the Robert Solow and Brundtland Comission concepts (Sen, Amartya. A ideia de justiça, p. 286). Robert Solow sees sustainability as the requirement to leave to the next generation "whatever it takes to achieve a standard of living at least as good as ours and to care for the next generation in the same way." SOLOW, Robert. An almost practical step toward sustainability (Washington: Resources for the Future, 1992. Apud. Sen, Amartya. A ideia de justiça, p. 284).

[15] Amartya Sen summarizes his analysis as follows: "To use a medieval distinction, we are not only "patients" whose needs deserve consideration, but also "agents" whose freedom to decide what to value and how to seek it can extend far beyond our own interests and needs. The meaning of our life cannot be placed in the box of our standards of living or the satisfaction of our needs. The prominent needs of the patient, no matter how important they may be, cannot eclipse the vital relevance of the agent's reasoned values." (Sen, Amartya. A ideia de justiça, p. 286).

the optimization of the use of global natural resources[16] in order to meet the expectations of future generations. One of the possible actions is to enable production and consumption more efficiently, with less cost and waste, avoiding the extraction/use of nonrenewable natural resources.

The energy transition from fossil to renewables is the result of sustainable development enforcement. From this environmental perspective the prosumer can be considered a trigger to implement this energy transition. Then, his legal protection is enshrined with environmental protection and also with human rights.

As presented below, three countries were chosen by the criteria of higher population density and larger territory located in Latin America, namely, Argentina, Brazil, and Colombia, using different concepts of consumer of electricity.

7.3.1 Brazil

Article 2 of the Brazilian Consumer Code (CDC) provides that "consumers are all individuals or legal entities that purchase or use products or services as final recipients".[17] From the absence of a definition of the meaning of "final recipient" in article 2, it was in charge of scholars to verify whether the concept would be expanded or restricted, that is, whether a simple withdrawal from the market of the good or service would be sufficient criterion to be covered by the Code. Some lines of interpretation have been drawn up to adjust the application of the law, among them: the finalist,[18] the maximalist,[19] and more recently, since 2002, after the Civil Code is in force, the depth finalist, which will be the focus of this study.

The depth finalist interpretation is among the finalist and maximalist interpretations. It is based on two criteria: (1) the extension of the concept of consumer by equivalence (by standard) is an exceptional measure under the Brazilian Consumer Code regime; (2) it is an essential requirement for this conceptual extension and, through the legal equivalence (art. 29, CDC), the recognition of the vulnerability of the party that intends to be considered a consumer. The vulnerability criterion is used to apply or exclude the application of the law to hypersufficiency cases or companies with significant economic size (MARQUES, Contratos no Código de Defesa do Consumidor, p. 366).

---

[16] On the principle of prudent, rational and sustainable use of natural resources, see Heffron et al. (2018).

[17] On the evolution of the legal definition of consumer, involving objective conceptions (act of consuming) and subjective (considerations about the person and the conditions of consumption): Bourgoignie (1988).

[18] The finalist interpretation of art. 2 of the CDC is based on the notion of factual and economic final recipient of good or service, that is, the consumer is the one who acquires or uses the product or service to fulfill his economic function and withdraws it from the market for his family or his own satisfaction. There is no profit purpose or to serve as serve as a raw material. It is the nonprofessional. It is highlighted that a relationship between two nonprofessionals is a relationship between equals and in this case the Civil Code must be applied (MARQUES, Contratos no Código de Defesa do Consumidor, pp. 302–303).

[19] The maximalist interpretation includes both the consumers of art. 2, CDC (finalist interpretation) and the bystander consumers (art. 2, sole paragraph, art. 17 and art. 29, CDC). It considers the rules of the Consumer Code as a new regulation of the Brazilian consumer market, not only to solely protect the nonprofessional consumer. It is enough that the consumer is a factual final recipient (purchases or uses a product or service), and not necessarily an economic final recipient (withdraws from the consumer market). This interpretation extends the scope of the rule. An example is a farmer who acquires fertilizer for planting; a company hiring a credit or transport service (MARQUES, Contratos no Código de Defesa do Consumidor, pp. 303–304).

The consumer, therefore, may be an individual or legal entity, since the fulfillment of the condition of being in an unfavorable position in the contract is verified, that is, more vulnerable in relation to the supplier.

With respect to decentralized generation of electricity, Normative Resolution (REN) no. 482/2012 (updated by REN 687/2015), of the Brazilian National Agency of Electricity (ANEEL),[20] provides that microgenerators are those with installed capacity less than or equal to 75 kW, and minigenerators are those whose generating plants have 75 kW and less or equal to 5 MW.[21]

Then, the concept of user-generator or consumer-generator to designate the prosumer does not exist expressly in the Brazilian legislation. There is only a citation in the art. 6 of the REN no. 482/2012 providing that "consumers responsible for a consuming unit (...) may adhere to the power compensation system."

Regarding the system, it should be clarified that Brazil chose to compensate[22] (see National Renewable Energy Laboratory, US Department of State, 2018) the electricity generated within the consuming unit. It could be compensated also with another consuming unit since it is proven the same ownership in the same distribution area, called remote self-consumption.[23,24]

Being able to generate and consume renewable energy can be considered as a means to provide well-being[25] and as a tool for an energy transition. In the legal perspective the prosumer must be still protected by the law given that he still is a consumer, but now more active, and due to the need to keep the legal warranties and to avoid risks when starting to self-generate electricity. The activity of the consumer does not affect the consumer concept already established and is essential to provide an effective energy transition, as long as he or she does not act with professional, commercial, or usual purpose.[26]

---

[20] ANEEL. REN no. 482/2012. Available from: <http://www2.aneel.gov.br/cedoc/ren2012482.pdf> (accessed 25.08.18.).

[21] ANEEL. REN no. 482/2012. Art. 2, I e II, REN no. 482/2012, respectively.

[22] ANEEL. REN no. 482/2012. Art. 2, III, REN no. 482/2012.

[23] ANEEL. REN no. 482/2012. Art. 2, VIII, REN no. 482/2012.

[24] The compensation system has the following advantages: (1) promote energy security by reducing dependence on the provision of public electricity services when generating—and consuming—energy in a decentralized manner; (2) enable the use of renewable sources and reduce the emission of greenhouse gases, thus being a positive incentive in the application of the COP 21 agreement, sustainable development goals 7 and 12 and the UN Guidelines for Consumer Protection/2015; and (3) increase access to energy in interconnected systems (grid), thus making it possible to promote the sustainability of development and ensure freedoms. Those advantages can be also be considered on a remuneration system (feed-in tariff).

[25] The Nobel Prize winner in Economics Amartya Sen is an exponent of the study of the criteria that lead to the identification of whether a given society can be considered developed, not based only on income criteria, or on economic graphs, but basing his thesis that development consists in the elimination of deprivations of liberty that limit the choices and opportunities of people to ponderously exercise their condition of agent. Sen emphasizes five types of freedom, considered instrumental, as (1) political freedoms, (2) economic facilities, (3) social opportunities, (4) guarantees of transparency, and (5) protective security. These rights and opportunities help to promote a person's overall capacity and can complement each other. Instrumental freedoms interrelate, connecting with each other, and contribute to the increase of human freedom in general (Sen, 2010, p. 10 e 25).

[26] For a deeper understanding, the extension of the consumer protection to the prosumer could be by the concept interpretation and/or by the contract relationship. See Bassani (2019).

7.3.2 Argentina

Argentine legislation provides the adoption of the decentralized generation modality through Law no. 27.424/2017. Unlike Brazil, article 3 of the Argentine law provides for the denomination of the modality and the subjects, such as the customer/user-generator and the distributor, transcribed below:

> ARTICLE 3 - For the purposes of this law, it is called:
>
> (. . .).
>
> h) Distributed generation: the generation of electricity from renewable sources by **users of the public distribution service who are connected to the network of the service provider** and meet the technical requirements established by the regulation to inject surpluses from self-consumption into said public network;
>
> (. . .);
>
> j) **User-generator**: the user of the public distribution service who has the equipment for generating energy from renewable sources in the terms of subsection h) above and who meets the technical requirements for injecting surpluses from self-consumption into said network under the terms established by this law and its regulation. It does not include large users or self-generators of the wholesale electricity market (InfoLEG Argentina, 2017).

Initially, the notion of decentralized generation set out in the law limits it to renewable sources, which excludes from the scheme the possibility for prosumers to use cogeneration through gas microturbines (Norma Martínez and Margarita Porcelli, 2018, p. 189), which could be a way to implement the energy transition. But the main point is the concept of customer-generator (*Usuario-generador*) referring to the customer who is qualified to connect to the grid for the generation of electricity for self-consumption, expressly excluding large customers or self-generators from the power wholesaler (*autogeneradores del mercado eléctrico mayorista*) provided by Law no. 27.191, which amended Law no. 26.190 (Norma Martínez and Margarita Porcelli, 2018, p. 189).

It is interesting to note that, according to Adriana Norma Martínez and Adriana Margarita Porcelli, the Argentinian Law "puts all consumers in the center of the scene, who would now become **generators or "prosumidores"**, that is to say, home scale energy generation through solar panels, making part of the electricity consumed be produced by the same user and the surplus can be sold to large distributors" (Norma Martínez and Margarita Porcelli, 2018, p. 189).

Then, it appears that the law establishes the right to generate electricity for self-consumption or to send the surplus to the grid. However, it is still lacking specific regulation, making it difficult to identify whether the country will adopt a remuneration (sale) or compensation system. The definition in article 3 of Law no. 27.424/2017 exposes the preference for self-consumption by using the customer-generator or prosumer definition. Nonetheless, it is considered necessary to await further studies and such regulation for further confirmation.

In this context the classification of the customer-generator as a consumer becomes clearer when checking the Argentine consumer legislation. Article 1092 of the Argentine Civil and Commercial Code provides the concept of consumer relationship and consumer, similar to the Brazilian concept as final recipient of the product and service, thus provided:

> ARTICLE 1092. Consumption relationship. Consumer
>
> Consumer relationship is the legal attachment between a supplier and a consumer. A consumer is a human or legal person who acquires or uses, free of charge or onerously, goods or services as the final recipient, for his own benefit or for the benefit of his family or social group. (. . .).
>
> ARTICLE 1093. Consumer contract
>
> Consumer contract is a contract between a consumer or end user and a human or legal person who acts professionally or occasionally or with a

company producing goods or providing services, public or private, *whose object is the acquisition, use or enjoyment of goods or services by consumers or users, for their private, family or social use.*

Rubén Stiglitz, when commenting on art. 1092 and 1093 of the Argentine Civil and Commercial Code, states that

> The regulation of the consumer contract requires a legal definition that depends on the characterization of the consumer relationship. (. . .). A consumer is a person who, without being a party to a consumer relationship, as a consequence of or on the occasion thereof, acquires or uses goods or services free of charge or for a consideration as the final recipient of his family or social group. The consumer contract is characterized by the fact that one of its parts is a final consumer of goods or the user of services (art. 1093 CCyC). It can be, indistinctly, the individual or legal person located when the economic circuit is exhausted and that puts an end, through consumption or use, to the economic life of the good or service. It is indistinct whether the use or utilization of goods and services is made for personal or family use, that is, for private use. The CCyC has included this criterion in art. 1093 CCyC in which it states that "the consumer contract is the one entered into between a *consumer or end user* with a human or legal person who acts professionally or occasionally or with a company producing goods or providing services, public or private, whose object is the acquisition, use or enjoyment of goods or services by *consumers or users,* for their private, family or social use) (Stiglitz, 2015).

Therefore the public service customer/user, even adding functions such as generating electricity, can be classified as a more active consumer or a prosumer, reaching the same conclusion when analyzing the Brazilian perspective, keeping its legal protection.

7.3.3 Colombia

Colombia does not provide a customer-generator or prosumer definition in the decentralized generation of electricity context. But Law no. 1715/2014 provides in art. 5, items 1 and 5, some relevant definitions:

> Article 5. Definitions. For the purposes of interpreting and applying the present law, it is understood:
>
> 1. Self-generation. That activity carried out by natural or juridical persons that mainly produce electricity, to attend to their own needs. In the event that surpluses of electricity are generated from such activity, these may be delivered to the grid, under the terms established by the Energy and Gas Regulation Commission (CREG) for such purpose (. . .).
>
> 18. Distributed Generation (DG). It is the production of electricity, near the centers of consumption, connected to a Local Distribution System (LDS). The capacity of the distributed generation will be defined according to the capacity of the system where it is going to be connected, according to the terms of the connection code and the other provisions that CREG defines for this purpose.

If the self-generation activity is carried out by individuals or companies, it is therefore possible to call them self-generators, in the sense of self-consumption, allowing the sending of surplus to the network connected to the Local Distribution System,[27] according to rules to be established by the Energy and Gas Regulatory Commission.[28] It is interesting to note that the emissions for implementation of a distributed generation project may not exceed those of projects with conventional sources.[29]

---

[27] Sistema de Distribución Local—SDL.

[28] Comisión de Regulación de Energía y Gas—CREG.

[29] Colombia (2019a). Artículo 44. Emisiones y vertidos de las instalaciones de FNCE. Los límites de emisiones o vertimientos establecidos para las instalaciones de FNCE, en ningún caso podrán ser más rigurosos que los límites establecidos en el caso menos exigente aplicado a fuentes de energía convencionales. En particular, el Gobierno Nacional desarrollará una normativa específica que regule las emisiones y los vertimientos de las instalaciones que utilicen recursos renovables de acuerdo a sus características específicas. Available from: <http://www.fedebiocombustibles.com/files/1715.pdf> (accessed 01.04.19.).

According to article 8 of the Law, countless hypotheses for implementation are provided, through compensation (net metering) or remuneration (FIT) systems. It also states other mechanisms to promote the adoption of the decentralized generation system, as presented next:

> Article 8: **Promotion of small and large scale self-generation and distributed generation.** The National Government shall promote small and large-scale self-generation and distributed generation by means of the following mechanisms: **a) Delivery of surpluses.** Autogenerators at small and large scale are authorized to deliver their surpluses to the distribution and/or transportation network. (. . .) b) **Bidirectional metering systems and simplified mechanisms for connection and delivery of surpluses to small-scale autogenerators.** Small-scale autogenerators may use low-cost bidirectional meters to settle their consumption and deliveries to the grid, as well as simple connection and surplus delivery procedures to make it possible for such mechanisms to be implemented, among others, by residential users; c) **Sale of energy by distributed generators.** The energy generated by distributed generators will be remunerated taking into account the benefits it brings to the distribution system where it is connected, among which the avoided losses, the useful life of the distribution assets, the support of reactive energy, etc. (. . .); d) **Sale of energy credits.** Those auto-generators that for the surplus of energy delivered to the distribution network become creditors of the energy credits referred to in literal a) of this article, may negotiate such credits and the rights inherent thereto with natural or legal third parties, according to the rules that the CREG defines for such purpose; e) **Mass disclosure programs** (. . .); f) **Focused dissemination programs** (. . .) (Colombia, 2014).

Despite not being possible to identify any data about the implementation of all the measures above, it is interesting to point out the provisions aiming to establish a broad range of possibilities to the prosumer, identifying a hybrid system (between compensation and remuneration systems). The freedom of choice of the prosumer is valued through those legal options. Still, the need to disseminate information on a mass and microscale is extremely relevant for a faster adoption of the decentralized generation systems.

Colombian Law provides in Law no. 142/1994 the definition of a consumer or customer/user in the public service sector as any natural or legal person who benefits from the provision of a public service, as well as the owner of the property where the service is provided or as a direct recipient of the service.

Article 5, item 5, of Law no. 1480/2011, called Consumer Protection Statute (*Estatuto de Protección al Consumidor*), defines the consumer as "any natural or legal person, as final recipient, who acquires, enjoys or uses a given product, whatever its nature, to satisfy a personal, private, family or commercial and domestic need when it is not intrinsically linked to his economic activity, shall be understood as including the user in the concept of the consumer".[30] From this definition, it can be deduced that consumers are those who withdraw goods or services from the market through their destruction or exhaustion by use, and who are not connected to their professional activity (Colombia, 2019).

But according to the Colombian understanding, consumers or users will not be considered those who "acquire, store, use or consume goods or services for the purpose of integrating them into production and/or transformation processes (in which case they will be considered producers), or marketing (in which case they will be considered suppliers)" (Colombia, 2019).

---

[30] From the original: "toda persona natural o jurídica, que como destinatario final, adquiera, disfrute o utilice un determinado producto, cualquiera que sea su naturaleza para la satisfacción de una necesidad propia, privada, familiar o doméstica y empresarial cuando no esté ligada intrínsecamente a su actividad económica. Se entenderá incluido en el concepto de consumidor el de usuario."

In a first moment, this affirmative would eliminate the protection of the prosumer; however, it is forgotten that the consumer is increasingly active in the consumption relationship, which suggests that if the production or storage is for nonprofessional, usual and commercial purposes, its protection by the Statute could be maintained.

7.4 Conclusion

Given that the emissions of greenhouse gases by fossil fuels is one of the most harmful causes of climate change, there is an urgent need to accelerate energy transition. Therefore there is no other path to human existence but to change the sources of power generation for a sustainable consumption.

The prosumer can be a catalyst of this energy transition. Not a new actor, but an active consumer. A consumer produces electricity for themselves, sending—selling or not—with freedom of choice, the surplus to the grid, all based on renewable sources. The North American examples illustrate the prosumer as a new category; nonetheless, the consumer in the United States has fewer legal protections than in South American countries or Europe. If there is a lack of protection for the consumer, the prosumer would also be unprotected. The challenge is to provide legal protection for this new activity.

All the three South American countries Brazil, Argentina, and Colombia protect the consumer in the usual consumer contract of electricity, and they have legal provisions about decentralized generation of electricity. The figures of customer-generator, user-generator, or even just consumer were used to point out the prosumer, which can be an initial milestone to frame the prosumer in the consumer concept.

The application of the theory of the depth finalism interpretation in the Brazilian case, together with the absence of the professional, commercial, or usual purposes, and the activity criterion of the consumer, enables to extend the consumer protection to the prosumer.

Therefore the prosumer could be the core of a new way to generate and consume power, pushing away the traditional actors based on polluting methods to provide electricity to society. Nonetheless, support is still needed to achieve this transition, which could be by positive incentives but also by legal protection.

References

Arendt, H., 2016. Tradução Roberto Raposo A condição humana, 13th ed. Forense Universitária, Rio de Janeiro.

InfoLEG Argentina, 2017. Régimen de fomento a la generación distribuida de energía renovable integrada a la red eléctrica pública—ley 27424. Available from: <http://servicios.infoleg.gob.ar/infolegInternet/anexos/305000-309999/305179/texact.htm> (accessed 19.03.19.).

Bassani, M.L., 2019. A proteção do prossumidor na geração distribuída de energia elétrica (231 f. Tese (Doutorado em Direito)). Universidade Federal do Rio Grande do Sul, Faculdade de Direito, Programa de Pós-Graduação em Direito, Porto Alegre.

Beaudrillard, J., 2001. A sociedade de consumo, 70. Editora, Lisboa.

Bianchi, M., 1998. The Active Consumer: Novelty and Surprise in Consumer Choice. Routledge, London.

Bourgoignie, T., 1988. Élements pour une théorie du droit de la consommation: au regard des dévelopments du droit belge et du droit de la Communauté économique europeenne. Story Scientia, Bruxelles, pp. 45–61.

Colombia, 2014. LEY 1715 DE 2014 (Mayo 13) Por medio de la cual se regula la integración de las energías renovables no convencionales al Sistema Energético Nacional. Available from: <http://www.fedebiocombustibles.com/files/1715.pdf> (accessed 01.04.19.).

Colombia, 2019. Available from: <http://www.redconsumidor.gov.co/publicaciones/quien_es_un_consumidor_pub> (accessed 01.03.19.).

Conant, M.A., Gold, F.R., 1981. A geopolítica energética. Tradução: Ronaldo Sergio de Biasi. Biblioteca do Exército, Rio de Janeiro, p. 20.

Cseres, K.J., 2018. The active energy consumer in EU Law. Eur. J. Risk Regul. 9 (2), 227–244.

DA SILVA, Solange Teles et al., 2018. Solar energy and the dawn of 'Solar Cities' in Brazil. In: Fontoura C., J A. et al. (Orgs.), Energy Law and Regulation in Brazil. Springer International Publishing, Cham, pp. 183–212.

European Union, 2016. Proposal for a Directive of the European Parliament and of the Council on Common Rules for the Internal Market in Electricity (Recast) COM/2016/0864 final — 2016/0380 (COD). Available from: <http://eur-lex.europa.eu/legal-content/EN/TXT/?uri = CELEX:52016PC0864>.

Ferrey, S., 2011. Sale of electricity. In: Gerrard, M.B. (Ed.), The Law of Clean Energy: Efficiency and Renewables. American Bar Association, Chicago, pp. 217—240.

Dalloz, 2017. Code Civil. Dalloz, França, p. 1313.

Guimarães, L.N., 2019. Challenges to the promotion of distributed energy resources in Latin America: a Brazilian case study. In: Sioshansi, F. (Ed.), Consumer, Prosumer, Prosumager: How Service Innovations Will Disrupt the Utility Business Model. Academic Press, London, pp. 235—258.

Heffron, R.J., Rønne, A., Tomain, J.P., Bradbrook, A., Talus, K., 2018. A treatise for energy law. J. World Energy Law Bus. 11, 34—48, p. 43—44.

Jacobs, S.B., 2016. The energy prosumer. Ecol. Law Q. 43 (1), 519.

Labussière, O., Nadaï, A., 2018. Energy transitions and potentials for democratic change. In: Labussière, O., Nadaï, A. (Eds.), Energy Transitions: A Socio-Technical Inquiry. Springer International Publishing, Cham, pp. 319—334. (Orgs.).

Lavrijssen, S., Carrillo Parra, A., 2017. Radical prosumer innovations in the electricity sector and the impact on prosumer regulation. Sustainability 9 (7), 1—21.

Leal-Arcas, R., Lesniewska, F., Proedrou, F., 2018. Prosumers: new actors in EU energy security. In: Amtenbrink, F., Prévost, D., Wessel, R.A. (Eds.), Netherlands Yearbook of International Law 2017: Shifting Forms and Levels of Cooperation in International Economic Law: Structural Developments in Trade, Investment and Financial Regulation. T.M.C. Asser Press, The Hague, pp. 139—172. (Orgs.).

Marques, C.L., 2016. Contratos no Código de Defesa do Consumidor: o novo regime das relações contratuais, eighth ed. Revista dos Tribunais, São Paulo.

Marques, C.L., Miragem, B., 2014. O novo direito privado e a proteção dos vulneráveis, second ed. Revista dos Tribunais, São Paulo.

Micklitz, H.-W., 2013. Do consumers and businesses need a new architecture of consumer law? a thought provoking impulse. Yearbook of Eur. Law 32 (1), 266—367.

Micklitz, H.-W., 2016. The consumer: marketised, fragmentised, constitutionalised. In: Leczykiewicz, D., Weatherill, S. (Eds.), The Images of the Consumer in EU Law: Legislation, Free Movement and Competition Law. Hart Publishing, Oxford, pp. 21—42.

Miragem, B., 2014. Consumo sustentável e desenvolvimento: por uma agenda comum do direito do consumidor e do direito ambiental. Revista do Ministério Público do Rio Grande do Sul (74), 229—244. jan./abr.

National Renewable Energy Laboratory, US Department of State, 2018. A Policymaker's Guide to Feed-in Tariff Policy Design. Available from: <http://www.nrel.gov/docs/fy10osti/44849.pdf> (accessed 10.02.18.).

Norma Martínez, A., Margarita Porcelli, A., 2018. Análisis del marco legislativo argentino sobre el régimen de fomento a la generación distribuida de energía renovable integrada a la red pública. Lex Social: Revista de Derechos Sociales 8 (2), 179—198.

Petit, V., 2019. The New World of Utilities: A Historical Transition Towards a New Energy System. Springer International Publishing, Cham, p. 193.

Prieur, M., 2015. 2. tir Droit de l'Environnement, droit durable. Bruylant, Bruxelles.

Reisch, L.A., Micklitz, H.-W., 2006. Consumers and deregulation of the electricity market in Germany. J. Consum. Policy 29 (4), 399—415.

Roberts, J., May 2016. Prosumer Rights: Options for a Legal Framework Post-2020. Client-Earth/Greenpeace, pp. 14—17. Available from: <https://www.documents.clientearth.org/wp-content/uploads/library/2016-06-03-prosumer-rights-options-for-an-eu-legal-framework-post-2020-coll-en.pdf> (accessed 25.11.18.).

Šajn, N., European Union, 2018. European Parliamentary Research Service. Electricity 'Prosumers'. p. 2. Available from: <https://www.europarl.europa.eu/RegData/etudes/BRIE/2017/603890/EPRS_BRI(2017)603890_EN.pdf> (accessed 30.11.18.).

Schill, W.-P., Zerrahn, A., Kunz, F., 2017. Prosumage of solar electricity: pros, cons, and the system perspective. Econ. Energy Environ. Policy 6 (1), 7—32. Available from: <https://doi.org/10.5547/2160-5890.6.1.wsch> and <https://www.diw.de/documents/publikationen/73/diw_01.c.552031.de/dp1637.pdf>. (accessed 14. 08.18.). pp. 4.

Schrijver, N., 2007. The Evolution of Sustainable Development in International Law: Inception, Meaning and Status. Collected Courses of the Hague Academy of International Law, 327. Martinus Nijhoff Publishers, The Hague, pp. 217—412.

Sen, A., 2010. Desenvolvimento como liberdade. Companhia das Letras, São Paulo.

Sen, A., 2011. A ideia de justiça. Tradução por Denise Bottmann e Ricardo Doninelli Mendes. 3. reimp. Companhia das Letras, São Paulo.

Stiglitz, R., 2015. Contratos de consumo. In: Caramelo, G., Picasso, S., Herrera, M. (Orgs.), Código civil y comercial de la Nación comentado. Libro Tercero Artículos 724 a 1250, first ed. Tomo III. Ciudad Autónoma de Buenos Aires, Infojus. p. 489. Available from: <http://www.saij.gob.ar/docs-f/codigo-comentado/CCyC_TOMO_3_FINAL_completo_digital.pdf> (accessed 31.03.19.).

The State of Washington Bill Text, USA, 2015. Washington 64th Legislature. Regular Session. House Bill 1096— Relating to Promoting a More Efficient and Reliable Electric Distribution System. Available from: 2015 Bill Text WA H.B. 1096. (LEXIS) (accessed 14.08.18.).

New York Public Service Commission, USA. May 2016. Issued and effective. CASE 14-M-0101. In: Proceeding on Motion of the Commission in Regard to Reforming the Energy Vision. Available from: 2016 N.Y. PUC LEXIS 274, 329 P.U.R. 4th 1 (N.Y.P.S.C. May 19, 2016) (LEXIS) (accessed 14.08.18.).

Volkova, I.O., Salnikova, E.A., Gitelman, L.M., 2018. Active consumers in the Russian electric power industry: barriers and opportunities. In: Syngellakis, S., Brebbia, C. (Eds.), Challenges and Solutions in the Russian Energy Sector, Innovation and Discovery in Russian Science and Engineering. Springer, Cham, pp. 27–35.

Brazilian Superior Court of Justice decisions

AgRg no AREsp 735.249/SC, Rel. Ministro RICARDO VILLAS BÔAS CUEVA, TERCEIRA TURMA, julgado em 15/12/2015, DJe 04/02/2016. Available from: <https://ww2.stj.jus.br/processo/revista/documento/mediado/?componente = ITA&sequencial = 1475957&num_registro = 201501562810&data = 20160204&formato = PDF> (accessed 09.09.18.).

Recurso Ordinário em Mandado de Segurança (RMS) n. 27.512/BA, relatado pela Ministra Nancy Andrighi. DJe 23/09/2009. Available from: <http://www.stj.jus.br/SCON/jurisprudencia/toc.jsp?processo = 27512&&b = ACOR&thesaurus = JURIDICO&p = true > (accessed 28.11.17.).

REsp 1195642/RJ, Rel. Ministra Nancy Andrighi, Terceira Turma, julgado em 13/11/2012, DJe 21/11/2012. REsp 1.196.951-PI, DJe 9/4/2012, e REsp 1.027.165-ES, DJe 14/6/2011. REsp 1.195.642-RJ, Rel. Min. Nancy Andrighi, julgado em 13/11/2012. Available from: <http://www.stj.jus.br/SCON/jurisprudencia/toc.jsp?livre = 201000943916.REG> (accessed 09.09.18.).

STJ. REsp 1599535/RS, Rel. Ministra NANCY ANDRIGHI, TERCEIRA TURMA, julgado em 14/03/2017, DJe 21/03/2017. Available from: <http://www.stj.jus.br/SCON/jurisprudencia/toc.jsp?livre = 201601246153.REG> (accessed 09.09.18.).

Further reading

InfoLEG Argentina, 2014. Codigo Civil y Comercial de la Nación. Ley n. 26.994. Available from: <http://servicios.infoleg.gob.ar/infolegInternet/anexos/235000-239999/235975/texact.htm#20> (accessed 13.01.19.).

Bassani, M.L., Ferreira, L.V., 2017. The energy access in rural or isolated areas in Brazil: a viability review. In: BRANT, Leonardo Nemer Caldeira et al. (Org.), Sustainable Development and Energy Matrix in Latin America: The Universal Clean Energy Accessibility, first ed. Konrad Adenauer Stiftung/Cedin, Belo Horizonte, v. 1, pp. 353–379.

ANEEL, Brazil, 2016. Micro e minigeração distribuída: sistema de compensação de energia elétrica, second ed. ANEEL, Brasília. Available from: <http://www.aneel.gov.br/biblioteca/downloads/livros/caderno-tematico-microeminigeracao.pdf>.

ANEEL, Brazil, 2012. Resolução n. 482/2012 e Resolução 687/2015. ANEEL, Brasil.

Brazil. Federal Constitution of 1988. Available at: http://www.planalto.gov.br/ccivil_03/constituicao/constituicao.htm.

Brazil, 1990. Consumer Protection Code. Federal Law n. 8.078/1990. Available at http://www.planalto.gov.br/ccivil_03/leis/l8078.htm.

Elias, N., 2000. The Civilization Process. Blackwell, Oxford.

Islas-Carmona, J.O., 2008. El prosumidor: El actor comunicativo de la sociedad de la ubicuidad. Palabra Clave 11, 29–39.

Kotler, P., 2010. The prosumer movement: a new challenge for marketers. In: Blättel-Mink, B.; Hellmann, K.-U. (Hrsg.), Prosumer Revisited: Zur Aktualität einer Debatte. VS Verlag für Sozialwissenschaften; GWV Fachverlage GmbH, Wiesbaden.

Ritzer, G., Dean, P., Jurgenson, N., 2012. The coming of age of the prosumer. Am. Behav. Sci. 56 (4), 379–398.

Seran, S., Izvercian, M., 2014. Prosumer engagement in innovation strategies: the prosumer creativity and focus model. Manage. Decis. 52 (10), 1968–1980.

Toffler, A., 2014. A terceira onda: a morte do industrialismo e o nascimento de uma nova civilização, 32th ed. Record, Rio de Janeiro.

Challenges of South American energy transition: energy efficiency and distributed generation

Maria Alejandra Ise[1,2,3], Silvina Cecilia Carrizo[1,2,3,4,5,6] and Marie Forget[7]

[1]National Scientific and Technical Research Council (CONICET), Buenos Aires, Argentina [2]National University of Northwest Buenos Aires (UNNOBA), Buenos Aires, Argentina [3]Centre of Studies on Territory, Energy and Environment (TEAM), Buenos Aires, Argentina [4]National University of La Plata (UNLP), La Plata, Argentina [5]School of Architecture and Urbanism (FAU), La Plata, Argentina [6]Centre of Urban and Territorial Research (CIUT), La Plata, Argentina [7]Laboratory on Environments, Dynamics and Mountain Territories (EDYTEM), Mixed Research Unit (UMR) 5204, University of Savoie Mont Blanc (USMB), Chambéry, France

8.1 Introduction

In South America the diversification of national energy mixes and the strengthening of energy systems have been an issue of public interest for at least two decades. Recurring energy deficits,[1] poverty, and climate change concerns trigger a sustainable transition. In the process of energy transition, renewable resources gain preeminence and begin to take up larger portions of the energy mix. This transition involves not only a shift from fossil to renewable energy but also a change in the way people use and produce energy.

The vast majority of the electricity produced in South America is produced in large generating plants that work mainly on hydropower and fossil fuels. These power plants generate electric energy that is transported, through high-voltage transmission systems, to remote locations. Significant losses occur along the way. Thus two major challenges may be identified: energy efficiency (EE) and distributed generation (DG).

---

[1]   In some areas energy shortages, blackouts and voltage drops are frequent. In other areas, distant from large urban centers, public networks are nonexistent, and people do not have access to modern energy services.

In the first place, it is necessary to make energy systems more efficient, as a means of ensuring universal access to adequate energy services, while at the same time making a sustainable use of available resources. EE is understood as a renewable clean resource, as it involves reducing energy requirements while achieving the desired energy output. It is particularly necessary in a context of energy poverty. Most countries in the region are implementing programs and regulations, which call for the installation of low-consumption equipment and promote efficient practices, mainly in residential and industrial sectors.

Second, DG is gaining more and more recognition. In contrast with large generating stations, DG involves producing and consuming energy on-site and/or providing support to a distribution network, connected to the grid at distribution-level voltages (IEA, 2002). DG, as a form of energy production for consumption in situ, is highly dependent on renewable sources and first and foremost on EE. In fact, EE and renewable energies-DG systems are closely interrelated, and they both tend to valorize territorial resources.

In order to kick-start the transition toward a sustainable, efficient, and DG system, governments take measures to allow electric service users to generate part or all of their energy needs using renewable sources. Different actors—public, private, and civil society—take actions to implement renewable energies projects. In this context, new ways of producing and consuming energy emerge: renewables gain importance, consumers become more rational and environment-conscious, and local resources are valorized.

The use of renewable resources to diversify the energy mix has been practiced in all South American countries, with different levels of success, and through different means (Carrizo et al., 2009, p. 65). However, national trajectories toward EE and DG are relatively recent. Most countries are still in the initial phase; some have already obtained interesting results,

while others have had no practical experiences or legal development yet (Curien, 2000; Carrizo, 2010, 398 p).

This chapter addresses the situation of South American countries in the realm of DG and EE. Public policies and incentive mechanisms are analyzed for each country, as well as the impacts derived from their implementation.

This chapter is organized in three sections, this introduction and two others:

- Section 8.2 describes energy challenges in South America and public policies that promote EE and DG.
- Section 8.3 focuses on the study of the most advanced experiences.

The objective is to identify promotion mechanisms for EE and for DG and to study the regulatory, social, economic, and technological challenges to sustainability in South America. This chapter argues that EE and DG reinforce each other and may contribute to the reduction of the vulnerability of people living in poverty conditions.

8.2 South America in transition: challenges to be faced

South America has abundant and diverse energy resources: oil, gas, coal, uranium, hydro, geothermal, wind, sun, biomass and hydropower, which is the main electricity source. There are countries rich in hydrocarbons, such as Ecuador, Brazil, and Venezuela. Argentina has vast reserves of nonconventional hydrocarbons that place the country as the second-world reserve in shale gas and the fourth in shale oil (US Energy Information Administration, 2013). Renewable resources are also abundant: the north of Argentina and Chile, as well as Bolivia and Peru are particularly well endowed with solar energy, with an annual average solar irradiance of 5.5 kW h/m^2/day (Grossi Gallegos, 2005, pp. 85−92). Brazil is a world leader in

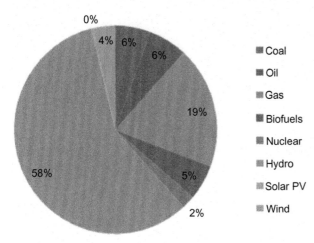

FIGURE 8.1   South American electric mix. Source: *Based on data from IEA, 2018a. Electricity Access Database. World Energy Outlook (IEA, 2018a); IEA, 2018b. Key World Energy Statistics. IEA Publications, France; IEA, 2019. Statistics Global Energy data at your fingertips. Available from: <https://www.iea.org/statistics> (accessed 19.06.19.).*

sugarcane ethanol production, and both Brazil and Uruguay have interesting potential for wind power.

Despite the wealth in natural resources, the South American energy mix shows low levels of diversification. What is more, in South America, there are areas where (1) energy supply is insufficient and frequent shortages occur; (2) there is no network infrastructure, and people do not have access to modern energy services.

Valorizing locally available renewable resources could contribute to the reduction of energy shortages and to the increase in energy security in the countries, thus improving the well-being of vulnerable populations. EE and DG work together in that direction: the former by reducing demand, and the latter by providing or improving energy services to all populations, regardless of their location. Efficiency and distributed renewable energy production tend to reduce poverty and inequities in two main ways: (1) a rational use of energy ensures that a lower amount of energy is consumed and that,

as a result, there is more offer to meet demand and (2) electric power can still be generated where there is no public network or service provider, using, for example, the sun or wind.

8.2.1 Renewable diversification and universal access

In Latin America the successful expansion of electricity access (95% of the population has electricity) favored an increase in energy demand, which was also driven by economic growth, urbanization, and higher living standards. Energy production in South America represents 5.47% of the world total (13,764 Mtoe) (IEA, 2018b). It stands out as the world region with the highest share of renewable energies in the electric mix, thanks to hydropower[2] (Fig. 8.1). Consequently, on a regional scale, levels of CO_2,[3] are acceptable, in spite of high level of energy consumption relative to population and GDP (Acquatella, 2008).

---

[2]   Traditionally and internationally, large-scale hydroelectricity has been considered a renewable source, even though some countries only accept small-scale hydropower as renewable. For example, in Argentina, renewable hydropower is the electricity generated in hydroelectric power stations of a capacity of less than 50 MW.

[3]   Average emissions are of 2.9 t of CO_2 equivalent per capita excluding emissions from land use change and forestry (García Ochoa, 2014).

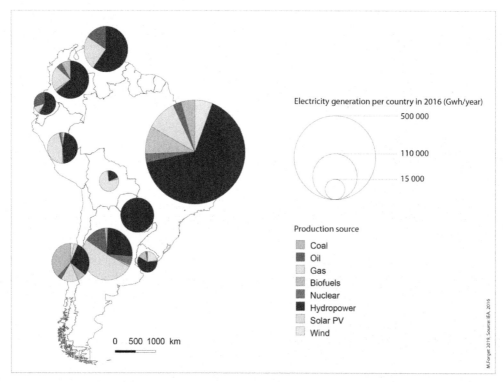

FIGURE 8.2 Electricity generation (GW h) in 2016 by country. Source: *Based on data from IEA, 2018a. Electricity Access Database. World Energy Outlook; IEA, 2018b. Key World Energy Statistics. IEA Publications, France; IEA, 2019. Statistics Global Energy data at your fingertips. Available from: <https://www.iea.org/statistics> (accessed 19.06.19.).*

However, even in countries with abundant hydropower,[4] such as Argentina, Brazil, and Paraguay, the primary energy mixes are still dependent on hydrocarbons (Fig. 8.2). There are countries that export oil and gas, countries able to cover national needs with their own production, and some countries that rely on oil and gas imports. Hydrocarbons are used for power generation in Venezuela and Ecuador (OPEC member countries). Natural gas is dominant in power generation mainly in Venezuela, Peru, and Argentina, where there are large gas deposits, and has increased in Brazil. Recent years have seen growth in nonconventional renewables.

Energy deficits and socio-technological and environmental challenges encourage South American countries to carry forward an energy transition (Carrizo and Velut, 2018, pp. 167–188). Public programs and incentives to renewable energies projects are given through different mechanisms (calls for

---

[4] The Itaipu Dam bordering Brazil and Paraguay is the world's second largest hydropower plant in installed capacity, after the Three Gorges Dam in China. The capacity is owned in equal shares by the two countries. Paraguay covers over three-quarters of its electricity from its generating units and exports the rest to Brazil. Brazilian hydroelectricity imports from Paraguay have remained stable [relative to tonnes of oil equivalent (TPES)] and represent one of the largest electricity exchanges in the world.

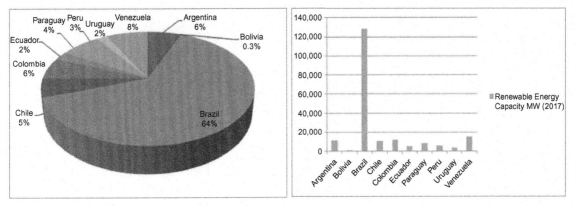

FIGURE 8.3 (A) Percentage of renewable installed capacity in South American countries. (B) Total renewable energy installed capacity in South America by country in MW (2017)[5]. Source: *Based on data from IRENA, 2018. Renewable Capacity Statistics 2018. International Renewable Energy Agency (IRENA), Abu Dhabi (IRENA, 2018).*

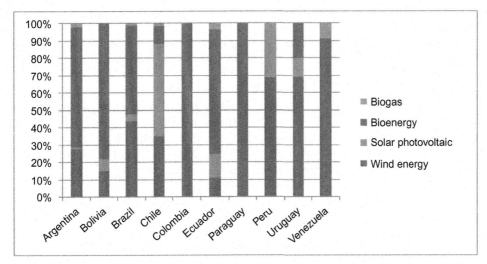

FIGURE 8.4 South American countries' installed capacity per renewable type.[5] Source: *Based on data from IRENA, 2018. Renewable Capacity Statistics 2018. International Renewable Energy Agency (IRENA), Abu Dhabi.*

tenders, quotas, special fees, and tax exemptions). Between 2005 and 2009, Brazil concentrated more than 70% of the investment in renewables. In 2015 investment in Brazil represented a bit over 40% of the total investment in South America. That was equal to 7 billion USD (Bloomberg New Energy Finance, 2016). Since 2010, the investment gap between countries of the region has been decreasing. As a result, renewable installed

---

[5] This figure does not account for hydroelectricity as international statistics do not distinguish between small and large hydroelectric power plants.

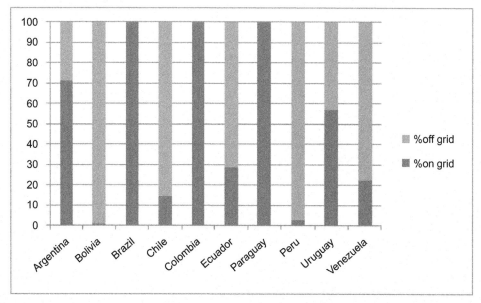

FIGURE 8.5   Percentage of total renewable installed capacity on-grid and off-grid in South American countries. *Source: Based on data from IRENA, 2018. Renewable Capacity Statistics 2018. International Renewable Energy Agency (IRENA), Abu Dhabi.*

capacity has augmented in the region (Figs. 8.3 and 8.4).

In line with the Sustainable Development Goals,[6] South American countries move forward in the energy transition, at the same time that they strive to defeat energy poverty, developing renewable energies both on- and off-grid (Fig. 8.5).

Energy poverty affects vulnerable populations of different countries and territories in different ways (Chevalier, 2009). Energy poverty—understood as the lack of adequate energy services—includes lack of energy access (which entails lack of basic services, such as lighting, cooking, or transport) and/or poor quality service, for example, in cases of unsafe illegal electricity connections. In Latin America, 95% of the population has access to electricity.[7] Approximately a third of those connections are illegal. In these contexts, safety risks and low density of charge tend to be the norm.

In countries such as Bolivia, Peru, and Paraguay, even if they have energy surpluses, per capita energy and electricity consumption levels are lower than in countries where there are deficits in production. On a global scale the relation between energy intensity, defined as the quotient between energy consumed per unit of GDP, and per capita electricity consumption for each country shows similar inequities (Fig. 8.6).

[6]   The Sustainable Development Goals, set by the United Nations General Assembly in 2015, are 17 global goals intended to tackle all forms of poverty, inequalities, and climate change. Among other goals related to sustainability, goal 7 calls for affordable and clean energy.

[7]   In the world, nearly 990 million people do not have access to electricity, 20 million of which are located in Central and South America (IEA, 2018a).

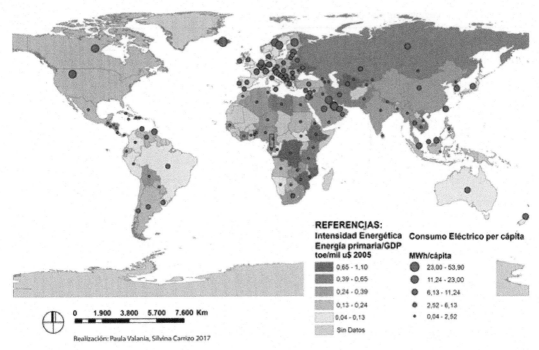

FIGURE 8.6 Energy intensity and per capita electricity consumption per country. Source: *From Valania, P., Carrizo, S., Geografía energética en transición. In:* Llegar con Energía, *Travelling Exhibition (Valania and Carrizo, 2017), based on data from IEA, 2014.*

As shown in Fig. 8.6, in general, countries of Africa and India have minimal per capita electricity consumption levels, below 1.5 MW h/year. Other countries, as the nordic ones, have higher per capita consumption levels and lower energy intensity, that is, to say they are those which use less energy to generate their GDP. In South America, Argentina, for example, uses 0.17 toe/thousand USD of GDP and the per capita consumption is of 3.05 MW h

EE and the use of renewable energies are tools that contribute to energy resilience and to the universalization of services. All countries of the region have implemented ad hoc programs to universalize energy access. For example, in Venezuela, since 2005, the Program Sembrando Luz (Sowing Light), implemented by the Foundation for Electric Development, has installed more than 3100 renewable energy and hybrid systems in remote and indigenous community areas (Table 8.1). In Peru the Master Plan for Rural Electrification was responsible for providing electric service to populations far from networks. In these last two countries the public programs mentioned account for the majority of the renewable installed capacity, which is not connected to the grid.

8.2.2 Energy efficiency and distributed generation policies

Higher consumption does not necessarily equal higher quality of life or productive modernization, or social and economic development. But expanding modern energy services does contribute, directly and indirectly, to eradicate poverty. It also serves to improve networks and to foster regional cooperation

TABLE 8.1   National programs for the expansion of energy services.

Country	Program	Objective
Argentina	PERMER	Renewable energy for rural population
Bolivia	Universalization Plan Bolivia with Energy 2010	2015: Universal urban access
		2025: Universal rural access
	Smokeless cookstove for a better lifestyle	100,000 households
	Bolivian plan of rural electrification Electricity to live with dignity	
Brazil	Light for all	Electricity for families in rural areas
Colombia	Energy plan for nonconnected areas	Energy for 550,000 habitants
Chile	Rural Electrification Program	Electricity for 140,00 households
Ecuador	Rural energization with renewable energies program	Electricity for 44,000 households
Peru	Master Plan for Rural Electrification	Electricity for 34,000 towns—362,000 households
	Sowing smokeless cookstoves	
		47,000 (in 2009)
	Massive solar energy project	Photovoltaic systems for 150,000 rural households
Uruguay	Rural electrification program	Photovoltaic systems for rural
		Population
Venezuela	• Sowing light • Electricity tables	Photovoltaic systems for population 10 km far from networks

From:Velut, S., Carrizo, S. 2013. Les défis de l'integrationénergétiqueenAmerique latine. In: Regional Conference of the International Geographical Union. International Geographical Union, Kyoto (Velut and Carrizo, 2013).

and integration in the energy sector. South American countries increase their energy consumption levels due to the fact that more people get access to energy and have increasing energy demands. Nevertheless, there are many opportunities to introduce EE measures, given their relatively high energy intensity. National states are trying to promote changes by introducing specific legislation (Table 8.2).

Respecting DG, Bolivia, Paraguay, and Venezuela, unlike other South American countries, have not yet developed legal frameworks or practical experiences.

• In Bolivia, 98% of urban population and 75% of rural population have access to electricity as of 2015.[8] The PLABER—Plan Bolivia de Electrificación Rural, launched in 2002, and the Program *Electricidadparavivir con dignidad*[9] (Supreme Decree 29635/2008) have contributed to the extension of electricity networks and the densification of photovoltaic systems. The World Bank

---

[8] Bolivia has a population of approximately 10 million inhabitants, 4.2 million of which (46%) live in rural areas (ENDE, 2016). In 2012, 67.9% of rural households were using biomass (firewood or dung) as fuel for cooking (INE, 2012).

[9] "Bolivian plan of rural electrification" and "Electricity to live with dignity."

TABLE 8.2 National legislation promoting energy efficiency.

Country	Law/Decree	Objective
Argentina	Decree 140/2007	Declares of national interest and priority the rational and efficient use of energy and approves the guidelines of the PRONUREE
Brazil	Law 9991/2000	Establishes the obligation of electricity companies to invest in research and development and in energy efficiency and sets minimum percentages of their gains to be destined to energy efficiency programs
	Law 10295/2001	Establishes maximum energy consumption levels, or minimum energy efficiency levels, of machines and appliances that are manufactured and sold
Colombia	Law 697/2001	Declares the rational and efficient use of energy a matter of social and public interest and of national convenience
Ecuador	Decree 1681/2009	Instructs the State to promote energy efficiency and clean and healthy practices and technologies. Establishes the creation of Energy Efficiency Committees in every entity of the public administration
Peru	Law 27345/2000	Declares of national interest the promotion of energy efficiency
Uruguay	Law 18597/ 2009	Establishes the institutional, legal, and financial framework for energy efficiency. It established the formulation of the National Plan for Energy Efficiency and the definition of the Avoided Energy Target
Venezuela	Law 39823/2011	It has the goal of promoting and guiding the rational and efficient use of energy in the processes of production, generation, transformation, transport, distribution and commercialization, as well as in the final use of energy

PRONUREE, National Program of Rational and Efficient Use of Energy.

financed the project called "Decentralized Infrastructure for Rural Transformation." The implementation of State-managed alternative energy generation projects has been promoted. A mechanism for the remuneration of electricity generated from renewable sources and injected into the National Interconnected System has been established (Supreme Decree 2048/2014).

- Paraguay listed as priority the use of renewable resources and the improvement of EE (Decree 6092/2016), but the country has no law encouraging renewable energy projects. In 2017 a decree vetoed a draft bill "for the promotion of nonconventional renewable sources and its special regimes," which proposed to authorize the injection of energy surpluses into the grid. In 2018 a new draft bill to promote renewable energies started being debated. Although Paraguay is one of the main world producers and exporters of hydroelectricity, its energy system is frequently unable to meet demand due to lack of investments in transmission infrastructure.

- Venezuela[10] issued in 2007 the National Registry of Renewable Energy (Resolution 77) and the Homeland Plan 2013—19 which

[10] In Venezuela the State, through CORPOELEC, is the only participant of the National Electric System, maintaining a vertically integrated monopoly (Organic Law of the Electric System and Service, from 2010). In 2007 the Renewable Energy Committee and Wind Energy Subcommittee were created.

promoted the change in the power generation matrix through expansion of mini hydroelectric power plants, with an emphasis on isolated indigenous communities and border communities. There is no evidence of significant progress toward the development of a DG system.

In countries where there is legislation on DG, most policies implemented to promote DG within the electricity market have allowed it only from renewable resources (Hernandez et al., 2015). These policies include regulations on how generated and consumed energy is measured and billed, as well as regulations on technical safety of installations. Three schemes are globally used to regulate DG:

- Feed in tariff scheme (FIT): Relies on a purchase obligation and a stable tariff payment per unit of produced electricity, which is guaranteed over a long period— usually 15–20 years (Yamamoto, 2012). The main advantage of a FIT scheme is its long-term predictability, which encourages investment (Poullikas, 2013).
- Net metering scheme: Energy consumed from the grid or injected into it is valued at the same price. Either a single bidirectional meter or two unidirectional ones record the prosumer's electricity consumption. The meter runs backwards when the generated electricity exceeds the demand and excess energy is fed into the grid. If, at the end of a billing period, the amount of energy imported exceeds the amount of energy exported, the prosumer must pay the electric company for the net amount of energy consumed. If the exported energy exceeds the energy imported, then the prosumer receives no compensation[11] (Hugues and Bell, 2006).

- Net billing scheme: The energy consumed from the grid and the energy generated and exported to the grid are valued at two different prices. Exchanges between the prosumer and the grid are measured and informed in monetary terms. The prosumer must pay for all the energy consumed and the electric company must buy all the energy generated in a billing period (Mir-Artigues, 2013; Hugues and Bell, 2006).

Through these three different mechanisms, most South American countries have implemented policies that enable individual users to install renewable energy generation equipment and interact with the grid (Table 8.3). Several experiences are in progress.

8.3 National trajectories and experiences in progress

In South America, some countries have taken further steps toward an efficient, distributed, and sustainable energy system. Among them, Brazil, Chile, and Uruguay can be considered pioneers as they were the first to implement programs of DG and EE with results that are worthy of attention. In Argentina, Perú, Colombia, and Ecuador, some experiences have started to emerge.

8.3.1 Pioneering experiences

Microgeneration, shared generation and citizen generation, are local expressions in Uruguay, Brazil, and Chile that represent different national trajectories in DG. Regarding EE, each of these countries has taken a different approach to promote it.

---

[11] This is a simple net metering scheme. Variations include net metering with buy-back, in which the prosumer is paid for the excess energy exported to the grid; net metering with rolling credit, in which excess electricity generated is used as a credit in the next billing period; or a combination of both buy-back and rolling credit.

TABLE 8.3 Countries and adopted distributed generation schemes.

Country	Year	Installed capacity	Maximum capacity allowed per installation
FIT			
Uruguay	2010	16.78 MW	Same amount of energy that is consumed from the grid on an annual basis
Net metering			
Brazil	2012	944 MW	75 kW (microgeneration)
			75 kW−5 MW (minigeneration)
Ecuador	2018	−	300 kW residential/1000 kW commercial and industrial
Net billing			
Chile	2014	24 MW	300 kW
Argentina	2017	−	Equal to the contracted capacity for energy demand, ranging between 3 kW and 2 MW
Colombia	2018	−	100 kW for distributed generators, 1 MW for small-scale self-generators, and 5 MW for large-scale self-generators

8.3.1.1 *Uruguay leads the way through "microgeneration"*

Uruguay is the first country in South America to implement on-grid DG. The country adopted a feed in tariff scheme in 2010. It also established an ambitious EE plan in 2015.

Efforts to make nonconventional renewables more attractive have been made in Uruguay since 2008. Renewable energies in Uruguay were expected to contribute with 50% of primary energy, and nonconventional energies were expected to provide 15% of electricity by 2015 [Ministry of Industry, Energy and Mining (MIEM)]. In 2016 hydroelectric power accounted for 59.2% of the electricity mix, while nonconventional renewable sources were responsible for 37.4% (IEA, 2019). Investment in renewables started to increase, mostly in wind energy, incentivized by the fact that UTE—Administración Nacional de Usinas y Trasmisiones Eléctricas[12]—guarantees 20-year power purchase agreements at a stable price.

In 2009 Law 18597 declared of national interest the efficient use of energy.[13] The concept, as defined by this Law, also includes the substitution of conventional energy sources for the use of nonconventional renewable resources. In 2015 the National Plan for EE established the goal of 1690 ktep of "avoided energy" in the period 2015−24. The decrease in energy consumption, according to the Plan, would be deeper in residential and transport sectors. The Plan includes actions such as (1) information campaigns and educational programs; (2) National Energy Efficiency Labeling System; (3) inclusion of energy managers in companies and public buildings; (4) evaluation

---

[12] National Administration of Power Stations and Electric Transmissions.

[13] This law defined the efficient use of energy as "all changes that result in an economically convenient decrease in the amount of energy needed to produce one unit of product, or to meet the energy requirements of the services that people use, ensuring the same or higher quality level."

of energy performance of buildings; and (5) financial instruments. A specific credit line for EE grants nonrefundable funds that cover two-thirds of the cost[14] of diagnostic studies or energy audits of a building, process, or system. An ESCO registered at the National Energy Department should be hired when applying for funding.

Regarding DG, since 2010, any user connected to the low-voltage distribution grid is able to install renewable energy generation equipment (decree 173). The prosumer should possess a bidirectional meter, installed by UTE, which records energy flows in both directions, separately. UTE is obliged to purchase all the energy delivered by the prosumer for a period of 10 years. Prosumers must pay for generation equipment, but they have tax benefits,[15] and they do not have to pay charges for their use of electricity networks. In addition, a special tariff applies to prosumers of the lowest scales: the energy they generate and feed into the grid is purchased by UTE at the price of the rate of the immediate upper scale.[16] The power of the prosumer cannot exceed the power contracted as a subscriber, that is, 100 or 150 kW, depending on the level of supply voltage, which is of 230 and 400 V, respectively (UTE, 2019a). Since 2017, the maximum amount of energy that a prosumers is allowed to feed into the grid is the same amount of energy that is consumed from the grid on an annual basis (Resolution 42/2017).

Since 2010, 537 users have become prosumers: private and public companies, governmental bodies, businesses, non-for-profit organizations, and residential homes—in minority. The majority installed photovoltaic systems; only six opted for wind energy generators. They account for 16.78 MW of installed DG capacity. A total of 41% of them are located in Montevideo and its surrounding metropolitan area (UTE, 2019b).

To achieve wide-spread DG development in Uruguay, some issues will need to be addressed regarding market development and economic incentives. Even though decree 173/2010 was originally conceived for residential users, they are the ones who opted for DG the least. Instead, businesses and industries are the predominant prosumers. Prices of the technology have seen significant reduction, but renewable generation equipment remain expensive for residential users who, in general, do not see direct benefit in becoming prosumers. Financial instruments aimed at making the initial investment affordable for home users are essential for DG to be a viable option for this sector.

8.3.1.2 Brazil, unprecedented regional distributed generation deployment

Brazil is the main energy consumer in the region, and it has been interested in diversifying its energy mix since the 1970s. Brazil produces hydrocarbons, with the activity being dominated by the State-controlled company PETROBRAS, which is also promoting renewable energies and is one of the biggest bioplayers with sugarcane ethanol production.[17]

In Brazil, several EE initiatives have been undertaken. In 2000 Brazil issued Law 9991 to

---

[14] The National Energy Department reimburses a maximum amount of USD3300, without taxes.

[15] In 2009 decree No. 354 exonerated renewable energy generation activities from paying income taxes for 90%, 60%, and 40% of the income until the years 2017, 2020, and 2023, respectively.

[16] For example, microgenerators who are users subscribed to the Basic Residential Consumption tariff will be paid for their energy at the price of the Simple Residential Tariff, while users included in the Simple Residential Tariff, for the first scale (0−100 kW h), will be paid the price of the second scale (101−600 kW h).

[17] For an analysis of Brazilian biofuel regulation refer to Chapter 13, Competitive renewables as the key to energy transition—RenovaBio: the Brazilian biofuel regulation.

regulate EE. This Law introduced the obligation of distribution utilities to invest 0.75% of their net operating revenue in research and development of the electricity sector and 0.25% in EE programs. In 2011 Law 10295 introduced the National Policy for the Conservation and Rational Use of Energy and established minimum efficiency levels of machinerie and equipment manufactured or sold in the country. To implement this policy a Committee in charge of indicators and levels of EE was created, integrated by authorities of the Ministry of Mines and Energy, ANEEL,[18] public university, and citizens. The National Energy Plan 2030 is the first official document that contains long-term goals for the Energy sector. It includes EE in its studies and constitutes the basis for the National Plan of EE. The National Energy Plan 2050 highlights the importance of long-term EE actions and includes DG, considering it a way of meeting electric demand. In this way, Brazil appears as one of the few cases where public policies took note of the relationship between DG and EE.

In Brazil, DG was first authorized in 2004 only for power generation by undertakings of an installed capacity up to 30 MW, connected to the distribution utility's network (Law 10848/2004). The consideration of prosumers within the concept of DG did not occur until 2012, when the Program for the Development of Distributed Generation was launched (Guimarães, 2019, p. 245). In this context, in 2012 ANEEL authorized micro-and minigeneration (Resolution ANEEL 482/2012). The scheme established is net metering with rolling credit. In this scheme the surplus energy delivered to the network is compensated to the user via a credit, in kW h, to be used in bill reduction within the following months. The costs associated with infrastructure and the meter are assigned to the prosumer,[19] while the distribution utility is responsible for operation and maintenance. Initially, minigenerators had to have a capacity between 100 kW and 1 MW and microgenerators a capacity up to 100 kW. In 2015 ANEEL's resolution was revised (Resolution ANEEL 687/2015), and some changes were introduced: (1) the capacity allowed for minigenerators was raised to the range between 75 kW and 5 MW, and for microgenerators it was lowered to a maximum of 75 kW[20]; (2) the validity of the energy credits was extended from 36 to 60 months; (3) it was allowed to use the credits to offset the bill of other households of the same owner (remote self-consumption); and (4) the figure of "shared generation" was created, allowing a consortium of people or a cooperative society to install a distributed micro- or minigeneration system.[21]

As of 2019, 78,284 DG systems have already been installed, which added 944 MW to the National System. The vast majority of distributed systems are photovoltaic (99.6%), which are mainly installed by commercial and residential users. Thermoelectric plants, which in large part generate from biogas and are located in rural areas, account for 0.20% of the installations. Hydroelectric plants—in general installed by commercial and industrial users—and wind power systems represent 0.1% and 0.07% of the installations, respectively (ANEEL, 2019).

Brazil still has great untapped potential for DG. Two major issues currently appear as

---

[18] National Agency of Electric Energy.

[19] The user generator must also pay the distribution company the concept of availability cost.

[20] ANEEL established simplified procedures for generators up to 75 kW, which include standardized application forms and the reduction of connection time to 34 days. In addition, as of 2017, users are able to apply and follow the state of their request online.

[21] The generated energy is discounted from partners bills.

relevant to future DG deployment: on the prosumer's side, the lack of an hourly tariff that compensates generation accordingly and lack of proper acknowledgment of the benefits of DG (in situ generation and avoided environmental costs); on the distribution utility's side, the issue of remuneration for the use of the network. On this point, a new regulatory revision is currently being discussed inside ANEEL. Initiated through Public Consultation 10/2018, it focuses on economic regulations to avoid cross subsidies from users who do not install distribution generation equipment, (who are therefore completely dependent on the network) to prosumers. A new regulation is expected to see the light of day in 2020.

8.3.1.3 Chile, a plethora of policies in energy efficiency and citizen generation

In Chile, where hydrocarbons and possibilities of developing nuclear or hydroelectric energy are scarce, efficiency measures and the use of alternative energies are largely promoted.

Several programs intend to improve energy services. For example, families are able to apply to have photovoltaic modules installed in their homes. With a minimum of 30 applying families, the Ministry of Energy calls for bids to determine which company provides the equipment, the installation and 5 years maintenance of home photovoltaic systems. Another example is the massive incorporation of solar collectors for water heating (Law 20365/2009). In 2017, more than 100.000 households had solar collectors for water heating (Ministerio de Energía, 2018). Additionally a mechanism was established to facilitate access

to a subsidy for thermal conditioning of households built before 2007. As a complementary measure, energy evaluation and rating of households are done by assessing the global energy performance. Every new house must comply with construction standards, such as roof insulation. As of 2018, 29,100 houses had been evaluated.

Since 2014, in Chile, self-consumption and the sale of surplus electric power have been allowed to regulated users[22] who install nonconventional renewable energy generation equipment (Net Billing Law 20571/2012 and Law 21118/2018). Compensation is made through a net billing mechanism. Self-consumption projects and community systems are both allowed. The so called "citizen generation" started out in 2015 with a maximum limit of 100 kW of installed capacity. As of 2018 the limit increased to 300 kW.

To accomplish wide-spread DG the National State set the goal of building a vast staff of qualified installers of DG systems. To that end, the Ministry of Energy has offered free of charge seminars on the legal framework of DG, connection procedures, and good practices of photovoltaic system installation. There is an official registry of installers, providers, and consultants that works as a search engine for interested citizens.[23]

Chile has great potential for deeper development of DG. The increase of the maximum allowed capacity (from 100 to 300 kW) was made in the hopes of enticing commercial and industrial users with the prospects of even greater reductions in their electric bills. As a result, DG now increases at a monthly rate of 1.1 MW

---

[22] The Law of Electric Service in Chile (Decree-Law 4/1982) distinguishes between two types of final users: (1) regulated users: those with a connected capacity of 500 kW or less; (2) free users: those with a connected capacity of more than 5000 kW. Users of connected capacity between 500 and 5000 kW are able to choose whether to be free or regulated. According to the law of rate fairness (Law 20928/2016), no user pays for energy a difference bigger than 10% between different communes.

[23] The State contemplates as well: (1) efficient cogeneration systems, which are allowed to inject their excess energy into the grid under a net billing scheme; and (2) small microgenerators with excess energy up to a limit of 9 MW.

(Valencia, 2019). This shows that DG has had broad acceptance. More than 4000 DG systems have been declared between January 2015 and October 2018, which equals a total installed capacity of 24 MW (Superintendencia de Electricidad y Combustibles, 2019). Users are concentrated in Santiago, Valparaíso, Concepción, and in reconstruction areas. In a not so distant future, adjustments in network infrastructure and integration of smart technologies will be necessary to accompany this development.

8.3.2 Emerging initiatives

As previously addressed, some countries in South America have already taken steps in the transition toward sustainable distributed and efficient energy systems, which are worth looking into. Other countries are starting to include DG and EE in their public agendas and taking initial actions. This last group of countries could derive substantial benefit from considering successful experiences to replicate or mistakes to avoid. Their trajectories are, in any case, interesting to follow.

8.3.2.1 Argentina, birth of a national law

In Argentina, there was no nationwide law regarding DG until the end of 2017. However, this was no obstacle to DG development as some provinces issued their own regulations, exercising their constitutional right to rule on power distribution. Three provinces serve as examples:

Mendoza enabled DG in 2006 (Law 7549) and in 2015, adopted a system where only one bill is issued containing information about the amount of energy consumed from the grid and injected into it by the prosumer. Energy credits are awarded to the prosumer, which can be used to offset payment in subsequent bills. The prosumer must pay the electric company the equivalent of a consumption of 45 kW h (Resolution EPRE[24] 19/2015). The province created a *mercado a término*, to allow prosumers to buy and sell energy to each other, and established incentives and benefits to encourage DG and the development of smart grids.

Salta established a net metering system (Law 7824/2014). The first 2 years after the connection the prosumer sells all the energy generated and pays for all the energy consumed. After 2 years, generated energy is used to offset the energy consumed. Residential users are allowed to install a maximum of 30 kW; industrial and commercial users connected to the low-voltage network, a maximum of 150 kW, and of 300 kW, if connected to the medium voltage network (Resolution ENRESP[25] 1315/2014).

The oldest and most successful experience is "Prosumidores," a program conceived by Santa Fe province (Resolution EPE[26] 442/2013). The program began operating in 2016 and was relaunched in 2018. The scheme followed is a feed in tariff. Prosumers, who can install equipment for a total capacity that equals or is lower than 80% of their average annual consumption, are paid an incentive price from a scale from ARS\$0.56 to ARS \$7.32/kW h.[27]

On a national scale, in 2017 the State set the goal of reaching 1000 MW of distributed renewable generation capacity in 2030 (Law 27424/2017 and Decree 986/2018). Net billing with rolling credit is the scheme

---

[24] Provincial electric regulator.

[25] Regulatory Entity of Public Services.

[26] Provincial Energy Utility.

[27] The maximum capacity allowed and the tariff paid depend on the kind of prosumer: (1) homes; (2) businesses; (3) clubs and social organizations; (4) residential conglomerates; (5) rural prosumers; and (6) maximal prosumers. More information available at: <https://www.santafe.gob.ar/ms/prosumidores/> (accessed 08.05.19.).

adopted. In a National Registry (RENUGER[28]), renewable energy prosumers are categorized into three types: small user-generators with an installed capacity up to 3 kW, medium user-generators—installed capacity between 3 and 300 kW—and big user-generator—installed capacity between 300 kW and 2 MW (Resolution S.E. 314/2018).

So far, nine provinces, of a total of 23, have adhered to the national framework: Catamarca, Chaco, Chubut, Córdoba, La Rioja, Mendoza, San Juan, Tierra del Fuego, and Tucumán. This shows that the country is still in the early stages and that plenty of issues remain to be discussed. The majority of the provinces have not adhered to the national framework, not only because they see energy distribution as their legal turf, but also because they find conflict with their own regulations. Another challenge for DG in Argentina is the high costs of the equipment, especially for residential users. For small and medium companies, DG seems relatively more attractive as they see a triple benefit: economic savings, lower energy demand, and positive corporate image as a company that cares for the environment and social welfare.

Regarding EE, in 2007 a national law declared of national interest and priority the rational and efficient use of energy and approved the guidelines of the National Program of Rational and Efficient Use of Energy. Efforts are being made to promote energy audits and efficiency labeling of households. However, practical experiences are few and isolated, and no connection with DG promotion can be found.

8.3.2.2 Peru: distribution for universal services

Of major concern in Peru is the universalization of energy access. In 2015, 6% of the population—mostly located in rural areas—did not have access to electricity (World Bank, 2019). In 2008 a program for rural electrification—Plan Maestro de Electrificación Rural con Energía Renovable en la República del Peru[29]—was launched. It envisaged providing electric energy to 360,000 households by means of renewable energies. In 2013 an auction was held for the installation and maintenance of 150,000 solar home systems in rural areas.[30] In 2016 the Ministry of Energy and Mines approved the National Plan for Rural Electrification 2016–25. The General Department of Rural Electrification carries out other projects aimed at rural populations.

The efficient use of energy, diversification of sources, and the geographical expansion of generation are promoted. For example, public entities must perform energy audits and equipment must be labeled according to their EE. Nonconventional renewable energy has increased significantly, reaching 3.7%, of electricity generated in 2016: wind energy accounted for 2%, biofuels for 1.2%, and solar 0.5%.[31]

DG has been on the public agenda since 2006. That year a law established to measures that favored DG: (1) the sale of excess energy in the spot market, which would be assigned to "bigger transmission generators"; and (2) the use of distribution networks to inject energy (Law 28832/2006). Since 2015, there is a legal instrument that authorizes electric generation for

---

[28] National Registry of User-Generators of Renewable Energies.

[29] Master Plan of Rural Electrification with Renewable Energy in Peru.

[30] For an analysis of Peru's Massive Solar Energy Program refer to Chapter 6, The Challenges and Contradictions of Peru's *Proyecto Masivo de Energía Solar*.

[31] The vast majority of the electricity in Peru in 2016 was produced in hydropower stations (46.5%) and in thermal stations (45.8%). Source: IEA <https://www.iea.org/statistics>.

self-consumption and the connection of renewable energy generation equipment to the public grid (Decree 1221/2015). Nevertheless, it has not been regulated, so DG is not yet in practice.

8.3.2.3 Colombia: a means for efficiency

In Colombia, Law 697/2001 declared the rational and efficient use of energy as a matter of national interest. Within this legal framework, PROURE—Program for the Rational and Efficient use of Energy—was created. In the Action Plan 2017–22, DG is seen as a means of improving EE as it reduces losses associated with energy distribution (Resolution MME 4 1286/2016).

DG was first promoted, together with large and small-scale self-generation and the use of renewable energies, by Law 1715/2014. In 2018 Resolution CREG[32] 30/2018 established the following:

- Small self-generators[33]—up to 1 MW of installed capacity—who use renewable sources are allowed to deliver their energy surpluses, using a bidirectional meter and receiving energy credits as compensation. Surpluses that surpass the amount of energy consumed during the billing period are valued at retail price. Surpluses equal or lower than the energy consumed in the billing period are compensated with the energy consumed. For every kW h, there is a charge: for the cost of energy trading (for

self-generators of up to 0.1 MW) and for the cost of energy trading and system service (for self-generators of more than 0.1 MW);
- Distributed generators—those of an installed capacity of 100 kW or less—may choose between two options to trade their energy surplus: sell them on the terms of Resolution CREG 086/1996[34] or sell them directly to the integrated commercializer.[35] The sale price is calculated at retail price plus an additional that is paid in acknowledgment of the advantages of DG.

8.3.2.4 Ecuador, recent substantial steps

In 2011 the conditions, preferential prices, and form of dispatch for electric energy generated from renewable sources were established (Resolution. CONELEC[36] 004/11), thus setting the context for the development of renewable energies projects. In 2016 Ecuador's electric energy mix was composed of 58% hydropower, 29.7% oil, and 10.1% gas. Nonconventional renewables accounted for 2.2% of electricity generation (IEA, 2019).

Regulations on DG and EE are very recent in Ecuador. No national law concerning DG has yet been sanctioned, but a regulation was issued in 2018 by ARCONEL.[37] Since then, users are allowed to use photovoltaic systems for microgeneration up to an installed capacity of 100 kW. Energy surpluses can be injected into the public network and are compensated

---

[32] Energy and Gas Regulation Commission.

[33] A self-generator is defined as "the natural or legal person that produces electrical energy only to meet his or her own needs and only uses the public network as a backup" (Resolution CREG 084/1996).

[34] According to this Resolution, generators of up to 10 MW are allowed: (1) to sell energy to the energy trader of a regulated market at retail price; or (2) to sell energy to nonregulated users, generators or energy traders of nonregulated markets at a price that is agreed between the parties.

[35] In Colombia the trader is the service provider who does not intervene in the activity of distribution, while the integrated trader is the service provider who is integrated with the network operator.

[36] National Council of Electricity. It was replaced in 2015 by ARCONEL, Agency for the Regulation, and Control of Electricity.

[37] As a public agency, it has the power to regulate technical, economic, and operational aspects of electric services.

through energy credits, which are used to off-set payments and are valid for 2 years (Regulation ARCONEL 42/2018). Soon after this, the maximum installed capacity was raised to 300 kW for residential users and up to 1 MW for industrial and commercial users (Resolution ARCONEL 57/2018).

Regarding EE, in early 2019, the Organic Law of Energy Efficiency was sanctioned, establishing the legal framework for the National System of Energy Efficiency.

8.4 Conclusion

EE and DG are two increasingly valued resources in South American countries. National trajectories differ in the speed of the changes and, especially, in the significance given to them: they can either be seen as a tool to reduce system deficits and to make systems more efficient; as a means of universalizing energy services and reducing energy poverty; or as an alternative to diversify energy sources and reduce CO_2 emissions.

Three groups of countries may be identified: the ones with the longest experience, where DG has been possible for 3 years or more, and where the challenge is to maintain what has been achieved and unlock the full potential; the ones that have only recently issued laws and regulations that enabled DG and are still, therefore, going through the initial phases of implementation; and the ones which have not implemented DG yet. EE has been addressed in all countries of the region, though in different ways. There's still a long way to go regarding efficient building techniques and the implementation of labeling systems that are effectively enforced. Information availability and the inclusion of EE practices in educational programs are other steps to be taken.

A project for the use of renewable energies in DG that aspires to sustainability must take into account EE, as it contributes to the reduction of energy poverty and environmental damage, and to the improvement of habitat conditions and energy security. The synergistic relationship between DG and EE should be further exploited and taken into account when designing programs or issuing laws aimed at either one of them. The cases of Brazil, Colombia, and Uruguay are examples in the region where this has been attempted.

In addition to considering EE, in order to unfold South America's full potential in DG, some regulatory and technological issues will have to be addressed. One of them could be rewarding the environment-friendly aspect of DG and its contribution to the reduction of transmission costs. Another one is the incorporation of smart technologies to accompany advances in DG. Smart grids could lead to better informed consumers who would be freer to make decisions and opt for a more rational and efficient use of energy. All of these issues are paramount to the goal of continuing the expansion of DG and EE as a means of ensuring an effective energy transition to sustainability in South America.

References

Acquatella, J., 2008. Energía y cambio climático: oportunidades para una política energética integrada en América Latina y el Caribe. CEPAL, Santiago de Chile.

ANEEL, 2019. Geracao Distribuida. Available from: <http://www2.aneel.gov.br/scg/gd/GD_Fonte.asp> (accessed 24.05.19.).

Bloomberg New Energy Finance, 2016.

Carrizo, S., 2010. Hydrocarbures en Argentine: réseaux, territoires, intégration. Éditions universitaires européennes, Sarrebruck.

Carrizo, S., Ramousse, D., Velut, S., 2009. Biocombustibles en Argentina, Brasil y Colombia: avances y limitaciones. Geograficando 5 (5), 63–82.

Carrizo, S., Velut, S., 2018. Energy transitions and regional integration in South America. In: Cargnin, Rückert, Lemos (Eds.), Territorial Planning and La Plata Basin Borders, 1. Editora Letra, Porto Alegre.

Chevalier, J.M., 2009. Les nouveaux défis de l'énergie: Climat, économie, géopolitique. Economica, Paris.

Curien, N., 2000. Economie des réseaux. La decouverte, Paris.

Ende (2016): Bolivia. Programa de Electrificación Rural (BO-L117). Componente II. Incremento de la capacidad de transmisión hacia áreas rurales. Programa de Gestión Ambiental y Social (PGAS). Junio 2016. ENDE Publications, available from: https://www.ende.bo/public/publicaciones/pdf/pgas-per-ii-bo-l-1117-componente-ii-junio.-docx.pdf (accessed 8.10.19).

Guimarães, L.N., 2019. Challenges to the promotion of distributed energy resources in Latin America: a Brazilian case study. In: Sioshansi, F. (Ed.), Consumer, Prosumer, Prosumager. How Service Innovations Will Disrupt the Utility Business Model. Academic Press, London, pp. 235–258.

García Ochoa, Rigoberto (2014): Pobreza energética en América Latina. CEPAL, Santiago de Chile, 2014.

Grossi Gallegos, H., 2005. Disponibilidad y características de la radiación solar en Latinoamérica. In: Blesa, M., Galvez, B. (Eds.), Tecnologías solares para la desinfección y descontaminación del agua. Escuela de Posgrado, Universidad Nacional de San Martín, Buenos Aires, pp. 85–92.

Hernandez, J., Santamaría, F., Trujillo, C., 2015. Impacts of regulation in the development of distributed generation. Electr. J. 28 (7), 83–95.

Hugues, L., Bell, J., 2006. Compensating customer-generators: a taxonomy describing methods of compensating customer-generators for electricity supplied to the grid. Energy Policy 34, 1532–1539.

IEA, 2002. In: Fraser, P. (Ed.), Distributed Generation in Liberalised Electricity Markets. IEA Publications, Paris. Available from: <http://library.umac.mo/ebooks/b13623175.pdf> (accessed 18.06.19.).

IEA, 2018a. Electricity Access Database. World Energy Outlook.

IEA, 2018b. Key World Energy Statistics. IEA Publications, France.

IEA, 2014. Statistics. Atlas of Energy. Available from: <energyatlas.iea.org> (accessed 10.10.19).

IEA, 2019. Statistics Global Energy data at your fingertips. Available from: <https://www.iea.org/statistics> (accessed 19.06.19.).

INE, 2012, Bolivia: disponibilidad de energía eléctrica y principal combustible o energético utilizado para cocinar en la vivienda de los hogares, por departamento, según área geográfica, censos de 1992, 2001, y 2012. Available from: <https://www.ine.gob.bo/index.php/educacion-5/vivienda-y-servicios-basicos-3> (accessed 19.06.19.).

IRENA, 2018. Renewable Capacity Statistics 2018. International Renewable Energy Agency (IRENA), Abu Dhabi.

Ministerio de Energía, 2018. Gobierno de Chile. Instrumentos de fomento para sistemas solares térmicos. Available from: <http://www.minenergia.cl/sst/wp-content/uploads/2016/05/franquicia-tributaria-2018.pdf> (accessed 19.06.19.).

Mir-Artigues, P., 2013. The Spanish regulation of the photovoltaic demand-side generation. Energy Policy 63, 664–673.

Poullikas, A., 2013. A comparative assessment of net metering and feed in tariff schemes for residential PV systems. Sustain. Energy Technol. Assess. 3, 1–8.

Superintendencia de Electricidad y Combustibles, 2019. Resumen Anual de la Industria Energética 2018. Available from: <http://www.sec.cl/pls/portal/docs/PAGE/SEC2005/ACERCA_DE_SEC/INFORME_SEC/ANUARIO-SEC-2018.PDF> (accessed on 08.05.19.).

US Energy Information Administration, 2013. Technically Recoverable Shale Oil and Shale Gas Resources: An Assessment of 137 Shale Formations in 41 Countries Outside the United States. US Department of Energy, Washington, DC. Available from: <https://www.eia.gov/analysis/studies/worldshalegas/pdf/overview.pdf> (accessed 18.06.19.).

UTE, 2019a. Microgeneración. Available from: <https://portal.ute.com.uy/clientes/redes-inteligentes/microgeneracion> (accessed 01.05.19.).

UTE, 2019b. Fuentes de generación. Available from: <https://portal.ute.com.uy/institucional/infraestructura/fuentes-de-generacion> (accessed 01.05.19.).

Valania, P., Carrizo, S., 2017. Geografía energética en transición. In: Llegar con Energía, Travelling Exhibition.

Valencia, R., 2019. La apuesta de la generación distribuida. Electricidad 227, 3–8. Available from: <http://www.revistaei.cl/reportajes/la-apuesta-la-generacion-distribuida/#> (accessed 24.05.19.).

Velut, S., Carrizo, S., 2013. Les defis de l'integrationenergetique en Amerique latine. International Geographical Union, Kyoto, Regional Conference of the International Geographical Union.

World Bank, 2019. Acceso a la electricidad. Available from: <https://datos.bancomundial.org/indicador/EG.ELC.ACCS.ZS> (accessed 20.05.19.).

Yamamoto, Y., 2012. Pricing electricity from residential photovoltaic systems: a comparison of feed-in tariffs, net metering and net purchase and sale. Solar Energy 86, 2678–2685.

Urban microgrids: benefits, challenges, and business models*

Marco Aurelio Lenzi Castro

Electricity Regulatory Specialist, ANEEL (The Brazilian Agency for Electric Energy Regulation), Brasilia, Brazil

9.1 Introduction

The energy sector worldwide is facing a rapid changing from the traditional large-centralized generation, long transmission lines, and unidirectional energy flow through the distribution lines to highly decentralized small generation near the loads and bidirectional flow through transmission and distribution lines.

The current energy transition implies the use of renewable energy sources, such as solar, wind, and biomass instead of the traditional fossil fuels, mainly coal, natural gas and diesel, to reduce the greenhouse gas emissions, which has been called decarbonization of the energy matrix.

In addition, this energy transition has been demanding an increase in the energy efficiency of residential buildings, offices, industry facilities, and public buildings by using electric devices with low energy consumption, reducing energy losses in cooling and heating systems, as well as increasing the level of automation to maximize the benefits for customers, as the primary objective, and for the environment, as a positive externality.

Those measures can be combined with distributed energy resources (DER), such as solar photovoltaics (PV) and energy storage systems, to achieve a higher level of energy efficiency and become a net-zero energy building.[1]

With the support of renewable energy public policies in countries such as Germany, Australia, Spain, China, and the United States (IRENA, 2018), as well as to huge investments from the private sector to finance the research and development (R&D) of new materials, improvement in engineering design, optimization software, and industrial process, the cost of DER has been experiencing a sharp reduction through the last decade, especially the

---

* The views and opinions expressed in this chapter are the author's own and do not necessarily reflect ANEEL's point of view.

[1] A net-zero energy building produces locally approximately the same amount of energy that it consumes during a year period by using renewable energy sources, becoming carbon neutral.

The Regulation and Policy of Latin American Energy Transitions
DOI: https://doi.org/10.1016/B978-0-12-819521-5.00009-7

cost of solar PV[2] (for small and large-scale power plants) and wind turbines (utility scale).

Thus this innovative and dynamic framework opened an opportunity for consumers to take responsibility for their own power supply and reduce their energy costs. They became prosumers and export their energy surpluses to the grid, making use of the incentives created by local public policies and regulatory rules that allow utilities to purchase that surplus energy[3] or reduce consumers' next energy bill,[4] depending on the country where they are applied.

In order to enable and manage the connection of those new resources the distribution grid has been gradually updated with digital devices, such as smart metering, automatic reclosers, protection relays, among others, associated with a huge flow of data, collected from customers, street transformers, and substations to become smarter grids.

Since consumers are demanding better services from their local utilities (higher reliability, resilience, lower energy costs, and environmental sustainability), along with the inevitable trend of decentralized and decarbonized energy supply and digitalized grid, it has been observed a growing movement towards microgrids projects, which is small smart grids able to operate connected or isolated from the main network, with enough DER to supply their own consumers.

The concept of remote microgrids is not new and has been used in many countries to provide electricity to isolated communities or in rural areas. Those grids are usually powered by diesel-based generators or hybrid systems (diesel-based along with one renewable energy source).

The innovation arises on microgrids installed in urban areas and connected to the main network, allowing consumers to manage their own generation, increasing the resilience of their energy supply by operating the microgrid in isolated mode during energy outages in the main network, reducing their energy costs, and improving the quality of the distribution network (ENEA, 2017).

This chapter is organized in eight sections in addition to the introduction:

- Section 9.2 describes how an urban microgrid works.
- Section 9.3 presents the main types and key drivers for microgrid projects.
- Section 9.4 highlights the main benefits that an urban microgrid can provide.
- Section 9.5 brings some examples of microgrids in different countries.
- Section 9.6 discusses the main challenges that such projects face (technical economic and regulatory).
- Section 9.7 presents different business models.
- Section 9.8 assesses the opportunities and regulatory challenges for microgrid projects in Brazil.
- Finally, the conclusion of this chapter is offered.

9.2 How does an urban microgrid work?

An urban microgrid can operate connected or isolated from the main grid (utility grid), according to the objectives based on which it was designed (Hirsch et al., 2018). This decision

---

[2] The levelized cost of electricity for solar photovoltaics has been reduced by 73% from 2010 to 2017 (IRENA, 2018).

[3] Feed-in tariff is a regulated tariff that is paid for electricity injected into the distribution network by small-scale renewable energy generator.

[4] Net metering is a regulatory scheme which allows consumers to pay only the positive difference between its consumption and the energy injected by its small-scale renewable generator into the distribution grid.

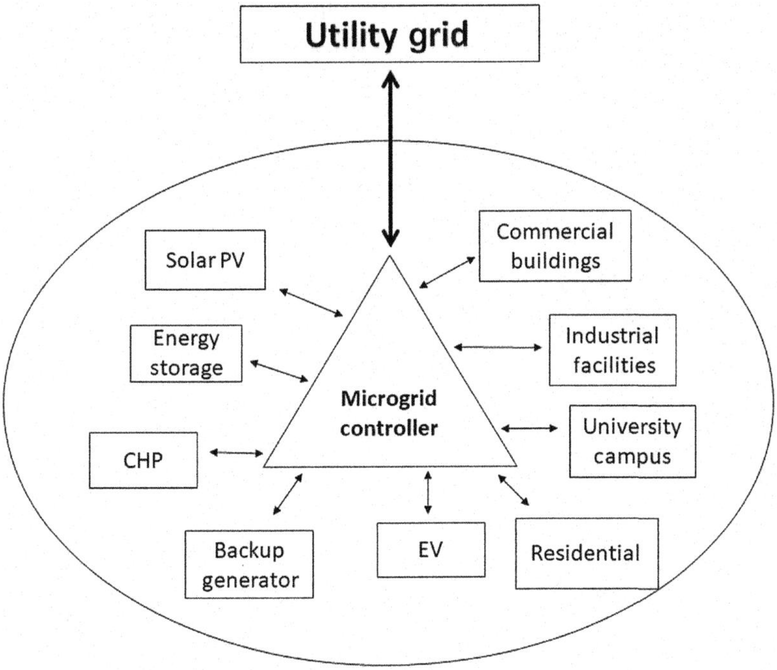

FIGURE 9.1 Microgrid schematic diagram.

is managed by the microgrid controller, and different aspects can be considered, such as the cost of the electricity during peak hours, the availability of the main grid, reduction of carbon emissions, increase in power quality, reliability and resilience of energy supply, export of power-generation surpluses, and provision of ancillary services.[5] Fig. 9.1 illustrates a schematic diagram of an urban microgrid.

A microgrid can balance the local power-generation resources and demand from its loads, while keeping the voltage and frequency inside the standard boundaries.

Those power plants generally use renewable energy resources, such as solar, wind, biomass, and can be conjugated with a combined heat and power (CHP) plant that converts natural gas or biogas to electricity and thermal energy with high energy efficiency (Mariam et al., 2016).

In addition, a battery energy storage systems, with lithium-ion or vanadium flow batteries, for instance, can be integrated into the microgrid to provide flexibility and quick response during peak times or outages in the utility grid, as well as to support the grid with ancillary services (Martin-Martínez et al., 2016).

The main component is the microgrid controller,[6] which is in charge of managing the energy resources to supply the loads, considering all technical and economical requirements to provide a reliable, safe, and cost-effective power supply (Kaur et al., 2016).

In order to achieve these goals the controller monitors technical parameters from both utility grid and microgrid, such as the voltage, current, and frequency, as well as manages the power demand and energy consumption from the microgrid loads.

---

[5] Such as voltage and frequency control, demand response, and supply energy to utility's customers during an outage.

[6] Software that optimizes the operation of the microgrid.

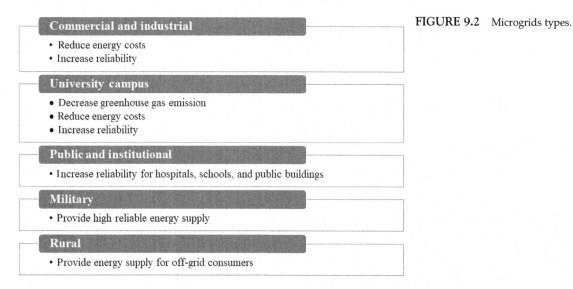

FIGURE 9.2   Microgrids types.

If an outage is detected in the utility grid the microgrid will be immediately disconnected and starts an island operation, where the load will be supplied only by its local DER.

9.3 Types of microgrids and key drivers

The most common types of microgrids can be aggregated in the following categories, as well as the main key drivers to deploy each type of project, as presented in Fig. 9.2:

The following reasons can justify most of urban microgrid projects (Hirsch et al., 2018):

- Increase energy reliability and resilience.
- Reduce energy costs.
- Environmental sustainability.

Increasing the reliability and resilience of the energy supply can be considered the main driver to build an urban microgrid, allowing critical loads, such as military bases, hospitals, industrial facilities, and university laboratories, along with other types of consumers (commercial and residential) that want a better network

service, to keep their ongoing activities during an outage in the utility grid.

The energy reliability might not be an issue in developed countries, such as Australia, especially in large cities where the distribution grids have higher level of automation and even self-healing schemes, but in small cities and rural areas, the grid does not have the same strength.

However, in developing countries, such as Brazil, the distribution networks are congested, with long feeders, voltage regulation issues, and high number of outages during the year (short and long periods).

However, extreme weather events, such as hurricanes, bushfires and shower storms might happen worldwide, and microgrids are able to increase the resilience of the energy supply, avoiding long periods of outages that normally follow such events.

Therefore those scenarios highlight opportunities to deploy urban microgrid projects and improve the reliability and resilience of the energy supply.

Another driver to build a microgrid is the reduction of the energy costs by using the local

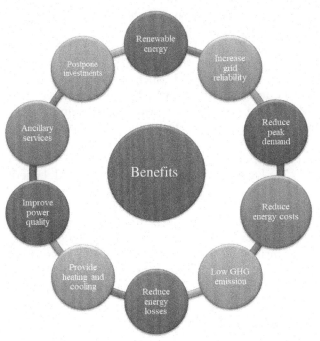

FIGURE 9.3 Main benefits from urban microgrids.

DER available inside the microgrid, such as solar energy, biomass, CHP, and energy storage, which can provide cheaper electricity and thermal energy, as well as avoid paying high peak-hour tariff to utilities.

This condition applies in both developed and developing countries, as the cost of renewable energy is decreasing worldwide, in a different pace in each country, and the cost of delivering energy to customers are rising for various reasons, such as the replacement of aging infrastructure and retirement of old and cheap coal power plants.

The third main key driver is the environmental sustainability, which is related to reduce the carbon footprint of the buildings, offices, industrial, commercial, and government facilities by increasing the energy efficiency of energy appliances, improving the cooling and heat systems, as well as use as much as renewable energy as possible, instead of fossil fuel energy sources.

9.4 Potential benefits from urban microgrids

There are numerous benefits (KEMA, 2014) that microgrids can provide depending on their design and how they will be operated (isolated mode in remote areas or grid connected), reaching not only their participants but also the main grid and society as well, as presented in Fig. 9.3:

The benefit that is first associated to urban microgrids is increasing the reliability of the energy supply, as mentioned earlier in Section 9.3, as the main driver for those projects, once they can operate in island mode if an outage is detected in the utility grid. Therefore the loads connected to an urban microgrid, such as hospitals, industries facilities, university premises, public building, or even residential units, would not stay without electricity during an outage, and they would be supplied by the local generation and energy storage.

After the fault is cleared and voltage and frequency return to standard levels in the utility grid, the microgrid could be reconnected and the loads would be supplied again by the main grid (Hirsch et al., 2018).

The reduction of peak demand and energy costs are directly related, once the microgrid controller can be set to start the island operation to avoid the high peak-hour tariff from utilities.

This action can benefit the utility grid because it reduces the loading of feeders and transformers exactly when they are more demanded, contributing to improve the voltage level at that part of the grid and even postpone investments from utilities to strengthen the network, depending on the size of the load and its impact on the grid.

On the other hand, microgrid consumers also benefits from this operation, once they reduce their energy costs by avoiding the highest tariffs of the day. In addition, they might join a demand response program (MIT, 2016) from utility, if available, and receive a financial compensation for reducing their demand at peak times during extreme heat waves in the summer, for instance.

The use of renewable energy to supply the loads within the microgrid, associated to a CHP plant that provides heating and cooling to fulfill the thermal energy needs of consumers, can reduce the greenhouse gas emissions and the carbon footprint of facilities, as well as the energy costs, which benefits microgrid consumers, utilities, and overall society.

As the energy is produced near the load, a positive externality is the reduction of the energy losses in the distribution grid, because the flow of energy will be reduced outside the microgrid, decreasing the thermal losses in the cables and transformers, which is a benefit for utilities and all consumers as less energy would be purchased to supply the loads (Hirsch et al., 2018).

Besides all benefits already listed, an urban microgrid can also provide ancillary services to the main grid, such as voltage and frequency control by providing reactive and active power to the grid and receiving a revenue for helping the utility to keep their network running within the technical standards (ENEA, 2017).

Lastly, microgrid consumers could benefit from a higher power quality (Palizban et al., 2014), which is crucial for sensitive loads such as industrial engines, university laboratories, and data centers, because the energy generated inside the microgrid can be set to have minimal harmonic distortions (by adjusting the inverter or installing harmonic filters), the voltage and frequency variations can also be managed by the microgrid controller.

9.5 Examples of urban microgrids

There are many projects in universities, hospitals, and military bases already developed worldwide, with an expected global market increase from 3.2 GW in 2019 to 15.8 GW by 2027 (including all types of microgrids, as described in Section 9.3), where the United States accounted for almost 35% of this market in 2018 (Microgrid Knowledge, 2019), or USD 2.5 billion.

The New York State Energy Research and Development Authority (NYSERDA) launched a competition in 2015[7] to stimulate the deployment of microgrids in local communities, with funding to support feasibility studies, technical designs, and building stage. After a competitive process, 11 projects were selected to receive up to USD 1 million.

---

[7] The NY Prize Competition. Information available from: <https://www.nyserda.ny.gov/All-Programs/Programs/NY-Prize> (accessed 02.03.19.).

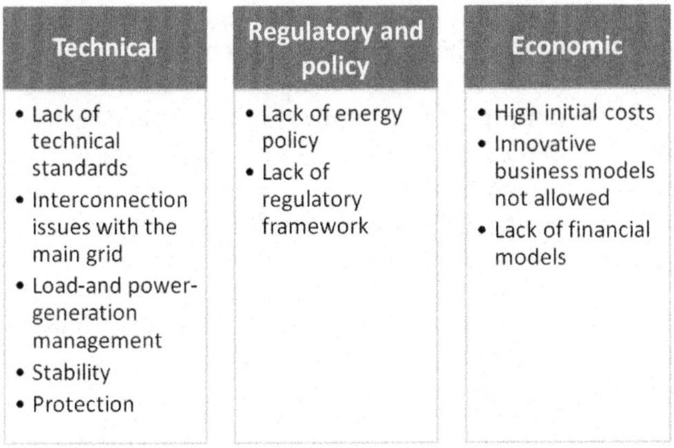

Technical	Regulatory and policy	Economic
• Lack of technical standards • Interconnection issues with the main grid • Load-and power-generation management • Stability • Protection	• Lack of energy policy • Lack of regulatory framework	• High initial costs • Innovative business models not allowed • Lack of financial models

FIGURE 9.4 Challenges for microgrid projects.

Outside the United States, Europe[8] and Asia[9] are the main regions with microgrid projects, while Australia[10] is leading the way in Oceania, and Canada has also projects in North America (Martin-Martínez et al., 2016). In Brazil, most of the microgrids are installed in small communities in rural areas that are not connected to the main transmission system (Bellido et al., 2018), but there are few pilot projects in University,[11] residential condominium (Rodrigues et al., 2017), and in rural areas that are connected to the main grid.[12]

9.6 Challenges for urban microgrids

Despite all the benefits that an urban microgrid can provide, there are several challenges that prevent the deployment of projects without receiving subsidies from governments or funding for R&D programs. The main challenges can be classified in three categories:

technical, regulatory and policy, and economic, as presented in Fig. 9.4.

9.6.1 Technical

As one of the main characteristics of a microgrid, the operation in the island mode requires specific technical procedures that must be followed by the microgrid operator during the transition from grid connected to the island mode and the reconnection to the main grid, in order to avoid or reduce voltage and frequency variations inside the microgrid and outside, near its boundaries (Soshinskaya et al., 2014).

The lack of local technical standards specific for microgrids interconnections stands out, and so far, IEEE Std 1547.4-2011 (IEEE, 2011) is the only international reference (Bellido et al., 2018) for microgrids operating in island mode.

The disconnection and reconnection of a large load (such as an industry) or power

---

[8] Mainly Germany, Spain, Greece, and England.

[9] Japan, China, and South Korea.

[10] Monash University microgrid is one example of projects in Australia. More examples available from: <https://arena.gov.au/?s = microgrid> (accessed 06.07.19.).

[11] University of Santa Catarina.

[12] Partnership between Itaipu and Copel. Available from: <https://www.itaipu.gov.br/sala-de-imprensa/noticia/itaipu-e-copel-desenvolvem-projeto-inedito-de-microgrid-no-oeste-paranaense> (accessed 06.07.19.).

plants might result in voltage and frequency variations that can damage consumer's electric devices or incorrectly trip distribution lines protections.

Another issue occurs if the power plants inside the microgrid keep injecting active power into the grid during a fault, which is a risk for utility's works that are reestablishing the energy service.

However, inside the microgrid, the challenge is balancing the load and local generation during the island operation to keep the stability of the grid and set correctly the protections for the new short-current circuit that are lower than the one while connected to the main grid (Hirsch et al., 2018).

Despite the importance of the technical issues that are being addressed by researchers, global electric and electronic manufactures, as well as in pilot projects in different countries, the lack of regulatory and energy policy to support the deployment of microgrid might be claimed the highest barrier along with the economic aspects, because an advanced regulatory framework could provide the guidelines, a suitable tariff structure, and allow innovative business models that reduce the uncertainty for investors and support microgrids projects.

9.6.2 Regulatory and policy

The regulatory framework is the main challenge (ENEA, 2017) that an urban microgrid project faces nowadays in many countries, because there are gaps in their energy policies and regulatory frameworks, which increase the uncertainty surrounding microgrid projects, preventing potential investors and financial institutions to embrace them. Therefore the main topics that should be addressed by the regulatory authorities to minimize those issues would be:

- establishing a clear definition of microgrids, as how they will be classified and treated under the regulatory rules;

- updating the interconnection standards and setting the technical requirements to allow urban microgrids to operate in island mode;
- defining a fair remuneration for the services that could be provided by microgrids to the main grid;
- designing a suitable tariff structure to stimulate customers to invest in local generation and energy efficiency plans; and
- allow innovative business models, such as those described in Section 9.6.

9.6.3 Economic

The initial costs to design and build an urban microgrid are still high due to the level of complexity of the project that should mitigate the current technical challenges described in Section 9.8.5.1.

In addition, there are still few companies in the market worldwide able to implement a suitable and flexible digital platform to control and manage the flow of data and energy circulating from the main grid to the microgrid, and vice versa, assuring that all safety standards are met and maximizing the performance of the local DER.

Another economic issue is the level of uncertainty about what types of business models would be allowed by the regulatory framework, which increases the risks and becomes a barrier for banks to finance such projects.

Therefore it is crucial to have clear rules providing the guidelines for investors about the business models, besides R&D resources and subsidized funding to support microgrids projects.

9.7 Business models for urban microgrids

There are different businesses models (Bellido et al., 2018; Kema, 2014) that can be

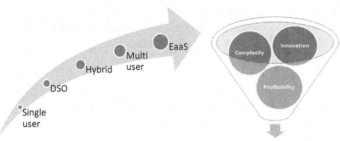

FIGURE 9.5 Microgrid business models.

Business models

applied to an urban microgrid, and their names vary according to the author of the study. Fig. 9.5 illustrates the most known models and how the levels of innovation and complexity increase from one type to another.

9.7.1 Single user

The single user is the most common model applied nowadays, since it faces the least regulatory barriers because investment and operation are done by the owner of the microgrid assets, which is also the unique participant (Bellido et al., 2018).

In this case, all loads and generation are controlled by just one participant, and the energy produced inside the microgrid is used for self-consumption or to export to the grid, depending on the regulatory framework of the country.

This is the simplest model with the lowest level of complexity, as the microgrid is restricted to supply just one consumer with different loads, such as a university campus, but with the lowest level innovation, because the owner bear all the costs, operational issues and risks, in addition to the regular fee to use the main grid while operating in the grid connected mode.

9.7.2 Utility microgrid

For this model (ENEA, 2017), utility owns and operates the lines and the distributed

generation inside the microgrid according to its interest, such as reducing peak demand or injecting active power to control the voltage on specific areas surrounding the microgrid.

An advantage for the participants of the microgrid is avoiding the high investment and operational costs. Different consumers can be integrated into the original microgrid without a special permission to extend the lines in a public area.

Participants pay an extra fee to benefit from a higher level of reliability, and other consumers, outside the microgrid, might have an increase in their energy rates to remunerate the new utility's assets.

The complexity of this model is that in many countries, the distribution utilities are not allowed to own generation assets, as a consequence of regulatory reforms that have segregated generation, transmission, and distribution activities and assets.

9.7.3 Hybrid

Within the hybrid model the microgrid is owned and operated by the utility, and participants are in charge of DER, which could also be object of a contract with a third party to install and operate the generators. In this scenario, participants avoid part of the investment and operational costs and could use their generators to supply the loads.

While operating in island mode, the small generators could sell energy to the other participants of the microgrid, if allowed by the regulatory framework of that country.

Similar to utility model, new consumers could be integrated into the original microgrid and participants pay an extra fee for a higher reliability. This model demands a strong coordination among utility and participants to allow a safe and efficient operation of the microgrid, especially during the island mode. The energy rates of all consumers might be increased to remunerate the new assets installed by the utility.

9.7.4 Multiuser

The multiuser model relies on a third party (private company, such as an ESCO[13]) with enough experience, knowledge and financial resources to design, build, own, and operate the microgrid and DER that serve different consumers and charge a fee in exchange for its service. The assets could also be transferred to the participants after the investment's costs are recovered by the third party.

Similar to the utility model, participants of multiuser model do not bear the capital costs or the technical challenges to operate the grid but pay a regular fee to cover the maintenance costs and also the access the main grid while operating in the grid connected mode.

This is one the most innovative model and presents the highest level of complexity, because they need a regulatory framework that allows one consumer with its own generation to sell energy to other (peer-to-peer energy trading), and the microgrid expands its poles and wires to integrate new participants within the concession area of a traditional utility.

In addition, a strong coordination among the participants is needed to balance the local-energy production and consumption as well as

managing their own billing system designed to charge the energy trading, which increases the complexity of those models.

Finally, there is no impact on the energy rates of consumers outside the microgrid, because the utility does not have to install new assets to provide a higher reliability or increase the power quality for the microgrid participants.

9.7.5 Energy as a service

Within the energy as a service (EaaS) model (Microgrid Knowledge, 2018), a third party provides not only the microgrid project, assets, operation, and maintenance, but also a full energy advising on energy management tailored for the needs and aims of each microgrid, including energy efficiency and wholesale market, as well as financial support for the project.

This model works with a structure similar to a power-purchase agreement for a long-term period, where consumers pay a regular fee for the hired services, which might include indicators to assess the results achieved by the microgrid provider versus the indicator that agreed upon in the contract and adjust the fees based on their performance.

One challenge for this model is adapting the regulatory rules to allow such microgrid providers to operate within the utility's monopoly area.

9.7.6 Peer-to-peer energy trading

A peer-to-peer energy trading (Mengelkamp et al., 2017) is not a new type of microgrid business model, but it could be applied to hybrid, multiuser, and EaaS models when associated with blockchain technology,[14] allowing one or more consumers with DER to sell energy directly to others inside the same microgrid, while keeping the record of all

---

[13] Energy Service Company.

[14] Keeps a record of the all transactions carried out, thus being able to charge or pay for the services used and supplied.

transactions carried out and enabling the payment for the exact services provided.

Besides, this innovative approach can reduce the energy costs for consumers and increase the profitability of the microgrid participants.

9.8 Scenarios for urban microgrids: the case of Brazil

The electricity regulatory framework in Brazil does not have specific rules that clearly define urban microgrids, how they can be implemented, and what kind of business models would be allowed. Therefore it is important to have a glance at the policy and regulatory framework of the country, how the electricity matrix looks like, and the overall power quality that consumers face before assessing the scenarios for microgrids in Brazil.

9.8.1 Policy and regulatory framework

Fig. 9.6 presents a schematic of the Brazilian electric power institutional structure, representing the main public and private organizations that are in charge of managing the energy sector.

As illustrated in Fig. 9.6, the energy policy is set by the President of Brazil, supported by the Ministry of Mines and Energy (MME), the National Council for Energy Policy, and Electricity Industry Monitoring Committee. The National Congress has also the competency to establish the energy policy through the issuance of new laws.

The long-term energy planning is performed by the Power Research Company that provides technical support for MME, who is in charge by law to set the expansion plan for the energy sector.

The Brazilian Agency for Electric Energy Regulation—ANEEL is responsible for the regulation and overseeing of the electricity market and utilities, with the support of State Agencies, implementing the public policy and calculating the tariff for consumers. The wholesale market is managed by the Chamber for the Commercialization of Electrical Energy (CCEE), while the centralized dispatch of large power plants ($>50\,MW$) and the

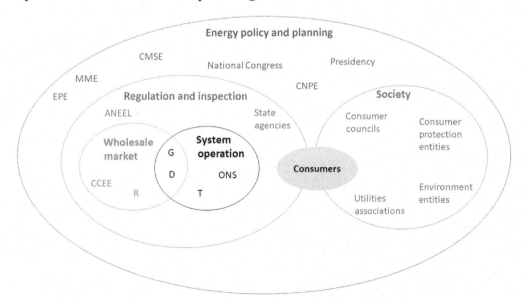

FIGURE 9.6   Brazilian Policy and Regulatory Framework.

management of the interconnect transmission lines are performed by the Independent System Operator (ONS).

However, there are various stakeholder associations, consumer councils and environment entities that play important role in this complex energy policy framework with the main objective to provide an affordable and reliable power supply to consumers.

9.8.2 Brazil's electricity matrix

Fig. 9.7 shows the energy resources available in Brazil to produce electricity (ANEEL, 2018a), where it can be observed that roughly 83% of the power plants use renewable energy, and hydro plays the major role with 64%, where large hydropower plants (> 50 MW) represent 94.5% of the market share, medium (between 3 MW and 50 MW) have 4.8%, and small (<3 MW) claims 0.7%.

Fossil fuel power plants come in second place and are important to supply energy when the hydropower plants are operating with low levels in their reservoirs, as well as during peak times in summer and provide power support to some week distribution networks.

Biomass (mainly sugarcane bagasse) and wind have almost the same percentage of participation in the electricity matrix. While the number of biomass power plants has increased slowly during the last decade, wind power plants have been growing fast and consistently since 2009, rising from 606 MW to 14.7 GW at the end of 2018.

Solar energy is growing fast in Brazil since 2016, driven by utility scale power plants, which represented in December of 2018 roughly 1% of the electric matrix (2 GW), as well as by the exponential growing of small-scale power plants installed in residential, commercial, industrial, and public buildings, reaching 565 MW at the end of 2018.

According to EPE (2019), it is expected that the participation of centralized and small-scale solar power plants will increase up to 3% of the electricity matrix until 2027, and the whole range of renewable energy sources will still represent 83% of the total electricity production in 2027.

Therefore the use of renewable energy and reduction of greenhouse gas emission is not a strong driver to deploy microgrid projects in Brazil.

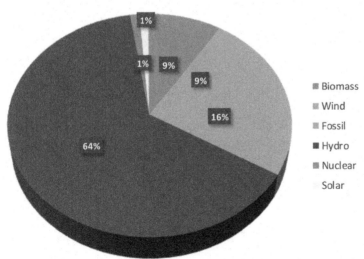

FIGURE 9.7 Brazilian electricity matrix in 2018.

Biomass
Wind
Fossil
Hydro
Nuclear
Solar

9.8.3 Overall quality of electricity supply

Brazil has 209 million inhabitants and 80 million electricity customers, supplied by 63 concessionaires and over 100 small electricity distribution companies that cover all the 27 states of the country.

According to ANEEL (2018a,b), the average number of hours that a consumer in Brazil stayed in 2017 without electricity was 14.35, which is higher than the limit of 12.77 hours for that year established by ANEEL, and there were an average of 8.20 failures (longer than 3 minutes) for the same period, lower than the average limit of 9.72 failures established by the regulatory agency.

In order to have an idea if those indicators represent a good or poor service to consumers, it is interesting to know the same indicators in Australia, for instance. According to Energy Networks Australia (2018), consumers face an average of 3.33 hours of outages (planned and unplanned) and an average of 1.5 interruptions of service per year. Thus the SAIDI[15] and SAIFI[16] indicators in Brazil are 4.3 and 6.5 times higher than in Australia, showing that power quality is an issue in Brazil.

In addition, according to the Brazilian Electricity Distribution Procedures—PRODIST,[17] Module 8—Power Quality, utilities must reimburse their consumers if they provide poor quality supply, which are measured by some indicators and compared with the regulatory limits. As a result, in 2017 utilities have returned nearly USD 66 million to their consumers throughout the year, giving credits in their energy bills.

Therefore increasing the reliability and quality of power supply are latent topics for consumers and utilities in Brazil, and urban microgrids could be part of the solution.

9.8.4 Main regulatory barriers in Brazil

As highlighted previously, there is a lack of regulatory and policy rules to guide the deployment of urban microgrids in Brazil, but it does not mean such projects are not allowed. However, they would face higher economic and regulatory barriers that reduce the attractiveness for potential investors with the current regulatory framework, as shown later.

9.8.4.1 Right of monopoly

Utilities have the right of monopoly to supply electricity for all customers that are located inside their concession area.[18] This exclusivity right was established in each concession contract that were signed by MME, ANEEL, and utilities, granting an average period of 30 years to provide electricity services to households, businesses, industries, public sector, and rural properties within a specific geographic area.

Therefore only utilities can install poles and wires in public areas to supply their own consumers. However, according to the Brazilian Federal Law 9074/1995, Article 15, and Decree 2003/1996, Article 23, an independent generator might be granted the permission to supply energy and connect its private network to consumers that belong to the same industrial or commercial complex if they also receive steam from the cogeneration process.

---

[15] The system's average interruption duration index.

[16] The system's average interruption frequency index.

[17] The technical code for the distribution network (up to 138 kV) that utilities, generators, and consumers must follow.

[18] There are 63 concession areas in Brazil.

9.8.4.2 Tariff structure

Similar to other developing countries, low voltage consumers in Brazil (mainly residential and commercial units) pay a flat value for their consumption ($/kWh), that is, the energy has the same value along the day and the both fixed[19] and variable[20] costs are charged as a function of the energy consumption (kWh), which is called volumetric tariff.

Within a volumetric tariff the utility's revenue depends directly on the amount of energy it sells to cover the fixed and variable costs. If a group of consumers are gathered to form a microgrid with their own generation, which means reducing their consumption from the main grid, the immediate consequence is a negative impact on utility's revenue.

Therefore in order to restore the original balance on the utility's revenue, regulators would have to increase the tariff for all the consumers, even for those without their own generation, which creates a negative impact for the all consumers.

However, the current tariff structure in Brazil establishes a peak demand period[21] of 3 h/day, usually between 5:00 and 8:00 p.m. or 6:00 and 9:00 p.m., whereas in Australia, it is from 7:00 a.m. to 7:00 p.m., 12 h/day.

This cost structure in Brazil does not represents the real peak demand of the electric system, especially during the summer,[22] and does not reward the contribution of solar generation and battery-storage system to reduce the demand commercial customers.

9.8.4.3 Export and sell electricity

An important source of revenue for an urban microgrid is exporting their excess of energy to the main grid, which could be done during off-peak times (in the case of using solar energy) or peak hours by discharging their energy storage system (usually batteries). Another option is operating in island mode to avoid high energy prices from utility.

In Brazil, consumers can have their own distributed generation[23] system and export the energy surpluses to the grid under a net metering scheme,[24] but they are not allowed to sell energy directly to other consumers or receive financial compensations from utility.

Therefore despite the benefit of having significant reduction in their electricity bills, the current rules prevent consumers to make profits from their own small power plants, and this would impact the revenues for urban microgrids.

In addition, generators or consumers with their own power plants are not allowed to supply electricity directly to consumers without using utilities' network.

---

[19] Fixed costs ($/kW) are due to cover the maintenance and operation cost of the network.

[20] Variable costs are charged to cover the price of energy in different periods of the day (peak, off-peak, and intermediate), which was purchased in advance by utility from independent energy producer or state-owned power plants, depending on the energy market rules.

[21] At this time, the cost of energy is much higher, approximately 50%, due to more expensive power plants running to supply the peak demand.

[22] 90,525 MW was the maximum demand in Brazil registered by ONS at January 31, 2019, 3:50 p.m. Available from: <http://sdro.ons.org.br/SDRO/DIARIO/index.htm> (accessed 06.07.19.).

[23] Renewable energy sources, such as solar PV, and cogeneration with high energy efficiency with up to 5 MW of installed capacity.

[24] ANEEL's Normative Resolution 482/2012 allows consumers to inject the excess of their own generation back to the grid to be used within the following 60 months to reduce their electricity bills.

9.8.4.4 Choice of retailer

Just a few types of consumers[25] are eligible to choose their own retailer and most of them must purchase electricity only from their local utility, including residential, small commercial, and industrial units. Those restrictions would initially reduce the attractiveness of the microgrid projects because their consumers may not be able to meet the regulatory requirements to join the so-called free electricity market and change their energy supplier.

9.8.4.5 Ancillary services

The current regulatory rules[26] allow only generators to be remunerated for providing ancillary services to the grid,[27] and urban microgrids do not fit into this classification, which prevent their participants to offer those services to support the network and receive additional revenue.

9.8.4.6 Interconnection procedures

The regulatory rules and the technical standards from utilities impose some restrictions for customers with distributed generation to operate in island mode during a short outage or even a long period without utilities services, forcing the disconnection of the small power plant and leaving the customer without energy supply. The main reason is the risk to reenergize a distribution line, while utility's workers are still fixing the problem and cause a serious accident.

9.8.5 Some paths for urban microgrids in Brazil

Despite the regulatory barriers listed on Section 9.7.4, there are some alternatives to overcome some restrictions and still follow the current rules, as well as some changes that could be implemented to allow innovative business models for microgrids.

9.8.5.1 Keeping the status quo

Within the current regulatory scenario, there are a few tools that could be used to support microgrid projects and allow two types of business models: the single user model and a variance of the hybrid model (using the concept of shared generation, as explained later).

Since the electricity reform in Brazil from 1995,[28] distribution utilities are not allowed to own generation assets, and for that reason, the DSO[29] business model would not be a feasible option in Brazil without a deep change in the legal framework.

9.8.5.1.1 Research and development resources from utilities

According to Brazilian Federal Law 9991/2000, utilities must invest up to 0.5% of their net operating revenue in R&D projects, following the regulatory rules[30] stablished by ANEEL, representing roughly USD 150 million per year.[31]

Therefore utilities have to deploy R&D projects related to their core business, which could be done in partnership with universities,

---

[25] Consumers with load above 500 kW and below 2 MW (new limits from 2020 according to MME's Decree 514/2018) can only purchase electricity from renewable energy source or from local utility, while consumers with load higher than 2 MW are free to choose their own retailer.

[26] ANEEL's Normative Resolution 697/2015 and Module 14 of the Interconnect Network Procedures.

[27] Frequency control, reactive support, black start, and special-protection scheme.

[28] The Brazilian electricity market reform started by Federal Law 9074/1995.

[29] Distribution system operator.

[30] R&D Procedures approved by ANEEL's Normative Resolution 754/2016.

[31] Source: <http://www.aneel.gov.br/programa-de-p-d> (accessed 04.03.19.).

research centers, or other utilities, and those resources could be an important funding source for microgrid projects, as well as capacity building for all participants and sharing their findings with other stakeholders.

9.8.5.1.2 Private networks

There are few exceptions in the regulatory framework that allow consumers to have their own network. According to Decree 5163/2004 and ANEEL's Normative Resolution 229/2006 (ANEEL, 2006), consumers can own private networks only if all assets are installed inside their properties, without crossing public roads or areas, such as industrial complexes, and private residential and commercial condominiums.

Outside those boundaries, ANEEL can authorize consumers to keep their own network under certain conditions.[32] This alternative currently allows different consumers to share a private substation by connecting their own transmission lines to supply their loads.[33] With such permission, those lines could cross public roads and other's properties.

In such a scenario, there are two options for consumers with private networks, such as those connected to urban microgrids: the lines should not cross any public thoroughfare or consumer must apply for a special permission from the regulator to keep control of them. Otherwise, the utility will incorporate those assets, manage the energy supply, and reimburse consumers for their investment on the grid, following the criteria established by ANEEL.

9.8.5.1.3 Shared generation

ANEEL's Normative Resolution 482/2012[34] (ANEEL, 2012) allows different consumers to share the same power plant and use the energy to reduce their electricity bills. One option is available for residential, commercial, or industrial consumers with their unit within a condominium. In this case the private network should not cross any public street or third-party properties that do not join the same condominium. Such network could be improved and become a microgrid by installing a battery system, a microgrid controller, and other devices.

In another alternative, customers should join a consortium[35] or a cooperative[36] to benefit from the energy produced by power plants of those entities and reduce their electricity bills. The main difference between this option and the first one is that consumers could be placed in different cities inside the same concession area of the utility to be able to share the energy from the same power plant.

In this case, as the loads are not connected to the same network, it cannot be considered a microgrid.

---

[32] Consumer must demonstrate that the private network is the least cost option to supply its loads or it has a specific technical characteristic, such as nonstandard voltage level or higher reliability to supply sensitive loads, as well as comply with safety standards.

[33] One example is the Authorizing Resolution 6660/2017 that granted permission to consortium Três Corações to keep four private distribution lines with 13.8 kV that interconnect the four industrial premises that integrate this consortium and cross public roads.

[34] ANEEL is revising the DG rules during 2019, and this chapter follows the current version that was updated by the Normative Resolution 687/2015.

[35] Private organization of legal entities that follows the appropriate Brazilian legislation.

[36] Private organization that was built to provide services to their associates, following the Brazilian legislation for such entities.

9.8.5.2 *Improving the regulatory framework*

In order to allow innovative business models, such as multiuser and EaaS, or simply reduce the risks for investors and maximize the benefits that an urban microgrid can provide to the main grid, some nontrivial changes could be done by the regulator, and others might demand updates in Presidential Decrees and Federal Laws, which are more challenging.

9.8.5.2.1 Microgrid definition

It is critical to have a clear definition of microgrid, its boundaries, what services can be provided and how it will be classified under the regulatory rules. For instance, if a microgrid is defined as a public utility that can provide services to any customer by building its distribution lines through public area, it will be treated as a regular utility and follow the same technical, economical, and commercial standards, including compliance with all customer rights before the Regulatory Agency.

However, the rules might restrict the area where the microgrid can provide its services, such as inside a defined perimeter (university campus, military area, industrial complex, commercial center, or even residential suburbs), without extending its poles and wires into public areas, where an electricity utility has its own assets already installed to supply its customers.

ANEEL could update the Module 3—access to the distribution system of the PRODIST to include an appropriate definition for microgrids.

9.8.5.2.2 Operation in island mode

In order to overcome this risk of accidents during the operation of the microgrid in island mode, the regulatory rules should establish the safety parameters for a protection scheme inside the microgrid to avoid the reconnection to the main grid, while the voltage at the point of common coupling does not return to the standard value and sustain it for a certain period of time. This topic could be addressed by ANEEL in a revision of the current version of Module 3 of PRODIST.

9.8.5.2.3 Services' payment

Urban microgrids can provide different services to its customers and to the grid, such as ancillary services, and the regulatory rules should address how these services would be remunerated, according to the business models that were adopted in each project.

Therefore those services should be valued according to the benefits that they can provide to the main grid, and ANEEL could establish, for instance, if the microgrid would be remunerated by the local utility through a fixed rate or a market price.

Thus a possible path is the revision of ANEEL's Normative Resolution 697/2015 (ANEEL, 2015) and Module 14 of the Interconnect Network Procedures, both under ANEEL's leadership.

9.8.5.2.4 Design a new tariff structure

In several countries with a liberalized electricity market, such as Australia, United Kingdom, and United States,[37] regulators have designed different types of tariffs that could be mandatory or optional for each class of consumers.

Most of them charge a fixed rate from consumers[38] (including residential and commercial), as well as variable rates for the energy supplied according to the time it is consumed,[39] which could also be purchased from different retailors.

---

[37] Some states in the United States have a liberalized electricity market, such as California and New York.

[38] Binomial tariff.

[39] Time of use tariff.

Within this tariff structure, utilities can recover their fixed costs even if customers have their own generation or are connected into a microgrid and reduce their energy consumption from the main grid. In this case, regulators do not need to increase the utility tariff because the initial equilibrium between costs and revenues was not changed.

Following this path, ANEEL started a process of public hearing[40] in 2018 to discuss the outcomes of a Regulatory Impact Analysis study performed by the agency to assess possible scenarios to implement a binomial tariff for low voltage consumers. According to the Agency Regulatory Agenda 2019–20,[41] the final decision of the Board of Directors about the new tariff structure is scheduled to be done until December of 2019.

Therefore in order to create an incentive to microgrid projects, the electricity tariff should be designed in such a way that values the energy that is exported to the main grid and, at the same time, do not impact the revenue of the utility or increase the tariff for other customers, which is a challenge for the regulator.

9.8.5.2.5 Expansion of the electricity free market

In order to boost innovative business models for urban microgrids, such as multiuser and EaaS, it is crucial to allow consumers to sell the energy produced in their own unit to others that are connected to the same microgrid (peer-to-peer trading), as well as choose their own energy supplier, such as the owner of the microgrid or another retailer.

However, such disruptions in the traditional electricity market in Brazil would demand changes in Federal Laws,[42] Decrees, and several ANEEL's normative resolutions.

Therefore a lot of effort should be done to implement those reforms and the journey would certainly be challenging, but this move has already started in Brazil, and it looks like sooner or later consumers will have more options to manage their energy supply and those changes would reduce the barriers for microgrids as well.

9.9 Conclusion

This chapter presented the main benefits that an urban microgrid can provide to its participants, utilities, and the overall society.

Urban microgrids could be suitable for developed countries with high grid reliability, energy matrix heavily based on fossil fuels but with expensive energy costs, such as Australia, as well as for developing countries, such as Brazil, with longer and more often power outages and most of electricity production from renewable energy sources.

Despite the potential benefits, there are technical, regulatory, and economic challenges that those projects must overcome to be a feasible, profitable, and sustainable alternative for consumers. The technical issues are being addressed by pilot projects in different countries, in partnership with university researchers and global industry manufactures, and with the advancement of smart grid technologies, it might be expected to mitigate those issues shortly.

The current electricity regulatory scenario in Brazil was not designed to embrace the microgrids projects, but a path to navigate within those restrictions and allow the simplest business models were provided, as well as possible changes that could be done in the regulatory

---

[40]   ANEEL Public Hearing 59/2018.

[41]   Approved by ANEEL Decree 5571/2019.

[42]   Such as Federal Laws 9074/1995 and 10,848/2004 and Decree 5163/2004.

framework to allow the most innovative models, such as multiuser and EaaS.

One of the key improvements is redesigning the tariff structure in Brazil. First, adjusting the peak demand period to the reality of each concession area, which would increase the economic benefits that solar generation and energy storage could provide to customers connected to an urban microgrid, as well to the main grid, by reducing the demand at such periods.

Second, establishing tariffs without cross subsidies, allowing utilities recover their fixed costs and providing the right economic signal for consumers, such as investing in energy efficiency and distributed generation.

Third, additional benefits that urban microgrid can provide to the main grid, such as ancillary services and demand respond, should also be addressed by the new tariff design to stimulate customers to support the grid operation and receive an appropriate revenue.

Finally, urban microgrids can play an important role during the current energy transition that is changing the way energy is produced and consumed worldwide, with higher energy efficiency, use of renewable energy sources and storage, as well as increasing the reliability and resilience of the electricity services for customers.

References

ANEEL, 2006. Brazilian Electricity Regulatory Agency. Normative Resolution 229/2006.

ANEEL, 2012. Brazilian Electricity Regulatory Agency. Normative Resolution 482/2012.

ANEEL, 2015. Brazilian Electricity Regulatory Agency. Normative Resolution 697/2015.

ANEEL, 2016. Brazilian Electricity Regulatory Agency. Normative Resolution 754/2016.

ANEEL, 2018a. Brazilian Electricity Regulatory Agency. Management Information Report – December 2018. Available from: <http://www.aneel.gov.br/informacoes-gerenciais> (accessed 29.03.19.).

ANEEL, 2018b. Brazilian Electricity Distribution Procedures – PRODIST. Module 8 – Power Quality.

Bellido, M., Rosa, L., Pereira, M., Falcao, D., Ribeiro, S., 2018. Barriers, challenges and opportunities for microgrid implementation: the case of Federal University of Rio de Janeiro. J. Cleaner Prod. 188, 203–216.

ENEA, 2017. Urban Microgrids – Overview, Challenges and Opportunities. Report.

Energy Networks Australia, 2018. Behind the News: Network Reliability. Available from: <https://www.energynetworks.com.au/news/energy-insider/behind-news-network-reliability> (accessed 05.03.19.).

EPE, 2019. Brazilian Power Research Company. Plano Decenal de Expansão de Energia 2027.

Hirsch, A., Paraga, Y., Guerrero, J., 2018. Microgrids: a review of technologies, key drivers, and outstanding issues. Renew. Sustain. Energy Rev. J. 90, 402–411.

IEEE, 2011. IEEE Std 1547.4-2011: Guide for Design, Operation, and Integration of Distributed Resource Island Systems with Electric Power Systems.

IRENA, 2018. Renewable Power Generation Costs in 2017. Available from: <http://www.irena.org/publications/2018/Jan/Renewable-power-generation-costs-in-2017> (accessed 03.06.19.).

Kaur, A., Kaushal, J., Basak, P., 2016. A Review on Microgrid Central Controller. Elsevier.

Kema, D., 2014. Microgrids–Benefits, Models, Barriers and Suggested Policy Initiatives for the Commonwealth of Massachusetts.

Mariam, L., Basu, M., Conlon, M.F., 2016. Microgrid: Architecture, Policy and Future Trends.

Martin-Martínez, F., Sánchez-Miralles, A., Rivier, M., 2016. A Literature Review of Microgrids: A Functional Layer Based Classification.

Mengelkamp, E., Gärttner, J., Rock, K., Kessler, S., Orsini, L., Weinhardt, C., 2017. Designing microgrid energy markets – a case study: the Brooklyn microgrid. Appl. Energy J. 210, 870–880.

Microgrid Knowledge, 2018. The Financial Decision-Makers Guide to Energy-as-a-Service Microgrids.

Microgrid Knowledge, 2019. Microgrids "Shining Light" for US in World Energy Markets: Report. Available from: <https://microgridknowledge.com/microgrid-market-aee-report/> (accessed 05.05.19.).

MIT, 2016. Utility of the future. MIT Energy Initiative. Available from: <http://energy.mit.edu/wp-content/uploads/2016/12/Utility-of-the-Future-Full-Report.pdf> (accessed 02.03.19.).

Palizban, O., Kauhaniemia, K., Guerrero, J.M., 2014. Microgrids in active network management – part II: System operation, power quality and protection. Renew. Sustain. Energy Rev. 36, 440–451.

Rodrigues. N., Pica, C.Q., Martins, M.A.I., Leites, T.A., Souza, B.P., 2017. Feasibility Study of a Realistic Microgrid: Analysis of the Pilot Microgrid in Fortaleza – Brazil.

Soshinskaya, M., Crijns-Graus, W.H.J., Guerrero, J., Vasquez, J.C., 2014. Microgrids: experiences, barriers and success factors. Renew. Sustain. Energy Rev. J. 40, 659−672.

Further reading

ABEEOLICA, 2019. Associação Brasileira de Energia Eólica. Evolution of Wind Power Plant Capacity. Available from: <http://abeeolica.org.br/noticias/energia-eolica-chega-a-1471-gw-de-capacidade-instalada/> (accessed 15.02.19.).

ANEEL, 2017. Brazilian Electricity Distribution Procedures − PRODIST. Module 3 − Access to the Distribution System.

Dohn, R.L., 2011. The Business Case for Microgrids. White Paper Siemens.

EPRI, 2016. Microgrids: Expanding Applications, Implementations and Business Structures. Smart Electric Power Alliance and Electric Power Research Institute White Paper.

Morris, G.Y., Abbey, C., Joos, G., Marnay, C., 2011. A Framework for the Evaluation of the Cost and Benefits of Microgrids.

10

Regulatory aspects of energy efficiency in Uruguay

Ana Laura Rodríguez D'Espada
Technological University (UTEC), Montevideo, Uruguay

10.1 Introduction

The current energy transition in Latin America is generally linked to the development of new sources of electricity generation, the diversification of the energy supplier, and the new role of the consumer as an energy producer. This transition is rarely related to the new behavioral paradigm linked to the design and execution of energy efficiency policies, whose purpose is to influence human behavior in order to perpetuate a behavioral change toward the efficient and not wasteful use of energy.

The objective of this chapter is to identify the instruments of public and legal policy used to encourage and involve the different sectors (industry, commerce, and government) and the population in the change of behavior toward an efficient energy consumption.

With this purpose the present chapter studies the national regulatory framework and, in particular, the main aspects of the energy policy, Law No. 18597/2009, the National Energy Efficiency Plan 2012—24, and the public policy instruments designed to print the change of behavior toward the fulfillment of the avoided energy consumption goal for the year 2024.

10.2 Regulatory background and historical context

Nationally, the first regulatory background related to energy efficiency can be found in the 1980s through the Decree 561/1980 entitled "energy policy."[1]

Concerned about the worldwide energy situation due to the gradual decrease of nonrenewable energy sources and the challenges to

---

[1] This regulation was passed during the de facto regime presided by Aparicio Méndez Manfredini (1976—81). During this regime the National Electricity Law (Decree Law No. 14694) and the Organic Law of the National Administration of Power Plants and Electric Transmissions (UTE) (Decree Law No. 15031) were also enacted, both of which are fundamental regulations for the electricity sector and remain in force.

substitute them with other sources, the de facto regime decided to take general measures in order to meet the needs of the energy demand without affecting development plans and minimizing adverse effects.

The decree entrusted the Ministry of Industry, Energy and Mining (MIEM) to devise measures leading to a more rational use of energy by designing rules leading to:

1. optimizing energy efficiency in existing buildings;
2. improving energy efficiency in household appliances, specifically, food refrigeration units;
3. projecting new buildings that could be energetically sufficient in terms of temperature, lightning, and ventilation; and
4. suggesting rules to control the installation of thermal conditioning.

In addition, the abovementioned decree created different study groups aiming at:

1. studying whether it was convenient to encourage nontraditional forms of environmental and water heating, especially among them, those designed to capture solar energy and suggesting rules to regulate their implementation;
2. planning rules in order to achieve a higher energy efficiency in household appliances, specifically, food refrigeration units;
3. planning measures leading to a more rational use of energy in urban areas;

4. analyzing the implementation of future public, freight, and passenger transport units with traction based on electrical energy or other fuels; and
5. studying the disposal of urban waste and presenting a concrete plan, nationwide, to treat them based on the need to optimize their energetic potential.[2]

Once democracy had been restored, the next legal record in terms of energy policies cannot be found until 1994. As enacted by Law No. 16517/1994, the Uruguayan State ratifies the United Nations Framework Convention on Climate Change[3] passed by the United Nations Conference on Environment and Development.

The main objective of the convention is to stabilize the concentrations of greenhouse effect gases in the atmosphere, which could cause dangerous anthropogenic interference, within a suitable time frame to allow ecosystems to adapt to climate change in order to prevent threats to food production and to prolong a sustainable economic development of the countries.

On that occasion, the States Parties acknowledged that in order to achieve the objectives, especially in developing countries, it was necessary to increase energy consumption:

> taking into account the possibilities for achieving greater energy efficiency and for controlling greenhouse gas emissions in general, including through the application of new technologies on terms which make such an application economically and socially beneficial[4]

---

[2] On November 30, 1980, less than 30 days after the Energy Policy was published, the *1980 Plebiscite* took place. The referendum, whose objective was to create a constitutional government to legitimize the de facto government and substitute the 1967 Constitution, was rejected by the electorate, unleashing a slow process toward democracy restoration. A few months later, on September 1, 1981, Aparicio Méndez was replaced by General Gregorio Álvarez, former head of the Army (1981–85) who resigned on February 12, 1985 with the restoration of a democratic regime. <https://es.wikipedia.org/wiki/Gregorio_%C3%81lvarez>.

[3] Convention adopted in New York on May 9, 1992, and signed in Rio de Janeiro on June 11, 1992.

[4] Paragraph 22 of the text of the international standard. <https://www.impo.com.uy/bases/leyes-internacional/16517-1994>.

In 2000 by Law No. 17279/2000, Uruguay incorporates the Kyoto Protocol (1997) to its own national legal system. In order to fulfill the commitments of limiting and reducing greenhouse gases emissions, States Parties agreed to design policies and measures to promote energy efficiency in the relevant sectors of their national economies.[5]

Finally, in 2006, according to Law No. 18102/2006, Uruguay incorporates to its legal system the framework agreement on regional energetic complementation among States Parties of the MERCOSUR and associated states. In this agreement the States Parties[6] commit to encourage:

> the organization of exchange and technical development programs to promote the rational and efficient use of conventional energy, energy efficiency, renewable energy, environment preservation and the harmonization of security and quality levels among member states

This section described the regulatory background related to energy efficiency leading to Law No. 18.597/2009 ["Efficient Use of Energy Law" (EE Law)] passed in 2009, which is further discussed in Section 10.4.

10.3 The national energy policy

Cajarville Peluffo (2008) defined *policy* as a direction or orientation activity that, through encouragement and coordination, determines objectives and goals to be achieved and, generally, the means to achieve them.

According to Enrique Sayagués Laso's classification of the duties that the state has, leading general and sectorial policies are essential ones. Essential duties are those which can be considered inherent to its nature, which cannot be conceived without the direct exercise of the state in their fulfillment. In the exercise of these duties the administration is endowed with more intense legal powers than with the fulfillment of other duties (Sayagues Laso, 1987, p. 56).

The 1967 Constitution, still in force, conceived and designed an executive power that acts as a leader and a political guide. As Cajarville Peluffo (2011, p. 56) suggests, one of the basic concepts behind the 1967 Constitution was to confer the executive power the authority and responsibility to become a central organ of leadership and political guidance, which it assumes in contemporary times.

Based on this constitutional background, article 198 of the Decree Law No. 14416/1975 creates the MIEM and the Administration of energy policy within it, which had as its executive unit the National Bureau of Energy (DNE). According to the aforementioned law, this unit would be responsible for responding to the national energy policy in terms of fuels and electric energy, including nuclear energy.

Article 4 of the Decree Law No. 14694/1977 (National Electricity Law) provides that the executive power is responsible for all matters related to the formulation and control of electric energy policies, especially in relation to the necessary authorizations for the use and preservation of primary sources used in the production of electric energy.

Thus the MIEM is responsible for designing and instrumenting government policies regarding the industry, energy, mining, and telecommunication sectors, as well as small and medium businesses, which are meant to transform and strengthen the national productive system, its energy framework, and the

---

[5] By means of Law No. 17732/2003, the Stockholm Convention on preexisting organic pollutants is incorporated into the national legal system, which refers to the consumption and nature of the raw materials used and their energy efficiency (see Decree No. 375/2006 declaring the application and execution of the "National Implementation Plan of the Stockholm Convention" to be of national interest).

[6] Colombia, Chile, Ecuador, Venezuela, Uruguay, Paraguay, Brasil, and Argentina.

communications system to allow a sustainable development with social justice within a regional integration framework and inserted into the globalized world.

In this context the Decree 561/1980 is passed, the first record of a long-term national energy policy. Its design was strongly supported by the desire to make viable the supply of natural gas from Argentina, the interconnectivity of electrical systems with Argentina and Brazil, and the reinterpretation of the seismic and geological studies in the continental platform, which could eventually lead to a possible hydrocarbon exploration.

Although the intentions of the energy policy were clearly based on the contents of the aforementioned decree, it must be highlighted that, in 1980, the Uruguayan government was already considering alternative sources to generate electricity.

In fact, the 1980 Decree entrusted the MIEM with the preparation of a plan to accelerate the study of alternative sources of energy such as solar, wind, biomass, and geothermal together with the instrumentation of measures to make them viable.

However, such plan, as well as the National Energy Model[7] to design the national energy policy, had to deal with the 1980 Plebiscite that sealed the fate of the de facto government making their decisions illegitimate for the people.

After the restoration of democracy, it was not until 2008 that a new global and national energy policy surfaced with the approval, by the executive power, of the energy policy 2005–30 that had been designed by the DNE, a decentralized body of the MIEM. This policy was endorsed by all political parties with parliament representation in 2010, therefore becoming a state policy.

The Energy Policy 2005–30 comprises objectives, goals, strategic guidelines, and actions to develop in four main theme areas that are as follows:

1. offer,
2. institutional,
3. social, and
4. demand.

Taking into consideration the importance of the area and the scope of this chapter, focus will be addressed to the demand area. The main objectives in terms of demand are

- to promote energy efficiency in all the sectors of national activity (industry, construction, transportation, agriculture, households, etc.) and for all energy uses (lighting, household appliances, vehicles, etc.) through a better use of energy resources, without reducing the levels of production, comfort and care to the people's daily needs, and promoting a cultural change regarding consumers' habits through the educational system, both formal and informal.

In terms of more specific objectives, it is expected the country will[8]:

1. have the necessary regulations and tax structure to promote energy efficiency, especially in terms of insulation materials for construction, lighting, vehicles, and other energy-consuming equipment;
2. have the financial means to promote technological and process modifications, both at an industrial and residential level, to improve efficiency in energy use; and
3. particularly for the transportation sector, achieve an energy perspective in terms of

---

[7] Article 8 of Decree 561/980 authorizes the Ministry of Industry, Energy and Mining to take measures leading to the urgent implementation of the National Energy Model, in order to have, as soon as possible, the basic technical-economic instrument to rationally project the future energy policy.

[8] Energy Policy 2005–30 <http://www.eficienciaenergetica.gub.uy/documents/20182/22528/Pol%C3%ADtica + Energ %C3%A9tica + 2005-2030/841defd5-0b57-43fc-be56-94342af619a0>.

transport state policies in order to promote changes in methodologies, means, and sources, which could improve energy efficiency in the sector.

In order to ensure the fulfillment of this objectives, the executive power devised a series of short-term (2015), mid-term (2020), and long-term (2030) objectives. In terms of energy efficiency the goals set were as follows:

Short-term goals (2015)	Mid-term goals (2020)	Long-terms goals (2030)
• A culture of energy efficiency will permeate society • The country will have national companies producing energy resources and supplies and developing energy-efficient processes	• Reduce energy consumption in a 20% compared to the trend scenario through a combination of actions promoting energy efficiency • The country will have companies that will act as regional leaders in the production of energy resources and supplies and developing processes, which will foster energy efficiency	• The country will have saved at least 10 billion dollars since 2010 by substituting energy sources and promoting energy efficiency, compared to the trend scenario • The country will have world leading companies producing energy resources and supplies and developing processes, which will foster energy efficiency

Even though this policy was approved in 2008, the EE Law[9] was passed in 2009. During this transition the paradigm of the policy shifted from energy efficiency to energy saving. At the time, Uruguay's electricity matrix was 100% hydrothermal and, as a consequence, very dependent on climate phenomena, alternating drought periods with floods in an evolutionary scenario of favorable energy demand.

Thus while the process of setting up new energy generation sources (nonconventional renewables) was set in motion, the strategy was to promote energy saving.

Decree 212/2008 passed a new Electrical Energy Saving Plan,[10] which consisted of prohibitions to energy consumption in the public sector[11] and recommendations to the commercial and residential sectors to apply measures to reduce their energy consumption, such as drastically reducing the power consumption of signs and buildings or any other decorative use of lights.

This saving plan was extended[12] by Decree 236/2008 partially suspended by Decree 418/2008 and completely eliminated by Decree 221/2010.

To understand these measures, it is necessary to understand the national energy context.

---

[9] Law No. 18597/2009.

[10] New with respect to previous plans approved by Decree No. 116/005 of 17/MAR/05 and its amendment No. 120/005 of 28/MAR/05.

[11] The working hours shall not extend beyond 6 p.m.; the lighting of facades and luminous signs shall be suspended; the lighting of the premises shall be reduced to a minimum; electrical equipment shall not be left in standby mode once the working day has ended; the use of air-conditioning and electric heating equipment, etc. shall be suspended.

[12] New mandatory measures were established: (1) For all users of energy supplied by the National Administration of Power Plants and Electric Transmissions: (a) disable half of the elevators per building body in cases where there are two or more units per body; (b) disable 50% of escalators; (c) do not maintain illuminated porches, facades, entrance halls of buildings, gardens or front retreats, for decorative purposes; except for lighting essential for security reasons. (2) For users of commercial, service and similar premises: (a) keep the electrical display equipment for sale off; (b) keep 50% of the interior lighting off; (c) keep 50% of the parking lot lighting off; (d) keep light signs off, with the exception of the open pharmacy signs; (e) keep 50% of the lights off in each shop window or similar. (3) Public performances and cultural, artistic, and sporting activities in the open air, which require electricity for lighting, shall be held exclusively during daylight hours.

At that time, Uruguay had a hydrothermal energy matrix that depended heavily on water resources. Therefore the adoption of saving measures, their suspension, and their elimination are explained by the evolution of the hydrological situation of the Rio Negro and Rio Uruguay river basins, where the country's four hydroelectric generation plants are located.

Depending on the state of the levels and the contributions to the reservoirs of the hydroelectric power plants, measures had to be taken to ensure the supply. These measures were embodied in the savings plans commented ut supra.

When the hydrological situation of the basins improved, it became possible to partially suspend the savings measures without this constituting a risk for the electricity supply.

The implementation of these savings plans showed valid and effective instruments to encourage the development of structural changes in the use of energy resources at the national level.

Thus the Institutional Energy Plan for executive power agencies[13] approved by Decree No. 527/2008 was created. Unlike the savings plans that responded to situations of supply risk, the Institutional Energy Plan is created as a preventive measure whose conceptual basis, unlike the savings plans, is energy efficiency, that is, the rational and efficient use of energy.

This denotes a change of paradigm and behavior.

Uruguay moved from the concept of energy saving to respond to supply crisis to the concept of efficient use of energy as a preventive and mitigating measure for shortage scenarios.

Although through this plan the concept of energy efficiency is incorporated into the national energy policy, a new saving plan was approved by Decree 154/2012. However, in this opportunity, the savings plan was transitory; in fact it was called "Interim Energy Saving Plan for the Public Sector" and attended a very concrete and transitory situation.[14]

After that, no other savings plans were implemented, presumably because the Energy Policy 2005–30 and its energy matrix modifications toward the use of renewable energy started to show their first results.[15]

In addition, the decrees that historically had promoted the daylight saving time to save energy during certain seasons, which had their first record in 1974 and its last provision dates from 2006,[16] were no longer implemented since 2015, with the implementation of Decree 178/2015.

It was argued that the circumstances that led to the decision of using daylight saving time measures had changed. It must be noted that by 2015, the year when the decree was passed, the EE Law was 6 years old. Consequently, the first short-term goal of the Energy Policy 2005–30 in terms of

---

[13]   The Plan establishes that all agencies of the Executive Power are obliged to develop and implement internal plans for the rational and efficient use of energy. Electricity saving goals will be set: the Plan orders a reduction in the purchase and then prohibits the purchase of incandescent lamps; it prohibits the purchase of T12 fluorescent tubes, among other measures.

[14]   Water supplies to the Salto Grande reservoir reached historic lows, the difficulties spread throughout the region, and it was not possible to ensure a constant flow of electricity exchange with neighboring countries; in addition, there was also an unusual growth in the electrical power demanded, higher than the year-on-year increases of recent years. <https://www.impo.com.uy/bases/decretos/154-2012>.

[15]   The relative weight of renewable sources increased from 38% in the period 2004–08 to 52% in the 5-year period 2009–13. By diversifying the matrix, dependence on climate and hydrological factors decreased. <http://observatoriosocial.mides.gub.uy/Nuevo_Test/midesv2/adjContents/adjcont455.pdf>.

[16]   Decree No. 311/2006.

modifications of the energy matrix toward the use of nonconventional renewables had been achieved.

Another note must be made about the implementation of the Energy Policy 2005–30 in terms of demand. Even though the transition toward a cleaner energy matrix with the incorporation of nonconventional renewables was developed due to the support of the executive power and the enactment of administrative general acts (decrees), in terms of energy efficiency, the transformation had a legal basis. The support of other transversal policies[17] also helped in transforming the energy matrix together with international commitments affirmed by Uruguay in terms of climate change (much later in time), which will be further addressed.

10.4 Law No. 18597/2009

The most relevant aspects of Law No. 18597/2009[18] are commented hereunder.

Article 1 of Law No. 18597, called EE Law declares the national interest of the efficient use of energy, in order to contribute to (1) national economy competitiveness; (2) the country's sustainable development; and (3) the reduction of greenhouse gas emissions.

With the EE Law the terms granted to the executive power for the elaboration of the Energy Efficiency Plan, its minimum content, and the periodicity of its revisions, among other aspects that are discussed below, are raised to legal rank.

The second article of EE Law provides a glossary for the purposes of the application and interpretation of the law. In that sense the efficient use of energy is defined as all the changes that cause a convenient reduction of the amount of necessary energy to produce a product or to satisfy the energy requirements of services, providing an equal or higher quality level and a reduction of the negative environmental impacts connected with the generation, transmission, distribution, and

---

[17]   Decree No. 310/2017 approving the National Climate Change Policy. Strategic and programmatic instrument prepared by the National System for Response to Climate Change and Variability.

[18]   Other legal and regulatory rules which are included in the current national regulatory framework in terms of energy efficiency are:

- Article 821 and 822 of Law No. 18719 incorporated subparagraphs 11 and 20 to article 1 Title 11 of TOCAF. They refer to taxing management of combustion vehicles and other equipment and artifacts with low energy efficiency.
- Law No. 18743 approves the Statute of the International Renewable Energy Agency (IRENA), signed in Bonn, Germany, on 2009.
- Law No. 18871 approves the Energy Security Treaty between Uruguay and Venezuela whose main objective is to develop an integration and cooperation process between both countries in order to promote a rational use of energy, leading to maximum savings and energy efficiency avoiding the waste of energy resources.
- Article 304 of Law No. 18996 exonerates FUDAEE from several taxes.
- Article 679 of Law No. 19355 approves an annual sum to be taken from general government revenues and sent to provincial governments to incentivize the maintenance of the executed plans and the development of those which seek to achieve an efficient use of energy.
- Law No. 19.646 approves the Free Trade Agreement between Uruguay and Chile which includes the promotion of regulatory measures and cooperation dialog to improve efficiency and effectivity in the development of regulations, technical regulations, and evaluation procedures. Cooperation in environmental areas will involve the design of special programs in areas such as renewable energy and energy efficiency.

consumption of energy. In addition, the concept of efficient energy use also incorporated the substitution of traditional forms of energy for nonconventional renewables that allow the diversification of the energy matrix and the reduction of polluting gas emissions.

At a glance, EE Law, in its first part, shows a classical conception of energy efficiency, while it incorporates a modern conception in its second part by including the concept of energy efficiency to entrepreneur activities meant to substitute the final customers use of traditional forms of energy for nonconventional renewable forms of energy that allow the diversification of the energy matrix and the reduction of polluting gas emissions.

Microgeneration or self-consumption businesses, those leading to the self-supply of electricity, must be considered under the light of EE Law as energy efficiency measures (MEE) and therefore benefitting from promotional instruments included in the law and regulatory decrees.

Under Chapter 2, "Jurisdiction and plan" (articles 3–8), legislators focus on the organic and institutional aspects of the Energy Efficiency Plan.

Regarding institutional aspects, article 3 entrusts the MIEM with the creation of the Energy Efficiency Unit (UEE), whose responsibilities and competences are established by the law.

The unit was created in order to establish the policies, rules, and infrastructure to enforce EE Law by creating the regulatory, technical, economic, and financial structure required to develop short-, mid-, and long-term policies, both economically and environmentally viable. Therefore the UEE could ensure a sustainable development, knowledge, and the raise of awareness among the population about the

efficient use of energy and the benefits associated with the responsible use of resources, as well as the dissemination of information about available energy sources and the impacts associated with their use.[19]

However, after EE Law was passed, article 403 of Law No. 18719/2010 stated that the DNE is responsible for promoting the efficient use of energy, making no reference to the UEE. This leads to the interpretation that the UEE was tacitly repealed. This interpretation can be confirmed today when observing the organizational scheme of the DNE where there is a new area for energy efficiency whose competences are the ones that were originally assigned to the UEE (Díaz Inverso, 2015).

Articles 5–8 of the EE Law modify some of the articles of Law No. 17.598[20] adding to the regulator some of the competences related to energy efficiency which were not included in the text of the same law.

Articles 8 and 9 focus on the coordination of sectorial policies to have an impact in terms of energy efficiency, particularly in terms of housing, transportation, and industrial development. This design is considered one of the objectives of the EE Law.

A similar consideration is made about provincial governments that will establish the minimum requirements of energy efficiency use for new buildings as well as in terms of public transport and street lighting, following the guidelines and rules of energy efficiency and environment established at a regional level and in coordination with the MIEM and the Ministry of Transport and Public Works, respectively.

Chapter 3, "Information to the public" (articles 11–15), legislates about the role of companies that provide energy services, both public and private, which must inform and encourage

---

[19] Article 3 of Law No. 18597/2009.

[20] Law on the creation of the Energy and Water Services Regulatory Unit (URSEA).

the efficient use of energy among their customers and by providing efficient consumption equipment using different instruments and commercial promotions associated with energy services billing.

Chapter 4, "About the mechanisms for the certification, promotion and financing of the efficient use of energy," (articles 16–23) legislates about the Efficient Energy Certificates (CEE) and the Uruguayan trust of saving and energy efficiency (FUDAEE).

More specifically, article 19, paragraph J, is the only disposition which makes reference to energy saving by describing FUDAEE's responsibilities: the administration of a contingency fund to be used during an energy supply crisis in order to finance plans to save energy and implement emergency operations in the energy market to secure supply continuity.

Chapter 5, "Of the elimination of taxing and administrative barriers for energy efficiency" (articles 24–28), enables the executive power to eliminate such barriers and to modify the Ordered Text of Accounting and Financial Administration (TOCAF), enabling the administration to hire public or private energy service companies (ESCOs) directly, as long as they are registered under MIEM and develop their activities under the Performance-based Remuneration Contracts, in which the investment is fully or partially funded by the ESCO.

Finally, the EE Law describes sanctions for those which use energy inefficiently or in a lavish way in contexts of supply crisis and establishes the executive power will rule over these cases.

The EE Law was implemented and detailed by, among others, Decree 86/2012, which approves the FUDAEE[21] and Decree 211/2015, which approves the National Plan of Energy Efficiency 2012–24.

10.5 The National Plan of Energy Efficiency

Article 4 from EE Law entrusted the MIEM with the design of the National Plan of Energy Efficiency (EE Plan) in coordination with other ministries and institutions in order to be approved by the executive power.

The minimum content for the EE Plan derives from paragraphs A to H from the aforementioned article and the action lines in terms of energy efficiency predefined by the Energy Efficiency Policy 2005–30, previously discussed in this chapter. The EE Plan must contain[22] the energy consumption avoidance goal for the whole period and the annual energy avoidance targets for each period.

The EE Plan was approved by Decree 211/2015 and included a total avoided energy target of 1.690 ktep between 2012 and 2024 (Graph 10.1), in a context of growing and sustained evolution of the consumption of electric energy (Graph 10.2).

Besides setting the goal for energy avoidance, EE Plan contains all the instruments to achieve those goals. EE Plan identifies general instruments which are transversal to all sectors and sector-specific instruments aimed at a specific group of the public, which is segmented from the rest due to their consumption characteristics.[23] In this chapter, focus is addressed on the general instruments.[24]

---

[21] Trust Agreement entered into by the Ministry of Economy and Finances (MEF) and MIEM on December 29, 2011.

[22] Subparagraph H of Article 4 of Law No. 18597/2009.

[23] See National Energy Efficiency Plan 2015–24. <www.eficienciaenergetica.gub.uy>.

[24] Those that are transversal to various sectors of activity and differ from those instruments of particular scope intended for a segmented sector of activity.

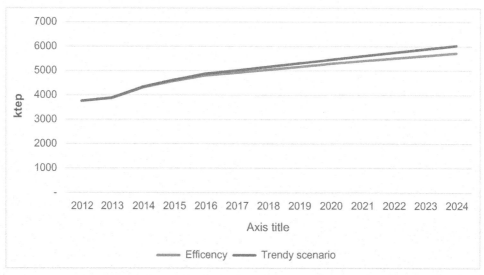

GRAPH 10.1   Energy consumption in a trendy scenario versus efficiency scenario. Source: *Based on DNE.*

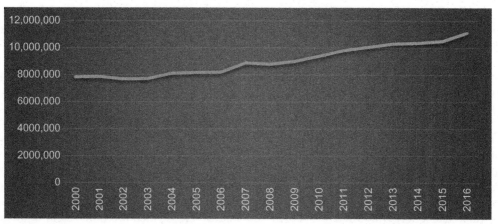

GRAPH 10.2   Final total energy consumption. Annual net electricity demand (MWh) (2000–16). Source: *Based on ADME.*

10.5.1 The legal and institutional framework

Energy efficiency is clearly a key component in a long-term energy policy. Having a solid, stable, transparent, and long-term legal and institutional framework is a very favorable, if not indispensable, condition to fulfill a transformation as the one pretended in this scenario. In this aspect, results are observable.

From a regulatory point of view the Energy Policy 2005–30, the EE Law, the international commitments to mitigate climate change, and the regulatory decrees are all into the same line. From an institutional point of view the leadership of the energy policy, an organic and adequate structure, the involvement of decentralized structures (provincial governments), academic organizations (universities), and the private sector (consultants, ESCOs, banks, and other structures) enable the functioning of FUDAEE as the main financial tool for the promotion of energy efficiency.

10.5.2 Access to information, education, and cultural change

According to the EE Plan, some of the actions leading to raising cultural awareness and democratizing the access to information about energy efficiency require more in-depth educational activities in topics such as energy, renewable energy, and energy efficiency by incorporating them into elementary and secondary schools curricula; designing communication campaigns in massive media; design and promote informational materials; and optimize web tools and simulators, among other actions.

On this matter, it must be highlighted that the DNE launched a contest called "EE in Educational Institutions and UTU[25]" which brings energy efficiency closer to students and teachers in formal education through a "School Divulgation" program in which UTE (National Electricity Provider) provides schools with talks about the importance of energy efficiency and the use of energy in the preservation of the environment for school children. In higher education levels the creation of the "Renewable Energy Engineering Degree" program at the Technological University of Uruguay (UTEC) must be noticed. Within its curriculum, this program expects to offer a full unit dedicated to energy efficiency.[26] Other higher education programs include postgraduate programs at University of the Republic (UDELAR) and other private universities.

In addition, since 2009, MIEM publicly awards with the National Prize of Energy Efficiency those organizations which have made important efforts to achieve goals in relation to saving and using energy efficiently and who deserve to be noticed and make visible by means of spreading their achievements in the area and their institutional commitment to the environment and efficient consumption. Over 60 initiatives have received this award in the last 10 years.

Regarding communication campaigns the topic has been present in different media such as radio and television through awareness-raising campaigns as well as in the access to web simulators and informational brochures about the EE Plan and the available instruments for its promotion.

Nonetheless, based on Decree 1/2019, which approves the 4th Open Government National Action Plan 2018–20, there is currently no public systematized information available, which could help in monitoring the evolution of the energy efficiency policies. Because of this, the national government sets a new goal for the end of 2019 to process, analyze, and return results through a report and by 2020 to publish the new data in an open format.

10.5.3 National Labelling System

Within the framework of the Labelling and Normalization of Energy Efficiency Program,[27] a National Labelling System (SNE) was implemented in order to classify products and equipment according to their energy efficiency in a simple and graphical way for a rapid interpretation from consumers.

According to Decree 429/2009, all equipment and artifacts in the national territory which consume energy from any source must

---

[25] Polytechnic School of Uruguay.

[26] On this subject see Mathisson et al. (2018).

[27] Institutionalized by Law No. 18597/2009 and Decree 429/009.

be evaluated and labeled according to the UNIT standards.

Decree 359/2011 states that MIEM will determine the methods and time limits for the incorporations of equipment to the SNE.

Since part of the equipment that is included in the SNE has been imported, Decree 116/2011 regulated the requirements for their access to the country through the Regulatory Unit of Energy and Water Services and the National Customs Bureau.

According to Uruguay Report 2017,[28] in 2016, air-conditioning units and heating bombs (for domestic use or similar up to 6 kW of refrigeration capacity) were included in the SNE. Previously, in 2011, the SNE had incorporated fluorescent lamps[29] and, in 2012, the electric water heaters and electric refrigerators for domestic use.[30]

10.5.4 Economic and financial instruments of promotion

The main mechanism to make the transformation toward a more efficient use of energy viable is undoubtedly the FUDAEE.[31]

In the case of FUDAEE the trustors, the ones who transmit the funds (property rights or other real and personal rights) for their administration, are the MIEM and the Ministry of Economy and Finance (MEF).

The trustee, which is in charge of administering the funds, is the National Development Corporation (CND), which is a public nongovernmental entity enabled as a trustee agent by the Central Bank of Uruguay. The funds the CND administer constitute an independent asset, that is, they cannot be confused with the assets of the trustee.[32]

With the funds transmitted by the ministries, the CND will finance actions leading to the achievement of the avoided energy goals set as one of the objectives of the EE Plan.

The transmitted assets or funds destined to the promotion of energy efficiency are found in article 21 of EE Law and, in summary, they come from the following sources:

1. Contribution of the energy services providers[33]—0.13% of their total annual energy sales (after 5 years this contribution can be increased up to 0.25%) through monthly payments based on the projected annual sales of the company. A total of 30% of this contribution can be materialized through energy efficiency certificates.
2. A total of 1% of the total investments in new fossil-fuel generators whose purpose is the trading of most of the energy generated to third parties (excluding cogeneration).
3. Funds from fines applied to energy users for inefficient and lavish practices.

---

[28] <http://200.40.96.180/images/ReporteUruguay2017.pdf>.

[29] Previously under transitional regime approved by Decree 428/009.

[30] Previously under transitional regime approved by Decrees 430/009 and 329/010.

[31] Article 1 of Law No. 17703/2003 defines the trust as the legal arrangement through which the trust owns a group of property rights and other real and personal rights, which are transmitted from the trustor to the trustee for them to be administered accordingly to the instructions contained in the trust, in benefit of a person (trustee), which is designed by himself, to be returned at the end of a term or condition to the trustor.

[32] Laws No.15.785/1985; 17.703 /2003 and 18.597/2009.

[33] Contributors: National Administration of Power Plants and Electric Transmissions (UTE), National Fuel, Alcohol and Portland Administration (ANCAP), Montevideo Gas and Conecta S.A.

4. Funds from international donations or loans, or other external sources which do not have a specific destination and are explicitly destined to be used for the promotion of energy efficiency and the reduction of greenhouse gas emissions.

5. State contributions determined by the executive power according to its budget regulations regarding promotion, savings, and efficient use of energy.

6. Funds from differential tax rates to inefficient equipment.

Once the trustee, the CND in this case, receives the fund, it must be administered according to the fiduciary instructions sent by the trustors (the ministries). These instructions are contained in the FUDAEE Operations Manual approved by Decree 86/2012. They describe administration and finance aspects of the instruments that will turn into the concrete actions leading to the achievement of the avoided energy goals.

These actions are described in the following subsections.

10.5.4.1 The energy efficiency trust

As with FUDAEE, the energy efficiency trust (FEE)[34] is structured as a trust. In this case the fund is managed by the National Guarantee System (SiGa). Companies that are interested in developing energy efficiency products and who lack investment credit can get access to a guarantee, which will back the requested loan to banks affiliated to the SiGa system.

These guarantees can cover up to 60% of the total financed amount. The maximum amount will depend on whether the technical endorsement is provided by a Category A ESCO, in which case the amount of indexed units is 824,000, or a Category B ESCO, in which the amount is 277,000 indexed units.[35]

In addition, the loan cannot exceed the 80% of the total investment for the project and the credit term cannot exceed 6 years. Finally, users who are interested in getting access to a guarantee must provide, at least, a minimum of 20% of the investment amount and the project must fulfill the energy efficiency condition (benefit−cost ratio less than 1).[36]

10.5.4.2 Energy efficiency certificates

The emission of CEE seeks to reward and incentivize energy saving generated by projects that implemented MEE. The conditions to get access to these certificates are described in article 16 of EE Law. It established that CEE can be provided to all energy users or energy service providers who present energy efficiency use projects implemented at least during the year prior to the application for the certificate and who have already been through one annual evaluation of results by a certified agent in energy saving.

The certified agent certifies the amount of energy avoided during the implementation of a project using the international performance measurement and verification protocol, which is applied compulsorily for measures with total savings equal or higher than 500 tep.

Once the saving has been certified and the legal and regulatory requirements are met, the MIEM issues the certificate.

According to EE Law, these CEE have a value in energy units, which equals the total number of energy avoided units during the

---

[34] Article 19 subparagraph (B) of Law No. 18597/2009.

[35] The value of the Indexed Unit as of May 2019 is $4.1723 (Uruguayan Pesos).

[36] To May 2019 the financial institutions which have joined this system are: Itaú Bank; Bank of the Eastern Republic of Uruguay; Bandes; BBVA; HSBC; Santander Bank; Scotiabank; Acac; Fucap y Fucerep. <http://www.eficienciaenergetica.gub.uy/fideicomiso-de-eficiencia-energetica-fee->

project's *life*, which results from the sum of the energy saving estimated during the implementation of the project based on technical parameters and the ponderation of energy avoided that the MIEM defines in each convocation.

The administration of certificates transactions and the responsibility for ensuring the transparency of the CEE market are responsibilities of FUDAEE; whose funds are also destined, among other activities, to finance the audit and control of the CEE.[37]

According to article 20 of EE Law, the MIEM must define the annual program of market operations and the reference prices for CEE transactions by FUDAEE in order to allow the fulfillment of the energy avoided goals outlined by the EE Plan.

Other legal regulations in this subject are contained in Decree 46/2016 that approves the use of the CEE Manual and the especial conditions of each CEE convocation approved by MIEM.

According to the energy efficiency area report[38] regarding the results of the second convocation in 2017, after evaluating the process, 248 MEE implemented by 59 companies and public institutions were approved. They were distributed into 188 different locations, 18 provinces, 3 of them received a total benefit of US$46,381,316.

10.5.4.3 Technical–financial assistance lines

This modality of financial aid reimburses two-thirds of the total costs for feasibility studies as well as other studies required for the preparation of projects that may increase energy efficiency.

The maximum amount to be reimbursed is US$3300 (without tax). Final users who require a detailed study from an ESCO, including the characterization of all uses and sources of energy in their facilities and which promotes the use of MEE in terms of use and main sources can apply to technical–financial assistance lines (LAEE).

According to the report "Results of the Second Convocation (2a LAEE),"[39] 47 studies carried out by 12 different ESCOs were benefited by LAEE. Ten of the companies and institutions which benefited from this assistance were small and medium businesses that were reimbursed a total sum of US$130,357.

10.5.4.4 Promotion and investment protection law

Law No. 16906/1998 (Investment and Industrial Promotion Law) declares of national interest the protection of national and foreign investments in the national territory. Article 11 states that companies whose investment projects are promoted by the executive power can get access to the benefits established by the abovementioned Law.

Under Decree 354/2009 the executive power declared the following activities, among others, as protected by article 11:

1. transformation of solar energy into thermal energy;
2. conversion of equipment and/or incorporation of processes that lead to the efficient use of energy;
3. services provided by ESCOs registered in the National Bureau of Nuclear Energy and Technology and qualified as Category A; and

---

[37] See article 19 of Law No. 18597/2009.

[38] <http://www.eficienciaenergetica.gub.uy/documents/20182/56464/Informe + cierre + CEE + 2017_v02.5.pdf/73c78fef-3dc1-4a4f-a10a-36fa9fdbc0d8>.

[39] <http://www.eficienciaenergetica.gub.uy/documents/20182/56464/Informe + cierre + 2%C2%AA + LAEE_v1.pdf/49a668dc-88f1-41b3-9ad2-d98a9c326db2>.

4. the national manufacture of machinery and equipment destined the activities mentioned above.

The decree also provides for a specific tax exemption regime for these activities, which translates into exemptions with variable and scalable percentages in the economic activities income tax (IRAE).

Law No. 16906 was regulated by different decrees.[40] Presently, the current regulation is the Decree 143/2018, which incorporates cooperatives to the list of subjects who can benefit from the tax exemptions.[41]

This decree is more extensive than Decree 354/2009 in terms of fiscal benefits since it allows projects to obtain tax exemptions not only form IRAE but from other taxes, such as the estate tax on movable property, as long as they are not benefited by other instruments for their entire lifespan; exemptions of rates; and taxes over imports of movable property and materials destined to civil works and the return of added value taxes for purchasing materials and services destined to civil work.

Micro and small businesses,[42] which meet the requirements described in article 27 of the aforementioned decree, will also receive an additional 20% IRAE benefit when they meet the general criteria requirements.

According to information from the MEF, in 2018, 597 projects were benefited by Law No. 16906, which included energy efficiency projects and, of which, 70% were awarded to small and medium businesses (PYMES).

According to the cleaner production investment indicator (P + L), the approved projects plan to invest US$423.1 million with the intention of increasing global efficiency and reducing risks to humans and the environment.[43]

10.5.5 Support programs for small and medium businesses

The EE Plan promotes consultancy by specialists in energy efficiency who act as shared energy managers, that is, they advise a group of companies. Other forms of action include support for professional development for business owners, fostering relationships with provincial governments, generate and spread information, and identify specific forms of consumption by type of equipment or standard installations in each field in order to enable businesses to find areas of improvement.[44]

10.5.6 Energy management

Developing human resources for the application of the ISO standard no. 50001 on energy management[45] will contribute to the achievement of the objectives. For this reason, it is essential to raise awareness among business owners about the convenience of having an "energy manager" for the control, monitoring, planning, and execution of actions leading to an improvement in the use of energy.

For PYMES whose income volume does not justify the incorporation of a manager, the

---

[40] Previously, Decrees No. 455/2007 and 2/2012.

[41] Previously, only IRAE taxpayers could avail themselves of this investment promotion scheme.

[42] Definition provided by Decree 504/007.

[43] <https://www.mef.gub.uy/innovaportal/file/26286/2/informe-anual-de-proyectos-presentados-ano-2018-.pdf>.

[44] See Energy Efficiency Plan on page 20.

[45] ISO 50.001 is an international standard developed by the International Organization for Standardization (ISO) based on the continuous improvement management system model, as well as ISO 9.001 (Quality Management) and ISO 14.001 (Environmental Management). This standard helps organizations integrate energy management into their overall management system, reducing energy costs and greenhouse gas emissions.

generation of resources to share best practices and even promoting the financing of technical assistance through different mechanisms, such as LAEE is encouraged.

10.5.7 Strengthening the capacity of consultants and providers

The EE plan includes different actions leading to improve the performance of consultants and providers. Among them, there are actions which seek to improve the registration and categorization of ESCOs and energy consultants as well as efficient equipment providers. These actions encourage their training and further specialization, project dissemination, generation of alliances, and working connections, among others.

Regarding actions that are already in practice, the updated registry of ESCOs and energy consultants is currently carried out by the DNE as well as the annual convocations for categorizing ESCOs into A or B categories: the training of certified evaluators in the Standard ISO 50001 as well as the certifying agents for the measure of energy efficiency. In addition, the legislation is promoting contracts with ESCO under the Performance-based Remuneration Contracts scheme based on article 25 of the EE Law, article 59 of Law No. 18834/2011 and its Regulatory Decree 289/2015, and paragraph 26, article 33 of TOCAF.

10.5.8 Buildings energy performance

Improving buildings' energy performance was proposed as another key element to reduce energy demand. The MIEM is moving forward toward the creation of a labeling system for houses. In 2014 the UNIT committee for energy efficiency was created to design technical standards for the evaluation and certification of energy performance in buildings.

According to MIEM,[46] there are currently 27 published standards and 8 drafts. In addition, in 2015, 5 typical meteorological years were generated for five different locations in Uruguay. This tool, among others available, allows to evaluate and model the thermal behavior of buildings. Finally, in 2017, an initiative was launched to support energy efficiency in houses by providing loans to improve housing through MEE. For this purpose, 80 architects were trained to perform these duties.

10.5.9 Other promotion instruments

Among other instruments the possibility of resorting to international financing was contemplated by defining promotion measure and supporting energy efficiency projects as part of the Uruguay's national commitments under the Paris Agreement and the Sustainable Development Objectives 5, 7, 8, 9, 10, 12, 13, and 17.

The first nationally determined contribution (NDC) was approved by Decree 310/2017 in which the government commits to reduce the emission of gases through the implementation of actions that will promote energy efficiency at residential, industrial, and transportation levels. More specifically, the country is developing regulations and incentives to promote the use of energy-efficient building materials as well as low-consumption lightning systems and modifications to the public transport system.

When Uruguay communicated its first Nationally Determined Contribution (NCD) in the framework of the Paris Agreement,[47] it listed the main mitigation measures being

---

[46] <http://www.eficienciaenergetica.gub.uy/edificaciones>.

[47] Adopted in Paris, French Republic in 2015, signed by Uruguay in New York, United States of America, on April 22, 2016, and incorporated into the national legal system through Law No. 19.439/2016.

implemented or to be implemented in order to achieve the unconditional objectives of mitigation. Among them, in term of energy efficiency, Uruguay committed to[48]:

1. replacement of 4 million incandescent lamps in the residential sector with efficient equipment and 30% of LED luminaires for street lighting in 2025;
2. mandatory energy efficiency labeling for household equipment by 2025;
3. regulations of energy efficiency labeling for houses and new building in 2025;
4. implementation of a pilot program in Montevideo to improve energy efficiency in houses and the CEE and the mandatory labeling of internal combustion light vehicles in 2025;
5. incorporation of the Touristic Green Seal in 4% of tourist accommodation establishments; and
6. establishing a laboratory for studying energy efficiency and gas emissions by 2025.

As can be seen, when Uruguay ratifies the Paris Convention, it already had a path traveled in terms of energy efficiency since Law 18.597/2009.

At the time of incorporating the aforementioned agreement into the national legal system, Law 18.597 had been in force for 7 years and had been effectively implemented, which together with other measures, related to the energy sector and included in Energy Policy 2030, also gave Uruguay a reasonable and favorable position in the subsequent work of defining the first NDC and of identifying the mitigation measures implemented and to be implemented in compliance with such objectives, the so-called Nationally Appropriate Mitigation Actions that had to be communicated to the convention within the framework of the Paris Convention.

Uruguay therefore finds harmony between international requirements and its regulatory framework, which allows it to direct the same efforts toward compliance with the NDC and its own national goals.

10.6 Conclusion

Uruguay has a clear, comprehensive, and legally hierarchical regulatory framework for energy efficiency.

Through the Energy Policy 2030 horizon for the demand theme area, Law No. 18597/2009 and the National Energy Efficiency Plan 2012–25, Uruguay anticipated the requirements of the international community for compliance with national environmental commitments to mitigate the effects of climate change.

Uruguay successfully underwent a paradigm shift in energy use.

As analyzed in this chapter, Uruguay went from designing energy saving plans to mitigate the effects of supply risk caused by climatic phenomena derived from situations of force majeure and to generating a policy of rational and efficient use of energy, as a preventive measure, in the long term, aimed at modifying human behavior, which is what any public policy should aim at.

The public policy toward the efficient use of energy was successfully incorporated into the national legal system. This incorporation was not limited to generating programmatic regulations but also considerably stopped in the design of a legal engineering that would facilitate the achievement of the agreed goals through a diversity of tools and mechanisms that cut across all sectors of activity, generating regulatory incentives, and committing citizens from an initial education, to social awareness.

Although common and transversal mechanisms were studied in all consumption sectors,

---

[48] <file:///D:/Datos/Descargas/Uruguay_Primera_Contribucion_Determinada_a_nivel_Nacional%20(2).pdf>.

the implementation of MEE has an uneven development depending on the sector of activity to which they are directed (industrial, commercial, public, etc.). However, incentives, particularly financial ones, are already reflecting results aimed at meeting the avoided energy target.

As a challenge identified by the current government, there is still no systematized public information to monitor the evolution of energy efficiency policies in Uruguay and to identify opportunities for the development of new projects.

In this sense, if the commitment assumed in the 4th National Action Plan of Open Government of Uruguay 2018–20 is fulfilled, in the first semester of 2019, there will be available a national georeferenced map with the different energy efficiency projects developed as well as the promotion instruments used and their location, so as to finish in the first semester of 2020 with the publication of data in open format, which allow democratizing the results of the implementation of the National Energy Efficiency Policy.

The stability of the energy policy, which became a state policy, the National Energy Efficiency Plan by 2024 with predefined periodic reviews, the institutional and legal stability of the country, and an early and advanced paradigm shift generated that the various actors, both public and private sector, contribute responsibly to this energy transformation toward sustainability and care for the environment in line with the objectives and commitments assumed by Uruguay at the international level.

References

Cajarville Peluffo, J., 2011. "El Poder Ejecutivo como conductor de Políticas Sectoriales en la Legislación Uruguaya" en: "Sobre Derecho Administrativo". [The Executive Power as a guide to sectoral policies in

Uruguayan legislation] in "Sobre Derecho Administrativo" [Administrative Law], Administrative Law Studies. FCU, Montevideo, 2008, vol. II.

Cajarville Peluffo, J., 2008. "El Poder Ejecutivo como conductor de las políticas sectoriales en la legislación uruguaya" [The Executive Power as a guide to sectoral policies in Uruguayan legislation] in "Sobre Derecho Administrativo" [Administrative Law], second ed. FCU, Montevideo. vol. I.

Díaz Inverso, R., 2015. "Marco jurídico de la Eficiencia Energética: Concepto y sujetos" en "Derecho de la Energía" [Legal Framework for Energy Efficiency: Concept and Subjects in Energy Law] Ed. University of Montevideo.

Mathisson, J.F., Coleto, M., Calcango, M., Suárez, G., Aguiar, M., 2018. Nuevas carreras en energías renovables en Uruguay; Ingeniería y Tecnólogo. Avances en energías renovables y medio ambiente [New degrees in renewable energies in Uruguay: engineering and technologist. Advances in renewable energies and the environment]. Argentina.

Ministry of Economy and Finances, 2018. Submitted Projects. Investment Promotion Scheme (Decrees 455/007, 002/012 and 143/018) and Sectorial Regimes (Decrees 329/016 and 110/016).

Sayagues Laso, E., 1987. Tratado de Derecho Administrativo [Treaty on Administrative Law] vol. I. fifth ed. FCU.

Further reading

Compiled Text of Regulations for the Electrical Sector, 2019. <http://www.ursea.gub.uy/Inicio/Energia_Electrica/Marco_Normativo/TCO/>.

Rodríguez D'Espada AL, y otro, 2014. "Energías Renovables. ¿Cómo lograr el cambio de paradigma?"[Renewable Energies: How to achieve a paradigm shift]. Law Review of the University of Montevideo. vol. 13, number 26.

Eastern Republic of Uruguay, 2017. First Nationally Determined Contribution to the Paris Agreement. <file:///D:/Datos/Descargas/Uruguay_Primera_Contribucion_Determinada_a_nivel_Nacional%20(2).pdf>.

Energy Efficiency Certificates (CEE), 2017. Convocation Results. <http://www.eficienciaenergetica.gub.uy/documents/20182/56464/Informe + cierre + CEE + 2017_v02.5.pdf/73c78fef-3dc1-4a4f-a10a-36fa9fdbc0d8>.

Energy Efficiency. <http://www.eficienciaenergetica.gub.uy/documents/20182/22528/Pol%C3%ADtica + Energ%C3%A9tica + 2005-2030/841defd5-0b57-43fc-be56-94342af619a0>.

<http://www.eficienciaenergetica.gub.uy/edificaciones-Building-Energy-Performance>.

<https://www.mef.gub.uy/innovaportal/file/26286/2/ informe-anual-de-proyectos-presentados-ano-2018-. pdf>.

Impo. Official Gazette. <www.impo.com.uy>.

Ministry of Industry, Energy and Mining, 2008. Energy Policy 2005—2030. </www.eficienciaenergetica.gub.uy/>.

National Plan of Energy Efficiency, 2015. <http://www. eficienciaenergetica.gub.uy/documents/20182/29276/S %C3%ADntesis_Plan_Nacional_de_EE.pdf/508a0c1d-a566-4d1d-b0a0-89601ca0ffbf>.

Uruguay, 2015. Report. <http://observatoriosocial.mides.gub. uy/Nuevo_Test/midesv2/adjContenidos/adjcont455.pdf>.

Uruguay, 2017. Report. <http://200.40.96.180/images/ ReporteUruguay2017.pdf>.

Wikipedia. <https://es.wikipedia.org/wiki/Gregorio_% C3%81lvarez>.

The Ecuadorian energy matrix: *from a fossil-fuel dependency to a renewable production*

Pablo Morales and Edisson Morales

Morales & Asociados Law Firm, Quito, Ecuador

11.1 Introduction

Energy transition is defined as a transformation in the energy sector (Hauffe et al., 2014); however, in the past, energy transition defined the need to interconnect local systems to a global path. Nowadays, when talking about energy transition, it refers to the abandonment of fossil-fuel dependency in the energy generation (IRENA, 2019).

Energy transition during the years has been carried out by technology and the idea of making its production more efficient, especially by reducing the ecological footprint. This fact has had a positive impact in the environment and it is why today the idea of energy transition is to abandon carbon and oil-based energy. Historically, Ecuador had an oil dependency to generate energy, therefore this chapter will analyze and determine what changes has Ecuador implemented regarding energy transition.

The actions procured by the government will be studied.

The review of the Ecuadorian energy sector will be divided into three parts, always presenting the legal architecture of Ecuadorian regulatory policy in the electricity market. The first part will show how energy is introduced in the country, how the primary system began to work, and how much oil was needed to produce energy; all of these in a phase called developmentalism or structuralism.[1] Second, the transition to market liberalization. And lastly, neo-developmentalism will be analyzed considering the actions that Ecuador has taken since 2008.

The developmentalism process means the elaboration of a legal structure, where laws and rules regarding energy matters are issued by the national governmental monopoly. The electricity journey in Ecuador started in 1897, when, in Loja, the first electricity utility *Luz y*

---

[1] Definition used by the United Nations Economic Commission for Latin America and the Caribbean.

The Regulation and Policy of Latin American Energy Transitions
DOI: https://doi.org/10.1016/B978-0-12-819521-5.00011-5

Fuerza began operations. After that, some of the biggest cities in the country followed. While electricity as a necessity grew, the State had to invest in the sector; that is why in 1926 the government decided to contract the company American & Foreign Power Co. to supply electricity to Guayaquil. During the 1940s, the entire infrastructure—built mainly by private investment—became a responsibility of the local governments, only a small public investment was introduced—mainly in small hydroelectric power plants.

This initial phase was characterized for the lack of planning and adequate policies (CONELEC, 2015b). Until the 1950s the electricity sector was rather small and could only cover the needs of 17% of the population (Echeverria, 2006, p. II-2). Finally, in 1961 the government made the decision to create an electricity institute to regulate and develop the sector.

The liberalization process and the introduction of economic regulation began in Ecuador in October 1996 with the issue of the legal framework "Electricity Sector Law" LRSE.[2] This law was meant to transform the electricity sector, introducing the participation of private investment and the creation of public institutions to regulate the entire sector. Due to the introduction of transparency and nondiscrimination as basic principles of the market players' behavior, by the year 1999 investment in the sector grew 23% (Neira, 2003, p. 4). This meant that regulation was adequate. Nevertheless, a large number of deficiencies continued and oil dependency too.

Neo-developmentalism began in Ecuador in 2008 with the introduction of the new constitution and the return of the State as center of the energy sector development and regulation. Since then, regulation has been State centered and focuses on departing from oil dependency and attracting foreign investment to generate clean energy.

Sector-specific regulation will be the basis of this chapter. A review of the competences given to the regulatory authority over the years is vital in understanding how the market shortcomings are fought through a precise set of regulatory tools. First, the behavioral tools dealing with regulated access and tariffs. Second, the structural tools tackling issues stemming from vertical integration. This overview will show how the Ecuadorian regulatory authority shall behave and which faculties need to be sharpened to obtain better and faster results.

11.2 Developmentalism

During the first few years of the electricity industry in Ecuador, there was a lack of special sector regulation, and electricity utilities became a decision from local governments. The first electricity utility was *Luz y Fuerza*, a company established in Loja, Ecuador in 1897 (CONELEC, 2015b, p. 26). Small undertakings were created to cover regional necessities and growth. In the 1950s only 17% of the population had access to electricity, which was generated mainly by private undertakings. At that time the State only had a small investment in hydroelectric power plants (Echeverria, 2006, pp. II-2). This scenario changed by the international obligations acquired in 1961 by the "Carta de Punta del Este" (Garcia, 1967) that aimed at generating economic growth. This document stated the need to create supervision institutes for the energy sector (CONELEC, 2015b, p. 27). In that same year the executive order no. 24 (Decreto Ley de Emergencia, 1961) was issued to create the Ecuadorian Institute of Electrification (INECEL) with the intention to regulate and plan the development of the sector. It also had the faculty to set tariffs, build new infrastructure, and operate all

---

[2] Ley de Régimen del Sector Eléctrico "LRSE", 1996. Congreso Nacional, Quito.

the levels of the supply chain (CONELEC, 2015b, p. 25).

Therefore when INECEL was created, a vertically integrated monopoly was formed, because this institute could regulate tariffs, plan the sector's future, and also build hydroelectric and thermoelectric power plants. The change in the entire structure was a result of the State's need of developing this sector, to interconnect, regulate, and serve electricity utilities as soon as possible.

In 1964 the scenario for the electric industry was segmented by small electric power plants. Approximately 1100 electric power plants generated almost 190 MW. Then, in 1967 the number of electric power plants increased to 1218. Out of these electric power plants, 60% were thermoelectric power plants. These numbers show the constant growth of the industry and the need for specific regulations. It also made Ecuador's dependency on fossil fuels to generate electricity visible.

State developmentalism grew in the 1970s with the military dictatorship of General Guillermo Rodríguez Lara and the oil boom. The income from the oil boom in the 1970s and 1980s allowed INECEL to create the Interconnected National System, the purpose of which was to integrate, standardize, and socialize the service (CONELEC, 2015b, p. 29).

At that time the creation of a vertically integrated monopoly was thought beneficial for the market, since it could bring:

- synergies and reduction of common costs existing in the generation, transmission, and distribution of electricity;
- better structure coordination and planning;
- scale economies on every level of the supply chain;
- cross subsidization of less profitable levels with monopoly rents;
- harmonic and planned investment in future generation; and

- positive externalities on the network reliability.

Considering the windfall profits coming from the oil exploitation—during the National Revolutionary government—the Electrification Basic Law (LBE for its acronym in Spanish) was issued on September 10, 1973 (Electrification Basic Law (LBE), 1973). This act created—in its Article 22—a public fund based on the petroleum rent. The fund was constituted by 47% of the income arising from of oil sales, which helped INECEL build hydroelectric power plants that aimed at electrifying rural counties (CONELEC, 2015b, p. 29).

During developmentalism the electric sector was declared as a public utility service of national interest[3] and was managed by INECEL. Following these thoughts, Article 3 of the LBE declared that only the State—through INECEL—can generate, transmit, distribute, and trade electricity.

In the same act, Articles 8 and 12 gave some prerogatives to INECEL and its directory. Among others, their competences were to:

1. program, coordinate, execute, and supervise the development of all the phases of electrification in the country, in accordance with the national energy policy;
2. inventory the country's energy resources for the purpose of producing electricity;
3. project, finance, construct, acquire, and operate works for the generation, transmission, and distribution of electrical energy, as well as the facilities and equipment necessary for its operation;
4. manage and obtain financial resources;
5. approve and modify the internal regulations of INECEL;
6. approve the tariffs for electric power services;

---

[3]   Article 1. LBE.

7. approve bidding rules, establishing formulas, and special price adjustments; and

8. authorize contracting.

In the mid-1980s the oil boom came to an end and its prices went down. It affected the economy and the *aggressive external debt* (Origen de la deuda externa comercial, 1976–1982) as a fiscal policy also came to an end. Having established a vertically integrated monopoly, INECEL kept aside the operation of independent generators (Ibarburu, 2001, p. 4). Empiric evidence shows that during the 35 years of INECEL's operation, the entire sector became part of the political agenda and was managed without real technical and economical understanding. The tariffs were politically managed and normally did not cover the total cost of the undertakings. The granting of subsidies was necessary to maintain the social electricity tariffs. The State was not able to maintain the network, and during 1992 the country suffered severe electricity shortages.

The existence of a vertically integrated monopoly, low oil prices, and external debt led to a shift in the political perspective regarding electricity. Therefore the sector was declared in emergency[4] and a global consultancy was ordered for the companies ENDESA and Price Waterhouse. This new perspective came from the neoliberal trend, imposed by the World Bank and the International Monetary Fund—Ecuador's major money lenders.

11.3 Liberalization

Considering that during developmentalism there was a lack of competition in the market,

the liberalization trend aimed at changes in the previous model, introducing new policies and competitors for the electricity market. Various countries around the globe did this, Chile (1982) and England (1990), for example, and their successful reforms derived in the creation of a new regulatory model. In addition, the European Community introduced in 1988 various proposals for liberalizing the sector. However, there was strong opposition by some member states. Finally, in 1996 after years of lobbying, Directive 96/92/EC (1997) was approved.[5]

According to Joskow (2008), there is a "standard recipe" to apply in the electricity policy (Ibarburu, 2001, p. 7):

- Separate the sector's different markets. Differentiate them into generation of electricity, transmission of electricity through high-voltage grids, distribution through lower voltage grids, and supply to end users. The differentiation can be greater, depending on the market's characteristics.
- Generate competition in markets not characterized by natural monopoly (subadditivity). In this case, introducing competition mainly on the generation granting equal treatment for newcomers and private and public investment. This should lead to the creation of an electricity market, where this commodity can be traded on the basis of supply and demand only.
- Strong regulation for the markets characterized as natural monopolies. Granting an objective and nondiscriminatory third-party access. All generators, wholesalers, retailers,

---

[4] Constitutional prerogative that allows the president to take immediate action on a given sector without following normal legal rules.

[5] This directive established a gradual and partial opening of member states electricity markets. Consumers and producers had the opportunity to negotiate the purchase and sale of electricity freely.

distributors, and final consumers will have equal treatment and no discrimination is possible.

- Tariffs are viewed with a more economic approach. The regulator can apply a cost plus or price cap if it wants to incentivize investment or apply incentive regulation to generate more efficiency.

Following these directives, Ecuador issued on October 1996 the LRSE, where the old conception of regulating the electricity sector was transformed. The most revolutionary idea for the Ecuadorian market was introduced in this market opening through the possibility to delegate the various activities of the sector to private companies by the State.

LRSE created CONELEC (Consejo Nacional de Electricidad), to substitute INECEL. The creation of CONELEC changed the regulatory model, abandoned the vertical monopoly, and reformed the sector structure. At that time, CONELEC was just the National Regulatory Authority leaving aside the generation, transition, distribution, and commercialization. In these circumstances, INECEL[6] had to disappear and the process of its liquidation began on March 31, 1999.

The first step toward liberalization was the corporatization of the different undertakings. This means not only changing their legal status but also a complete change of business vision. The electricity enterprises now have to operate according to commercial principles. Moreover, the new undertakings would have to pay taxes and market-based interest rates, produce commercially competitive returns of capital, apply procurement, as well as to manage their own budgets, loans, and labor employment (Bacon, 2001, p. 5).

11.3.1 Sector structure

With this change in perspective the electricity sector's structure changed. It was formed by:

- National Regulatory Authority: CONELEC and CENACE (Centro Nacional de Control de Energía);
- generation undertakings;
- transmission network operator; and
- distribution and retail undertakings.

11.3.1.1 Regulatory authorities

The sector was divided into four levels: generation, transmission, distribution, and retail. It was strongly regulated by CONELEC and technically and commercially supervised by CENACE.

CONELEC was the National Regulatory Authority, controlling the behavior of all market participants. It was in charge of delegating the various activities in the production chain. In addition, it had to elaborate the national plan of electrification, which is mandatory for public undertakings and referential for private ones.

According to the LRSE, all its actions were in line with the principles of decentralization, unbundling, efficiency, and deregulation. And it should not involve in any business activity. Finally, it must also approve the tariffs for transmission services and for end users (retail).

CENACE, on the other hand, was a private corporation with a technical purpose. Its members are representatives of all the generation, transmission, and distribution companies, as well as of large consumers. The transparent and technical management of the

---

[6] This enormous company was the main shareholder of almost all the electricity companies in the country and it was in charge of regulating and managing the entire sector.

TABLE 11.1   Generation of electricity and import made from 2002 to 2008.

Year Concept	Unit	2002	2003	2004	2005	2006	2007	2008
Energy generated	GWh	11,887.56	11,546.13	12,584.85	13,404.02	15,115.85	17,336.65	18,608.53
Imported from Colombia	GWh	56.30	1119.61	1641.61	1716.01	1570.47	860.87	500.16
Imported from Peru	GWh	0	0	0	7.44	0	0	0
Total	GWh	11,943.86	12,665.74	14,226.46	15,127.47	16,686.32	18,197.52	19,108.69

information had to deliver an adequate view of the operation.

11.3.1.2 Generation

From 1999 to 2001 the generation market was formed by six companies[7] that provided electricity to the network and seven self-generators (CONELEC, 2015a, p. 35). Then, in 2001 only five companies generated electricity.[8]

11.3.1.3 Transmission

According to the legal provisions of LRSE, the transmission network operator renders the public service of transmitting the electricity from the production sites to the consumption points. The network is named TNS (Transmission National System). The transmission of electricity is a legal monopoly carried out by a private entity, TRANSELECTRIC S.A., whose only shareholder is the State. The company is in charge of the design, construction, operation, and maintenance of TNS.

11.3.1.4 Distribution and retail

Distribution and retail of electricity were done through private companies owned by the State, municipalities or provincial government, and other private investors as main shareholders. Clearly, the undertakings are regionally placed. This means that they have a monopoly to distribute and retail energy in a given area. There are 11 companies[9] rendering this service and only 1 of them is privately owned.[10]

11.3.2 Fossil dependency

The analysis period is from 2002 to 2008, considering that the generation companies in this period are still even. Moreover, the analysis was conducted till 2008 because by the end that year, a new Constitution began to govern Ecuador, which changed the panorama.

From 2002 to 2008, the generation of electricity increased from 11,887.56 to 18,608.53 GWh as shown in Table 11.1.[11]

[7] Hidropaute S.A., Hidroagoyán S.A., Hidropucará S.A., Termoesmeraldas S.A., Termopichincha S.A., Electroguayas S.A.

[8] Hidropucará S.A. was absorbed by Hidroagoyán S.A.

[9] Empresa Eléctrica Ambato, Regional Centro Norte S.A.; Empresa Eléctrica Azogues C.A.; Empresa Eléctrica Regional Centro Sur C.A.; Empresa Eléctrica Provincial Cotopaxi S.A.; Empresa Eléctrica Provincial Galápagos S.A.; Empresa Eléctrica Regional Norte S.A.; Empresa Eléctrica Quito S.A.; Empresa Eléctrica Riobamba S.A.; Empresa Eléctrica Regional del Sur S.A.; Corporación Nacional de Electricidad CNEL S.A.

[10] Empresa Eléctrica del Ecuador Inc.

[11] All the statistics shown in Tables 11.1 and 11.2 are taken from the CONELEC statistics of the Ecuadorian Electricity Sector (CONELEC, 2012).

TABLE 11.2 Generation of electricity (MW).

Energy resource	Power plant	2002	2003	2004	2005	2006	2007	2008
Renewable	Hydric	1745.81	1745.93	1745.94	1808.54	1800.64	2057.29	2057.29
	Biomass	0	0	35.00	64.80	73.80	73.80	106.90
	Eolic	0	0	0	0	0	2.40	2.40
	Solar	0	0	0	0.02	0.02	0.02	0.02
Total		1745.81	1745.93	1780.84	1874.45	2133.54	2133.54	2166.51
Nonrenewable	Thermic[a]	1657.95	1733.50	1756.91	1881.23	2187.19	2343.45	2374.36

[a]It is important to clarify why thermal and fossil fueled are related. To generate electricity in a thermal power plant, it is necessary to burn fossil fuels; therefore in this chapter "thermal" shall be understood as fossil fueled.

Table 11.2 shows the amount of electricity produced from fossil materials.

Analyzing the data shown, Ecuador depended on fossil fuels. From 2002 to 2008, Ecuador generated 13,580.62 MW out of renewable resources, and in the same period, 13,934.59 MW were generated from fossil resources.

11.3.3 The fall of liberalization

It is unquestionable that the change made through liberalization enabled the market to grow. This can be verified by the percentage of the nation's GDP that it covers. Normally, the electricity sector investment grows 5% per year, but it increased significantly more in 1999 when the sector was deregulated and liberalization began: its growth was 23% (Neira, 2003, p. 2). This proved that liberalization began with the "right foot" and advanced in the proper direction. However, the path to liberalization was interrupted for two reasons. First, the liberalization process did not have the expected transience; second, due to the judicial decisions made.

When INECEL closed, all of its generation and transmission assets and shares were transferred to the Solidarity Fund.[12] Also, INECEL was divided into six companies responsible for generating electricity, one transmission company, and some retail companies, which got INECEL's liabilities (Roa, p. 5). In that sense, although regulation for the sector changed, it still had the same actors.

By 2002 the energy sector was composed of 13 generating companies, but only 4 were privately owned. Only Transelectric (Solidarity Fund owned) was in charge of transmission. There were 20 retail companies owned by Solidarity Fund and municipalities (State). In other words, private investment was significantly small (Roa, pp. 6–8).

Moreover, the liberalization process failed due to the judicial decisions made. Article 28 of the LRSE allowed the sale of 51% of State companies' shares in an attempt to attract private investments. However, the political arena was chaotic and the privatization process was completely stopped by the numerous interest groups that fought and lobbied against such sales. This dispute ended after the Constitutional Court (Ecuador Constitutional Court, 2000) declared the unconstitutionality of the articles that allowed the sale of such percentages of the companies' shares. The main argument for this was the extremely privatized conception of the reform.

[12] Fondo de Solidaridad, a company owned by the state.

This meant that all decisions made by the government to implement the law were unsuccessful. The sale of most distributor companies to different investors throughout the assessment of Salomon Smith was a complete failure. The government was really keen in maintaining this policy but the result was too extreme for all others. The possibility of privatization was halted via political and judicial lobbying. This was probably one of the most severe blows to the liberalization process.

The institutional architecture of the liberalized market did not work due to minimal private investment, low tariffs that did not cover the expenses of distribution companies, and also the INECEL's debts that were transferred to the new companies. Hence, some changes had to be made. In 2006 Congress created the Fund for Investment in the Electricity and Oil Sector (FEISEH) in an attempt to create an easier path to finance the high-priority generation projects (Pelaéz-Samaniego et al., 2007, p. 4179). This new legislative instrument was meant to use the surplus of oil extraction in the electricity sector. Nevertheless, the funds it created were not enough and the sector's troubles, such as lack of investment, low tariffs, and tariffs debt, continued.[13]

The National Government and the Ministry of Energy—through CONELEC—in December 2007 (Consejo Nacional de Electricidad, 2009–2020) transformed the entire sector's concept. The government considered that it should play a more active role—especially in the way this business is run—and that a deregulated market did not work. The notion of interventionism was stated in the 2008 Constitution.[14]

11.4 Neo-developmentalism

The starting point for neo-developmentalism is the constitutional process between 2007 and 2008. Before the implementation of the new Constitution, Congress at that time—through acts called "mandatos constituyentes"[15]—regulated the energy sector. The most relevant of these acts for the energy sector was Mandato Constituyente 15. This act established that CONELEC had unlimited competence to regulate tariffs; to aid investment for generation, transmission, and distribution of electricity; and to finance programs to electrify rural areas.

Following that, the constitution was issued and its Article 313 provided the guidelines to exploit the energy sector. It establishes that the State reserves for itself the right to manage, regulate, and control strategic sectors. The same article determines that energy in all of its forms is a strategic sector. Article 314 points out that the State is responsible for providing electricity; and Article 315 allows the creation of public companies to provide public services and manage strategic sectors.

As stated by CONELEC, the new Constitution as well as subsequent regulations introduced a new operation scheme for the wholesale electricity market, which set free competition aside and changed market regulations. Thus distributors needed a contract with the State to participate in commercial transactions.[16]

---

[13] In 2006 there were blackouts in the months of July, August, and October. Available from: <https://www.eluniverso.com/2006/08/02/0001/9/8C86208481364D5384EFF8768A8D35E6.html>; <https://lahora.com.ec/noticia/449455/apagones-en-varios-sitios>; and <http://www.ecuadorinmediato.com/index.php?module=Noticias&func=news_user_view&id=43161> (accessed 28.06.19.).

[14] In 2008 after a referéndum Ecuador decided to change their Constitution.

[15] Constitutional mandate issued by the president.

[16] CONELEC, 2016. Memoria Institucional 1996–2015. p. 40.

The new regulations changed the sector's structure once again. CONELEC acquired the faculty to regulate, control, and plan the electric sector. In addition, the structure changed regarding distribution and generation. In January 2009, 2 companies were created: CNEL S.A. (Corporación Nacional de Electricidad S.A.), which absorbed 10 distribution companies,[17] and CELEC S.A. (Corporación Eléctrica del Ecuador S.A.), which absorbed 5[18] generation companies and the only transmission company.[19]

Lastly, in 2010 generation activities and the transmission network operator were bundled into CELEC EP.[20] This new company is responsible for the production of cheap and reliable energy and for guaranteeing nondiscriminatory access for other generators, distributors, and large customers.

While these regulations were applied, the political model boosted public investment in the construction of new hydroelectric power plants. In 2012 Ecuador had 31 hydroelectric power plants[21] producing 2255.94 MW; and in 2013 14 new hydroelectric plants[22] were under construction, producing 3021.21 MW (CONELEC, 2012–2021).

11.4.1 Current sector structure

Currently, the sector's structure has changed. Energy sector is regulated by the Constitution, laws, executive acts, and also municipal acts.

The constitution points out that the State is obliged to use nonpolluting energies[23] and must promote the development of renewable energies.[24] In addition, it states that energy is a strategic sector for the Ecuadorian economy; therefore the State will regulate, control, and manage this sector.[25]

Energy is also a public service[26] and is regulated by the Electric Power Public Service Law LOSPEE.[27] Article 9 of LOSPEE establishes that the institutional structure is:

1. Ministry of Energy and Nonrenewable Natural Resources,

---

[17] Empresa Eléctrica Esmeraldas S.A., Empresa Eléctrica Regional Manabí S.A. (Emelmanabí), Empresa Eléctrica Santo Domingo S.A., Empresa Eléctrica Regional Guayas-Los Ríos S.A., Empresa Eléctrica Los Ríos S.A., Empresa Eléctrica Milagro C.A., Empresa Eléctrica Península de Santa Elena S.A., Empresa Eléctrica El Oro S.A., Empresa Eléctrica Bolívar S.A., Empresa Eléctrica Regional Sucumbíos S.A.

[18] Hidroagoyán S.A., Hidropaute S.A., Termoesmeraldas S.A., Termopichincha S.A., Electroguayas S.A.

[19] Transelectric S.A.

[20] Empresa Pública Estratégica Corporación Eléctrica del Ecuador.

[21] Paute, Agoyan, Pucará, Daule Peripa, Mazar, San Francisco, Sibimbe, Ocaña, Hidroabanico, Calope, Esperanza, Poza Honda, Loreto, Saymirin, Saucay, Buens Aires, Cumbayá, El Cármen, Guango polo, Nayón, Papallacta, Recuperadora, Perlabi, Rio Blanco, Vindobona, Alao, Ambi, Carlos Mora, Corazón, Uluchil-2, Pasochoa.

[22] San José del Tambo, Guangopolo II, Mazar-Dudas, Esmeraldas II, Saymirín V, Chorrillos, Topo, Victoria, San José de Minas, Manduriacu, Paute-Sopladora, Toachi-Pilatón, San Bartolo, Delsitanisagua, Quijos, Minas-San Francisco, Coca Codo Sinclair.

[23] Constitución del Ecuador. Article 15. Op cit.

[24] Constitución del Ecuador. Article 413. Op cit.

[25] Constitución del Ecuador. Article 313. Op cit.

[26] Article 1 Ley Orgánica del Servicio Público de Energía Eléctrica. Registro Oficial Suplemento 418 de 16 de enero de 2015.

[27] Ley Orgánica Del Servicio Público De Energía Eléctrica. Fecha de publicación: 16-ene.-2015. Registro Oficial Suplemento 418.

2. Electric Control and Regulation Agency (ARCONEL),
3. Electric National Operator (CENACE), and
4. Specialized Institutes.

Therefore the sector's structure is:

1. Planning Ministry of Energy and Nonrenewable Natural Resources,
2. Regulation: ARCONEL,
3. Generation: CELEC EP,[28]
4. Transmission: CELEC and CENACE, and
5. Distribution and Retail: CELEC and CNEL EP.[29]

11.4.1.1 *Planning*

The Ministry of Energy and Nonrenewable Natural Resources is the public entity in charge of planning sectoral policies.[30] Among its competences, it has to establish policies for the sector, evaluate regulations issued by ARCONEL, grant licenses to participate in the electric sector, and also plan the use of renewable energies.

11.4.1.2 *Regulation*

The entity responsible for regulation and control is ARCONEL that was created as a technical entity; it is also in charge of issuing technical regulations for the electric sector,[31] as well as controlling energy utilities.[32]

11.4.1.3 *Generation*

LOSPEE established that the generation market is open to public and private companies, which acquire licenses to participate in the market.[33] It also states that new-generation projects must follow guidelines and policies set forth in the National Development Plan. CELEC EP is a public company that owns the majority of electric power plants in Ecuador.

11.4.1.4 *Transmission*

LOSPEE establishes that the State must provide the transmission service through a public company, and only exceptionally private companies may be authorized to participate in the transmission phase.[34]

CELEC EP is the public company responsible for TNS. CENACE is the technical body that operates the TNS. Essentially, this entity is in charge of transmitting electricity to distribution and retail companies. CENACE also coordinates electricity import and export.

11.4.1.5 *Distribution and retail*

Public and private companies may participate in this phase.[35] The State owns two public companies, CELEC EP and CNEL EP, which are part of this market. CNEL is the largest electricity public company, which provides electricity for almost all the territory, with few exceptions.[36] Private and public companies can

---

[28]   Decreto Ejecutivo 220. Registro Oficial 128 del 11 de febrero de 2010. Article 2 CENEL EP has the objective of generate, transmit, distribution, retail, import, and export energy.

[29]   Decreto Ejecutivo 1459. Registro Oficial 922 del 28 de marzo de 2013. Article 2 CNEL EP has the objective of distribution and retail energy.

[30]   Article 11 LOSPEE. Op cit.

[31]   Article 14 LOSPEE. Op cit.

[32]   Article 15. LOSPEE. Op cit.

[33]   Article 40. LOSPEE. Op cit.

[34]   Article 42. LOSPEE. Op cit.

[35]   Article 43. LOSPEE. Op cit.

[36]   Quito, Ambato, Cuenca, Loja, and Riobamba have their own provider.

be part of this market; however, they never compete against each other, since they are authorized to provide energy for specific areas.

11.4.2 Departing from fossil dependency

As outlined before, Ecuador has made large public investments to change the energy matrix in recent years; as part of the National Development Plan.[37]

The very first plan (Plan Nacional de Desarrollo, 2007−2010, p. 272) emphasized on the need to create a sustainable power system by using renewable resources departing from fossil-fuel use. To accomplish it the government proposed to build new hydroelectric power plants and potentiate new electric generation systems, such as solar, wind, and biomass. This is very feasible since the country has a large number of rivers with great potential and it is located in the equator, which means 12 hours of daily sunlight year-round.

Subsequently, the government adopted an economic model based on international debt to build power plants. China is now the largest investor in the sector (Nathanson, 2017). Considering these new investments, the government decided to change the energy matrix primarily by constructing new hydroelectric power plants.

Table 11.3 (CONELEC, 2012) shows that Ecuador has begun to change its energy matrix, by reducing the use of fossil fuels in the electricity matrix and replacing it with hydroelectric power.

Using data from the table provided, it is possible to identify the percentages of electricity generation each year and determine how the energy matrix is changing; the difference is shown in Graphics 11.1 and 11.2.

During the past decade, Ecuador has begun to change its energy matrix and depart from its fossil-fuel dependency.[38]

11.4.2.1 A commitment toward hydroelectric energy

The change in the energy matrix that leaves aside fossil dependency started in 2008, when the new Constitution of Ecuador was issued. This new Constitution's fundamental principle is the protection of nature[39] and natural resources.[40] Another principle establishes the obligation of the State to pursue economic development by ensuring and respecting nature.[41] Following these directives, SENPLADES[42] issued various acts related to the energy sector.

For instead the first development plan (Plan Nacional de Desarrollo, 2007−2010, p. 272) pointed out the need to set up a sustainable power system by using renewable resources. It also established that hydrographic resources must be exploited.[43]

Policy 4.5 of the development plan stated that Ecuador had to develop renewable energy by establishing new hydroelectric power plants and wind systems.[44]

---

[37] Plan Nacional de Desarrollo.

[38] About this topic, refer to Chapter 8, Challenges of South American energy transition: energy efficiency and distributed generation.

[39] Constitución del Ecuador, 2008. Asamblea Constituyente. Montecristi. Article 71−74.

[40] Constitución del Ecuador. Op cit. Article 395−399.

[41] Constitución del Ecuador. Op cit. Article 395.

[42] Secretaría Nacional de Planificación y Desarrollo.

[43] Política 4.3 Plan Nacional de Desarrollo (2007−2010), p. 158.

[44] Política 4.5 Plan Nacional de Desarrollo (2007−2010), p. 160.

TABLE 11.3   Electricity generation (MW).

Year	Energy resource					Nonrenewable
	Renewable					
	Hydro	Biomass	Wind	Solar	Total	Thermal
2008	2057.29	106.9	2.4	0.02	2166.51	2374.36
2009	2059.25	106.8	2.4	0.02	2168.47	2541.90
2010	2242.42	101.3	2.4	0.02	2346.13	2796.55
2011	2234.41	101.3	2.4	0.07	2338.18	2893.75
2012	2263.89	101.3	2.4	0.08	2367.67	3086.73
2013	2263.89	101.3	18.9	3.9	2387.99	3108.23
2014	2248.09	144.3	21.15	26.41	2439.95	3291.58
2015	2407.61	144.3	21.15	25.54	2598.60	3406.38
2016[a]	4446.36	144.3	21.15	26.48	4638.29	3586.14
2017[b]	4515.96	144.4	21.15	25.59	4707.10	3321.19
2018[c]	5072.41	144.3	21.15	27.63	5265.49	3394.35

[a]*Estadísticas Anual y Multianual del Sector Eléctrico Ecuatoriano (2016).*
[b]*Estadísticas Anual y Multianual del Sector Eléctrico Ecuatoriano (2017). ARCONEL. https://www.regulacionelectrica.gob.ec.*
[c]*Balance Nacional de Energía (2018).*

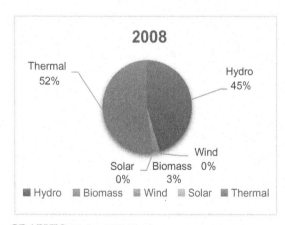

GRAPHIC 11.1   2008 Matrix.

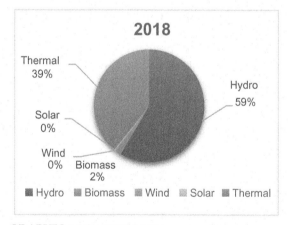

GRAPHIC 11.2   2018 Matrix.

The choice of hydroelectric power plants makes sense considering the hydric resources Ecuador has 31 hydric systems that represent 51.41% of the national territory.

Regarding that matter the government stated that the energy policy[45] is based on the valuation and internalization of hydric potential, which may eliminate the use of polluting energies (Ministerio del Ambiente, 2017).

The government had implemented policies and regulations to boost the development of renewable energies. The most relevant policies are the National Plan for Good Living (Plan Nacional del Buen Vivir, 2017) and the National Strategy for Climate Change.[46] These—in general terms—established the goal to increase electricity generation using renewable energies from 60% to 90%, by 2021.

To ensure the construction of new hydropower plants, the government took two courses of action: (1) it created public companies to manage electricity generation[47] and (2) gave certain projects in concession.[48]

Other courses of action were to develop the use of alternative energies, mainly to promote the use of solar and wind energy, including the following regulations: (1) Nonconventional Energy Law,[49] (2) Institutional Framework for Environmental Incentives,[50] and (3) Rules for Investments of Production Law.[51]

11.5 Conclusion

The Ecuadorian electricity market has substantial problems. The biggest one, without a doubt, is the political inability to apply the rules given by law. Even, when the legal framework is complete and contains the guiding principles of transparency, objectivity, nondiscrimination, and competition, somehow government decisions are able to evade the law and create chaos in the market.

This is probably why the historical characteristic of a vertically integrated structure in the electricity market has made its return. The transmission system operator is a legal monopoly that also controls the majority of produced energy. The policy is to confer more power for direct State intervention, not only as a regulator but also as the protagonist of the market. Following our most recent legal reform, the market is organized to seek sovereign investment, neglecting direct private investment.

The 1996 attempt to liberalize the process through economic regulation, by introducing the participation of private investment and creating public institutions to regulate the entire sector was an utter fail. The more rational response to the problems coming from the vertical integration of State-owned companies, worldwide, is the creation of a regulated

---

[45] Regarding the relation between energy planning and energy transition. Refer to Chapter 4, Energy (and climate) challenges in South America: what planning for what transition?

[46] Estrategia Nacional de Cambio Climático.

[47] Empresa Pública Estratégica Hidroeléctrica Coca Codo Sinclair. Decreto Ejecutivo 370. Registro Oficial 206 del 3 de junio de 2010. Empresa Pública Hidroeléctrica Zamora Chinchipe Hidrozachin. Ordenanza Provincial. Registro Oficial 287 del 27 de septiembre de 2010.

[48] Mazar-Dudas. Quijos.

[49] Ley de Energías No Convencionales. Registro Oficial 223 del 16 de abril de 1982.

[50] Marco Institucional Para Incentivos Ambientales. Registro Oficial Edición Especial 387 del 4 de noviembre de 2015.

[51] Reglamento de Inversiones Del Código Orgánico de la Producción. Registro Oficial Suplemento 450 del 17 de mayo de 2011.

market, with its respective nuances. However, stating that healthy energy markets have followed this path is not necessarily a positive indicator.

Economic analyses prove that, if applied correctly, this is a coherent way for achieving government goals for the sector. Beginning with vertical disintegration, the liberalization process focuses on the introduction of competition in the market. In this case, it is essential to make the distinction between the monopolistic and nonmonopolistic levels of production. These ideas did not come out randomly but rather based on economic modeling and by questioning its characteristics and rational behavior. The same applies if we address the new corporate structure, breaking down with the slow and expensive administration of bureaucrats.

Historically, a major swing between two opposing viewpoints has occurred: on the one hand, a State-owned and controlled industry, and on the other, a liberalized industry. These changes are commonly found in our legislation, being unsustainable for any market participant except for the State. The very basic principles of rule of law and a predictable behavior are basic elements for a healthy industry.

To tackle this issue, it is essential to form a technical consensus at the legal and economic levels and, after that, reach a consensus at a political level. Once the discussion is settled at the scientific level, it can reach the political level, where a long-term decision should be made in order to achieve any real result stemming from the implementation of a given theory. All changes, implemented since 1996, were never socialized and a fear of privatization has always existed, supported by the idea that private investors only want to maximize their profits. Social activists halted the proper implementation of a well-designed liberalization program, falling within a political agenda that brought total failure to the system.

It is important to note that electricity, according to the Constitution, is a strategic sector in the Ecuadorian economy. This categorization is the foundation for the sector's new architecture. Given that the standard recipe failed and now the government's vertically integrated monopolistic approach is failing too, there is a need for a more technical approach.

Due to this, a well-studied mix of planning and sector-specific regulation is necessary. The main issue present is the lack of investment in generation, other than that made by the State. When there are windfall profits coming from oil price, this might seem compulsory, yet international investment is and will also be needed.

National regulators should keep in mind that any direct intervention from the State will strengthen specific public ventures, coming either from low rate or nonreimbursable loans, as well as other means. No private investor will deem this as favorable ground. Moreover, no investor will endow its economic resources in favor of the country's economy but rather to favor its own interest. This behavior is normal and should be encouraged by the State, although being cautious that there is competition on even grounds and that return on investment is reasonable. This is the crux of future discussions, whether the economy will need the aid of foreigner investment or not.

Today, the political agenda is proglobalization and prointernational trade, which was radically different 2 years ago. The previous government decided to withdraw 13 bilateral investment agreements, since the Constitution forbids the implementation of foreign jurisdiction within the State. Meanwhile, the Ministry of Production alleged that they would find a way to encourage foreign investment and that these claims would take some time before being enforceable. Until this day, there is no clear perspective regarding this or the previous government's decisions.

It is important to rethink the sector's structure using available legal tools and political constraints. Based on the analysis from this study, the need for a strong regulatory authority is undeniable. This authority should be independent and not subject to government decisions. The presence of externalities that should to be internalized—or at least solved—puts the future of this sector in the hands of an efficient National Regulation Agency. Management of public service obligations, the supply of universal services, safety standards, environmental protection, and investment should be efficient and subject to frequent comparison between its results and those of regulators within the region and other international groups.

It is important to attain—through behavioral tools—real third-party access regulation and tariff supervision. Tariffs should have a more economic approach. The regulator should apply incentive regulation to improve efficiency. Yet more important, the National Regulation Agency shall compel market participants to pay the full amount on-time. The main goal must be to avoid deficit from tariffs.

This will result in different outcomes. The first one is that efficient companies will behave rationally concerning their investments and management. The second is that inefficient ventures will go out of business if they continue to have the same behavior. This may sound punitive; however, this will clarify where the problems lie and will lead to finding solutions. Another possible outcome is that bundling as many ventures as possible will result in mixed results that cover up these problems, delaying their solutions.

It is unfortunate to find out that there is no real understanding of the consequences arising from the merger and creation of CELEC EP and CENEL.

Since 2008 Ecuador has attempted to change its energy matrix; however, the route seems to be mistaken. Public policies have been created to change the energy matrix by promoting the construction of hydroelectric power plants; however, out of eight hydroelectric power plants only three are operating. That is why Ecuador continues to depend on thermal energy. Added to this, other types of energy, such as solar and wind, have been used minimally.[52] State investments have focused mainly on hydroelectric power plants.

In conclusion, despite Ecuador's efforts to issue regulations and policies to change the energy matrix, it has not been able to abandon fossil-fueled energy. The main reason is the fact that the State only promotes and implements policies regarding the hydroelectric power production. What the State should tend is to diversify the energy matrix, promoting another type of renewable energy. Also, it is advisable to take measures of saving and managing energy waste, such as the implementation of smart grids. Another measure could be to promote the self-generation of energy for consumption.

References

Bacon, R.W., 2001. Global Electric Power Reform, Privatization and Liberalization of the Electric Power Industry in Developing Countries. The World Bank, Washington, DC.
Balance Nacional de Energia, 2018. ARCONEL. <https://www.regulacionelectrica.gob.ec/balance-nacional/>.
CONELEC, 2012–2021. Plan Maestro de Electrificación.
CONELEC, 2012. Folleto Multianual. Estadísticas del Sector Eléctrico Ecuatoriano. CONELEC, Quito, Ecuador.
CONELEC, 2015a. Memoria Institucional 1996–2015. ARCONEL, Ecuador.
CONELEC, 2015b. Memoria Institucional 1996–2015. Available from: <https://www.regulacionelectrica.gob.ec/biblioteca/>.

[52] Regarding the implementation of solar projects in Latin America, refer to Chapter 6, The challenges and contradictions of Peru's *Proyecto Masivo de Energía Solar*.

Decreto Ley de Emergencia, 1961. No. 24. Registro Oficial 227.

Directive 96/92/EC, 1997. Directive 96/92/EC of the European Parliament and of the Council of 19 December 1996 Concerning Common Rules for the Internal Market in Electricity.

Garcia, P., 1967. Document Prepared by OAS Members that Stated the Necessity to Develop Latin America. Available from: <http://revistas.bancomext.gob.mx/rce/magazines/702/6/RCE_6.pdf>.

Echeverria, D., 2006. Aplicación del módulo MODPIN en la planificación de la generación de energía eléctrica en el Ecuador. Escuela Politécnica Nacional.

Ecuador Constitucional Court. Judgment 193-2000-TP. 2000. Registro Oficial No. S-234.

Executive Act. Electrification Basic Law (LBE), 1973. Decreto Supremo 1042. Registro Oficial 387.

Agencia de Regulación y Control de Electricidad. Estadísticas Anual y Multianual del Sector Eléctrico Ecuaoriano, 2016. Available from: <https://www.regulacionelectrica.gob.ec/wp-content/uploads/downloads/2017/08/Estad%C3%ADstica-anual-y-multianual-sector-eléctrico-2016.pdf>.

Estadísticas Anual y Multianual del Sector Eléctrico Ecuatoriano, 2017. Available from: <https://www.regulacionelectrica.gob.ec>.

Hauffe, J., Bode, A., Neumann, D., Haslauer, F., 2014. Global Energy Transitions. Weltenergierat, Germany, accessed 15.04.19.) <https://www.worldenergy.org/wp-content/uploads/2014/12/Global-Energy-Transitions-2014_EfG2014-mit-AT-Kearney.pdf>.

Ibarburu, M., 2001. Regulación de los mercados eléctricos competitivos. Universidad de la República.

IRENA. Energy transition. <https://www.irena.org/energytransition> (accessed 15.04.19.).

Joskow, P., 2008. Challenges for Creating a Comprehensive National Electricity Policy. Technology Policy Institute.

Ministerio del Ambiente. Estrategia Nacional De La Biodiversidad Y Su Plan De Accion. 2017 Acuerdo Ministerial 125. Registro Oficial Edición Especial 41.

Nathanson, M., 2017. Hidroélectricas: una mirada a la inversion china en la amazonia. Available from: <https://gk.city/2017/12/10/hidroelectricas-en-ecuador/>.

Neira, E., 2003. Diagnostico del Sector Eléctrico Ecuatoriano. BCE, Quito.

Origen de la deuda externa comercial, 1976—1982. Comisión para la auditoría general del endeudamiento público. <http://www.auditoriadeuda.org.ec/index.php?option=com_content&view=article&catid=46&id=58&Itemid=56> (page visited 06.07.10.).

Pelaéz-Samaniego, M.R., García-Perez, M., Cortez, L.A.B., Oscullo, J., Olmedo, G., 2007. Energy Sector in Ecuador: Current Status. Elsevier — Energy Policy. Available from: <www.elsevier.com/locate/enpol> (accessed 28.06.19).

Consejo Nacional de Electricidad, 2009—2020. Plan Maestro de Electrificación.

SENPLADES, Plan Nacional de Desarrollo, 2007-2010. Available from: <http://www.planificacion.gob.ec/wp-content/uploads/downloads/2013/09/Plan-Nacional-Desarrollo-2007-2010.pdf>.

SENPLADES, Plan Nacional del Buen Vivir, 2017. Resolución 2. Registro Oficial Suplemento 71.

Roa, T.(S.F.). La liberalización del sector eléctrico en América Latina. Los casos de Ecuador y Perú. Available from: <https://totumasymaracas.files.wordpress.com/2010/01/sector-electrico-en-ecuador-y-peru_2009_tatiana.pdf>.

Environmental impact assessment and public participation of geothermal energy projects: the cases of Chile, Costa Rica, Colombia, and Mexico

Miguel Saldivia Olave[1,2] *and Sofía Vargas-Payera*[2]

[1]University of Cambridge, Cambridge, United Kingdom [2]Andean Geothermal Center of Excellence (CEGA), University of Chile, Santiago, Chile

12.1 Introduction

After 100 years of geothermal explorations in South America, in 2017 the first geothermal power plant was inaugurated. Cerro Pabellón, a binary cycle plant located in the north of Chile has a capacity of 48 MW, and it is the world's first large-scale facility of this kind to be built at 4500 m above sea level. This slow development differs from what happened in Central America, where the first geothermal plant started to operate in 1975. In Latin America, there is a huge potential [30,800 MWe (Think GeoEnergy, 2017)] for this clean energy due the presence of subduction of the Cocos plate beneath the landmasses of Central America, and the Nazca and Antarctic plates beneath South America (Muñoz and Stern, 1988). However, the development of geothermal energy has been a back-and-forth process, and the direct use is still underdeveloped.

Despite the potential, in Latin America the geothermal energy lacks some relevant considerations regarding the sustainable development. The 2030 Agenda for Sustainable Development (UNGA Resolution A/RES/70/1, 2015), signed for most countries of United Nations, including those from Latin America, includes the development of sustainable energy as part of the Sustainable Development Goals (SDGs) (UNGA Resolution A/RES/70/1, 2015). A question arises regarding what is sustainable energy. The scope of this answer is still unclear, but it is possible to affirm that it refers to the energy source that contributes to the sustainable development. Nonetheless, this chapter does not intent to discuss the definition of sustainable development (Holder and Lee, 2007, pp. 217–263; Cordonier Segger and

Khalfan, 2004), one of the most contested concepts in environmental law (Lee, 2005, p. 46). However, in order to understand the relationship between the energy transition and the sustainable development, it is useful to analyze the compliance of this process with some principles of international law applicable to sustainability. Some of these principles are contained in international instruments, such as the New Delhi Declaration,[1] which include the following:

1. the sustainable use of natural resources;
2. the principle of equity and the eradication of poverty;
3. the principle of common but differentiated responsibilities;
4. the precautionary approach to human health, natural resources, and ecosystems;
5. the principle of public participation and access to information and justice;
6. the good governance; and
7. the principle of integration and interrelationship, in particular in relation to human rights and social, economic, and environmental objectives (ILA Resolution 3/ 2002, 2002).

Thus the overall purpose of this chapter is to review the development of the geothermal energy in light of the sustainable development's principles. The analysis then will consider two principles: the precautionary approach to natural resources and the public participation. In terms of methodology the chapter is focused on case studies of different development levels of geothermal activities in Latin America: Chile, Costa Rica, Colombia, and Mexico:

- The first section highlights the current scenario of this industry in the region and the huge potential of contribution to the energy transition.

- The second section continues by placing some legal provisions related with the sustainable development, in particular, with regards to the environmental protection and public participation.
- The third section explores the obligation of the environmental impact assessment (EIA) of geothermal activities in the countries abovementioned.
- In turn, the fourth section examines then how the geothermal activities comply with the principle of public participation, in the case of both local communities and indigenous peoples.
- Finally, fifth section sums up final remarks and challenges identified in the review of the cases studied.

12.2 Geothermal energy in Latin America

There is nothing new to affirm that the Latin American territory has an outstanding geothermal potential due to being located on the Pacific Ring of Fire, thus there is volcanic activity, which includes over 200 potentially active volcanoes (Muñoz and Stern, 1988), but the challenge has been to overcome several difficulties to use geothermal resources. Among the problems, legal and economic barriers have been critical, as well as the fact that high potential are often located in remote areas, logistically far from the largest centers of energy consumption, especially in Andean countries (Mejia et al., 2014).

Currently, the Latin American geothermal development is concentrated in Mexico, Costa Rica, Guatemala, Nicaragua, Salvador, Honduras, and Chile. Mexico and Central American countries have led the power generation in the territory. The first geothermal plant in Central America, Ahuachapan, was inaugurated in El

---

[1] The New Delhi Declaration was signed on April 6, 2002 during the 70th Conference of the International Law Association, held in New Delhi, India.

Salvador in 1975, followed by the Momotombo in Nicaragua in 1983[2]. From 1973 to 1978 the oil crises started to happen; as a result, several countries in Central America began to make efforts to diversify the energy matrix, in the light of unstable supplies for imported oil (see footnote 2). In the rest of the countries the development has been slow, which implies several up-and-down moments. The following section briefly describes the scenario of the geothermal development in selected countries.

12.2.1 Chile

The first attempts to use geothermal energy in Chile began in the 1920s, when an Italian group from Larderello explored El Tatio geothermal field[3]. Although those initial efforts were not successful, 100 years later in the same province, Ollague, the first geothermal plant of an Andean country, Cerro Pabellón, began to operate in 2017. In spite of the fact that the first and recent Chilean geothermal developments were located in the same region, the Chilean geothermal history is not linear.

In Chile, there are estimated values of power potential between 1000 and 2700 MWe (Aravena and Lahsen, 2012). Despite the potential, geothermal energy is underdeveloped. Several efforts to assess geothermal potential have been made by public and private institutions for the last 100 years.

By the end of 1990s and due to a national energy crisis, diversification of the energy matrix was an urgency. In this context, in 2000 the Geothermal Concession Law (Law 19657/2000) was enacted to promote the geothermal development. 10 years later, a national geothermal research center started to operate. The Andean Geothermal Center of Excellence (CEGA), financed with public funds, has been focused on generating and improving geothermal knowledge in Chile.

In Chile the electricity market is private and the State plays a regulatory function. The geothermal concessions system, ran by the Ministry of Energy, had a peak in 2016, and then those have been decreasing by several economic and legal reasons (Sanchez-Alfaro et al., 2015, pp. 1390–1401). In terms of low-enthalpy development, the documented installed heat pump capacity is 8.6 MWt with an energy use of 34 TJ/yr (Lahsen et al., 2015). The use of heat pumps started in 1996 when 51 units were installed in the southern part of the country (Lahsen et al., 2015). Since 2015, CEGA started to promote the direct use of geothermal energy as an alternative to firewood in the most polluted cities of Chile. This strategy has not been isolated. In 2019 the Ministry of Energy presented to the Congress a draft bill to modify the Geothermal Law in order to promote low-enthalpy projects.

12.2.2 Costa Rica

Similar to Chile, in Costa Rica an energy crisis led to a diversification of the energy matrix. The difference here was that this process took place 20 years before and since then the geothermal industry has been increasing. Costa Rica has an estimated geothermal power potential of 1000 MWe, but many of the best areas are in national parks and therefore off-limits for exploitation. Because geothermal resource is considered as public interest, Costa Rican Electricity Institute (ICE, for its acronym in Spanish) is exclusively in charge of carrying out investigation, exploration, and exploitation of geothermal resources, without the need for permits or concessions. In this setting, ICE has been exploring and exploiting geothermal fields since 1976. Currently, there are two

[2] DiPippo, R., 2016. Chapter 15 – Geothermal Power Plants in Central America and the Caribbean. In: Geothermal Power Plants, fourth ed., Butterworth-Heinemann, pp. 427–460, ISBN 9780081008799.

[3] Tocchi, 1923. Il Tatio. Larderello SpA, Ufficio Geologico. Unpublished document.

geothermal fields producing: Miravalles (160 MWe) and Pailas II (55 MWe).[4]

Similarly in other Latin American countries, the geothermal development is associated to volcanic activity, in this case the Cocos Plate subducts under the Caribbean Plate, however, unlike what happens in Chile or Colombia, for example, the geothermal resource is located close to urban areas, which brings the advantage of being closed to the electricity transmission lines, and the challenge of working closely to local communities.[5]

Since 2000 ICE has tried to promote geothermal direct use; however, there are currently no projects with low enthalpy in the country besides touristic balnearies.

12.2.3 Colombia

Taking into account the effects of climate change in Latin America, Colombia has made political efforts to include renewable energies because its energy matrix is almost completely based on hydropower. The promotion of geothermal power started in 1970s in Nevado del Ruiz volcano (Battocletti and Lawrence, 1999). Since then, Colombia has an advanced level of identification of its geothermal resource (Alfaro, 2015) due to the explorations made by the Colombian Geological Survey. However, despite its potential—by 2025, geothermal sources are expected to generate at least 1400 GWh of electric power per year (Salazar et al., 2017); currently, there is no geothermal energy production. Some of the reasons to explain that are (1) limited technical and scientific capacity, (2) the location of high-potential areas. They are located on isolated zones without connection to the National Interconnected System (SIN by its

acronym in Spanish) (Mejia et al., 2014) and (3) technical, social, and economic barriers such as costs of investment, lack of public and private coordination, and lack of industrial development (Gómez-Navarro and Ribó-Pérez, 2018).

Since 2017 the Colombian Geothermal Association has aimed to promote the use of geothermal energy in the country. This organization is facing the great challenge of promoting synergy among communities, government institutions, industry, and academia.

Today there are no geothermal power plants in Colombia; however, according to the Mining & Energy Planning Unit (UPME by its acronym in Spanish), there are two forward projects. The first one is a geothermal plant located at Nevado del Ruiz volcanic mountain, which could be in operation in 2020, and the other is a binational project with Ecuador, Chiles-Tufño-Cerro Negro, which could be inaugurated by 2025 (Salazar et al., 2017).

12.2.4 Mexico

Mexico has the longest geothermal tradition for electricity generation in the region. In the 1950s the Pathé Geothermal field became the first geothermal plant in the American continent. It was partially operational until 1973 (Cemiegeo). By 2015, Mexico had a geothermal installed capacity of 1.07 MW distributed into four public geothermal fields: Cerro Prieto (720 MW), Los Humeros (94 MW), Los Azufres (194 MW), and Las Tres Virgenes (10 MW) (Bertani, 2015, pp. 1−19). The Mexico's Federal Electricity Commission is the government-owned electricity utility. At the same time, there are two private projects owner by Grupo Dragon: Domo San Pedro and Volcán Ceboruco,[6] both located at Nayarit.

[4] Nietzen, F., Solís, L., 2015. Production-injection at the Costa Rica geothermal fields: miravalles and pailas. In: Proceedings World Geothermal Congress 2015 Melbourne, Australia, 19−25 April 2015.

[5] Interviewed to ICE member. Fieldwork, March 2019.

[6] Cemiegeo, Geotermia en México. Available from: <http://www.cemiegeo.org/index.php/geotermia-en-mexico> (accessed 25.03.19.).

Similar to Chile, in Mexico, there is a geothermal research center that aims to explore and study the geothermal resources. The Mexican Center for Innovation in Geothermal Energy is an alliance among academia and industry, in order to promote the development of geothermal energy for electricity generation and low-enthalpy projects (Romo-Jones and Kretzschmar, 2017).

12.2.5 Countries' legal frameworks

Regarding the general regulation of the geothermal energy in these countries, Chile and Mexico have specific legal frameworks, whereas Colombia and Costa Rica only have broader provisions in some energy regulations.

In the case of Chile the exploration and exploitation of geothermal energy is regulated by Law 19657 on Geothermal Energy Concessions promulgated on December 21, 1999. This regulation establishes a system of concessions to explore and exploit the resource for power generation purposes. In turn, in 2014 Mexico enacted the Geothermal Energy Law, a special regulation that also establishes a regime of permits and concessions.

Colombia, on the other hand, does not have a specific regulatory framework for the geothermal sector. In general terms, renewable natural resources are regulated by the National Code of Renewable Natural Resources and Environmental Protection (Decree-Law 2811 of 1974), which defines geothermal energy and includes it within the so-called renewable resources. This regulation establishes study permits for these resources and variable term concessions.

In Costa Rica, research, exploration, and exploitation of the country's geothermal resources is considered of public interest. This implies that the state reserves the exploitation of high-enthalpy geothermal resources associated with the generation of electrical energy. The exploration and exploitation activities are carried out by a state-owned company,Costa Rican Electricity Institute (ICE), which has the exclusive authority to take advantage of high-enthalpy geothermal resources without the need for permits or exploitation concessions. Therefore, private companies can only access low-enthalpy resources. In those cases, public service concessions are granted for the generation of electric power.

12.3 Energy and sustainable development

On September 25, 2015, 193 member states of the United Nations adopted the 2030 Agenda for Sustainable Development, an urgent call for global action for people, planet, and prosperity through a set of 17 SDGs.[7] This agenda included, for the first time in history, a global goal—the SDG 7—aiming "to ensure access to affordable, reliable, sustainable and modern energy for all" (see footnote 7).

In turn, the Paris Agreement, adopted in December 2015, is a landmark targeted to address climate change by reducing substantially the greenhouse gas (GHG) emissions in order to limit the global temperature increase in this century below 2°C above preindustrial levels (Rajamani and Guérin, 2017, p. 75). That goal depends substantially on the energy transition, since the energy sector is by far the largest source of GHGs, being responsible for more than two-thirds of worldwide emissions, with fossil fuels being the source of about 80% of those emissions and in particular electricity produced from fossil fuels accounting for 40% of them (International Energy Agency, 2013, p. 15).

In fact, the Intergovernmental Panel on Climate Change has identified the energy sector as "part of the problem and the solution" (Banet, 2017, p. 72). In turn, International Renewable Energy Agency (IRENA) published

---

[7] UNGA Resolution A/RES/70/1 (adopted September 25, 2015) (UNGA Resolution A/RES/70/1, 2015).

a report showing that clean energy could achieve 90% of the energy-related CO_2 emission reductions required to meet the main goals of the Paris Agreement.[8]

The sustainable development is a complex concept in the environmental law, introduced in 1980 in a joint report issued by the United Nations Environmental Programme, the World Wildlife Fund, and the International Union for Conservation of Nature (Dupuy and Viñuales, 2018, p. 91). From a legal perspective, this concept is explained in Principle 4 of the Rio Declaration, which states that "in order to achieve sustainable development, environmental protection shall constitute an integral part of the development process."[9]

Later, in 2002 the New Delhi Declaration established a series of principles related to the sustainable development. The analysis of this soft law instrument is helpful for assessing the contribution of laws and policies to the sustainable development (Cordonier Segger and Khalfan, 2004, p. 15). Although the New Delhi Declaration states seven different principles related to the sustainable development, this chapter will focus on two of them since they have a direct application to the development of energy projects.

12.3.1 The precautionary approach to human health, natural resources, and ecosystems

According to Principle 4 of the New Delhi Declaration[10]:

> a precautionary approach is central to sustainable development in that it commits States, international organizations and the civil society, particularly the

scientific and business communities, to avoid human activity which may cause significant harm to human health, natural resources or ecosystems, including in the light of scientific uncertainty.

The declaration also states that the sustainable development requires a precautionary approach with regard to human health; environmental protection and sustainable utilization of natural resources should include, among other conditions, the consideration of an Environmental Impact Assessment (EIA).[11]

In practice, this precautionary approach is applied by means of the EIA. This procedure was established by the Principle 17 of the Rio Declaration, which refers to an evaluation established in a national instrument for those activities likely to have a significant adverse impact on the environment and are subject to a decision of a competent national authority.[12] Generally, the EIA is regulated in domestic regulations and in some international treaties such as the Convention on Environmental Impact Assessment in a Transboundary Context (Espoo Convention) (Dupuy and Viñuales, 2018, p. 78).

In the case of geothermal energy the sustainable utilization of geothermal energy means that it shall be produced and used in a way that is compatible with the well-being of future generations and the environment (Shortall et al., 2015, p. 391). In practice the assessment of these potential environmental impacts is conducted by countries according to their domestic regulations.

12.3.2 The principle of public participation

In turn the New Delhi Declaration highlights that public participation is essential to

---

[8] IRENA (2017).

[9] Principle 4, Rio Declaration (1992) (Rio Declaration on Environment and Development. 1992).

[10] Paragraph 4.1., ILA Resolution 3/2002.

[11] Paragraph 4.2., ILA Resolution 3/2002.

[12] Principle 17, Rio Declaration (1992) (Rio Declaration on Environment and Development. 1992).

sustainable development and good governance and that it is a condition for responsive, transparent, and accountable governments as well a condition for the active engagement of equally responsive, transparent, and accountable civil society organizations, including industrial concerns and trade unions.[13] This principle comprises the effective protection of the human right to express opinions and ideas.[14] It also requires access to effective judicial or administrative procedures in the state where the measure has been taken to challenge such measures and to claim compensation.[15]

The scope of public participation in the environmental field is quite broad and covers different situations and instruments. For instance, according to the Aarhus Convention, the consultation process to citizens must include the following:

1. "decisions on specific activities"[16] (i.e., exploration or exploitation activities of a geothermal project);
2. "plans, programmes and policies relating to the environment"[17] (i.e., an energy plan for promoting renewables); and
3. "the preparation of executive regulations and/or generally binding normative instruments"[18] (i.e., a new regulation for obtaining a geothermal concession or permit).

12.4 Environmental protection of geothermal activities

In general terms, geothermal power generation creates a much lower GHG emission than most other technologies. Nonetheless, negative environmental effects cannot be excluded during geothermal power generation. These will differ according to the characteristics of the site, reservoir, and power plant (Rybach, 2003, p. 466). In order to evaluate those potential impacts the EIA of a project is crucial. The assessment reflects a practical application of the precautionary principle, and its main purpose is to identify those eventual effects and the corresponding mitigation measures.

This section examines the obligation of EIA as a legal mechanism for avoiding uncertainties related to negative impacts and guaranteeing the environmental protection in the case of geothermal activities in Latin America.

The EIA is the administrative procedure that determines whether the environmental impact of a project or activity conforms to domestic regulations. It comprises a process that provides a written statement with the environmental consequences of an activity, program, or policy (Sands, 2003, p. 800).

As a general rule, a project or activity listed in the environmental regulation of any country must be environmentally assessed. In the case of the geothermal energy the regulation regarding the environmental evaluation differs in some Latin American countries. For example, some of the environmental legislations include geothermal energy expressly, either for the exploration stage or for exploitation (CEGA, 2018, p. 22). In the case of the exploration stage, some laws mention drilling or exploration activities, but in other cases, there is no obligation for such project phase. In the case of exploitation, it refers to the power

---

[13] Paragraph 5.1., ILA Resolution 3/2002.

[14] Paragraph 5.2., ILA Resolution 3/2002.

[15] Paragraph 5.3., ILA Resolution 3/2002.

[16] Article 6, Aarhus Convention.

[17] Article 7, Aarhus Convention.

[18] Article 8, Aarhus Convention.

generation plants; however, in several cases, distinctions are made regarding the capacity of the plants (CEGA, 2018, p. 22). Still, there are even cases in which geothermal energy is not even mentioned as an activity subject to evaluation.

12.4.1 Chile

Geothermal energy is not mentioned explicitly in the Chilean environmental regulation. Eventually, a geothermal project would be obliged, indirectly, to undergo environmental evaluation only in two situations as follows:

1. under the provision of letter "c" of article 10 of Law 19300/1994, which refers to "power generating plants greater than 3 MW" and
2. under the provision of letter "p" of article 10 of Law 19300/1994, which refers to "execution of works, programs or activities in national parks, national reserves, natural monuments, reserves of virgin areas, sanctuaries of nature, marine parks, marine reserves or in any other areas placed under official protection."

The first option is applicable when the geothermal project is in the final stage of operation in which it is undoubtful that a power generation plant will be installed. This situation only contemplates electrical purposes for geothermal energy. But this scenario also presents a practical difficulty, since it may take several years before a concessionaire decides to build a power plant. Previously, studies and explorations must be carried out to determine if the geothermal resource is sufficient to generate energy. The law does not rule on this previous stage; then there is a loophole as to whether the exploration or prospecting should be environmentally assessed or not.

This scenario could change with the incorporation of geothermal surveys as an activity subject to the environmental assessment. This proposal to modify the environmental regulation is currently being discussed in the Chilean Congress. If it is approved, for the first time, geothermal energy would be expressly mentioned in the Chilean environmental legislation.

12.4.2 Colombia

Even though the Colombian environmental legislation does not mention expressly the geothermal energy, some provisions are applicable for both exploration and exploitation stages. The obligation of the EIA varies depending on the size and characteristics of the project. This evaluation can be carried out at national or regional level.

According to Decree 1076/2015 of the Environmental and Sustainable Development Sector of Colombia, a geothermal project must be environmentally assessed, at a national level, only when it considers the construction and operation of electric power generation plants with an installed capacity equal to or greater than 100 MW.[19]

In turn, a project must be environmentally assessed, at a regional level, when it considers the following:

1. the exploration activities and use of virtually polluting sources with installed capacity equal to or greater than 10 MW and less than 100 MW[20] and
2. the construction and operation of power plants with a capacity greater than or equal to 10 MW and less than 100 MW (see footnote 20).

[19]   Article 2.2.2.3.2.2 of Decree 1076/2015 of Colombia.

[20]   Article 2.2.2.3.2.3 of Decree 1076/2015 of Colombia.

In the regional level the regional corporations are entitled to grant an environmental license for activities carried out in the area of their regional jurisdiction.

12.4.3 Costa Rica

In the case of Costa Rica, there is no obligation to evaluate those projects related with exploration activities. Regarding the exploitation phase, the environmental regulations include the electricity generation based on geothermal source. The type of the assessment depends on the size of the power plant:

1. A project greater than or equal to 2000 kW must be evaluated as a category A.
2. A project smaller than 2000 kW and greater than 1000 kW under a category B1.
3. A project smaller than 1000 kW and greater than 100 kW under category B2.
4. A project smaller than 100 kW under a category C.[21]

In the Costa Rican case, the categories represent different environmental impacts and requirements for the assessment process.[22]

12.4.4 Mexico

The Law DOF 11-08-2014 of Geothermal Energy states that the Secretariat of Energy, which is public agency in charge of authorizing the geothermal concession, must coordinate with other agencies and competent bodies in environmental matters, so that the regulations in this matter are observed and to avoid damages to or deterioration of the environment, derived from exploration and exploitation activities of geothermal resources.

This provision also refers to the protection of the sustainability and integrity of the aquifers, during the exploration and exploitation of resources.[23] Notwithstanding the above, the requirements for an application of a geothermal concession do not expressly include an environmental authorization as a proof of the corresponding environmental assessment.[24]

In turn, the Regulations of the General Law of the Ecological Balance and the Protection of the Environment of Mexico establishes the obligation of EIA for energy projects under these circumstances:

1. construction of geothermal power plants, with the exception of generation plants, with a capacity of less than or equal to 0.5 MW, used to support residences, offices, and housing units[25] and
2. plants for cogeneration and self-supply of electric power greater than 3 MW.

There is no mention of exploration works.

12.4.5 Interim conclusions

This revision reveals different standards of environmental assessment in Latin American countries, as summarized in Table 12.1.

In the case of Chile, there is no explicit reference to geothermal activities in the environmental regulation. The existent legal obligation

---

[21] Annex of List of Activities subject to Environmental Impact Assessment of Costa Rica.

[22] According to article 6, Decree 31849/2013, Category A refers to projects with a high environmental impact, Category B1 refers to projects with a high moderate potential environmental impact, Category B2 refers to projects with a low moderate environmental potential impact and Category C refers to a low potential environmental impact.

[23] Article 50, Law of Geothermal Energy (2014) of Mexico.

[24] Articles 11−14, Regulations of Law of Geothermal Energy, 2014 of Mexico.

[25] Letter (K) No. I, Regulations of the General Law of the Ecological Balance and the Protection of the Environment in the matter of Environmental Impact Assessment of Mexico.

TABLE 12.1 Legal obligations of environmental assessment of geothermal activities.

Country	Obligation during the exploration phase	Obligation during the exploitation phase
Colombia	Depending on the capacity of the project, it is evaluated at a regional level when it comprises the exploration projects and use of virtually polluting sources with installed capacity equal to or greater than 10 MW and less than 100 MW[a]	Depending on the capacity of the project, it is evaluated at a national or regional level when it comprises: 1. The construction and operation of electric power generation plants with an installed capacity equal to or greater than 100 MW (assessment at a national level)[b] 2. The construction and operation of power plants with a capacity greater than or equal to 10 and less than 100 MW (assessment at a regional level)[a]
Costa Rica	No obligation	Generation of electricity from geothermal sources depending on its size: greater than or equal to 2000 kW (category A), less than 2000 kW and greater than 1000 kW (category B1), less than 1000 kW and greater than 100 kW (category B2), and less than 100 kW (category C)[c]
Chile	No obligation	Construction and operation of power plants greater than 3 MW[d]
Mexico	No obligation	1. Construction of geothermal power plants with the exception of generation plants with a capacity of less than or equal to 0.5 MW, used to support residences, offices, and housing units[e] 2. Plants for cogeneration and self-supply of electric power greater than 3 MW[e]

[a]*Article 2.2.2.3.2.3 of Decree 1076/2015 of Colombia.*
[b]*Article 2.2.2.3.2.2 of Decree 1076/2015 of Colombia.*
[c]*Annex of List of Activities subject to Environmental Impact Assessment of Costa Rica.*
[d]*Article 10 letter (c), Law 19300/1994 of General Bases of the Environment of Chile.*
[e]*Letter (K) No. I, Regulations of the General Law of the Ecological Balance and the Protection of the Environment in the matter of Environmental Impact Assessment of Mexico.*

refers only to electric power plants in general terms. In other words a geothermal project is obliged to environmental assessment only during the exploitation phase when it is generating electricity. Any exploration work, including drilling, is not subject to environmental evaluation. A similar situation is provided by the legislation in Costa Rica and Mexico. Colombia is the only country whose legislation considers exploration activities in the environmental regulation.

Therefore, the inclusion of the exploration phase of geothermal activities is a pending issue in relation with the environmental protection and the assessment and prevention of potential environmental effects. Finally, it is very important to emphasize that the obligation of environmental assessment is directly related to the right to citizen participation, as indicated in the following section of this chapter.

12.5 Public participation in energy projects

In terms of citizen participation, a similar scenario is observed in Latin American countries. In general, the public participation takes place when the EIA is carried out.

In Chile, for instance, the Environmental Evaluation Service (SEA by its acronym in Spanish) is the national agency that evaluates the EIA and organizes public consultations with the surrounding communities. In total, consulting companies may need up to 1 year to carry out an EIA and more than 200 days for the SEA. According to the Chilean laws, an EIA is required only if the project either generates more than 3 MW or if it comprises protected areas such as national reserves. In Chile, Mexico, Costa Rica, and Colombia, there is no obligation for geothermal companies to undertake environmental evaluation, except when the project is in an advanced stage, exploitation phase, or if the power plant is located in protected areas.

In terms of indigenous consultation, Chile, Mexico, Costa Rica, and Colombia have endorsed the International Labour Organization Indigenous and Tribal Peoples Convention 169, which is based on the compliance with some principles of international law applicable to the sustainable development and international social and environmental standards. However, the majority of the analyzed countries apply the international agreement according to its own criteria. Colombia, on the other hand, has incorporated an obligation of consultation for indigenous peoples.[26] Although in Chile there is no obligation in that matter, since 2016 the Ministry of Energy has carried out five indigenous consultations within the context of geothermal exploitation concession. The most important concerns highlighted are the consultation process and the EIA are part of the exploitation phase and groundwater rights.

12.6 Conclusion

The energy transition does not only refer to the technology change from fossil fuels to renewables sources, such as geothermal. The energy transition must be conducted accordingly to the principles of sustainable development. Actually, sustainable energy development is an emerging paradigm which involves reducing negative environmental impacts, whilst simultaneously increasing energy access, security, and the efficiency of energy use (Shortall et al., 2015, p. 392).

In light of the criteria of environmental protection and public participation, the development of geothermal energy in Latin America complies with some basic requirements of sustainable development. In relation to the EIA of geothermal energy projects, some of the legislations include geothermal energy expressly in their environmental laws. The difference in the regulations lies in the stage of the project subject to evaluation. Only in few cases geothermal energy must assess its impacts, associated with exploration and exploitation. In other cases, such as Chile and Costa Rica, geothermal energy should only be evaluated at the time of exploitation as a power generation plant (CEGA, 2018, p. 37). In this regard, taking into account that citizen participation is associated only to environmental regulation, and this process is ambiguous in several countries, citizen participation is also vague.

Notwithstanding the above, the comparative review of legislation at international level shows a much more advanced development regarding the regulation of the possible environmental impacts of geothermal energy. In some cases, environmental regulations are very detailed. As an example, in the case of Iceland, the law on EIA expressly mentions geothermal activity for both its exploration and

---

[26] CEGA (2018), Análisis de legislación comparada sobre energía geotérmica.

exploitation stage. In the case of exploration, it is included in the list of activities subjected to environmental assessment depending on whether the environmental impacts are significant, depending on the scope of the project, its nature and location. In these cases, it is mentioned (1) drilling of production wells and research wells in high-temperature geothermal regions and (2) geothermal drilling in low-temperature areas where the mineral or hot spring sources are on the surface or in close proximity.[27] In the case of exploitation, geothermal energy is mentioned within the projects that must obligatorily undergo environmental evaluation.[28]

Therefore, the pending challenges in countries such as Chile, Costa Rica, and Mexico is to move forward to a detailed regulation protecting the resources involved in the geothermal activities.

There are also other environmental issues not addressed in this chapter, for instance, the use of groundwater as part of the execution of a geothermal project. For example, Chile is one of the few cases in which these water rights are granted to the geothermal concessionaire by the sole regulation in the law, skipping any request and procedure. In the case of comparative experience, some legislation oblige the concessionaire to request a water concession separately. Other regulations establish criteria of reasonableness and solidarity for the use and granting of water rights (CEGA, 2018, p. 38). The protection of water resources in geothermal activities is, therefore, one of the many pending challenges in Latin America.

References

Alfaro, C., 2015. Improvement of perception of the geothermal energy as a potential source of electrical. In: Proceedings World Geothermal Congress 2015. Melbourne, Australia, 19–25 April, 2015.

Aravena, D., Lahsen, A., 2012. Assessment of exploitable geothermal resources using magmatic heat transfer method, Maule Region, Southern Volcanic Zone, Chile. Geotherm. Resourc. Council Trans. 36.

Banet, C., 2017. The Paris Agreement to the UNFCCC: Underlying dynamics and expected consequences for the energy sector. In: Roggenkamp, M.M., Banet, C. (Eds.), European Energy Law Report, vol. 11. Intersentia.

Battocletti, L., Lawrence, B., 1999. Geothermal resources in Latin America & the Caribbean. In: Report prepared for Sandia National Laboratories and the US Department of Energy, Office of Geothermal Technologies. Contract No. AS-0989.

Bertani, R., 2015. Geothermal power generation in the world 2010–2014 update report. In: Proc. World Geotherm. Congress.

CEGA, 2018. Informe Análisis de Legislación Comparada de Geotermia.

Cemiegeo, Geotermia en México. Available from: <http://www.cemiegeo.org/index.php/geotermia-en-mexico> (accessed 10.05.19.).

Cordonier Segger, M.-C., Khalfan, A., 2004. Sustainable Development Law: Principles, Practices and Prospects. OUP.

Dupuy, P.-M., Viñuales, J., 2018. International Environmental Law. CUP.

Gómez-Navarro, T., Ribó-Pérez, D., 2018. Assessing the obstacles to the participation of renewable energy sources in the electricity market of Colombia. Renew. Sustain. Energy Rev. 90. Available from: https://doi.org/10.1016/j.rser.2018.03.015.

Holder, J., Lee, M., 2007. Environmental Protection, Law and Policy. CUP.

ILA Resolution 3/2002, 2002. New Delhi Declaration of Principles of International Law Relating to Sustainable Development. Available from: <https://www.cambridge.org/core/services/aop-cambridge-core/content/view/S0165070X00000528> (last accessed 29.04.19.).

International Energy Agency, 2013. Redrawing the Energy-Climate Map – World Energy Outlook Special Report.

Lahsen, A., Rojas, J., Morata, D., Aravena, D., 2015. Geothermal Exploration in Chile: Country Update.

Lee, M., 2005. EU Environmental Law. Challenges, Change and Decision-Making. Hart.

Mejia, E., Rayo, L., Méndez, J., Echeverri, J., 2014. Geothermal development in Colombia. In: Presented at Short Course VI on Utilization of Low- and Medium-Enthalpy Geothermal Resources and Financial Aspects of Utilization, Organized by UNU-GTP and LaGeo,

---

[27]   Annex 2, Law 106/2000, on Environmental Impact Assessment of Iceland.

[28]   Article 5, Law 106/2000, on Environmental Impact Assessment of Iceland.

in Santa Tecla, El Salvador, March 23–29, 2014. Report retrieved <https://orkustofnun.is/gogn/unu-gtp-sc/UNU-GTP-SC-18-07.pdf> (accessed 18.05.19.).

Muñoz, J., Stern, C.R., 1988. The Quaternary volcanic belt of the southern continental margin of South America: transverse structural and petrochemical variations across the segment between 38S and 39S. J. South Am. Earth Sci. 1 (2).

Rajamani, L., Guérin, E., 2017. Central concepts in the Paris agreement and how they evolved. In: Klein, D., et al., (Eds.), The Paris Agreement on Climate Change. Analysis and Commentary. OUP.

Rio Declaration on Environment and Development, 1992. UN Doc. A/CONF.151/26.Rev.1.

Romo-Jones, J., Kretzschmar, T., 2017. The Mexican Center of Innovation in Geothermal Energy, CeMIE-Geo: Challenges and Opportunities. Procedia Earth Planet. Sci. 17. Available from: https://doi.org/10.1016/j.proeps.2017.01.013.

Rybach, L., 2003. Geothermal energy: sustainability and the environment. Geothermics 32.

Sanchez-Alfaro, P., et al., 2015. Geothermal barriers, policies and economics in Chile–Lessons for the Andes. Renew. Sustain. Energy Rev. 51.

Sands, P., 2003. Principles of International Environmental Law. CUP.

Salazar, S.S., Muñoz, Y., Ospino, A., 2017. Analysis of geothermal energy as an alternative source for electricity in Colombia. Geotherm. Energy 5 (1).

Shortall, R., Davidsdottir, B., Axelsson, G., 2015. Geothermal energy for sustainable development: A review of sustainability impacts and assessment frameworks. Renew. Sustain. Energy Rev. 44.

Think GeoEnergy, 2017. Presentation Geothermal Conference for Latin America & Caribbean (GEOLAC). Mexico City. Available from: <https://www.slideshare.net/thinkgeoenergy/global-geothermal-development-an-overview-september-2017-79449031> (accessed 18.04.19.).

UNGA Resolution A/RES/70/1, 2015. Transforming Our World: The 2030 Agenda for Sustainable Development.

Further reading

Aarhus Convention on Access to Information, 1998. Public Participation in Decision-Making and Access to Justice in Environmental Matters.

CEPAL, 2017. Estadísticas de producción de electricidad de los países del Sistema de la Integración Centroamericana (SICA).

Energy in Colombia, 2015. Country Update. Proceedings World Geothermal Congress 2015. Melbourne, Australia, 19–25 April.

Ministerio de Minería de Chile, 2008. Estado de la Geotermia en Chile.

SENER, 2016. Prospectiva de Energías Renovables 2016–2030.

Competitive renewables as the key to energy transition—RenovaBio: the Brazilian biofuel regulation

Danielle Mendes Thame Denny[1,2,3]

[1]University College London Faculty of Laws, London, United Kingdom [2]Fundação Armando Álvares Penteado, São Paulo, Brazil [3]Universidade Paulista, São Paulo, Brazil

13.1 Introduction

As previously written in Chapter 1, Is there a Latin American electricity transition? A snapshot of intraregional differences, Latin America is a region of the globe with plenty of natural resources. Among other energy sources, there is a considerable amount of land for biomass. Unfortunately, "Latin American and Caribbean (LAC) countries have often lacked the political will, the economic environment, and/or the technological know-how to explore its natural resources."[1] The untapped potential is massive, even if the necessary environmental restrictions are taken into consideration.

The LAC energy transition should be implemented by means of promotion of adequate investments in renewables. The region's economic development must be structured in a way to take advantage of the so-far unexplored energy potential while at the same time promoting social welfare and avoiding jeopardizing the environment.

Already 56.6% (see footnote 1) of the total power generation is delivered to the region's grid by means of renewable energy sources, and this number can increase consistently to 100% (see footnote 1) in the subcontinent. One alternative to reach this goal is by promoting public policies that increase renewables' competitiveness, particularly in sugarcane ethanol.

Sugarcane ethanol is considered almost liquefied solar energy, because of its highly positive energy balance (Goldemberg et al., 1978). This means that it produces much more energy than it is consumed during the production process, taking into consideration planting, fertilizing, transporting, and other production stages. Besides that, it has an immediate use,

---

[1] As quoted in Chapter 1, Is there a Latin American electricity transition? A snapshot of intraregional differences.

without needing previous infrastructure intervention.

This chapter analyzes the Brazilian biofuel regulation as an example of public policy oriented to promote LAC energy transition. Many instruments were recently created to increase competitiveness of the biofuel not only nationally but also internationally, therefore seeking economic, social, and environmental sustainable development.

One of the main problems LAC countries face regarding energy governance is the lack of regional integration. They cannot be "considered energy integrated, neither in the physical nor in the market-based aspects" (see footnote 1). The case of RenovaBio might highlight some mechanisms that can be politically structured to benefit the entire chain of renewables' production and distribution, despite the country analyzed.

This chapter is divided into four sections after Section 13.1:

- Section 13.2 provides an overview of the sugarcane industry in Brazil;
- Section 13.3 covers the Brazilian regulatory framework;
- Section 13.4 takes a broader approach to RenovaBio and examines the possibilities of the use of technology; and
- Section 13.5 concludes the chapter.

13.2 The sugarcane industry in Brazil

In Brazil, thanks mainly to public policies starting with Proálcool (Denny, 2018, p. 353),[2] there have been massive private investments in the production of ethanol from sugarcane, collaborating to launch the country into one of the leading positions in terms of renewable energy sources in its energy mix (ANEEL, Agência Nacional de Energia Elétrica, 2017; EPE, Empresa de Pesquisa Energética, Ministério das Minas e Energia MME, 2017; Maia and Viana, 2016).

Regulatory coordination was unfortunately scarce so, in the late 2000s, Brazil's ethanol producers had to face a crisis inflicted by the government's oil price policy, which resulted in artificially low gasoline prices designed to counteract inflation, but that also had a negative influence on ethanol's demand.[3]

This illustrates that despite the sugarcane industry being the fourth main exporting sector of the Brazilian agribusiness (MAPA, Ministério da Agricultura, Pecuária e Abastecimento, 2017), it is inserted in a conflictive regulatory context, characterized by economic interests, social justice, and environmental protection. At international level, the conflicts are set out mainly due to price volatility, multilateral

---

[2] Proálcool was the National Ethanol Program implemented in 1975, under the government of Ernesto Geisel, 2 years after the first oil crisis. In order to reduce oil imports—at that time, 90% of the gasoline consumed was imported—and offer an alternative market for sugar, the Proálcool was created through the Decree 76593/75, which initially contained the establishment of low-interest loans (about 20%), guaranteed the construction of new units of production, fixed the price of gasoline at a level. This last measure conferred a competitive advantage on ethanol and on Petrobras' investments in the distribution of ethanol throughout the country, in addition to officially advertising the use of ethanol, Denny (2018, p. 353). The result was an increase of over 500% in ethanol production between 1975 and 1979. The second phase of the program, following the second oil shock in 1979, aimed at the production of vehicles powered by hydrated ethanol. To stimulate the shift of consumer attitudes toward ethanol, several incentives were offered, such as fixing the price of ethanol below gasoline (65% in 1980 and 59% in 1982), reducing taxes on the sale of ethanol-powered cars and increasing taxes on gasoline-powered cars, in addition to reducing IPVA (the tax on the property of the car) of ethanol-powered cars. The rise of Proálcool came about in the 1980s, when pricing policy made ethanol much cheaper than gasoline, more than tripling ethanol production between 1979 and 1985 (Bastos, 2007, p. 11).

[3] As quoted in Chapter 3, Assessing the macroeconomic effects of sustainable energy transitions in Costa Rica and Chile: a multisectoral balance-of-payments-constrained growth approach.

accusations at international fora regarding labor, use of water, and menace to forest areas. On the other hand, ethanol offers great potential, since it brings several positive externalities not only to national economies but also to the environment and society. The subsection later depicts some of these positive externalities.

Regarding ethanol itself, it is an alcohol comprising a single type of molecule (Piacente et al., 2016), and when used as a fuel, it can be considered ecologically adequate, as the cultivation of its raw material reduces the amount of carbon dioxide released to the atmosphere by means of photosynthesis. Likewise, when used in combustion engines, it contributes to the reduction of greenhouse gas (GHG) emissions, in comparison with what would be released by the use of hydrocarbons (Ortiz and Rodrigues, 2006). Brazilian sugarcane ethanol corresponds to a potential 86% reduction in CO_2 emissions compared to the use of oil and the emissions related to land use change recovers in 2–8 years (Jaiswal et al., 2017).

This fuel can be produced from several raw materials containing starch or sugars (Macedo, 1993). There are mainly three categories that are used: saccharides, in which sugar (sucrose) gives rise to ethanol, such as sugarcane, sorghum, sugar beet, and fruit juices in general; starch, originating alcohol from corn, cassava, sorghum, potatoes, and babassu (mesocarp) (Macedo, 1993); and cellulosic raw materials, such as eucalyptus, quince, sawdust, sugarcane bagasse, babassu pericarp, rice husk, among others (Macedo, 1993).

Sugar cane is the most widely used raw material in Latin America, South Africa, India, and Australia for ethanol production (Piacente et al., 2016). In the United States and Canada, maize is the predominant product; in Europe, cellulosic raw materials and forest residues; and in Asia, cassava, wheat, and sorghum are used for ethanol production (Piacente et al., 2016). First-generation ethanol is considered those produced according to traditional methods, similar to the ones used for the production of ethylic beverages from the direct extraction of the juice from the saccharides and starches.

Second-generation ethanol (E2G) is produced from groups of cellulosic products. This is a method which allows for the full exploitation of the raw material, processed in two phases:

1. juice extraction from the saccharides or starches and traditional processing for the ethanol production and
2. cellulose, a by-product of the first phase independent of the type of raw material used, is processed in enzymatic reactors, making possible the extraction of E2G (Piacente et al., 2016, 7).

In order to analyze which raw material is preferable, besides its availability, the energy balance must be taken into account, because like any energy resource, ethanol production consumes energy (Oliveira et al., 2014). In LAC, this balance is very positive since sugarcane has a high efficiency of photosynthesis and is a crop of easy dissemination in the region's climate conditions (Souza et al., 2018). Nevertheless, it brings the potential to be even more efficient, sustainable, and constant if more technological methods are used and if wasted materials are reused (Souza et al., 2018). For this, appropriate policies to guarantee competitive prices, to overcome opportunity costs, and to stimulate investments should be put in practice (Souza et al., 2018).

13.2.1 Some positive externalities of biofuels in Brazil

Biofuels can help creating a new economy, which is clean and renewable, contributing to overcome the challenges of decreasing the use of hydrocarbons as a main energy source. Until new transport technologies or energy production is put in place, ethanol in Brazil, for example, seems to be one of the main current options, because of its several advantages.

The first one of them is of macroeconomic nature: each barrel of biofuel produced in Brazil corresponds to a barrel of oil that is not imported, making the balance of payments less deficient. In terms of trade balance the sugar and ethanol's production stands out among the five main export sectors of Brazilian agribusiness. As of May 2019, official data informed that the sugarcane industry is the fifth largest export sector of Brazilian agribusiness with a 5.4% share of Brazilian foreign sales. The other main products in 2019 were soy-related exports (39.2%), forest products (15.4%), meat (15.3%), coffee (5.5%), and sugarcane (5.4%) (MAPA, Ministério da Agricultura, Pecuária e Abastecimento, 2019).

The second advantage refers to public health. The mixture of ethanol to the gasoline substitutes the use of highly toxic additives, such as chumbo-tetraetila and MTBE (metil-térciobutil-éster) (Morceli, 2006, p. 27), which cause serious pulmonary and environmental problems, such as acid rain and aquifers' contamination.

The third one has environmental character. Biofuels are renewable, and they pollute much less than fossil fuels. Ethanol in Brazil is produced in areas of consolidated urbanization, more precisely in the state of São Paulo, causing very few pressure or impacts to more sensible biomes, such as the Amazon and Pantanal (Denny, 2018). The map of production shows that the farms are around 2000 or 2500 km away from these sensible environments (UNICA, União da Indústria de Cana-de-Açúcar, 2017).

Forecasts show that ethanol from Brazilian sugarcane has the potential to displace 3.8%— 13.7% of crude oil consumption and 1.5%—5.6% of net CO_2 emissions globally [as of 2014, and generate the equivalent of 3.63—12.77 million barrels of crude oil per day in 2045 (Jaiswal et al., 2017, 3), even considering the impacts on climate change, the limitations of conservation areas and some possible scenarios of land use and demand for agricultural products]. The analyses of scenarios are necessary because the total arable land in South and Central America is projected to shrink by 1%—27% due to climate change and population growth (Jaiswal et al., 2017, 3). This may correspond to a reduction of 13.7% of oil consumption and a reduction of 5.6% of CO_2 emissions globally, as of 2014 (Jaiswal et al., 2017, 3).

The fourth advantage comes with the possibility to generate electricity from the burning of the cane's pulp, providing not only the necessary energy for the production of ethanol and sugar but also a power surplus that can be injected into the distribution network, avoiding transmissions costs, since the sugar plants, to a large extent, are located near the power load in Brazil's Southeastern region.

Fifth, geopolitically, ethanol reduces the dependence on foreign crude oil and on exporters in regions of political instability. Besides that, in Brazil the installed capacity of existent refineries are close to maximal—in 2018 it was around 80% (Oddone and Agência Nacional do Petróleo ANP Gás Natural e Biocombustíveis, 2018, 12). Therefore an increase on the demand of any oil products, such as diesel or gasoline, would force prices up.

Sixth, liquid biofuels bring a social advantage, since they—specifically ethanol in Brazil—allow the creation of 795,000 (IRENA, International Renewable Energy Agency, 2019) job opportunities,[4] boost income in the

---

[4] The liquid biofuel workforce counts for 795,000, summing up the other renewable energy sources there are 832,000 jobs in the sector at end of 2018 in Brazil. The Centro de Estudos Avançados em Economia Aplicada estimates that the direct jobs in the sucroenergy industry correspond to around 8% of the jobs of the whole agribusiness sector in Brazil, and many more indirect jobs are created Cana Online (2019).

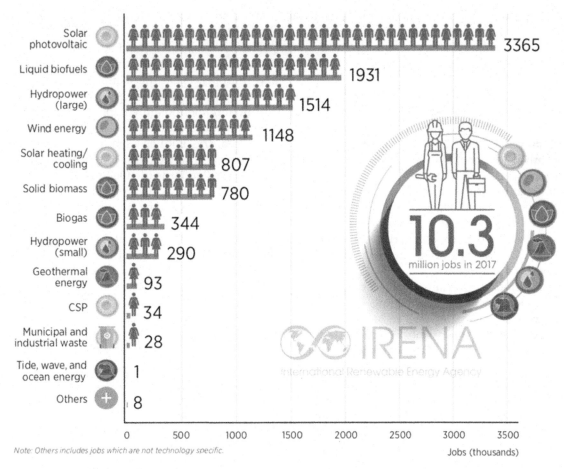

Note: Others includes jobs which are not technology specific.

FIGURE 13.1   Global renewable energy employment by technology. Source: *IRENA*.

community, increase salaries, provide formalization of contracts, and access to benefits such as medical assistance and basic food provision (Postal and da Silveira, 2018, 12). In addition, in 2013 an agreement among sugar mills' owners about the workforce was signed, in order to curb the degrading conditions of the rural seasonal workforce of the so-called boias frias (people who work in the sugarcane plantations).

As Fig. 13.1 depicts, liquid biofuels is one of the main sources of employment in the energy sector.

Regarding Brazil, ethanol production is even more advantageous as it represents an excellent opportunity to lead the transition to a low-carbon economy. With more than 45% of energy production coming from renewable sources (ANEEL, Agência Nacional de Energia Elétrica, 2017), Brazil has one of the world's cleanest energy mixes, using hydroelectric plants to generate electricity and ethanol as fuel in transportation. Sugar cane can produce ethanol to be used directly in vehicles,[5] or as

---

[5]   Hydrated ethanol.

an additive,[6] both mitigating the negative externalities caused by hydrocarbon fuels.

13.2.2 Why regulate biofuels?

Because of these positive externalities, the ethanol market has been object of incentive policies, creating supply and demand, and guaranteeing competitiveness. However, these incentives are neither constant nor coordinated, especially in Brazil.

Public policies have the ability to artificially increase the competitiveness of renewable energies and to stabilize domestic and international price volatility, reinforcing the environmental performance of sugarcane ethanol and its positive externalities.

Further on, government policies combined with clear market negotiations can push renewables ahead, leading to a "Developed Market Stage,"[7] where the fostering mechanisms can cease, given that renewables would be already price competitive, and consumers could choose them even in a free-market economy and a multiparty system.

In addition, action-oriented governance initiatives coordinated by various stakeholders can promote policy dialog and collaboration among key countries, organizations, universities, and the private sectors, with a common goal of accelerating development and implementation of low-carbon sustainable alternatives for transportation, chemicals, plastics, and other sectors.

This cross-sector coordination is far-reaching surpassing national borders, something very hard to achieve in this area. The environmental issue and the energy market have been left out of international trade negotiations, such as the World Trade Organization. International trade fora are still relatively reticent, despite the fact that distortive measures, national incentives, public policies, quotas, and subsidies are common. In some countries, such as United States and Europe, even the national stimulus to renewables can disguise protective measures and undermine foreign competition. When the environment is on the agenda of international trade fora, this is often done in dispute settlement panels or in technical committees discussing market standards (de Queiroz, 2005).

Even in such a disarranged international scenario, the sugarcane agroindustry can still be cited as one of the best examples of articulation globally (Shikida et al., 2011), thanks to the economic and technical discussions led by Brazilian-based UNICA (Union of Sugarcane Industry), in the defense of the interests of the ethanol industry, along with international market players such as banks, investors, international organizations, governments, and industry. But in a scenario where regulatory uncertainty is common, no matter how committed the international players are, in the absence of clear, transparent, and respected rules, governance and international business are hampered or even unfeasible.

Many countries[8] have adopted policies to foster ethanol in recent years, in order to force blending with hydrocarbon fuels, reducing carbon emissions and protecting environmental and human health (Biofuels Digest, 2018). For Brazil and other ethanol producers, these countries became potential export markets; regulations regarding ethanol production, therefore, turn into an urgent issue for competitiveness and international free trade. The aim to dominate the market, in a deregulated business environment, has the potential to divert

---

[6]   Anhydrous ethanol.

[7]   As quoted in Chapter 2, Using market mechanisms for a reliable energy transition in Latin America.

[8]   64 countries have targets or mandates. 14 countries in the Americas, 12 in Asia-Pac, 11 in Africa and the Indian Ocean, and 2 from non-EU countries in Europe Biofuels Digest (2018).

discussions away from sustainability toward protectionism. This eventually opens up space for scientific litigation, green washing practices, and implementation of antidumping measures (Denny, 2018).

A general articulation of public and private, international and national interests is needed to face these challenges and guide the economy toward a greener production. Moreover, creating solutions for low-carbon transport and renewable energy economy also help countries achieve their nationally determined contributions established in the Paris Agreement (UN, United Nations, 2015a) and the Sustainable Development Goals of the Agenda 2030 (UN, United Nations, 2015b).

As a matter of fact, baring these broad international targets for low-carbon transport fuels, such as sugarcane ethanol, is one of the fastest alternatives to reduce the carbon intensity of the transport sector (Milanez et al., 2017), ahead of a more comprehensive change in the fleet, infrastructure, and technology demanded by the other renewables sources.

Notwithstanding these qualities, renewable energy sources have no chance if the competitiveness is measured only by production price. Regulation is necessary especially to force investment, price the negative externalities, and demand sustainability standards.

13.2.3 Roadmap to innovations

Currently, sugarcane is the third largest Brazilian crop in planted area, behind only soy and corn (IBGE, Instituto Brasileiro de Geografia e Estatística, 2017, 8). In 2016 the sugarcane plantations already occupied 10.5 million ha, or 13.5% of the national total planted area in Brazil (IBGE, Instituto Brasileiro de Geografia e Estatística, 2017, 8). In addition, the incremental industry had revenues of about US$10 billion with the sugar-energy sector and indirect benefits. Regarding the carbon market, of the 68 Brazilian projects registered as Clean Development Mechanism at the United Nations Framework Convention on Climate Change (UNFCCC), 24 are from the sugar-energy sector, which generated an estimated reduction of 473.9 thousand tons of CO_2 and generated US$3.5 million in revenues (Sousa and Macedo, 2010, 31).

Energy efficiency has already improved a lot. Ethanol was half as efficient compared to gasoline when Brazil's National Ethanol Program[9] began in 1975 (Cortez, 2016), 2 years after the first oil crisis. Today, this difference in efficiency is around 25%, and new technologies such as biobutanol promise to decrease it further, possibly reaching 5%, according to Embrapa[10] (AGRA, Agribusiness Intelligence, 2017). New technologies can improve even further the efficiency of ethanol, by means of using the bagasse of the sugar cane in the production of cellulosic ethanol, known as E2G. The estimated production cost for this type of technology is around US$0.39/L, while in the traditional way, it is approximately US$0.30 (Milanez et al., 2015). Therefore without a proper regulation, it has no large-scale economic viability.

New varieties of sugarcane, which are more suitable for exploitation through this type of technology, have been studied. The best option so far is "cana-energia" (energy-sugarcane) (Milanez et al., 2015).[11] This type of plant could be planted in degraded areas with low

---

[9] Proálcool (Decree 76593/1975).

[10] Brazilian Agricultural Research Corporation is a state-owned research corporation affiliated with the Brazilian Ministry of Agriculture.

[11] This type of sugarcane is a variety obtained from the crossbreeding of the species *Saccharum officinarum*, most commonly used today, with a broad stem and rich in sugars and *Saccharum spontaneum*, a primitive species, more like a grass, that grows faster and is richer in fiber but has a thin stem and less sugars.

agricultural aptitude, increasing productivity per hectare and, because it cannot be used for human consumption, would not compete with agricultural land for the food production.

The techniques for plant breeding are divided into the traditional technique (breeding and selection) and the use of genetic engineering. In Brazil, there are three genetic improvement programs for sugarcane: one academic, one private, and one from the government.

The interuniversity network for the development of the sugar and alcohol sector (Ridesa) is a technical cooperation among 10 federal universities. The sugarcane technology center [Centro de Tecnologia Canavieira in Portuguese (CTC)] is a private company whose shareholders are from the sugar and alcohol industry. The Agronomic Institute of Campinas, a research institute of the agribusiness technology agency, is a body of the Secretariat of Agriculture and Supply of the State of São Paulo. In June 2017 the commercial use of the first genetically modified sugarcane was approved—the "CTC20"—which is resistant to borer (*broca* in Portuguese), the main pest that threatens production (FGV, Faculdade Getúlio Vargas, 2017, 47).

The National Bank for Economic and Social Development[12] together with the Funding Authority for Studies and Projects,[13] structured a joint initiative (BNDES-Finep) to support the technological innovation of the sugarcane industry sectors named PAISS (Joint Support Plan for Agricultural Technological Innovation in the Sugar and Energy Sector).[14] The aim of this joint initiative is to select business plans and projects that contemplate the development, production, and trade of new industrial technologies for the processing of sugarcane biomass. The purpose is to organize the entry of requests for financial support within the two institutions and to allow greater coordination of actions in order to promote and better integrate the available financial instruments.

PAISS is designed to stimulate the current installed capacity of E2G production—around 140 million liters per year—through low-cost funding offer and nonrefundable financial resources. Nevertheless, this technology is still far from wide commercial application (Jaiswal et al., 2017), since the price is not yet competitive and the production is not scalable. Until further technological innovations, the material that exceeded what was necessary to generate the steam needed for the distillation process—which could be destined for the E2G process—has been used so far only as fuel in the cogeneration stage of power production.

Besides that, the use of sugarcane bagasse for thermal and electric energy can grow. Currently, domestic demand for ethanol—anhydrous and hydrated—does not exceed 5% of the total energy consumption in Brazil, and ethanol and biodiesel together account for only 7.6% (EPE and MME, 2017, 37; MME, Ministério das Minas e Energia, 2017b). The reuse of sugarcane bagasse from which had been extracted sugar and alcohol for thermal and electric energy can increase this figures. Moreover, that use brings the potential to be optimized so that more of the bioelectricity surplus is exported to the distribution utility network. Considering all other biomass energy sources with commercial use—agroindustrial, liquid biofuels, forestry, animal waste, and urban solid waste—they still account for only 8.9% of the total Brazilian energy matrix, 76.8% of which comes from sugarcane bagasse (IEA, Instituto de Economia Agrícola, 2017, 1).

---

[12] A federal public company associated with Brazil's Ministry of Development, Industry, and Trade.

[13] An organization of the Brazilian federal government under the Ministry of Science of Technology.

[14] In Portuguese: *Plano de Apoio Conjunto à Inovação Tecnológica Agrícola no Setor Sucroenergético.*

In 2015–16, bioenergy production with sugarcane bagasse was 166.40 million tons of carbon equivalent, with 55.2% being produced in the State of São Paulo (IEA, Instituto de Economia Agrícola, 2017, 2). According to the 10-Year Energy Expansion Plan 2026, the technical potential of annual electricity generation for the distribution network using sugarcane biomass can reach the equivalent of almost two mills of the size of Itaipu (the biggest Brazilian hydropower plant and second biggest worldwide), generating 165 TWh/year until 2024 (EPE and MME, 2017, 37).

13.3 Brazilian regulatory framework

Law 13033 was enacted in September 2014, providing for the obligation to increase the mandatory percentage of anhydrous ethanol to gasoline by up to 27.5%, provided that its technical feasibility is verified. If the share of biofuels in the Brazilian fuels matrix is currently around 26%, this is largely due to the mandatory mixing percentage. In the case of gasoline the mixture of anhydrous ethanol can be between 18% and 27.5%. Since March 16, 2015 the mixture is at 27% (MME, 2017b, 35).

To guarantee the national supply of biofuels the Brazilian National Agency of Petroleum, Natural Gas and Biofuels (ANP in its Portuguese acronym) requires minimum stocks of anhydrous ethanol and constant proof of ability to meet the mixing mandate[15] (ANP, Agência Nacional do Petróleo, Gás Natural e Biocombustíveis, 2011). In order to supply the formation of monthly stocks in sufficient volume for the monthly commercialization of gasoline C (gasoline plus anhydrous ethanol), distributors have to purchase anhydrous ethanol in two ways: under a supply contract regime, whereby the distributor contracts with the anhydrous ethanol supplier 90% of the volume sold in the previous year, between May 1st of the previous year to April 30th of the current year of analyses, or under the direct purchase regime.

For producers, the legal regime is different. On January 31st and March 31st, cooperatives or ethanol trading companies must own stocks with a minimum of 25% and 8%, respectively, of its commercialization of anhydrous ethanol with the distributor in the previous calendar year. Moreover, if they have signed a contract with the distributor of at least 90% of the volume of anhydrous ethanol sold in the previous year, they are exempt from attending the 25% minimum stock criteria—still need to prove the 8% minimum stock on March 31st. The National Council for Energy Policy (CNPE in its Portuguese acronym)[16] (CNPE, Conselho Nacional de Política Energética, 2017a), extended these contracting limitations also for biofuels importers.

Other government actions include tax differentiation between hydrous ethanol and gasoline C[17] and the provision of specific financing lines for the industry, which induce its growth to a greater or lesser degree. In the end of 2018 the total amount of taxation over petrol in Brazil was 44%[18] (Fecombustíveis, 2018). The total taxation over diesel was 23%[19]

---

[15] These criteria for inventory formation are set forth in the ANP Resolution 67/2011.

[16] Resolution no. 11, dated 04/11/2017.

[17] The tax differentiation is basically on the following Brazilian taxes: CIDE, ICMS, PIS/COFINS.

[18] From this 44%, 25%–34% was State tax ICMS that varies around the country, R$0.79/L was PIS/COFINS (PIS is a mandatory employer contribution to an employee savings initiative and COFINS is a contribution to finance the social security system), and R$0.1/L relating to a carbon tax CIDE-combustível.

[19] From this 23%, 12%–25% was ICMS, R$0.35/L of PIS/COFINS, and no CIDE-combustível.

(Fecombustíveis, 2018). Ethanol on the other hand was taxed at 26%[20] (Fecombustíveis, 2018).

13.3.1 RenovaBio

The National Biofuels Policy Act (RenovaBio)[21,22] (CNPE, Conselho Nacional de Política Energética, 2017b) constitutes the institutional normative framework to articulate biofuel policies involving ethanol, biodiesel, biomethane, biokerosene, E2G, and others. It sets the legal environment to foster opportunities for production expansion and biofuels insertion into the national energy matrix, aiming also at making fuel supply more regular and contributing to the country's fuel market predictability of these energy sources. RenovaBio is also an answer to meet Brazilian commitments to the Paris Agreement (UN, United Nations, 2015a) under the UNFCCC.

Simulations regarding the investments that will be made in the sugar and ethanol industry with the RenovaBio show that it is expected the fundraising volume to grow almost fourfold until 2030 (NovaCana, 2017). The share of ethanol in the domestic market compared to gasoline and natural gas vehicle without the RenovaBio would be 31% by 2030, but, if fully implemented, the new RenovaBio measures could contribute to an ethanol share of 54% in the same period (NovaCana, 2017).

The main features of RenovaBio are the introduction of a market mechanism structured by the Biofuel Decarbonisation Credit (CBIO in its Portuguese acronym), as seen in Fig. 13.2, the establishment of 10-year targets[23] for reducing gas emissions and certification instruments for biofuel production in order to reflect how each producer or import agent contributes to the mitigation of GHG emissions in the atmosphere.

The CBIO is an instrument registered in an accounting form, for purposes of proving the individual target of the fuel distributor.[24] It will be a financial asset, that is, a carbon credit to be traded on the stock exchange and issued by the biofuel producer. The targets for reducing carbon intensity will be individualized by ANP for distributors in terms of its share in the fossil fuel market in relation to the total market for these fuels (MME, 2017b, 56). Each fuel distributor is responsible for complying with the target individualized by ANP. The invoice trade by the producer to fuel distributors makes possible to assess how and which agents have fulfilled the goal.

In more details the Brazilian national emission reduction targets for the fuel matrix were defined for a period of 10 years by CNPE[25] (CNPE, Conselho Nacional de Política

---

[20] From this 26%, 12%−30% was ICMS, R$0.13/L for the producer and R$0.11/L for the distributer of PIS/COFINS, and no CIDE-combustível.

[21] Congresso Nacional Brasil, 2017. Lei no. 13576/2017: Dispõe Sobre a Política Nacional de Biocombustíveis (RenovaBio) e Dá Outras Providências. Available from: <http://www.planalto.gov.br/ccivil_03/_ato2015-2018/2017/lei/L13576.htm>.

[22] Renovabio was enacted by CNPE Resolution 14/2017.

[23] Brazil's national energy policy council (CNPE) sets a goal of reducing the country's carbon emissions by 10% by 2028. Oil and gas regulator (ANP) has published a regulation to establish decarbonization goals for fuels distributors that establish the individual targets will be measured with decarbonization credits (CBIOs) and will take into account the amount of greenhouse gas emissions from fuels each distributor sold in the previous year.

[24] Brasil, Lei no. 13.576/2017: Dispõe sobre a Política Nacional de Biocombustíveis (RenovaBio) e dá outras providências.

[25] Resolution no. 5, June 2018.

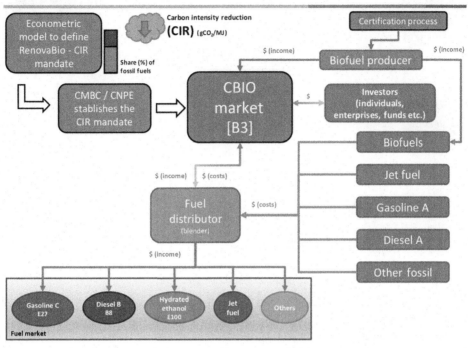

FIGURE 13.2   RenovaBio innovation. Source: *Data from MME, Ministério das Minas e Energia, 2018. RenovaBio: Price Is What You Pay. Value Is What You Get. Available from: <http://www.mme.gov.br/documents/10584/55980549/RenovaBio.pdf/> (accessed 15.06.19.) (MME, Ministério das Minas e Energia, 2018).*

Energética, 2018). These national targets set by CNPE are broken down into annual mandatory individualized targets for fuel distributors as they participate in the fossil fuel market. The participation in the RenovaBio is voluntary for the biofuel producer and importer. Those who want to join can claim a Certificate of Efficient Production of Biofuel, specifically granted to each biofuel production or importer unit.

The biofuel producer and importer, participants in the RenovaBio, are obliged to make available all the information necessary to calculate the Energy Efficiency Note and the fraction of the volume of eligible biofuel, including the generation, treatment, and

conversion of biomass to biofuel. Through the certification of the biofuel production, different grades will be assigned to each biofuel producer and importer, in an amount inversely proportional to the carbon intensity of the biofuel produced[26] (ANP, Agência Nacional do Petróleo, Gás Natural e Biocombustíveis, 2018). The note will accurately reflect the individual contribution of each producer and importer to the mitigation of a specific amount of GHGs in relation to their fossil substitute (in terms of tons of CO_2 equivalent).

Biofuels producers and importers wishing to join the program will contract firms accredited at the ANP to carry out the Biofuel

[26] ANP Resolution 758/2018.

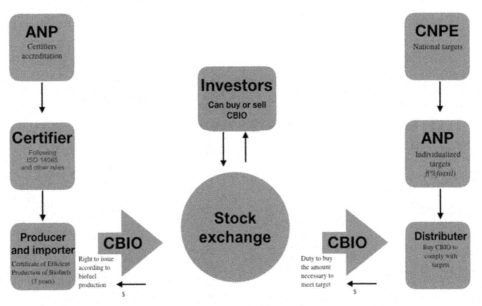

GRAPH 13.1   Renovabio operation scheme.[27] Source: *Own elaboration.*

Certification and validation of the Energy and Environmental Efficiency Note. The Certificate of Efficient Production of Biofuels is valid for 3 years, counted from the date of its approval by ANP, and may only be issued by the inspecting firm after approval of the process by the agency (Graph 13.1).

Fuel distributors must demonstrate compliance with mandatory individual targets through the purchase of CBIO; therefore it is a tradable financial asset, derived from the certification of the biofuels production process, based on the respective levels of efficiency achieved in relation to their emissions. Biofuels producers and importers that have voluntarily joined the program may sell the credits originated from the certified production. Fuel distributors will meet the mandatory

annual target by proving ownership of the sufficient amount of CBIOs in their portfolio.

13.3.2 Combustíveis Brasil

In Brazil the national supply of fuels is regarded as a public service[28] and consists of both fossil and biofuels. Therefore the government must ensure—also for biofuels—the regularity and continuity of supply, as well as protect the consumers' interests regarding price, quality and supply, among other objectives contained in the national energy policy.[29]

In order to do so the authorities have to supervise "supply, import, export, transport, transfer, storage, distribution, resale, and sale of biofuels" as well as "conformity assessment and certification" with quality standards.[30]

---

[27]   Brasil, Lei no. 13576/2017: Dispõe sobre a Política Nacional de Biocombustíveis (RenovaBio) e dá outras providências; MME (2017b).

[28]   Congresso Nacional Brasil, 1999. LEI No 9.847, DE 26 DE OUTUBRO DE 1999. Regulamenta Lei Do Petróleo.

[29]   Congresso Nacional Brasil, 1997. LEI No. 9.478, DE 6 DE AGOSTO DE 1997—Lei Do Petróleo.

[30]   *Brasil.*

Hence, and also to stimulate free competition by attracting new investment in the energy sector, Ministry of Mines and Energy, Energy Research Office, and ANP have been coordinating a broad dialog with the market around technical issues consolidated in a final report submitted to public consultation and to the CNPE.

This multisectoral and multilevel initiative is called Combustíveis Brasil ("Brazil Fuels" in Portuguese), and it proposes actions and measures to stimulate investments and diversify the fuel supply sector. The primary objective of this strategy is to redesign the fuel supply scenario by taking Petrobras off the center of operations. In addition, it intends to reassess the fuel logistics chain, to reposition economic agents beyond Petrobras, to contribute to greater clarity in the rules of trade contracts, and to stimulate arbitration (MME, 2017b). The Combustíveis Brasil initiative parallels the RenovaBio Program and corresponds to the strategy of increasing sustainability for the refining, transportation, and derivatives trade segments. It is intended to guarantee operations on viable economic bases, including the transfer of logistic-related costs inherent to the distribution activity to its buyers and the introduction of the limited supply points [Pontos de Fornecimento Limitados in Portuguese (PFL)] (MME, Ministério das Minas e Energia, 2017a).

The PFLs—when in force in the trade contracts with diesel and gasoline distributors—will foresee the possibility, for example, of reducing the supply of liquid fuels by up to 50% of the reference volume in the Brazilian cities of São Luís (MA), Belém (PA), and Araucária (PR) and up to 30% in Fortaleza (CE), provided there is a 60-day notice (MME, 2017b). This possible variation mainly reflects insufficient transport and distribution infrastructure and the lack of transparency regarding asset utilization information. The conditions of fuel storage and movement in the Northern and Northeastern regions of Brazil are the most critical in the country due to lower infrastructure development and higher expectations of demand growth (MME, 2017b, 29).

Considering this scenario, incentives need to be put in place to foster investments that distributors have to make, especially in these more critical regions, legally designing a clear and predictable regulatory environment. The historical monopoly of Petrobras and Transpetro resulted in a lack of technical capacity or adequate infrastructure from the private sector to meet this demand. At this point, multisectoral and multilevel alternatives, such as Combustíveis Brasil, are paramount to pave the way leading to technical planning, which, on its turn, can boost a gradual, articulated, and pioneering transition, in order to allow the adaptation of stakeholders to a more competitive scenario in the transport sector.

To a large extent, the CBIO created by RenovaBio can serve as a tool to enable these necessary and desirable private investments in the fuel distribution. Given that these credits derive from the Energy and Environmental Efficiency Notes associated with the biofuels—marking the difference between their carbon intensity and the carbon intensity of their fossil fuel substitute production—it indirectly also contributes to a better environmental efficiency of the Brazilian energy sector.

13.3.3 Certification and trade

This difference[31] will be established by a certification process, which analyzes the correction of the technical data fed into RenovaCalc, a support tool designed to calculate the carbon intensity of a biofuel (in $g/CO_2eq/MJ$), referring to the processes of biomass production and biofuel produced by the industry. The certification process is done by a third entity—

---

[31] Difference between their carbon intensity and the carbon intensity of their fossil fuel substitute production.

the inspecting or certifying company—and evaluated by audits to verify if it complied with the technical protocol of environmental performance evaluation based on the life cycle assessment (LCA)[32] (CONMETRO, Conselho Nacional de Metrologia Normalização e Qualidade Industrial, 2010).

The RenovaBio legal framework contemplates future regulation, establishing that the standard unit of the CBIO will be 1 t of CO_2 (carbon dioxide equivalent), allowing the credit to have fungibility and interchangeability with other economic sectors in the future, as well as with other decarbonization mechanisms further adopted in other areas of the oil business and automobile industries.

The producer or importer of biofuel will issue an accounting decarbonization credit, hiring a bookkeeper, bank, or financial institution (MME, 2017b, 62) in proportion to the volume of biofuel traded—produced or imported. These book-entry registrations are electronic processed, being guaranteed the safety, ease in circulation and reduction of costs that would involve in the case of printed certificates. At this point the use of technologies that allow for transparency, accurate monitoring, audits, and data comparison, such as distributed accounting technology [Distributed ledger technology (DLT)], which will be explained later, can contribute.

Each insurer is obliged to maintain a specialized team to verify the existence or not of the operation that served as guaranty to CBIO[33] (MME, 2017b, 62). Once these credits

have been entered, they can be traded over-the-counter, which is likely to be managed by B3 S.A. Brasil, Bolsa, Balcão S.A. ([B]³). Since March 2017 BM&FBOVESPA S.A. Securities, Commodities and Futures Exchange merged with Cetip S.A. Organized Markets forming this [B]³. "Today, besides being the only securities, commodities and futures exchange operating in Brazil, it is also the leading depository for fixed-income securities in Latin America and the leading clearinghouse for private assets in Brazil" (B3, 2017).

The costs of listing in the financial market must be negotiated individually between the producer/importer and the bookkeeper and collectively among the credit holders with the managers in the organized market.[34] Being the first ones more expensive than the last ones, they need a technical team to verify the ballast of the notes.[35]

As it can be seen, conditioning the public incentives to greater social and environmental efficiency and sustainability—legal features introduced by RenovaBio, Combustíveis Brasil, and CBIO—ends up stimulating governance strategies and ethanol production in accordance with sustainability standards. In general terms, in order to achieve the economic benefits provided for in these programs, producer, distributor, transporter, retailer, importer, and other stakeholders in the market must demonstrate compliance with this higher efficiency standards and social and environmental sustainability, which is achieved by means of the

---

[32]   This methodology has a strong scientific base, standardized by ISO 14040 and 14044.

[33]   Even with this estimated high operational costs, representatives of ANBIMA (Brazilian association of financial and capital market entities) expressed interest in becoming bookkeepers.

[34]   Organized over-the-counter market with entity accredited at the Brazilian Securities and Exchange Commission [Comissão de Valores Mobiliários (CVM)].

[35]   In contracts traded on B3 SA, for example, registered book-entry certificates (CPR) have the unit cost of registration/deposit of US$7.87, the cost of custody of this asset US$1.10 per lot, being the trading cost US$2.47.

adoption of compliance policies and sustainability standards.

13.3.4 Life cycle assessment program

The RenovaBio life cycle assessment program[36] is part of a system that uses a strong, internationally recognized methodology for organizing, reporting, and disseminating standardized information on life cycle inventories of industrial production. This methodology is the one to be used in the preparation of public and private inventories.[37]

This program aims to standardize information to better understand the environmental impacts associated with the entire product and service life cycle and thus serve for environmental management. It aims to identify and measure the "environmental impacts of materials, processes and products," so that with the information obtained, standardized with scientific bases, it is possible to "develop new products and detect improvements to be applied" (CONMETRO, 2010, 5).

The analysis during the life cycle of the environmental impacts of a product and service takes into consideration in addition to the materials the energy consumed by the productive processes. Converting all this into the equivalent carbon emissions can track the impact in the environment from the extraction of the natural resources used as inputs, including the processes of transformation, transportation, use, and final disposal.

Through this methodology, it is possible to identify what stages of the life cycle are most impacting and which should be the primary focus of process improvements and product development. Under the RenovaBio the life cycle assessment program is essential for the certification of biofuels within the biofuel production unit (mills) and also for importers. It will report

1. upstream of the agricultural process (inputs and infrastructure), with data from the "ecoinvent" database;
2. the agricultural process itself (the agricultural phase); and
3. the agroindustrial process (industrial phase).

Moreover, in order for the carbon intensity of the national biofuel to be comparable to that of the imported biofuel, GHG emissions from the distribution phase would also be accounted for (MME, 2017b, 41).

13.3.5 RenovaCalc

RenovaCalc is a data processing algorithm. The environmental performance conformity assessment protocol will have its documents and worksheets processed by RenovaCalc's support tools in following two ways:

1. with the "specific production profile," in which the primary production processes data are considered area under direct management of the plant or
2. with a "default production profile" (MME, 2017b, 41).

In the first, organizing data, completing RenovaCalc spreadsheets and auditing for certification will be more difficult. On the other hand, the "standard production profile," data supply and verification processes will be simpler and less expensive but will only correspond to the most common technological level at any given time, so the biofuel producer will not be able to distinguish favorably from its

---

[36] *Programa de Avaliação do Ciclo de Vida.*

[37] In Brazil, these standards are summarized in ABNT NBR ISO 14040: 2001 and in ABNT NBR ISO14044 and it is internationally stated by the International Life Cycle Partnerships for Sustainable World (UNEP/SETAC) and the International Organization for Standardization (ISO).

competitors, only demonstrate that it fulfills what is standard.

To obtain the environmental performance index of biofuels, by the life cycle assessment, the following steps are followed:

1. establishment of the reference unit (or functional unit 14 = 1 MJ of biofuel[38]) and the reference flow 15 of the LCA, which are considered in the following calculations;
2. sum of GHG emissions from the biofuel life cycle;
3. conversion of GHG to g CO_2eq, according to the characterization factors of each GHG (MME, 2017b, 41).

The RenovaCalc will allow the registration of data related to the parameters to be monitored and with this information it will perform the analysis of the evaluation steps. The result obtained will be the carbon intensity index in g CO_2eq/MJ of the determined biofuel under analysis.

The certification of the efficient production of biofuels envisaged in the RenovaBio will not evaluate the conformity of processes and components, but only the calculation of GHG emissions in the existing production process. So there is no prerequisite for certification; all biofuel producers may require it to be included in the RenovaBio and will therefore only submit their production process to a measurement according to the RenovaCalc. There is no minimum sustainability requirement for the biofuel producer to be included in RenovaBio since agents' production will only be measured, not selected. Besides that, certifiers previously registered and supervised by the Brazilian government will accredit this certification. ANP will supervise their actions and may impose administrative and pecuniary sanctions if they fail to meet the requirements set forth in the RenovaBio.

In addition, the data that support the composition of the energy-environmental efficiency note should be submitted to public consultation. It is important to emphasize that these requirements related to the disclosure of data on the process of constitution of the note of energy-environmental efficiency will have an important effect on mutual supervision between the producing plants of biofuels, aiding the identification of frauds. This mechanism is designed to avoid asymmetric information that could compromise the smoothness of the certification process and consequently the competitive game between the various stakeholders.

13.3.6 BVC (CanaSoft)

Another long-term standardized data processing initiative is the virtual sugarcane biorefinery (BVC) (CanaSoft), a computational simulation platform that makes it possible to technically evaluate—considering the three axes of sustainability: economic, environmental, and social—the viability of different guidelines and the integration of technologies in the sugarcane and other biomass production chain. It was developed by the integrated biorefineries evaluation division of the National Laboratory of Bioethanol Science and Technology, integrated to the National Center for Research in Energy and Materials (CNPEM). The object of this computational simulation is to assess the production of cellulosic ethanol, sustainable chemistry products whose elements degrade less the environment, alternatives of agricultural management for sugarcane and different strategies for better trade, and use of ethanol as biofuel.

In the agricultural area the evaluation model developed and used in the BVC (CanaSoft) incorporates parameters such as

---

[38] The researchers created this reference unit to facilitate the estimates, by this metric 14 U = 1 MJ of biofuel. They also adopted the reference flow 15 from the life cycle assessment program.

types of harvesting and planting, transportation stages, agricultural operations, machinery, labor, agrochemicals, fertilizers, among others. In the industrial sector, BVC uses mass and process energy balances to carry out the technical evaluation of different product options and technologies for biorefineries through process simulation software. In the area of logistics and final use, it is possible to analyze the influence of ethanol transport on the total impact of the product, among other factors. Indicators of social and environmental sustainability are calculated in the BVC through methodologies, such as input product matrix and life cycle analysis, respectively (CTBE, Laboratório Nacional de Ciência e Tecnologia do Bioetanol, 2017).

This technology is available to internal users of the CNPEM and to the external public of Brazil and abroad that had submitted a research proposal through the users' electronic portal and has been admitted by CNPEM, which in turn provides service, training, and support in the operation of their equipment to those with clearance to access it. While it has advanced in the processes of certification of the productivity of biofuels, the country prepares to improve the private systems to regularize such sources of energy, methodology already advanced in Europe.

13.4 RenovaBio broader applications and the use technology

RenovaBio "opens the doors for the participation of all economic agents that, regardless of their size, are efficient and able to sustainably supply biofuels" (Novato and Lacerda, 2017, 2). It can enhance the efficiency and sustainability of the renewable market, generate domestic employment, income, and socioeconomic development; the measurement tools can be used in other markets to redirect different production areas toward a more efficient

and sustainable path, especially if the measurements tools were programed using a technology that could be used by different agents efficiently, with transparency and reliability.

DLT brings the potential to increase network collaboration speed and efficiency with greater transparency and reliability. The main feature of this technology is to form distributed networks around cryptography and distributed consensus-based networks, being "a new type of database that allows multiple users to share information and can modify it in a secure and reliable way, even if they do not trust each other" (Denny and Paulo, 2019, 12).

Participants in a DLT keep the "general ledger" of the database updated in real time. In addition, the registered data can only be modified according to strict rules and if there is a general agreement. Participants can achieve consensus on changes in the state of this shared database, that is, validating transactions between users, without having to rely on the integrity of any of the participants or an authority (Denny and Paulo, 2019, 12). What each participant sees from the shared database matches the exact view of all other participants and thereby eliminates problems related to lack of payment receiving, duplication of files or corruption of the data. For this reason, DLT can facilitate the transfer of assets and data without the use of a reliable central authority.

In addition, participants become potential auditors; they can independently check what they see (i.e., the content of the database at a specific time) because what they access is the same for all other participants. This ensures that all participants have a consistent view of the state of the shared database. As a result, any incorrect alteration of the data (e.g., tampering by a malicious user) would be immediately detected, rejected, and unfeasible by all participants.

Another advantage is that costs are reduced because automatic business-to-business processing eliminates the overhead caused by the services of various intermediaries who are hired today to comply with bureaucracy. And the risk is mitigated because the DLT functions as an unchanging audit trail, greatly reducing the chances of tampering and collusion (Bueno and Veiga, 2018). This type of technology could be useful in RenovaBio especially because it would contribute to fill the information gap among participants and also to increase reciprocal trust.

13.5 Conclusion

Sugarcane used to produce energy has the potential to displace a significant amount of crude oil consumption and therefore mitigate the net CO_2 emission globally. There is no need to change the energy infrastructure, and it brings gains in human health, social benefits, and many other positive externalities. But the liquid biofuels on their own do not have enough competitiveness to beat fossil fuels. Therefore regulation must intervene to artificially decrease the relative prices and increase incentives.

A regional approach toward an integrated Latin American energy transition is only possible if a complementary regulatory paradigm is put in place to consider resources availability, cost-effectiveness, and technical reliability. And this depends on the construction of a regulatory, political, and economic common ground to ease the flow of investments into the region, guarantee a safe and clean energy transition at the same time that promotes regional economic development.

The Brazilian biofuel regulation contains some examples of policies to reach this goal, mostly creating certifications and market tools to increase the competitiveness of the renewables, benefiting the whole energy production and distribution chain, no matter the

nationality of the enterprise involved, therefore contributing to the Latin American energy transition as a whole.

References

AGRA, Agribusiness Intelligence, 2017. World Ethanol & Biofuels Report. Available from: <https://www.agra-net.com/agra/world-ethanol-and-biofuels-report/> (accessed 15.06.19.).

ANEEL, Agência Nacional de Energia Elétrica, 2017. Matriz de Energia Elétrica. Available from: <http://www2.aneel.gov.br/aplicacoes/capacidadebrasil/OperacaoCapacidadeBrasil.cfm> (accessed 15.06.19.).

ANP, Agência Nacional do Petróleo, Gás Natural e Biocombustíveis, 2011. Resolução 67/2011 ANP — estoques mínimos de etanol anidro e a comprovação constante de capacidade de atendimento do mandato de mistura.

ANP, Agência Nacional do Petróleo, Gás Natural e Biocombustíveis, 2018. Res. ANP no 758/2018: Regulamenta a certificação da produção ou importação eficiente de biocombustíveis de que trata o art. 18 da Lei no 13.576, de 26 de dezembro de 2017, e o credenciamento de firmas inspetora. Available from: <https://www.in.gov.br/materia/-/asset_publisher/Kujrw0TZC2Mb/content/id/52003440/do1-2018-11-27-resolucao-n-758-de-23-de-novembro-de-2018-52003305?utm_source=Comece+seu+dia+bem-informado&utm_campaign=f8bdc0a5ee-newsletter-2018_11_27&utm_medium=email&utm_term=0_5931171aac-f8bdc0a5ee-129476021&ct=t(newsletter-2018_11_27> (accessed 15.06.19.).

B3, 2017. B3 S.A. Brasil, Bolsa, Balcão S.A. ([B] 3) Profile and History. Available from: <https://ir.bmfbovespa.com.br/static/enu/perfil-historico.asp?idioma=enu> (accessed 15.06.19.).

Biofuels Digest, 2018. Biofuels Mandates Around the World 2018: Biofuels Digest. Available from: <https://www.biofuelsdigest.com/bdigest/2018/01/01/biofuels-mandates-around-the-world-2018/> (accessed 15.06.19.).

Bueno, R., Veiga, A.K., June 2018. Plataformas Blockchain. Características e Aplicações. In: Presented at the Blockchain developer PoliUSP, São Paulo. Available from: <http://www.bbchain.com.br> (accessed 15.06.19.).

Cana Online, October 4, 2019. O Setor Sucroenergético Se Mantém Como Um Grande Empregador e Gerador de Renda l CEISE Br. Available from: <http://www.ceisebr.com/conteudo/o-setor-sucroenergetico-se-mantem-

como-um-grande-empregador-e-gerador-de-renda.html> (accessed 15.06.19.).

CNPE, Conselho Nacional de Política Energética, 2017a. Resolução CNPE no. 11, de 11/04/2017 – diretrizes para a importação de biocombustíveis.

CNPE, Conselho Nacional de Política Energética, 2017b. Resolução CNPE no. 14/2017. Available from: <http://www.mme.gov.br/documents/10584/4489543/Resolução_CNPE + 14%2C + de + 2017.pdf/deab4620-df97-45d8-899e-cbb63a9f8ade> (accessed 15.06.19.).

CNPE, Conselho Nacional de Política Energética, 2018. Resolução CNPE no. 5/2018 – Estabelece as metas compulsórias anuais de redução de emissões de gases causadores do efeito estufa para a comercialização de combustíveis. Available from: <http://www.mme.gov.br/documents/10584/71068545/Resolução + no + 5_2018_CNPE.PDF/a46326ab-df5d-4d3f-ad52-b9f1ffc7ab1d> (accessed 15.06.19.).

CONMETRO, 2010. Conselho Nacional deMetrologia, Normalização e Qualidade Industrial. Resolução no 04, de 15 de dezembro de 2010 – Dispõe sobre a Aprovação do Programa Brasileiro de Avaliação do Ciclo de Vida – ISO 14040 e 14044. Available from: <http://www.inmetro.gov.br/legislacao/resc/pdf/RESC000236.pdf> (accessed 15.06.19.).

Cortez, L.A.B. (Ed.), 2016. Proálcool 40 Anos. Blucher FAPESP. Available from: <https://www.blucher.com.br/livro/detalhes/proalcool-40-anos-1196> (accessed 15.06.19.).

CTBE, 2017. Laboratório Nacional de Ciência e Tecnologia do Bioetanol. In: Biorrefinaria Virtual de Cana (BVC). CTBE. Available from: <http://ctbe.cnpem.br/biorrefinaria-virtual-cana-bvc/> (accessed 15.06.19.).

Denny, D.M.T., 2018. Agenda 2030 e governança ambiental: estudo de caso sobre etanol da cana de açúcar e padrões de sustentabilidade como bonsucro. UNISANTOS – Universidade Católica de Santos. Available from: <http://biblioteca.unisantos.br:8181/handle/tede/4581> (accessed 15.06.19.).

Denny, D.M.T., Paulo, R.F., 2019. Blockchain and the Agenda 2030. Blockchain Technology and Application. Nova Science Publishers, New York.

EPE, Empresa de Pesquisa Energética, Ministério das Minas e Energia MME, 2017. Plano Decenal de Expansão de Energia 2026/Ministério de Minas e Energia. Empresa de Pesquisa Energética. MME/EPE, Brasília. Available from: <http://www.epe.gov.br/pde/Paginas/default.aspx#intro> (accessed 15.06.19.).

Fecombustíveis, 2018. Federação Nacional do Comércio de Combustíveis e de Lubrificantes. In: Tributação Dez 2018. Available from: <http://www.fecombustiveis.org.br/revendedor/tributacao/> (accessed 15.06.19.).

FGV, Faculdade Getúlio Vargas, 2017. Cadernos de Biocombustíveis – FGV Energia.

Goldemberg, J., Da Silva, J.G., Serra, G.E., Moreira, J.R., Conçalves, J.C., 1978. Energy balance for ethyl alcohol production from crops. Science 201 (4359), 903–906. Available from: <https://doi.org/10.1126/science.201.4359.903> (accessed 15.06.19.).

IBGE, Instituto Brasileiro de Geografia e Estatística, 2017. A Geografia Da Cana-de-Açúcar. Brasília. Available from: <https://ww2.ibge.gov.br/apps/dinamica_agropecuaria/> (accessed 15.06.19.).

IEA, Instituto de Economia Agrícola, April 2017. Geração de Bioenergia de Biomassa Da Cana-de-Açúcar Nas Usinas Signatárias Ao Protocolo Agroambiental Paulista, Safra 2015/2016. Available from: <http://www.iea.sp.gov.br/ftpiea/AIA/AIA-19-2017.pdf> (accessed 15.06.19.).

IRENA, International Renewable Energy Agency, 2019. Renewable Energy and Jobs – Annual Review 2018. IRENA – International Renewable Energy Agency. Available from: <https://irena.org/-/media/Files/IRENA/Agency/Publication/2018/May/IRENA_RE_Jobs_Annual_Review_2018.pdf> (accessed 15.06.19.).

Jaiswal, D., De Souza, A.P., Larsen, S., LeBauer, D.S., Miguez, F.E., Sparovek, G., et al., 2017. Brazilian sugarcane ethanol as an expandable green alternative to crude oil use. Nat. Clim. Change 7 (11), 788. Available from: <https://doi.org/10.1038/nclimate3410> (accessed 15.06.19.).

Macedo, L., 1993. Alcool Etilico da Cachaca ao Cereal. Editora Icone, São Paulo, SP, Brasil.

Maia, E. A. F. de S., Viana, M.A., 2016. Seria o Brasil um líder na temática das energias renováveis? Fronteira: revista de iniciação científica em Relações Internacionais 15 (29 e 30), 69–86.

MAPA, Ministério da Agricultura, Pecuária e Abastecimento, 2017. Balança Comercial Do Agronegócio – Agosto/2017 – Secretaria de Relações Internacionais Do Agronegócio. Available from: <http://www.agricultura.gov.br/assuntos/relacoes-internacionais/estatisticas-de-comercio-exterior> (accessed 15.06.19.).

MAPA, Ministério da Agricultura, Pecuária e Abastecimento, 2019. Balança Comercial Do Agronegócio – Maio/2019 – Secretaria de Relações Internacionais Do Agronegócio. Available from: <http://www.agricultura.gov.br/assuntos/relacoes-internacionais/estatisticas-de-comercio-exterior> (accessed 15.06.19.).

Milanez, A. Yabe, D. Nyko, M.S. Valente, L.C. Sousa, A.M.F. L.J. Bonomi, C D F de J, M.D.B. Watanabe, et al., March 2015. De promessa a realidade: como o etanol celulósico pode revolucionar a indústria da cana-de-açúcar: uma avaliação do potencial competitivo e sugestões de política pública. Available from: <https://web.bndes.gov.br/bib/jspui/handle/1408/4283> (accessed 15.06.19.).

Milanez, A.Y., Mancuso, R.V., Godinho, R.D., Poppe, M.K., March 2017. O Acordo de Paris e a Transição Para o Setor de Transportes de Baixo Carbono: O Papel Da Plataforma Biofuturo. BNDES, Brasília. Available from: <https://web.bndes.gov.br/bib/jspui/bitstream/1408/11756/1/BS%2045%20O%20Acordo%20de%20Paris%20e%20a%20transição%20para%20o%20setor%20de%20transportes%20de%20baixo%20carbono%20%5B...%5D_P_BD.pdf> (accessed 15.06.19.).

MME, Ministério das Minas e Energia, 2017a. Combustível Brasil: Ministro Lança Iniciativa Para Repensar Refino e Abastecimento. Available from: <http://www.mme.gov.br/web/guest/pagina-inicial/outras-noticas/-/asset_publisher/32hLrOzMKwWb/content/combustivel-brasil-ministro-lanca-iniciativa-para-repensar-refino-e-abastecimento> (accessed 15.06.19.).

MME, Ministério das Minas e Energia, August 2017b. RenovaBio — Nota Explicativa Sobre a Proposta de Criação Da Política Nacional de Biocombustíveis.

MME, Ministério das Minas e Energia, 2018. RenovaBio: Price Is What You Pay. Value Is What You Get. Available from: <http://www.mme.gov.br/documents/10584/55980549/RenovaBio.pdf/> (accessed 15.06.19.).

Morceli, P., 2006. Futuro Para o Álcool Brasileiro. Revista de Política Agrícola, Brasília, DF, no. Ano XV-No 3, p. 19—27, Jul./Ago./Set. 2006.

NovaCana, 2017. Simulações Dos Impactos Do RenovaBio Para o Setor Sucroenergético. Available from: <https://www.novacana.com/n/etanol/politica/primeiras-simulacoes-impactos-renovabio-sucroenergetico-080617/> (accessed 15.06.19.).

Novato, M., Lacerda, M.I., 2017. RenovaBio-towards a new national biofuel policy and a truly sustainable world. Innov. Energy Res. 6 (2), 1—2. Available from: <https://doi.org/10.4172/2576-1463.1000164> (accessed 15.06.19..

Oddone, D., Agência Nacional do Petróleo ANP Gás Natural e Biocombustíveis, 2018. The Oil and Gas Industry in Brazil. ANP Agência Nacional do Petróleo, Gás Natural e Biocombustíveis, New York. Available from: <http://www.anp.gov.br/images/Palestras/Decio-Oddone_American-Council_June-2018.pdf> (accessed 15.06.19.).

Oliveira, K.B.M., Valdés Serra, J.C., Oliveira, L.M., 2014. Balanços energéticos para a produção de etanol para as diferentes matérias primas. Geoambiente Online Revista Eletrônica do Curso de Geografia UFG (22), . Available from: <https://www.researchgate.net/publication/314980007_BALANCOS_ENERGETICOS_DA_PRODUCAO_DE_ETANOL_PARA_DIFERENTES_MATERIAS_PRIMAS> (accessed 15.06.19.).

Ortiz, L., Rodrigues, D., 2006. Em Direção à Sustentabilidade Da Produção de Etanol de Cana de Açúcar No Brasil. Available from: <http://www.ecoa.org.br/arquivos/444052181.pdf> (accessed 15.06.19.).

Piacente, F.J., Silva, V. de C., Biaggi, D.E., 2016. Produção de Bioetanol a Partir Do Milho Estudo de Prospecção Tecnológica a Partir de Bases de Patentes. Blucher Eng. Proc. 3, 1211—1225. Available from: <http://www.proceedings.blucher.com.br/article-details/produo-de-bioetanol-a-partir-do-milho-estudo-de-prospeco-tecnolgica-a-partir-de-bases-de-patentes-25340> (accessed 15.06.19.).

Postal, A.C.M., da Silveira, J.M., 2018. Ten Years of Sugarcane Expansion in Brazil: An Analyis of Perception of Local Stakeholders About Social, Environmental and Economic Impacts. Available from: <https://www.researchgate.net/publication/322686792_PERCEPTION_OF_SUGARCANE_IMPACTS_ACCORDING_TO_LOCAL_STAKEHOLDERS> (accessed 15.06.19.).

de Queiroz, F.A., 2005. Meio Ambiente e Comércio Na Agenda Internacional: A Questão Ambiental Nas Negociações Da OMC e Dos Blocos Econômicos Regionais. Ambiente Sociedade 8 (2), 125—146. Available from: <https://doi.org/10.1590/S1414-753X2005000200007> (accessed 15.06.19.).

Shikida, P.F.A., de Azevedo, P.F., de Freitas Vian, C.E., 2011. Desafios Da Agroindústria Canavieira No Brasil Pós-Desregulamentação: Uma Análise Das Capacidades Tecnológicas. Revista de Economia e Sociologia Rural 49 (3), 599—628. Available from: <https://doi.org/10.1590/S0103-20032011000300004> (accessed 15.06.19.).

Sousa, E.L., Macedo, I. de C., 2010. Etanol e Bioeletricidade: A Cana-de-Açúcar No Futuro Da Matriz Energética. UNICA — União da Indústria de Cana-de-açúcar. Available from: <https://pt.scribd.com/document/37651191/Etanol-e-bioeletricidade-A-cana-de-acucar-na-matriz-energetica-brasileira> (accessed 15.06.19.).

Souza, S.P., Horta Nogueira, L.A., Martinez, J., Cortez, L.A.B., 2018. Sugarcane can afford a cleaner energy profile in Latin America & Caribbean. Renew. Energy 121, 164—172. Available from: <https://doi.org/10.1016/j.renene.2018.01.024> (accessed 15.06.19.).

UN, United Nations, December 2015a. Acordo de Paris. Available from: <https://treaties.un.org/doc/Publication/UNTS/No%20Volume/54113/A-54113-China-080000028047bf86.pdf> (accessed 15.06.19.).

UN, United Nations, 2015b. Agenda 2030. <http://www.un.org/ga/search/view_doc.asp?symbol = A/RES/70/1&Lang = E>.

UNICA, 2017. União da Indústria de Cana-de-Açúcar. In: Mapa Da Produção de Cana-de-Açúcar No Brasil. Available from: <http://www.unica.com.br/mapa-da-producao/> (accessed 15.06.19.).

The role and impacts of policies in hampering the biogas transition in Brazil[*]

Luiz Gustavo Silva de Oliveira[1,2]

[1]Copernicus Institute of Sustainable Development, Utrecht University, Utrecht, The Netherlands
[2]International Centre for Integrated assessment and Sustainable development, Maastricht University, Maastricht, The Netherlands

14.1 Introduction

Nowadays energy systems have been pressured by two huge forces: new technologies and its unsustainable character. Whereas new technologies, such as renewable energy and blockchain technologies, introduce varieties of energy system, being sourced by fossil fuels begets negative impacts at global and local scales. This situation demands radical changes in energy systems designs and operation. This debate, very often referred to as energy transition (Araújo, 2014; Fouquet, 2016), poses several complex questions to be solved.

The investigation and comparison of several energy transition cases suggest some issues such as the adequate conditions and how to implement energy transitions. Still, scholars and practitioners commonly underline the issues of temporality, regionality, scale, speed, technological features, socioeconomic landscape, cultural background, institutional structures, and actors involved (Araújo, 2014; Fouquet, 2016; Sovacool, 2016; Grubler, 2012; Smil, 2016; IRENA, 2019, 2018). Furthermore, although actions to promote these conditions take into consideration private actors and firms' strategies, analysts repeatedly accentuate the relevance of an adequate set of policies and regulations to establish conditions for energy transitions (Rogge et al., 2017).

The case of energy transition in Brazil illustrates not only the relevance of these conditions and the role of policies and regulations

[*] This chapter is a partial result of the PhD research on "Informing systemic policies to promote emerging technologies—Fostering the Brazilian biogas innovation system" at Utrecht University and funded by CNPq (Science Without Borders program).

but also how these conditions define the trajectories of different energy technologies. Brazil is often considered a case of success mainly because of its experiences in the biofuel industry[1] (IEA, 2013), moving from traditional to modern types of biomasses (Goldemberg and Coelho, 2004). Sugarcane energies, and more recently biodiesel, achieved high levels of legitimacy and market share, supported by the enactment of several policies and regulations, favorable resource conditions, coordination of research and development (R&D) as well as industrial activities with biofuel demand (Furtado et al., 2011; Rico and Sauer, 2015; Salles-Filho et al., 2017). Moreover, other Brazilian macro-conditions played an important role, such as the long history of sugarcane and the huge production of soybeans in the country.

On the other hand, biogas energies still struggle to achieve the same levels of legitimacy and market share. Despite having the long history of biogas experimentation and huge potential (EPE, 2016; Palhares, 2008; Bley, 2015; ABiogas, 2015, 2018), biogas energies occupy a very marginal share of Brazilian energy matrix (EPE, 2018). This fact can be attributed to a series of barriers specific to biogas projects (Jende et al., 2016) but can also be understood from a more systematic perspective accounting for factors at different levels (De Oliveira and Negro, 2019).

De Oliveira and Negro (2019), investigating the evolution of biogas field, discuss how the conditions for the development and diffusion of biogas technologies were conditioned by sectoral regulatory framework, specific regional conditions, and specific macro-events. Although the authors present a comprehensive account of the biogas transition in Brazil, including the role of policy contexts, they do not explain how biogas-specific problems interacted with the institutional context to

hamper the development of biogas field. This chapter discusses this issue.

For this end and following De Oliveira and Negro (2019), biogas field is analyzed as a sociotechnical system using the technological innovation system (TIS) framework. TIS framework focuses on investigating emerging technologies or technological fields accounting for how actors and their interactions under institutional and infrastructural settings fulfill system level processes (Bergek et al., 2008a). This systemic perspective allows identifying not only barriers at project level but also problematic factors at the level of organizational fields and sectors. Moreover, this chapter borrows concepts from policy mix literature to discuss how the evolution of the established policies and regulations influenced the Brazilian Biogas Innovation System (BBIS).

Thus

- Section 14.2 briefly introduces the concepts applied in the analyses as well as the methods;
- then, Section 14.3 discusses the evolution of biogas field to disclose its problematic factors and how the interrelationships between these factors and the policies are established; and
- lastly, Section 14.4 discusses how these phenomena entail the necessity for more coordination of new policies and regulations as well as what can be learned to other energy transitions in Latin America.

14.2 Problematic factors of technological innovation systems and policy mixes

Energy transitions comprise a complex set of processes that occur at distinct levels subjected to several conditions (Grubler, 2012; Smil, 2016; Cherp et al., 2017; Sovacool, 2016).

---

[1] Brazilian power sector is also a success case because of its almost full integration and high share of hydropower.

For instance, increasing the share of renewable energy technologies in power systems requires new practices of consumers, changes in institutional frameworks, the supply of new technologies, and the integration into power systems. Consequently, energy transitions can be explained by many different conceptual frameworks, depending on the main question asked (Sovacool and Hess, 2017). As this chapter is interested in explaining the interrelationship of problematic factors, policies, and regulation in biogas field in Brazil, it takes the sociotechnical system perspective of TIS (Bergek et al., 2008a) and borrows concepts from policy mix literature (Rogge et al., 2017).

14.2.1 What are systemic problems and blocking mechanisms

Systemic problem is a concept derived from the innovation system framework (Woolthuis et al., 2005; Wieczorek and Hekkert, 2012). Innovation systems studies understand innovations as a consequence of systemic interactions of different actors under certain institutional and material conditions (Edquist, 1997). Among the different approaches,[2] the TIS is the one which concentrates its analyses in specific technologies or technological fields (Bergek et al., 2008a). TIS framework states that configurations of system elements (actors, interactions, institutions, and infrastructures) enable or constrain the fulfillment of system processes (entrepreneurial activities, knowledge production and exchange, guidance of search, market creation, resource allocation, legitimacy, and positive externalities) (Hekkert et al., 2007; Bergek et al., 2008a,b). Moreover, TIS framework claims that system processes are blocked by either systemic problems (Wieczorek and Hekkert, 2012) or blocking mechanisms (Bergek et al., 2008a). Therefore

the development of a TIS across different phases is conditioned by these configuration and processes and the present systemic problems or blocking mechanisms.

Systemic problems and blocking mechanisms are similar concepts to explain problematic factors at system levels. Systemic problems are basically understood as negative attributes of system elements (Wieczorek and Hekkert, 2012). For instance, lack of capability of actors or inexistent institutions. In turn, blocking mechanisms can comprise all different sorts of negative influences, including those external to the TIS (Bergek et al., 2008a). The main goal of these concepts is to explain what hinders system development, so that it is possible to inform interventions such as policymaking.

14.2.2 Possibilities and limitations of systemic problems and blocking mechanisms explanations

Recent studies have highlighted that although systemic problems and blocking mechanisms are useful concepts to inform policymaking, they fall short in explaining important issues. First, part of these issues comes from the limitations of TIS framework in investigating contextual influences (external to the innovation system) systematically (Bergek et al., 2015). This fact is directly observed in the conceptualization of systemic problems, which does not explain contextual (exogenous) influences. For blocking mechanisms the lack of analysis of contextual influences does not lie in the conceptualization, which enables the discussion of these influences. It lies in the empirical use of the concept, in which hardly ever accounts for contextual influences and many times is used interchangeably with systemic problem concept. Second, the interaction of problematic factors at system level is not

[2] See Edquist (1997) and Weber and Truffer (2017).

discussed by these concepts. Often empirical studies find that interactions of problems are important factors for system hindrance (Kieft et al., 2016; Patana et al., 2013; Kriechbaum et al., 2018).

Following De Oliveira et al. (n.d.) and De Oliveira and Negro (2019), these issues can be tackled by specifying and studying the exact activities that actors engage.[3] It means that instead of only discussing the negative attributes of systemic problems, it is necessary to explain how these attributes lead to specific activities of actors that hamper the fulfillment of system processes. For example, the lack of technological capability of actors may manifest itself as poor project design or need to import knowledge. Also, by specifying activities, it is possible to examine in which contexts actors are embedded, and how these contexts influence their activities. For instance, actors in biogas field can be from power or sanitation sectors. These sectors have different institutional frameworks that enable or constrain actors differently.

Lastly, specifying activities also allow investigating which actors are involved in different activities and which activity precedes each other, that is, the interaction of problematic factors. For instance, the lack of technological capability of actors may lead to poor design, which in consequence may result in higher operational costs. In sum, for this chapter, systemic problems are negative attributes of structural elements of TIS, and blocking mechanisms are the way these problems manifest themselves to hinder the system processes. Moreover, these blocking mechanisms are influenced by both TIS-specific or TIS-contextual conditions.

14.2.3 Interrelations between technological innovation system problematic factors and policies

TIS analyses provide valuable insights about the development of emerging technologies to inform policymaking. One important insight is which institutions are more relevant for the specific technological field.[4] In other words, TIS analyses allow describing the effects of policies across different policy fields on a particular TIS. However, TIS analyses rarely discuss the interrelationship of these policies with TIS structures, processes, or problems.[5] One way of accounting for these interrelationships is to map the policy mix and its characteristics (Reichardt et al., 2016).

Policy mixes are "complex arrangements of multiple goals and means which, in many cases, have developed incrementally over many years" (Kern and Howlett, 2009:395). From this definition, it is possible to derive that policies are not single entities but composed of elements, goals, and means. Moreover, these goals and means range from strategic level to operational level (Howlett and Rayner 2013). Policy goals can be understood as the ideas, objectives, and requirements to address an issue (Howlett and Cashore, 2009; Howlett and Rayner, 2013). Policy means, or policy instruments, are understood as the techniques applied by the state to achieve such goals (Howlett, 2011; Howlett and Cashore, 2009; Howlett and Rayner, 2013).

Another important contribution of policy mix literature comes from the analysis of interactions of policy goals and instruments (Kern and Howlett, 2009; del Río and Mir-Artigues, 2014; Rogge and Reichardt, 2016). These

---

[3]   This understanding comes from the mechanism-based approach as in Beach and Pedersen (2016).

[4]   TIS analyses account for both formal and informal institutions (Markard et al., 2016). However, in this chapter, the focus is more on formal institutions—policies and regulations as in Streeck and Thelen (2005:10–16).

[5]   Important exceptions are Reichardt et al. (2016, 2017).

studies claim that by analyzing interactions between policy goals and instruments, which have specific design features and are embedded in particular contexts,[6] it is possible to identify some properties of policy mixes. Although there is debate on policy mix literature about these exact properties (Rogge et al., 2017), it is possible to say that policy coherence and consistency are widely accepted properties. The coherence of policy goals refers to the alignment of goals for an issue, while the consistency of policy instruments is the alignment of instruments toward a particular goal (Kern and Howlett, 2009; Howlett and Rayner, 2013).

Here, the most relevant for the analysis of this chapter is that these properties indicate the effects of established policies on actors and their activities. Put simply, if policy goals are not coherent, they tend to promote conflicting solutions. A common example is enacting energy efficiency policies and policies to promote energy-intensive industries at the same time in the same country. Still, if policy instruments are inconsistent, they will stimulate divergent behavior. One example is the implementation of financial instruments to promote distributed generation at the same time of the implementation of bureaucratic processes of connection of these generators to the grid. Therefore by identifying these properties in combination with system problems and blocking mechanisms, it is possible to discuss how policy mixes enable or reinforce problematic factors at system level.

14.2.4 Methodology

The discussion presented next comprises three main steps. First, it was necessary to map the evolution of biogas field as the BBIS. This step is already covered by De Oliveira and Negro (2019), and here it was only presented an overview of this transition dynamics. Second, the systemic problems and blocking mechanisms of BBIS had to be discussed. For this, it was applied an event history analysis as main method (Negro et al., 2007; Suurs and Hekkert, 2009) for the description of systemic problems and congruence method for the description of blocking mechanisms (Beach and Pedersen, 2016). Third, the interrelationship between systemic problems and blocking mechanisms with the policy mixes was the last step performed by coding policy goals and instruments and comparing them.

14.3 The role of policies and regulations in hindering the biogas transition in Brazil

Biogas activities in Brazil are rather diverse due to the several possible technological routes. Biogas production can be sourced by manure, wastewater, sludge, and organic matter of municipal solid waste and industrial residues (Coelho et al., 2018). In addition, biogas can be used in many different applications such as power generation and fuel substitution. Consequently, the technologies for biogas production and treatment derive from the choices of feedstocks and biogas uses. This diversity can be classified into three stages of a value chain: logistics and treatment of substrates (upstream), biogas production and treatment (midstream), and biogas logistics and use (downstream).

These stages permit to examine how each BBIS is influenced by different groups of policies. The upstream is mostly affected by environmental, sanitation, and agricultural policies. The midstream is the realm of biogas-specific policies and is influenced by trade and science, technology, and innovation policies. The downstream is conditioned by energy

---

[6] See Howlett (2011) for more details.

policies (power and fuels). The analysis next shows how BBIS systemic problems and blocking mechanisms and the evolution of sectoral policies create intricate problems for these activities, which indicate the points to intervene to foster the development of biogas field in Brazil.

14.3.1 Evolution of the Brazilian Biogas Innovation System

Biogas technologies7 in Brazil have long history of experimentation; the installation of the "Granja do Torto" biodigester in 1979 is often pointed out as the initial milestone (Palhares, 2008; Bley, 2015). From this point onward the BBIS started to be structured. The historical analysis demonstrates this structuration occurring throughout four main phases. *The first phase (1979—86) represented the kickoff of systematic efforts to develop biogas energies in the country*. It occurred mainly as a result of a combination of contextual pressures that allowed the emergence of biogas technologies as one of the possible solutions. The context of the expansion of biogas projects in China and India, the increasing pressures for improving urban and rural sanitation systems, and mainly the oil crises in the 1970 led to activities that comprised the first mechanism of BBIS structuration.

Basically, these activities were the convergence of visions among key actors (Ministry of Mines and Energy and state-level bodies) that biogas technologies were an interesting solution for the problems of oil dependence and environmental and sanitation problems. Noteworthily, these activities occurred in an environment where actors were engaged to seek for technological solutions that undermine the oil dependence. For instance, this is the same period of PROALCOOL program,

initial biodiesel initiatives, and creation of PROCEL. These initial activities fulfilled the guidance of search process that triggered the activities on the second mechanism.

Since biogas became a possible solution, key actors started to allocate resources to develop biogas projects. There was an important national funding line, resource allocation of several state-level bodies including of energy and sanitation utilities and some private companies. Specific knowledge on biogas technologies was provided by universities, rural extension organizations, utilities, and private companies. Consequently, several projects were developed, and this wave of projects triggered the third mechanism, the increase of legitimacy of biogas technologies. From the development of projects, especially in rural areas, farmers, municipalities utilities, universities, and private companies raised their expectations and trust in biogas technologies. It kept up the vision on biogas technologies as possible provider of alternative energy and environmental treatment. However, these high expectations and trust collapsed when the contexts change with the counter oil shocks and Brazilian macroeconomic crises.

Following the massive change in contexts, *the second phase (1986—2002) observed the tumble of BBIS*. The environment of economic crises led key actors to cutoff resource allocation mainly due to lack of financial capacity. This effect was widespread among national, state, and municipal level bodies and private companies. The 1980s decade is widely known as the lost decade. Consequently, although in the beginning of the phase some important landfill projects maintained their activities, on the beginning of 1990s, almost the totality of biogas projects ceased their activities.

The trust on biogas technologies, especially in rural areas, was also affected by many operational problems. Biogas projects did not live

---

[7] For a detailed account of the evolution of biogas field, please see De Oliveira and Negro (2019).

up the expectations of cost reductions, which yielded frustration with biogas technologies. In addition, this period was a period of intense changes of macro-institutional frameworks. This is illustrated by the redemocratization (1985–88), new constitution (1988), and the opening of economy and the liberal reforms (1990s). Although these changes created a turbulent environment for risky investments, such as technological innovations, they also established not only a more stable macroeconomic environment but mainly robust regulatory frameworks in sectors such as power, oil and gas, environment, science, technology, and innovation. This new institutional environment would allow biogas projects to blossom again in the following phase.

The third phase (2003–11) represented the return of biogas activities triggered by many CDM (Clean Development Mechanism) projects. Similarly to the first phase, the beginning of this phase observed a convergence of problematic situations and conditions that pointed at biogas as feasible solution. The explosion of urban population as well as the huge increase in livestock (mainly swine and poultry) and agriculture production from 1980s to 2000s emphasized the need for improved sanitation systems and environmental treatment options. In parallel, the increased relevance of climate change discussions in international arenas led to the establishment of CDM.[8]

These two circumstances enabled the alignment of visions around biogas technologies as solution once more. This occurred because CDM rewarded the degradation of methane,[9] which led to highly successful CDM projects in swine production and landfills and consequently initiatives and high-level discussions at national governmental bodies. Thus it was observed, once more, the sequence of activities

that guided the activities (alignment of visions around biogas solutions), resource allocation (leveraged by CDM), and increase of legitimacy.

This success continued until changes in the CDM rules and the international crisis in 2008. The former prohibited projects without energy recovery. The latter reduced the price of carbon and consequently the revenue of CDM projects. Also, several projects both in urban and rural areas still suffered with operational problems. Therefore similar to the first phase, initial sequence of the mechanisms was partially hindered.

However, given the new institutional environment (partly developed in the prior phase) and the favorable macroeconomic situation in the beginning of this phase, other mechanisms took place, favoring knowledge creation. Knowledge mostly resulted from learning-by-doing processes, which led to other type of experimentation with biogas technologies. Overcoming common problems on designing and operating biogas problems allowed the development of energy recovery projects in landfills, sanitation plants, and farms. Projects such as Ambev projects, Bandeirantes landfill, Nova Gerar, Itaipu project, and Barueri sanitation plant became important examples. These experimentations not only demonstrated the possibility of recovering biogas energies but also entailed a second wave of learning processes which would motivate important development in the following phase.

The fourth and current phase (2012–onward) is the first to notice a more active role of biogas players in structuring biogas field. Until this phase, most of the biogas activities occurred at project level without much coordination or concern to institutionalize biogas field. Virtually no specific biogas policy or regulation was observed before

---

[8] CDM was a resource allocation mechanism to develop projects that mitigate greenhouse gases in developing countries funded by developed countries.

[9] The methane GWP (global warming potential) is 28 times higher than carbon dioxide IPCC (2014).

2012. This picture changed because of the two waves of learning mechanisms in the previous phase, which counteracted the suspicion actors had on biogas technologies.

At the beginning of this phase (between 2012 and 2013), important biogas policies in São Paulo and Rio de Janeiro (RJ) states were established, CIBiogás and two biogas associations were created, an National Power Sector Regulatory Agency (ANEEL)[10] strategic R&D call was developed, and a national biogas program (PROBIOGÁS) was initiated. These activities aimed to develop and diffuse knowledge on biogas projects and policies to mitigate important struggle points of biogas sector (Jende et al., 2016). For instance, ANEEL strategic R&D focused on replicable projects, Itaipu project resulted in an important model of biogas project, ABiogás disclosed its policy proposal, and PROBIOGÁS published a series of technical studies.

This new accessible knowledge kept the expectations high on biogas technologies, however little influenced the resource allocation. First, it occurred because from 2012 onward Brazilian economy started to slow down. This scenario affected directly the investment in sanitation sectors and indirect in energy utilities due to market constrains. Sectoral crises, such as in the sugarcane sector, have also undermined the investment in innovations.

Second, the first national biogas regulation appeared only in 2015 (ANP regulation on biogas specification) and limited the production and use of biogas from urban residues.[11] This

fact hampered initiatives, especially in RJ state that based its policy on landfills. This issue was solved 2 years later,[12] but the worsening of the economic crisis in Brazil produced an adverse environment for investments. Despite this hostile setting, other biogas-specific policies[13] and a new biofuel law (RenovaBio[14]) were implemented. The state-level policies aim to foster biogas activities mostly exempting taxes, and the RenovaBio policy creates a carbon market for fuels sector.

14.3.2 Systemic problems and blocking mechanisms hindering the evolution of the Brazilian Biogas Innovation System

This evolution of biogas activities in Brazil demonstrates that players interested in biogas energies have experienced distinct types of problems throughout the different phases of development. Also, these problems were consequences of several factors, such as capabilities of actors, infrastructural conditions, and institutional settings. Describing these problems is the first step to understand the role of policies and regulations and possible interventions to foster biogas in Brazil.

14.3.2.1 Problematic factors of the first and second phases

After the initial successful experiences in the first phase, with at least 3000 biodigesters installed in rural areas (Palhares, 2008) and important landfill projects, the downturn came mainly due to contextual changes. However,

---

[10]　National power sector regulatory agency.

[11]　The ANP (National Oil, Natural gas and biofuel regulatory agency) resolution 8/2015 defined the specification for biomethane but excluded biomethane from urban residues, such as municipal solid waste. The Rio de Janeiro's policy was mostly based on recovering and treating landfill gas.

[12]　This issue was solved by the publication of the ANP 685/2017, which regulated biomethane from urban residues.

[13]　Ceará state, Rio Grande do Sul, and Santa Catarina are the most relevant.

[14]　Law no. 13576/2017.

these changes influenced BBIS mainly through structural couplings. Structural couplings are elements that make part of more than on system. Hence, they can transmit forces from one system to another. New technological fields heavily rely on these couplings since they still do not have the necessary elements.

This was the case of initial activities in BBIS that were conducted by several sectoral actors (utilities, governmental bodies). Also, sectoral policies and regulations were the main rules that guided the development of biogas activities. Therefore when the macroeconomic crises affected the capacity of sectoral actors, and several institutional reforms created short-term uncertainty, actors ceased the activities with biogas technologies.

Changes in contexts explain the major part of the failing of BBIS during the first and second phases. However, these changes do not explain the particularities of biogas issues. The first important systemic problem was the lack of biogas-specific policies and regulations, virtually not observed during the first phase. The successful experimentation and expansion of biodigester projects did not lead to formal regulations to govern the quality of process and biogas. Consequently, several projects presented problems in design and operation, which entailed a general frustration with biogas technologies.

This lack of specific regulations in combination with the centralized institutional energy framework and the low quality and accessibility of energy grid hampered the capture of the energy value of biogas. Entrepreneurs had virtually no options other than self-consumption for biogas energies. They also had hard time to access basic technologies, such as motor generators and compressors. These hurdles to create value with biogas energies reinforced the frustration with biogas projects.

Lastly, very few actors engaged in development of technical knowledge and interactions across regions and projects were also very few. This fact supported the little institutionalization in this period. In order to define regulation and standards for biogas projects and energies, actors had to codify technical knowledge and transmit this knowledge to other actors. Although there was some knowledge development, including specific technologies,[15] it was not observed the interactions to create a critical mass of biogas players.

14.3.2.2 Problematic factors of the third phase

In the third phase, it was observed the resurgence of biogas activities supported by and new sectoral institutional frameworks, better macroeconomic conditions, and the international support of CDM resources. As discussed in Section 14.3.1, this phase started with a positive wave of CDM projects followed by a wave of disappointment and finishing with a new positive cycle motivated by knowledge development. Here, the mechanisms hindering the development of biogas field were very similar to those observed in the first phase, caused by similar systemic problems.

The first successful wave of CDM projects focused mostly on methane degradation with a very low level of technological development. Second, it was still observed virtually no development of biogas-specific rules. Third, actors were still very isolated and few cross-sectoral and cross-governance level interactions occurred during this period. However, given the new context, they operated in a different fashion.

The focus on methane degradation of CDM projects overlooked the need of important technological advancements in biogas projects. Although knowledge on developing CDM

---

[15] For instance, UNESP advanced the biodigester technologies, Dedini developed technologies for vinasse and Sanepar, mangels, and MWM for using biogas in tractors.

projects and business models for biogas technologies came out as an important development of these projects, the little technological development and adaptation of biogas technologies could not address important technical issues, such as costs of biogas treatment and adaptation of biodigester scales. Therefore when the changes on CDM rules came out, actors either had to import high cost technologies or stop developing biogas projects.

Another important consequence of this focus on methane degradation was the low level of biogas-specific rules. As there was little commercialization of biogas electricity or biomethane, the specific rules developed concentrated on how to promote CDM projects. For instance, it was created an interministerial committee to evaluate the possibilities of expanding the CDM projects on landfills. What is more, similarly to the first phase, the loose regulation[16] of biogas projects allowed once more low-quality projects that entailed operational struggles particularly in rural areas. Different expectations also fed into the frustration on biogas projects (Querol et al., 2015).

Lastly, although there were relevant projects, they did not lead to the development of network or long-term relationships between players of different sectors or regions. These networks are crucial to the development and diffusion of emerging technologies (Musiolik et al., 2012). Hence, biogas projects were very case specific with few replicable models. In addition, knowledge on biogas technologies or projects did not reach high-level decision makers. For instance, during this phase, several bioenergy initiatives were promoted by national- and state-level governments, but most of them did not include biogas energies.[17]

14.3.2.3 Problematic factors of the fourth phase

In the last phase, although recently there has been important evolution in comparison to the past, for example, the creation of intermediary players (e.g., CIBiogás and Abiogás) and the PROBIOGÁS program that was in place from 2013 to 2017, biogas technologies still face difficulties, such as the uncertainty about costs and benefits of biogas projects and the lack of information about biogas solutions (Jende et al., 2016). These difficulties can be related to three main systemic problems: misalignment of funding conditions and requirement of biogas projects, the lack of knowledge on biogas technologies and projects by key actors, and interactions focused mostly on projects.

The misalignment of financial conditions and the need for biogas projects occurred mostly due to inadequacy of funding conditions and the required guarantees of biogas entrepreneurs. This mechanism hindered resource allocation across BBIS. Moreover, the level of stringency by which the resource allocation is hindered also depended on the contextual conditions posed by the implicit effects of macroeconomic policies and conditions. Macroeconomic conditions have a direct impact on how accessible or not are these resources. The austerity measures implemented by the government by the end of 2014 are illustration of these restrictive conditions.

The lack of knowledge on biogas technologies led to important blocking mechanisms of BBIS. First, it reinforced the different visions and expectations on biogas technologies and values biogas actors had during this period. Consequently, actors framed problems and possible solutions mostly according to their sectoral boundaries, leading to uncertain

---

[16]   Mostly regulated by environmental rules.

[17]   For instance, there was the establishment of agroenergy plans (2004 and 2005) and of the national biodiesel program (2004 and 2005).

future perspectives on biogas field. Second, little knowledge on biogas across actors increased costs of biogas projects due to the need of importation or to the expensive operational costs resulted from trial-and-error operations. Third, this lack of knowledge also resulted in suboptimal project designs and business models. These three mechanisms combined to influence the perception of biogas activities as complex and highly uncertain. Consequently, players avoided investing in biogas projects.

Lastly, new intermediary players and few networks, such as the BiogasFerti,[18] were not enough to overcome the project-based interaction in biogas field, which were also observed in the previous phase. This fact maintained the difficulties of actors to find relevant information and resources as well as spaces for exchange experiences and best practices. This problem tends to reinforce the problems of lack of knowledge and misalignment of funding conditions.

Furthermore, although these were the most relevant problematic factors, other important problems came into actions to reinforce these factors. First, the biogas-specific regional policies were not sufficient to create a national agenda for biogas field. This inexistence of a national agenda strongly supports the misalignment of funding conditions.[19] Second, the lack of political coordination, especially across sectors, creates an adverse environment for creation of networks or interactions that include actors from distinct sectors and levels.

Third, public bodies, both at national and state levels, given their rigid organization structures, have hard time to incorporate the new knowledge on biogas technologies. Hence, it is difficult to overcome the different

framings on problems and solutions of biogas field across these bodies. Fourth, the high costs of biogas projects due to importation and operational issues are inserted in an adverse economic context. This context brings into biogas field not only the effects of macroeconomic crises but also the low innovative character of Brazilian firms and the short-termism and rent-seeking features of Brazilian investors.

14.3.3 Interrelationships of systemic problems blocking mechanisms

The previous sections discussed the effects of intense institutional reforms in several sectors and policy fields on biogas activities. Understanding which conditions are these helps to explain the role of policies and regulations in the biogas transition but also supports informing policymaking. The later discussion focuses on the third and fourth phases because they discuss current problems. Moreover, the first phase and part of the second presented a completely different institutional setting, which makes the discussion less relevant. Also, in the second phase, there were very few activities in BBIS, and little time to observe the impacts of institutional reforms.

The influences of policies and regulations on BBIS activities for the third and fourth phases can be summarized into two categories: the influences of sectoral governance structures and the influences due to incoherence inconsistency of policy goals instruments. After the institutional reforms in the second phase, sectors and policy fields presented different governance structures. Most of the sectoral policies, such as environmental, sanitation, science, technology and innovation

---

[18] It is a network created to promote the use of digestate as biofertilizer and harmonize laboratorial procedures across biogas projects, among other goals. It includes Embrapa, CIBiogás, INT, INMETRO, among others players.

[19] For instance, for both ethanol and biodiesel, national policies or programs support the creation of specific funding conditions.

(STI), climate change, and natural gas policies, have shared governance between state and national levels.[20] However, power sector is mostly regulated at national level.

This fact introduces differences on how policies and regulations are carried out across policy fields. For instance, in power sector, the tariffs are established by the national regulatory agency (ANEEL) or by the players in a national free market. In contrast, state-level regulatory bodies are in charge of regulating natural gas tariffs and free markets. Moreover, these governance structures introduce different sorts of actors that are relevant to BBIS activities. They bring different capabilities, visions, and expectations, regional realities that are translated into different policy issues.

Problematic factors were highly interdependent on sectoral governance structures. This fact was directly observed for the problems of lack of specific biogas policies and regulations and lack of meaningful interactions. Policy studies have demonstrated the relevance of accessing the right policymakers for agenda-setting, advising, lobbying, and formulating policies (Howlett and Newman, 2010; Weible and Sabatier, 2017). The great variety of governmental bodies across sectors and governance levels that regulate biogas activities reinforces or maintains that lack of biogas policies and regulations. It occurs because interactions and knowledge exchange become more difficult.

Biogas players have a hard time to identify and prioritize which players to interact with within this landscape. First, these governmental bodies are in different stages of the policy-making process and across several sectors and regions. Second, in a situation in which biogas players still struggle to develop meaningful interactions and few players understand the whole life cycle of biogas projects, these governance structures hamper the identification and access to these governmental bodies. As governmental bodies are in charge to design and implement policies and regulations, the lack of interactions with biogas players is an important factor for their lack of knowledge on biogas projects and on future trajectories of biogas activities. Therefore biogas technologies keep out of their agenda (legitimacy).

It is not difficult to see that these governance structures have also influenced the incoherence and inconsistency of policies and regulations. Two important policy interactions that defined adverse conditions for biogas activities came from environmental and power sector policies and bioenergy policies. The positive wave of projects that occurred in the beginning of the third phase, by applying the CDM model, was also a consequence of state-level environmental policies that became stricter in regulating the disposition of residues in water resources.

These policies were enacted by the end of 1990s where the reforms in both power and natural gas sectors had not been completed yet. Therefore these environmental regulations aiming to reduce local pollution and the not clear institutional framework for distributed generation in the end of 1990s and beginning of 2000s provided the environment for CDM biogas projects without energy recovery. This situation would change after power sector reform, which established new regulations for decentralized energy, including tariffs' incentives.

Another important policy incoherence that indirectly hampered biogas activities came from the isolation of bioenergy policies. During 2004 and 2005, several initiatives to promote bioenergy in Brazil were enacted. The National Biodiesel Programme and social

---

[20] Some sectors are also subjected to governance structures at municipal levels, for instance, sanitation, environmental, and climate change sectors.

label[21] both in 2005 promoted the production and use of biodiesel. The two national agroenergy plans and the creation of Embrapa agroenergy between 2005 and 2006 defined targets and increased legitimacy of bioenergy in Brazil. However, these initiatives did not account for biogas activities.

The expansion on biogas projects in rural and urban areas and some regulatory initiatives as in state-level environmental bodies and municipalities were not enough to reach national governmental bodies. Consequently, biogas field did not experience the implementation of instruments, such as specific funding lines as the sugarcane and biodiesel fields. This incoherence between local, regional, and national levels goals reinforced the perception of biogas technologies as complex or not relevant and maintained the situation of misalignment of funding conditions.

Another interrelationship of policies and BBIS problems came from the inconsistent implementation of policy instruments across distinct sectors. First in the upstream side of the value chain, inconsistencies were observed in sanitation and waste management sectors. Sanitation treatment plants have historically been subsidized because of the relevance of sanitation sectors. Although these subsidies are relatively reasonable, they hamper biogas electricity[22] in the sector. It occurs mainly because the main business model for sanitation treatment plants is the power generation with self-consumption model.

In the waste management sector, Law 12305/2010 established the national waste management policy. This policy established that residues could not be disposed in landfills in the case of any other feasible destination. This goal can be understood as a consequence of several successful projects of landfill gas capture. In addition, the law established the end of dumps by 2014, which would lead municipalities to search for alternatives. Therefore there was an expectation of more biogas projects as part of the portfolio of solutions. However, the policy failed to implement instruments to provide resources (knowledge, skilled labor, and financial resources) to municipalities, and the goal was not accomplished.

In the downstream side, inconsistencies were observed in power sector and in the interaction of biogas and fuels sector policies. Power sector policies and regulations provided the main types of market creation mechanisms through the discount on tariffs of grid use, the net-metering scheme, and the auctions. However, the calibration of these instruments was not always aligned to the reality of biogas projects. In the case of the net-metering regulation (2012) the regulation defined the scale of the distributed generators more aligned to solar photovoltaic projects, up to 1 MW. Consequently, part of biogas projects, for example, landfill projects, was not applicable to this scheme.

In addition, 4 months after the initial publication, ANEEL's Normative Resolution 482/2012 was modified to remove the possibility of virtual net-metering. This modification aimed to avoid conflict with state-level bodies in charge of VAT-like taxes. This conflict came from the fact that states understood the met-metering scheme as power trade, therefore applicable to taxation (and prohibited by law to residential and small power consumers). These taxes completely undermined the incentive character of the net-metering scheme. This situation lasted until big states exempt their taxes, being followed by other states later on. In 2016 ANEEL reintroduced the possibility of

---

[21] Label created to stimulate the social inclusion of small producers in biodiesel supply chains.

[22] And for any energy efficiency project.

virtual net-metering and raised the allowed installed capacity. At the moment, this instrument is in discussion once more, which feeds uncertainty in some investors.

In the case of biogas-specific policies the main inconsistency came from the RJ state policy and ANP resolution that regulated biogas and biomethane. The RJ policy (2012) created an important market mechanism for biomethane[23] by making compulsory the purchase of biomethane up to 10% of gas market[24] in the state by the gas utilities. RJ state aimed to develop its potential of landfill gas. However, by this time, biogas and biomethane were not regulated by ANP, in charge of defining the regulations for fuels specification nationally. ANP published the regulation only in 2015 and excluded the possibility of using biomethane from urban residues. This fact hampered the development of RJ policy until 2017, when a new regulation included the possibility of using urban residues. This temporal inconsistency can be understood as a result of the problematic factors of lack of interactions and different frames on biogas technologies.

14.4 Conclusion

14.4.1 Problems and possibilities for biogas transition in Brazil

The discussion of systemic problems, blocking mechanisms, and their interrelationships with the complex policy context indicates that actions to foster biogas field can be promoted at different levels, targeting different actors and activities. First, it emphasizes the need of vertical and horizontal policy coordination.

This coordination goes beyond policies enacted by laws and decree. Several policy instruments are defined by regulatory and executive agencies through different types of legal documents, such as resolutions and ordinances. This coordination is not easy, given the reality of the governance structures in Brazil; however, it is likely to increase the effectiveness of new policies.

Second, this discussion broadens the possibilities for recommending fostering actions. These actions can target specific systemic problems, such as improving funding conditions or creating schemes for financial guarantees of biogas projects. They can also focus on specific activities caused by systemic problems. For example, to find information in biogas field is still difficult due to the lack of formal networks and project-based interactions. To stimulate the creation of platforms or formal network seems an interesting solution. Still, actions can focus on mitigating negative contextual influences, such as reducing rigidity on organizational of public bodies to increase knowledge absorption.

In the last phase, biogas players seem to have chosen the strategy of have closer connections with their specific local and regional bodies as individual players and use associations (intermediary players) to reach national bodies. This strategy makes sense because the general governance structure of the federative systems establishes the national policies governing regional policies. The state-level policies of Rio Grande do Sul (2017) and Santa Catarina (2018) as well as the ABiogás policy proposal (2015 and 2018) for a national program of biogas and biomethane can be understood as a consequence of these actions.

---

[23] In the law called renewable natural gas.

[24] It excluded the market of power generation, which represented by far the biggest market.

14.4.2 What can be learned for energy transitions in Latin America

This study on biogas transition in Brazil provides useful insights to pay attention for other energy transitions in Latin America. First, it becomes clear that although it is necessary to comprehend the specificities of actors and projects in energy transitions, more aggregated levels are highly relevant and systemic analyses are essential. More importantly, these analyses must go beyond techno-economic aspects to cover social, political, and cultural factors.

Purely techno-economic analyses fall short to understand the role of actors and their interactions and to explain characteristics of path-dependence and lock-in, which are critical to energy transitions. It is pivotal to investigate specific socioeconomic conditions, local realities, and institutional settings. For instance, several Latin American countries have also faced cycles of economic and political crises which creates conditions not necessarily captured on techno-economic analyses. Given the relevance of technologies, sociotechnical systems represent a powerful conceptual framework for this task.

Second, the emergence of new energy technologies and changes in energy systems occur embedded in spaces governed by various sorts of policies and regulations. The design features of these policies, the characteristics of policy goals and types of policy instruments, and the interactions of these policies, enable or constrain actors' behaviors in distinct fashions. Policy mix analyses are decisive because as emerging technologies are not totally suitable for market and institutional structures, the characteristics of established policy mixes may hamper fostering actions. This fact is emphasized for technologies like biogas in Brazil, which are influenced by many different policy fields. Policy mix analysis can suggest points of improvement in policy coordination and integration.

References

ABiogas, 2018. Proposta de Programa Nacional Do Biogás e Do Biometano. São Paulo. <http://www.abiogas.org.br/>.

ABiogas, 2015. Proposta de Programa Nacional Do Biogás e Do Biometano – PNBB. São Paulo. <http://media.wix.com/ugd/e3a792_c21f5cd0da-fe4a0997ef62fd84d94806.pdf>.

Araújo, K., 2014. The emerging field of energy transitions: progress, challenges, and opportunities. Energy Res. Soc. Sci. 1, 112–121. Available from: https://doi.org/10.1016/j.erss.2014.03.002.

Beach, D., Pedersen, R.B., 2016. Causal Case Study Methods: Foundations and Guidelines for Comparing, Matching, and Tracing. University of Michigan Press, <https://books.google.nl/books?id = Am7tDAAAQBAJ>.

Bergek, A., Hekkert, M., Jacobsson, S., Markard, J., Sandén, B., Truffer, B., 2015. Technological innovation systems in contexts: conceptualizing contextual structures and interaction dynamics. Environ. Innov. Societal Transit. 16, 51–64. Available from: https://doi.org/10.1016/j.eist.2015.07.003.

Bergek, A., Jacobsson, S., Carlsson, B., Lindmark, S., Rickne, A., 2008a. Analyzing the functional dynamics of technological innovation systems: a scheme of analysis. Res. Policy 37 (3), 407–429. Available from: https://doi.org/10.1016/j.respol.2007.12.003.

Bergek, A., Jacobsson, S., Sandén, B.A., 2008b. 'Legitimation' and 'development of positive externalities': two key processes in the formation phase of technological innovation systems. Technol. Anal. Strateg. Manage. 20 (5), 575–592.

Bley Jr., C., 2015. Biogás – a Energia Invisível. Planeta Sustentável, São Paulo. 9788567785042.

Cherp, A., Vinichenko, V., Jewell, J., Brutschin, E., Sovacool, B., 2017. Integrating techno-economic, sociotechnical and political perspectives on national energy transitions: a meta-theoretical framework. Energy Res. Soc. Sci. 37, 175–190. Available from: https://doi.org/10.1016/j.erss.2017.09.015.

Coelho, S.T., Garcilasso, V.P., Nunes Ferraz Jr., A.D., dos Santos, M.M., Joppert, C.L., 2018. IEE Intituto de Energia e Meio Ambiente da Universidade de São Paulo. In: Assumpção, I. (Ed.), Tecnologias de Produção e Uso de Biogás e Biometano. Synergia, São Paulo, <http://gbio.webhostusp.sti.usp.br/sites/default/files/anexosnoticias/livro-tecnologias-producao-uso-biogas-biometano.pdf>.

De Oliveira, L.G.S., Negro, S.O., 2019. Contextual structures and interaction dynamics in the Brazilian Biogas Innovation System. Renew. Sustain. Energy Rev. 107, 462–481. Available from: https://doi.org/10.1016/j.rser.2019.02.030. no.

De Oliveira, L.G.S., Subtil, J.L., Negro, S.O., n.d. Exploring blocking mechanisms and interdependence of systemic problems in technological innovation systems. Environ. Innov. Societal Transit.

del Río, P., Mir-Artigues, P., 2014. Combinations of support instruments for renewable electricity in Europe: a review. Renew. Sustain. Energy Rev. 40, 287–295. Available from: https://doi.org/10.1016/j.rser.2014.07.039.

Edquist, C., 1997. Systems of Innovation: Technologies, Institutions, and Organizations. <https://books.google.nl/books/about/Systems_of_Innovation.html?id=CvVASmavaDUC&pgis=1>.

EPE, 2016. Demanda de Energia 2050. In: Estudos Da Demanda de Energia. Rio de Janeiro. <http://www.epe.gov.br/Estudos/Documents/DEA 13-15 Demanda de Energia 2050.pdf>.

EPE, 2018. Empresa de Pesquisa Energética. In: Balanço Energético Nacional. Rio de Janeiro.

Fouquet, R., 2016. Historical energy transitions: speed, prices and system transformation. Energy Res. Soc. Sci. 22, 7–12. Available from: https://doi.org/10.1016/j.erss.2016.08.014.

Furtado, A.T., Scandiffio, M.I.G., Cortez, L.A.B., 2011. The Brazilian sugarcane innovation system. Energy Policy 39 (1), 156–166. Available from: https://doi.org/10.1016/j.enpol.2010.09.023.

Goldemberg, J., Coelho, S.T., 2004. Renewable energy—traditional biomass vs. modern biomass. Energy Policy 32 (6), 711–714. Available from: https://doi.org/10.1016/S0301-4215(02)00340-3.

Grubler, A., 2012. Energy transitions research: insights and cautionary tales. Energy Policy 50, 8–16. Available from: https://doi.org/10.1016/j.enpol.2012.02.070.

Hekkert, M.P., Roald, A.A.S., Negro, S.O., Kuhlmann, S., Smits, R.E.H.M., 2007. Functions of innovation systems: a new approach for analysing technological change. Technol. Forecasting Soc. Change 74 (4), 413–432. Available from: https://doi.org/10.1016/j.techfore.2006.03.002.

Howlett, M., 2011. Designing public policies: principles and instruments. Routledge Textbooks in Policy Studies. Routledge, <https://books.google.nl/books?id=2kNqRQAACAAJ>.

Howlett, M., Cashore, B., 2009. The dependent variable problem in the study of policy change: understanding policy change as a methodological problem. J. Comp. Policy Anal.: Res. Pract. 11 (1), 33–46. Available from: https://doi.org/10.1080/13876980802648144.

Howlett, M., Newman, J., 2010. Policy analysis and policy work in federal systems: policy advice and its contribution to evidence-based policy-making in multi-level governance systems. Policy Soc. 29 (2), 123–136.

Available from: https://doi.org/10.1016/j.polsoc.2010.03.004.

Howlett, M., Rayner, J., 2013. Patching vs packaging in policy formulation: assessing policy portfolio design. Politics Governance 1. Available from: https://doi.org/10.12924/pag2013.01020170.

IEA, 2013. World Energy Outlook 2013. OECD/IEA, Paris.

IPCC, 2014. Climate change 2014: synthesis report. In: Contribution of Working Groups I, II and III to the Fifth Assessment Report of the Intergovernmental Panel on Climate Change. Geneva. <http://www.ipcc.ch>.

IRENA, 2019. A New World—The Geopolitics of the Energy Transformation.

IRENA, 2018. Global Energy Transformation: A Roadmap to 2050, Executive Summary. <www.irena.org>.

Jende, O., et al., 2016. Barreiras e Propostas de Soluções Para o Mercado de Biogás No Brasil. In: Cardoso, H., Giersdorf, J., Costa Jr., L., Knopki, R., Santos, W. PROBIOGÁS, Brasília. <http://www.cidades.gov.br/images/stories/ArquivosSNSA/probiogas/Barreiras-Mercado-Biogas.pdf>.

Kern, F., Howlett, M., 2009. Implementing transition management as policy reforms: a case study of the Dutch Energy Sector. Policy Sci. 42 (4), 391–408. Available from: https://doi.org/10.1007/s11077-009-9099-x.

Kieft, A., Harmsen, R., Hekkert, M.P., 2016. Interactions between systemic problems in innovation systems: the case of energy-efficient houses in the Netherlands. Environ. Innov. Societal Transit. 24, 32–44. Available from: https://doi.org/10.1016/j.eist.2016.10.001.

Kriechbaum, M., Alan, C.B., Posch, A., 2018. Interaction patterns of systemic problems in distributed energy technology diffusion: a case study of photovoltaics in the Western Cape province of South Africa.. Technol. Anal. Strateg. Manage. . Available from: https://doi.org/10.1080/09537325.2018.1473851.

Markard, J., Wirth, S., Truffer, B., 2016. Institutional dynamics and technology legitimacy—a framework and a case study on biogas technology. Res. Policy 45 (1), 330–344. Available from: https://doi.org/10.1016/j.respol.2015.10.009.

Musiolik, J., Markard, J., Hekkert, M., 2012. Networks and network resources in technological innovation systems: towards a conceptual framework for system building. Technol. Forecasting Soc. Change 79 (6), 1032–1048. Available from: https://doi.org/10.1016/j.techfore.2012.01.003.

Negro, S.O., Hekkert, M.P., Smits, R.E., 2007. Explaining the failure of the Dutch innovation system for biomass digestion—a functional analysis. Energy Policy 35 (2), 925–938. Available from: https://doi.org/10.1016/j.enpol.2006.01.027.

Palhares, J.C.P., 2008. Biodigestão Anaeróbia de Dejetos de Suínos: Aprendendo Com o Passado Para Entender o Presente e Garantir o Futuro. <http://www.infobibos.com/Artigos/2008_1/Biodigestao/index.htm>.

Patana, A.S., Pihlajamaa, M., Polvinen, K., Carleton, T., Kanto, L., 2013. Inducement and blocking mechanisms in the Finnish life sciences innovation system. Foresight 15 (6), 428–445. <http://www.emeraldinsight.com.proxy.library.uu.nl/doi/pdfplus/10.1108/FS-10-2012-0081>.

Querol, M.P., Seppänen, L., Jackson Filho, J.M., 2015. Understanding the motivational perspectives of sustainability: a case of biogas production. Production 25 (2), 266–277. Available from: https://doi.org/10.1590/0103-6513.095212.

Reichardt, K., Negro, S.O., Rogge, K.S., Hekkert, M.P., 2016. Analyzing interdependencies between policy mixes and technological innovation systems: the case of offshore wind in Germany. Technol. Forecasting Soc. Change 106 (106), 11–21. Available from: https://doi.org/10.1016/j.techfore.2016.01.029.

Reichardt, K., Karoline, S.R., Negro, S.O., 2017. Unpacking policy processes for addressing systemic problems in technological innovation systems: the case of offshore wind in Germany. Renew. Sustain. Energy Rev. 80, 1217–1226. Available from: https://doi.org/10.1016/j.rser.2017.05.280.

Rico, J.A.P., Sauer, I.L., 2015. A review of Brazilian biodiesel experiences. Renew. Sustain. Energy Rev. 45, 513–529. Available from: https://doi.org/10.1016/j.rser.2015.01.028.

Rogge, K.S., Reichardt, K., 2016. Policy mixes for sustainability transitions: an extended concept and framework for analysis. Res. Policy 45 (8), 1620–1635. Available from: https://doi.org/10.1016/j.respol.2016.04.004.

Rogge, K.S., Kern, F., Howlett, M., 2017. Conceptual and empirical advances in analysing policy mixes for energy transitions. Energy Res. Soc. Sci. 33, 1–10. Available from: https://doi.org/10.1016/j.erss.2017.09.025.

Salles-Filho, S.L.M., De Castro, P.F.D., Bin, A., Edquist, C., Ferro, A.F.P., Corder, S., 2017. Perspectives for the Brazilian bioethanol sector: the innovation driver. Energy Policy 108, 70–77. Available from: https://doi.org/10.1016/j.enpol.2017.05.037.

Smil, V., 2016. Examining energy transitions: a dozen insights based on performance. Energy Res. Soc. Sci. 22. Available from: https://doi.org/10.1016/j.erss.2016.08.017.

Sovacool, B.K., 2016. How long will it take? Conceptualizing the temporal dynamics of energy transitions. Energy Res. Soc. Sci. 13, 202–215. Available from: https://doi.org/10.1016/j.erss.2015.12.020.

Sovacool, B.K., Hess, D.J., 2017. Ordering theories: typologies and conceptual frameworks for sociotechnical change. Soc. Stud. Sci. 1–48. Available from: https://doi.org/10.1177/0306312717709363.

Streeck, W., Thelen, K., 2005. Beyond Continuity – Institutional Change in Advanced Political Economies. Igarss 2014. Oxford University Press, <https://doi.org/10.1007/s13398-014-0173-7.2>.

Suurs, R.A.A., Hekkert, M.P., 2009. Cumulative causation in the formation of a technological innovation system: the case of biofuels in the Netherlands. Technol. Forecasting Soc. Change 76 (8), 1003–1020. Available from: https://doi.org/10.1016/j.techfore.2009.03.002.

Weber, K.M., Truffer, B., 2017. Moving innovation systems research to the next level: towards an integrative agenda. Oxford Rev. Econ. Policy 33 (1), 101–121. Available from: https://doi.org/10.1093/oxrep/grx002.

Weible, C.M., Sabatier, P.A., 2017. In: Weible, C.M., Sabatier, P.A. (Eds.), Theories of the Policy Process, fourth ed. Westview Press, New York.

Wieczorek, A.J., Hekkert, M.P., 2012. Systemic instruments for systemic innovation problems: a framework for policy makers and innovation scholars. Sci. Public Policy 39 (1), 74–87. Available from: https://doi.org/10.1093/scipol/scr008.

Woolthuis, R.K., Lankhuizen, M., Gilsing, V., 2005. A system failure framework for innovation policy design. Technovation 25 (6), 609–619. Available from: https://doi.org/10.1016/j.technovation.2003.11.002.

PART 3

Boosting energy transition: socio-technical issues

Foresight for Chile's energy transition—unleashing societal transformations

Joni Karjalainen, Noora Vähäkari and Sirkka Heinonen

Finland Futures Research Centre, University of Turku, Helsinki, Finland

15.1 Introduction

This chapter focuses on the energy transition in Chile and the increasing adoption of renewable energy sources as a future trajectory.

Energy transitions are long, deep affairs of socio-technical change, constrained by path dependencies, with considerable unclarity over their duration (Sovacool, 2016). Energy transitions have been understood in terms of changes in fuel sources or associated technologies, despite their profoundly systemic nature and expected potential to shape societies. More recently, systems-level approaches have been used in analyzing them (Geels, 2002, 2011; Geels and Schot, 2007). Furthermore, the frameworks of sustainability transitions, as a rapidly emerging and evolving field (Markard et al., 2012; Chang et al., 2017; Köhler et al., 2019), aim to offer responses to grand societal challenges and see the energy sector transition as a critical field of study. Such approaches aim to provide theories and models to understand potentially radical shifts to new kinds of socio-technical systems, while offering insights and tools that inform governance and policy-making efforts (Loorbach et al., 2017).

In spite of growing efforts for sustainability transitions, such performances often still focus on technological solutions, while disregarding local conditions, capacity, values, actor roles, or agency (Pesch, 2015; Wittmayer et al., 2017). Transformations of energy systems, in fact, consist of social processes that accompany shifts in energy technologies, affect the social outcomes, and stimulate and manage change (Miller et al., 2013). A focus on individual agency and capabilities as drivers of change places more emphasis on the local level than in institutional processes (Rauschmayer et al., 2015). An integrated approach to energy justice (Jasanoff, 2018; Williams and Doyon, 2019) is an example of efforts to understand how the ways the energy system and its actors are organized may change over time. The approaches of sustainability transitions endorse a long-term

The Regulation and Policy of Latin American Energy Transitions
DOI: https://doi.org/10.1016/B978-0-12-819521-5.00015-2

timeframe but have only rarely incorporated the analytical frameworks used in futures studies, as an academic niche and foresight practice.

The exploration of emerging directions is of great importance especially at times of turbulence and search for new national energy strategies. An analysis of energy futures can take various approaches to consider a range of factors, which may affect how energy systems are expected to change, even radically, over the next several decades. Awareness that the future may look little like the past is also useful for questioning asymmetries in power and presentist bias (Boston, 2017). Interestingly, in recent times an emerging body of energy futures work has surfaced in Latin America (Alvial-Palavicino and Opazo-Bunster, 2018; Herreras Martínez et al., 2015). Reaching to alternative futures challenges "used futures". Used futures is a term that describes visions and images of the future that are less relevant than they were in the past, perhaps in part because of their outdated assumptions (Milojević and Inayatullah, 2015).

The aim of the chapter is to explain how futures thinking can be used to analyze energy transitions and to explore and illustrate the potential benefits of using foresight methodologies and tools. The chapter focuses on Chile, which has shown ambition for a renewable energy transition at a regional level, and specifically probes the possibility of an energy transition to an increasingly renewable energy powered society in Chile, where the growing uptake of renewable energy is already informing the country's long-term energy strategy. An approach of deliberative foresight, as a critical and emancipatory future-oriented attitude, is specifically adopted to explore the power dynamics related to knowledge creation. The chapter also discusses and examines the sociocultural factors underpinning the opportunities embedded in a transition to renewable energy futures in Chile, which is often perceived to spearhead novel developments in Latin America.

After this section,

- Section 15.2 presents Chile's energy situation;
- Section 15.3 discusses the approaches of transition studies and foresight as related to energy;
- Section 15.4 opens up a case study of how foresight methodologies were used to imagine and analyze a future transition in Chile;
- Sections 15.5 and 15.6 present the results and discuss the relevance of the findings and the approach for the research agenda; and
- Section 15.7 concludes the findings and argumentation.

15.2 The promised land for renewable energy technologies

In a topography dominated by the Andes mountain range and a long coastline, Chile is often portrayed as a pioneer in Latin America owing to its level of human development, steady economic growth, and low economic risk (Benedikter and Siepmann, 2015). An open and transparent economy enables, for instance, low to zero tariffs for imported technology (Nasirov et al., 2015). However, Chile also has the highest income inequality of OECD countries and energy poverty is very high. Around 4000 people are estimated to die prematurely every year due to air pollution and 10 million people live in air-polluted cities (Schueftan et al., 2016; Gallardo et al., 2018).[1] Energy efficiency in commerce, housing, and industry is a

---

[1] In the small cities of South-Central Chile the burning of firewood for heating, cooking, and other uses is a source of particulate matter, so-called black carbon, and transport congestion affects the air quality in Santiago de Chile.

nation-wide challenge. Chile is a well-documented case of the influence of neoliberal economic ideas in policy planning (Manuschevich, 2016). In similar vein the country's electricity sector and its reforms are difficult to disassociate from the rest of the economy, including the mining sector (Pollitt, 2004). In recent decades, Chile's energy and electricity demand have increased (Nasirov et al., 2015)[2] and with that CO_2 emissions.[3]

Electricity generation in Chile was 70%−80% based on hydropower that is vulnerable to droughts and a central interconnected system until the 1990s (Bezerra et al., 2012). In the late 1990s the natural gas imports from Argentina started but soon halted when Argentina's economy dwindled, which caused a severe energy crisis in Chile. Between 2005 and 2015, the use of coal and natural gas as fuels in electricity generation increased, and this trend has continued. A dependence on fuel imports has also upheld high energy prices (Sanchez-Alfaro et al., 2015). As a response, governments in Chile have taken an active stance to address such challenges. A short-term focus has been on the increase of generation capacity and lowering the prices. Coal-based and large-scale hydro plants for electricity generation have been opposed on the grounds of social and environmental concerns (Alvial-Palavicino and Opazo-Bunster, 2018).

At a glance, Chile appears as the "promised land" for the developers of nonconventional renewable energy technologies, that is, all renewables excluding large hydropower (Santana, 2014; Nasirov et al., 2015). Chile has advantageous solar irradiance to make use of solar photovoltaics (PV) and concentrated solar power, especially in its northern parts at the Atacama Desert, as well as wind power and geothermal energy potential.[4] Changes to the Electricity Act, with other measures, have enabled the development of solar and wind power, and the Energy Agenda 2014 has introduced a series of short- and long-term reforms (Alvial-Palavicino and Opazo-Bunster, 2018). As the price of solar and wind power has fallen, they have won a growing share of competitive tenders, and more renewable energy projects are becoming online thanks to improved grid connections. Despite the effects to the electricity market, even high shares of variable electricity generation from renewable energy sources are possible. *Energía 2050* is the Government of Chile's long-term energy policy plan, which aims to act as a roadmap of interventions to guide a transition to a sustainable energy system (Gobierno De Chile, 2015). It sets a target of at least 70% renewable sources in electricity production by 2050, and there are debates on even a 100% RE target as a future option.[5]

Energía 2050, as Chile's long-term energy policy, based on a participative process and techno-economic scenarios, successfully challenged preconceptions regarding the cost of renewables and set an ambitious renewable energy target for the country. However, it is argued that it overlooked citizens' expectations, values, nonconformist views, and failed to propose alternative rules for governing the

---

[2] Electricity consumption in Chile has increased from 16.4 (1990) to 76.4 TWh (2016).

[3] See Chapter 3, Assessing the macroeconomic effects of sustainable energy transitions in Costa Rica and Chile: a multisectoral balance of payments constrained growth approach, this issue on Chile's economic structure and energy situation.

[4] See Chapter 12, Environmental impact assessment and public participation of geothermal energy projects: the cases of Chile, Costa Rica, Colombia, and Mexico.

[5] In 2016 out of the total electricity generated in Chile (79.3 TWh), wind energy (2.4 TWh), solar PV (2.6 TWh), biofuels (6.0 TWh), and hydropower (23.3 TWh) together contributed 43%. The rest was generated from coal (30.2 TWh), gas (11.8 TWh), and oil (2.9 TWh) (IEA).

electricity sector (Alvial-Palavicino and Opazo-Bunster, 2018). In fact, there are still numerous other issues holding back the full potential of Chile as a renewable energy country. Renewable energy technologies remain poorly known or opposed locally due to information barriers and lack of awareness (Cordoves-Sánchez and Vallejos-Romero, 2019; Haas et al., 2018; Vargas Payera, 2018). More recently, Chile has increased strategic efforts, especially in solar energy, to promote local and even regional solar-based industrial development, but there is a lack of human capital along the whole value chain of the projects and only a few local companies active in the sector (Haas et al., 2018). These attributes can be characterized as loose elements of a relatively weak innovation ecosystem nationally.

15.3 Theoretical framework

Sustainability transitions consist of a variety of research approaches used for analyzing systemic issues and their complexities with a purposive aim. Multiple tools and frameworks have been developed in the recent years to support the elaboration of interactions, interlinkages, and relations of actors and phenomena that guide such transitions (Köhler et al., 2019). Whereas historically energy transitions have often been described as processes of mostly technological development, an aim to purposefully deliver sustainable energy transitions magnifies the human facets of such endeavors, in recognition of the scale, scope, and the very nature of efforts to disengage from the global use of fossil fuels. Understanding these processes of change requires a multilevel and interdisciplinary analysis, and an awareness that even radical changes in the principles of energy systems, guiding frameworks, and actor roles might take place in the long term.

15.3.1 Energy transitions

Energy systems, as socio-technical entities consisting of technologies, practices, actors, and policies guiding them, are anticipated to face major structural changes in the future of nonlinear nature, affecting the way these and related systems are organized at various levels. A plausible explanatory framework is offered by the multilevel perspective (MLP), which draws on sociology, institutional theory, and innovation studies, as the basis and most popular of many sustainability transition frameworks. The MLP framework is used to describe development processes of technical innovations where an emerging technology or innovation (niche-level) is fighting to break into the mainstream. Incumbent structures are, and need to be, challenged for a niche structure to overcome the present regime—or a disruption of a structure to take place (Rotmans and Loorbach, 2009). The MLP presents how various factors in multiple levels stabilize, hinder, or reinforce an action or development. Breaking regime resistance requires multiple constraints by several lock-in systems to be overcome (regime-level). If successful, a technology or an innovation is expected to ultimately achieve a leader status or replace the existing regime. In the process, even the broader norms of the society may be shaped or major implications cast (landscape), as has been the case with telecommunications business and the effects of digitalization. MLP underlines the structures, disruptions, and nonlinearities that influence the process (Geels, 2010, 2011).[6] Dahle (2007) makes an interesting comparison between bottom-up revolutionary patterns and sustainability transitions.

Technologically, solar and wind power are emerging as the cheapest form of electricity and higher shares of renewable energy may be achievable (Lund, 2014; Agora Energiewende,

---

[6]   Geels (2004).

2015; Brown et al., 2018; Hansen et al., 2019). An emerging body of techno-economic studies assesses the potential for mass deployment of variable, renewable energy in Latin America (Barbosa et al., 2016; Gils et al., 2017; Aghahosseini et al., 2019). The future of the energy sector, as a whole, is increasingly driven by decentralization, decarbonization, digitalization, and electrification (WEF, 2018). In a renewable energy system, very high shares of solar and wind energy are used, complemented with other renewable energy sources, and the intermittency of solar and wind production is balanced. A detailed set of such principles is explained in what can also be called a neo-carbon energy system (Neo-Carbon Energy, 2017, 2018; Breyer et al., 2019).

As many sectors of the society as possible (heating, cooling, and transport) would be electrified, and short-term and seasonal energy storages used. The transport sector could be electrified either directly or indirectly with the aid of synthetic fuels from the electricity that is generated from a renewable energy system. An electrified energy system is entirely carbon neutral by the use of carbon dioxide from the air through power-to-X technologies. This relies on a method of using the electricity produced with renewables and combining it with carbon dioxide from the air to produce synthetic fuels, gases, liquids, materials, and products, even food proteins. Such carbon capture and utilization technologies that are becoming economically increasingly feasible, are chemical processes that act as bridging technologies, as they can transform electricity from renewables into synthetic materials, gases, and liquids. Exciting and promising as these directions may sound, not many studies yet explore such anticipated developments.

Here, an examination of sustainability transitions differs from conventional technological transitions (Geels, 2011). If conventional transitions are emergent, sustainability is purposive in its very nature. Nevertheless, the complex nature and definition of sustainability makes explaining a societal transformation process (e.g., toward sustainable development) difficult. Either way, explaining the growing adoption of renewable energy technologies cannot be limited to the analysis of technology diffusion but must recognize a broader range of factors, such as agency, governance (Patterson et al., 2017), and power (Smith et al., 2005). Here, an anticipation of the future (Miller et al., 2014; Miller, 2018; Poli, 2017) within complexity (Poli and Valerio, 2019) is advocated, which emphasizes the prominence and interplay of numerous technological (Kelly, 2016) as well as social forces, and is open to the potential for changes even in the metarules that drive the evolution of socio-technical systems (Schot and Kanger, 2018).

15.3.2 Futures studies and foresight as a co-creative and analytical approach

Future has excited human beings since ancient times, in fact, throughout their history. "Futures studies" means a systematic, multidisciplinary, and critical long-term analysis of complex future developments. The future cannot be predicted, but we are able to imagine, explore, and analyze future possibilities. Futures studies or futures research, as an intellectual tradition, has since the 1940s evolved and been formalized into scientific enquiry, with an increasingly rich epistemological, theoretical, and methodological offering.[7] Many futures studies practitioners emphasize the

---

[7] Masini (2006), Kuosa (2011), Tapio and Heinonen (2018), Son (2015). In some universities, Futures Studies has an academic status of a department and is considered a discipline, not just scientific enquiry in, for example, University of Turku.

ethical underpinning of the domain, in parts because thinking *of* the future, as alternatives, induces implications *for* the future. A classical goal of futures researchers is to contribute toward making the world a better place to live in, and to benefit people and the life-sustaining capacities of the Earth (Bell, 1997).

Foresight is a pragmatic approach within futures studies as structured participatory debate about the future of complex issues for enabling present-day decisions and mobilizing joint actions (Eu, 2014; EFP, 2019). Foresight processes tend to entail both critical and emancipatory as well as utilitarian or instrumental elements (Ahlqvist and Rhisiart, 2015). Many types of futures from possible, probable to preferred futures can be envisioned (Amara, 1981). Pitfalls of an emphasis on probabilities concern, on one hand, the negligence of emerging unlikely futures and, on the other, the omission of preferred futures. Dator (2009) suggests that the main task of futures thinking is to enable social change through aspirational futures. The opening up of futures or thinking in alternatives provides numerous analytical insights. Futures thinking operates beyond utopias or dystopias, which often have been the subject and interest of science-fiction writers. The purpose of foresight is to explore and open up possible trajectories, whereas transition management, in comparison, focuses on the decision-making and governance of transition processes (Hölscher et al., 2019).

Even if the future is still often looked at with a predominantly linear conception of reality, foresight has also emerged to shape Latin American thought. Foresight in Latin America has lately proven successful when the elements of creativity and design have been combined in innovative ways (Medina Vasquez, 2006; Mojica, 2010; Cordeiro, 2012). *Deliberative foresight* is an emerging niche inside futures studies and refers to future-oriented discussions for stakeholders, with a special interest in what could and should

happen, involving those whose futures are affected. Its aim is to explore the solutions and steps necessary to realize those aspirations as a jointly deliberated multiactor effort (Karjalainen and Heinonen, 2018). According to Caiati et al. (2019), anticipatory capacity and management of the future should be made part of the concerns and skills of all stakeholders from experts to ordinary citizens. Because complexity and uncertainty cloud the future, specific and diversified foresight tools are required to probe the unknown. There are various types of futures research methods. Wide horizon scanning is crucial in foresight, since important developments may be missed due to limited vision (Schoemaker, 2019). Both creativity and criticality can and should be entangled and elevated in participatory futuring engagements, as they produce more novel and varied ideas (Balcom Raleigh and Heinonen, 2018).

Specifically in the field of energy, foresight can be used with transition studies or alongside other types of approaches, such as energy modeling, market, and policy analysis. There have been foresight efforts in emerging economies, but they have predominantly assumed a narrow focus. A typical implementation has focused in hydrocarbon resources, performed by established large industry players and the state. As shown by experiences in Brazil, such foresight exercises have tended to be ad hoc, expert-led, probabilistic, and technology oriented (Proskuryakova, 2017). The limitation of such efforts is that they may ignore possibilities that fall outside the range of business as usual, sometimes even on purpose. Foresight can, however, be used to explore implausible energy futures to reveal the latent or hidden causalities of an increasingly complex and interconnected world, full of emergent potentials. This way, far-reaching changes and even entirely novel trajectories can be discovered, potentially unlocked, and opened up for broader discussions.

15.4 The steps of a foresight exercise in Chile

A deliberative mode of energy foresight was used to study renewable energy futures for Chile, explicitly taking note of the multilevel and multiactor nature of transitions, in a specific cultural context. The foresight process used evidence of ongoing changes at the global level and then emphasized uncertainties to explore related opportunities and challenges in Chile.

15.4.1 Analyzing a transition pioneer with transformative scenarios

Energy futures in Chile were explored in a large international research project, which sought to understand energy transitions toward a renewable energy-based system. The foresight part of the project also aimed to anticipate emerging cultural lifestyles and the socio-economic implications that could support or result from a new renewable energy-based system. The research combined various methods, assumed a global perspective, and investigated related drivers and obstacles. In the horizon scanning, where megatrends, trends, and other emerging issues are considered, weak signals were emphasized (Heinonen and Hiltunen, 2012). Weak signals are signs, indicative of novel phenomena and issues not yet well understood or in the present considered to be marginal, but which could strengthen in the future, therefore requiring identification and interpretation.

A research approach was assumed to recognize the massive pressures in the energy sector to understand an often-ignored possibility of even radical changes. Different types of manifestations and variations of a possible energy transition to renewable energy were opened up by constructing transformational scenarios (Dator, 2009; Heinonen et al., 2017c). Scenarios are a commonly used foresight tool for strategic conversation (Van Der Heijden, 2005). Numerous types of scenario approaches can be distinguished (Amer et al., 2013; Van Notten et al., 2003), but in this case, a semibackcasting approach was used to explore futures until 2050. The scenarios were generic in the sense that they could be adapted and tested through various ways. It was seen that opening up a wide range of alternative future opportunities could stimulate debate, help examine the potential of an entirely renewable energy—based system, and how such futures could be beneficial to different stakeholders. Chile was then chosen as one of the case study countries because of its efforts to adopt renewable energy, especially solar and wind energy, into its energy system. Analyzing the actions of a pioneering country was assumed to be relevant to anticipating regional and global regime changes.[8] This bears similarity to the study of niche experimentation in transitions approaches, but the reasons are different.

Futures research is interested in identifying pioneers and learning from them because this is assumed to generate futures knowledge. Analyzing pioneers is conceived as analogous to weak signals, in shedding light on the avenues opening up toward futures (Heinonen and Hiltunen, 2012). Pioneers are forerunners who are forced to operate in the midst of uncertainty and perform problem-solving to identify systemic bottlenecks. Understanding pioneers generates critical understanding about the role of key actors who drive transformations and create the futures. Pioneering countries are a forerunner type of key actors that drive cultural and technological change

---

[8] The Neo-Carbon Energy research project (2014–17) is one of the largest energy projects in Finland, funded by Tekes—the Finnish funding agency for innovation (subsequently Tekes was merged into Business Finland, a wider national innovation agency). The other case study countries were Argentina, Kenya, Tanzania, South Africa, Australia, and China.

(Heinonen, 2017). If pioneers are successful, their actions may provide leverage points for broader changes and even produce emancipatory effects (Heinonen and Karjalainen, 2019).

To explore the dynamics of the renewable energy transition, the key uncertainties explored by the initial scenarios had to be tested and contextualized. The scenario uncertainties were a peer-to-peer based logic and ecological consciousness. The peer-to-peer logic refers to constantly deepening, networked societal interaction, and an emerging culture of nonhierarchical interactions. Taken together, these are affecting how societies and sociotechnical systems are organized (Fattah, 2002; Bauwens, 2007; Benkler, 2017). Following the adoption of early networked approaches, the digital revolution has connected and unleashed citizens to communicate, coordinate, and collaborate in increasingly self-organized and informal ways. The peer-to-peer principles are also increasingly manifested in novel technologies that adopt such peer-to-peer models. Remarkably, the potential of information and communication technologies in the energy sector was already raised by Freeman (1992), and later in a revolutionary manner by Rifkin (2014). The convergence of peer-to-peer models, ICTs, and decentralized renewable energy are specifically discussed by Ruotsalainen et al. (2017), who also discuss inclusivity in social processes and the outcomes of technological development. Another uncertainty concerned ecological and climate change consciousness and how they would be expected to evolve in the future. However, this uncertainty received less attention in the research and in the Chilean case study, even if the deepening of ecological awareness was perceived to be important.

15.4.2 Futures Clinique

The actual research took place in three stages which includes the follows: desk research (including the scenario construction) before a field study, research work in Chile (where the scenarios were tested), and post-fieldwork phase. A key event in the research process was a Futures Clinique, as a special type of a futures workshop. Literature reviews were conducted pre- and post-Clinique.

In a workshop a group of people analyze a focal issue, debate it, and are expected to come up with solutions, proposals, or visions (Nygrén, 2019). A futures workshop, developed and introduced to futures studies by Robert Jungk, is a deliberative meeting, which tackles an issue that bears a long-term future trajectory (Jungk and Müllert, 1987). The participants typically represent a diversity of backgrounds, work together in small groups, and use specific methods and tools to discover futures topics. A Futures Clinique is a specially structured futures workshop, designed to anticipate or create radical futures that may differ significantly from the present and pays special attention to weak signals (Heinonen and Ruotsalainen, 2013). Futures Cliniques can be used as part of participatory processes such as transition arenas, which aim to empower various actors to take up roles and responsibilities in addressing sustainability issues (Hölscher et al., 2019).

A Futures Clinique, whose main principles are explained in Heinonen and Ruotsalainen (2013), was conducted at the Chilean Intellectual Property Institute INAPI in Santiago de Chile October 24, 2016 with 30 participants. The Futures Clinique was organized jointly by Finland Futures Research Centre of the University of Turku and Consejo Chileno de Prospectiva y Estrategia (CChPE). The participants represented a broad group of sectors (energy, innovation, mining, R&D, urban planning) and expertise (government, private sector, associations, research), of whom some, not all, already worked on Chile's energy futures.

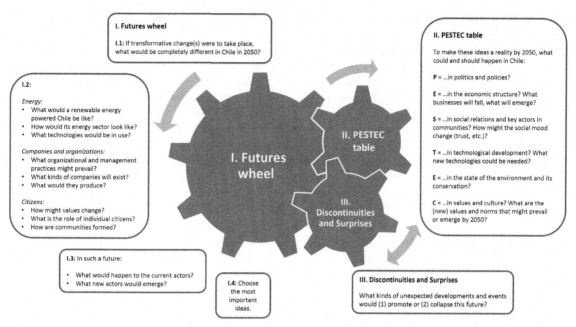

I. Futures wheel

I.1: If transformative change(s) were to take place, what would be completely different in Chile in 2050?

I.2:

Energy:
• What would a renewable energy powered Chile be like?
• How would its energy sector look like?
• What technologies would be in use?

Companies and organizations:
• What organizational and management practices might prevail?
• What kinds of companies will exist?
• What would they produce?

Citizens:
• How might values change?
• What is the role of individual citizens?
• How are communities formed?

I.3: In such a future:
• What would happen to the current actors?
• What new actors would emerge?

I.4: Choose the most important ideas.

I. Futures wheel

II. PESTEC table

III. Discontinuities and Surprises

II. PESTEC table

To make these ideas a reality by 2050, what could and should happen in Chile:

P = ...in politics and policies?

E = ...in the economic structure? What businesses will fall, what will emerge?

S = ...in social relations and key actors in communities? How might the social mood change (trust, etc.)?

T = ...in technological development? What new technologies could be needed?

E = ...in the state of the environment and its conservation?

C = ...in values and culture? What are the (new) values and norms that might prevail or emerge by 2050?

III. Discontinuities and Surprises

What kinds of unexpected developments and events would (1) promote or (2) collapse this future?

FIGURE 15.1 Three main phases of an adapted Futures Clinique process: futures wheel, PESTEC table, and discontinuities and surprises.

The Futures Clinique process, described in Fig. 15.1, advanced across three main phases: futures wheel, PESTEC table, and the anticipation of discontinuities. In the Futures Clinique the participants were posed with questions on the assumed transition to a renewable energy system from a socio-cultural perspective, and a timeline until the year 2050 was provided. The anticipated transformation, in its various forms, was explained and illustrated with the aid of four transformative scenarios in the form of a futures provocation. The summaries of the scenarios were presented to the participants before the moderated group work.

In the group work, the futures wheel, a user-friendly foresight method that resembles a mind map was at first used to collect ideas, discuss them, and anticipate their effects (Glenn, 2009; Bengston, 2016). The participants explored changes from the present to a possible future, with the help of a scenario given to them. The participants were asked to imagine the follows:

• What could be completely different in Chile in 2050 compared to the present, if the society was to develop like the group's scenario describes? (transformation)
  ○ How would a renewable energy powered Chile look like? (energy)
  ○ How might organizational practices, management and leadership styles, working life, and institutional power structures change? (organizational change)
  ○ How might citizens, customers, culture, and values change? (value change)

The most pertinent ideas were then analyzed by using a PESTEC table. The PESTEC table is an adapted tool of morphological analysis to break down issues under study into perspectives: political, economic, social, technological, environmental, and cultural/citizen/customer (PESTEC). The groups primarily focused on desirability, and an analysis of what would make the group's ideas a reality by 2050. At the end the participants also considered the role of surprising events (Heinonen et al., 2017a). This method pays attention to discontinuities and draws from the idea of black swan events, considered low in probability and high in impact, and therefore in principle, they are very difficult to anticipate or detect (Taleb, 2007). The results were presented for cross-fertilization. The moderators documented and synthesized the results (Heinonen et al., 2017b). Post-Clinique, the results were thematically analyzed by two researchers with a cluster analysis where emerging themes from the workshop data were identified and coded (Gibbs, 2007). The results are next presented to illustrate the foresight process and for further reflection.

15.5 Results

The results presented are derived from a thematic analysis and emphasize issues in relation to the decentralization of power and knowledge production that underpins a peer-to-peer based socio-technical vision. They are to be interpreted as matters regarding possible futures, their uncertainties, and discontinuities. In other words, these issues may open up as a part of Chile's energy transition.

15.5.1 Decentralization and regionalization of power

The growing adoption of renewable energy sources is connected to decentralization in its many forms. A decentralized energy paradigm and a peer-to-peer culture are bold aspirations.

For them to wholly take root in Chile, cultural change, political decisions, and support will be required. Legal and policy changes nationally have already enabled renewable energy uptake. In the future the principles of carbon neutrality or circular economy could further inform long-term visions, planning, and policies. The public sector might further invest in innovation, R&D, and create incentives that promote renewable energy sources and the novel principles. In this regard, deliberations of socio-cultural visions may prove to be important to complement economic studies. In recognition of the decentralized paradigm, national decision-making would perhaps have to increasingly support local and regional initiatives. One way to envision alternative decision-making configurations is a multilevel regulatory system that aligns with the practices of regional and horizontal governance.

If these developments were to continue, one possibility would be the emergence of an increasingly regionalized energy system, which would divide Chile into macroregions. This could require more collaboration in energy policy, decentralization of political power, but hopefully, it could also lead to a stronger appreciation of natural resources locally. A regionalized energy approach could even support cultural development and identity. In such processes of change, open governance and open data could alleviate people's concerns and make the public organizations responsive in crucial areas, such as environmental and risk management.

Specifically at a local level, policy or institutional innovations would likely be useful. An educated population would be increasingly capable of advancing the energy transition. Digital forms of education and interest groups could support citizen's learning on renewable energy technologies. Over time, local governments might even decide to set up binding tools that promote horizontal engagement, participatory policy-making, and choose to

TABLE 15.1 Group work metaphors.

Chile 2050—metaphors			
"Decentralized energy for the people"	"The power of energy"	"The bridge for a new mind-set"	"Integrated and diverse society"

Group work results from Futures Clinique.

establish new institutions, such as a Local Quality of Life Council. The metaphors, produced by the groups for Chile 2050, such as "The bridge for a new mind-set," which are illustrated in Table 15.1, reflect the proposals for a future where a decentralized energy paradigm intertwines with an increasingly educated population.

In such circumstances, empowered, responsible, and educated citizens, involved in decision-making, would have an ability to establish a culture of collaboration and assume a long-term view. A collaborative culture and an attitude of resilience would be useful both in innovation and in organizing against any adverse developments. As an aspiration, citizens from diverse groups were imagined to have an ability to test ideas, as individuals and collectives, present social boundaries. A new mind-set could not only be future oriented but also ensure that the past is appreciated. If collaboration was seen as imperative for innovation, a do-it-yourself mentality, in contrast, was seen far-fetched. The self-reliance of citizens was even associated with individualism.

The conversation dynamics illustrate how the promotion of decentralized approaches is enabled by education, human capital, and knowledge creation, all of which are also vital ingredients of social development.

15.5.2 Endogenous and exogenous knowledge creation

An aim to further advance a transition necessitates social engagement but to be a pioneer is also about the creation of new knowledge from facing the challenges. At present, Chile is presented as a renewable energy pioneer from a policy and market perspective. A focus on the market environment undermines often-raised questions in Chile of the social license of corporations, social-industry tensions, and undermines the significance and possibility of processes of change and innovation outside corporations. Being a knowledge and technology innovator, in the true meaning of the word, means an appreciation of experiments, reflection, and science. Therefore in alternative actor-network and knowledge configurations, inventions would be pushed by local demand.

Endogenous knowledge creation promotes the identification of the potential for innovations resulting from socio-technical change. In an alternative institutional landscape, knowledge on renewable energy would be nurtured. Education, training, and R&D initiatives would flourish, entrepreneurs would rise to the opportunity, and there would be conscious efforts for knowledge/technology transfer. Technology transfer refers to the appropriation of technologies, proactiveness in exploiting them, design of local applications and standards. Coordination and cooperation would be required to bridge and bond actors, including international experts, in a multidisciplinary way in order to align local needs and technological development. By 2050, human capital, local and foreign, would be in full force to create novel ideas and ensure the active diffusion and retention of information and knowledge. With changes in education, values might shift, bringing also other types of societal changes, such as the valuation of personal life before work and the appreciation of humanity.

In a vision of endogenous knowledge creation, entrepreneurs and innovators in citizen laboratories would act and harness open data, partner with colleges, local and regional universities, and establish university—industry initiatives. Entrepreneurs and professionals would be helped by bravery and a mind-set shift that tolerates risks, failure, and uncertainty. Multiple novel processes could even begin to diversify the productive structure of the economy. In addition to community-led developments, new energy actors and business models focused on sustainability, such as social enterprises, consultancies, and energy cooperatives, would create shared value as an outcome of the decentralized principles. Over time, local technology companies would be expected to emerge. In Chile the areas of electrification, energy storage, food, and green fuels could be particularly relevant.

The impacts of decentralization, the peer-to-peer logic, and environmental consciousness might also transform existing economic structures. The Atacama Desert, in addition to its solar resources, makes Chile home to the largest known lithium deposits in the world, essential in ICTs, electric cars, and other low-carbon technologies. If mining preserves its role in Chile's economy, perhaps novel practices such as deep-ecological mining would be adopted. On the other hand, in a scenario of exogenous knowledge creation, Chile would continue to rely on external innovation. In the case of the renewable energy transition, this would imply more of a performative act, with limited appropriation of renewable energy technologies and a lack of intellectual property rights in any sustainable business ideas. Such a future trajectory might be created, if transition processes were completely unidirectional. An external push trajectory that only focuses on technology, in its most extreme form, might even jeopardize cultural identity and undermine natural heritage.

15.5.3 Anticipating discontinuities and surprises preceding transformation

It is also possible to anticipate surprising events, which could reinforce, hinder, or even collapse a renewable energy-based transition in Chile, if realized, as shown in Table 15.2. Discontinuities and surprising events embody foresight relevance, since they may lead to disruption and subsequent transformation. The effects of climate change on fragile environments were an often-voiced concern, as they would undermine any other gains and bring unexpected and predominantly negative consequences in Chile and elsewhere.

The withering of traditional hierarchies might not only open a space for decentralization and grassroots activities but also leave an institutional vacuum. A breakthrough in energy storage technology could hasten the demand for Chile's lithium resources. If Chile manages to develop the lithium supply chain over time, this could even provide an opportunity for local manufacturing. The plans to advance lithium production in Chile have outraged citizens due to the involved company's reputation and the aggravation of pressures on Chile's scarce and already privatized water resources. Technological breakthroughs could have unforeseen impacts. Graphene as a synthetic material could potentially substitute Chile's current copper production, whereas developments in an alternative energy technology could hasten or undermine the present transition.

15.6 Discussion

The foresight process and the results illustrate issues related to the foresight—energy

TABLE 15.2 Black swans with interpretations of feedbacks for Chile.

Black swan	Reinforcing loop (+)	Hindering loop (−)
Climate change		Unexpected impacts canceling all other gains
		Affecting the use of renewable energy sources
		Turning Atacama Desert uninhabitable and unbearable to live
Autonomy and horizontal decision-making		Anarchy or chaos, if no novel institutional structures to replace old hierarchies
Lithium breakthrough	New source of economic growth for local manufacturing	Corporate malpractice
	Energy storage as a game-changer	Environmental and water scarcity in a neoliberal context
Graphene breakthrough		Triumph of graphene could act as a copper substitute and shock Chile's economy
Competition		Another country emerges as a renewable energy leader before Chile
Another energy breakthrough	Possibly expedites renewable energy transition	Possibly undermines (or collapses) renewable energy transition

Group work results from Futures Clinique.

research—public policy nexus, methodology as well as how transformative change is understood.

Chile's reputation as a pioneer owes to a conducive policy and market environment, which has also enabled increases in installed capacity of renewable energy. An aim to develop the value chain aligns with an appetite to increase the economic gains.[9] In order to develop the national innovation ecosystem, developing local firms' capabilities are not enough, because a supportive social infrastructure (users, societal groups, authorities, research institutes, etc.) is important in the development and use of innovations.[10] The potential benefits from the renewable energy transition are associated with collaborative institutional structures, purposeful development of human and social capital, and enhanced knowledge/technology transfer. Such changes accentuate dynamics of power and mind-set shifts. The proposals for decentralized or regionalized governance and recognition of discontinuities and surprises outline a further range of plausible outcomes.

The hypothesis of society-wide transformations was tested in particular with the peer-to-peer logic, and the proposals illustrate arrangements of inclusive and reciprocal participation, with supportive values, structures, and processes for such changes. In further work the role of ecological consciousness should be more closely studied (Kiessling et al., 2017). Class differences, inequalities, and labor

[9] See also Chapter 3, Assessing the macroeconomic effects of sustainable energy transitions in Costa Rica and Chile: a multisectoral balance of payments constrained growth approach, for the macroeconomic argument.

[10] Geels (2004).

market fragmentation uphold contestations of capital, power, and privatization. Such qualities may undermine the peer-to-peer logic and Chile's aspirations as a knowledge and technology innovator to create shared value from innovation activities. An aim to innovate underscores the collaborative dynamics of actors and capabilities to build a stronger innovation ecosystem. Confidence between actors in innovation processes and expediting knowledge retention seem to relate to the preferred futures of Chile's energy transition.

The processes of socio-technical change should recognize aims for a more integrated society, a need to appreciate diversity and promote a capacity for dialogue. Trust (or the lack of) has been directly associated with the acceptance of any new activity and technology in Latin America (Greenberg, 2014). Numerous fora on Latin American relations call to build the future together through mutually beneficial partnerships (Casanueva, 2013). Trust holds societies together, enables collaboration, and creates space for innovation ecosystems to take off (Murphy, 2006). In general, high-trust societies perform better (Van Lange, 2015). The mode of deliberative foresight emphasizes a proactive creation of novel ideas and opportunities, with a diversity of viewpoints, so that transition processes are shaped by problem identification and shared goal-setting. In fact, despite the majority of techno-economically oriented scholars, many futures researchers view that the future of the world depends more on culture than on technology (Medina Vasquez, 2006), which is why socio-technical change intertwines so closely with transforming values and visions. This also goes back to the calls of enhanced inclusion of values, agency, and capacities when studying sustainability transitions.

A core process for transitions is a shift in belief systems, ideologies, and public opinion (Geels, 2010). The purposiveness in sustainability transitions emphasizes outcomes that

necessitate transition-oriented dialogues. With foresight, it is possible to probe and create images of the futures, including scenarios as narratives, leading to possible futures. Foresight may be beneficial for understanding and opening up alternatives in times of transitions, and also helps to create new strategic choices in a particular field. Alvial-Palavicino and Opazo-Bunster (2018) view the introduction of foresight in Chile as a policy innovation, but a challenge of future-oriented practice is to overcome assumed linearity in processes and outcomes. Mojica (2010) suggests that the consideration of multiple realities requires overcoming the dominance of probabilistic thought. Consequently, more attention should be given to foreseeing discontinuities as leading toward transformation.

An explicit aim of foresight processes is to engage actors, generate insights and knowledge through collaborative processes, and to enable multiple stakeholders to learn how and why to use the future (Inayatullah, 2009; Rhisiart et al., 2015). As an exploratory method, the Futures Clinique purposefully allows even radical ideas to be opened up for investigation. In such processes, participants may engage, learn, and even develop a common perspective on change, similarly to transition management approaches. Such efforts can be systematically used to empower frontrunners to "(re-)define and take up roles in contributing to sustainability transitions" (Hölscher et al., 2019). In 2018 the energy stakeholders in Chile conducted further vision work as a participatory process with a set of cross-institutional workshops, background documentation, open discussion forums, feedback channels, and a revision and synthetization process to discuss the future of energy (Escenarios Energéticos, 2018).

Prevailing assumptions and cognitive schemas have to be challenged because dominant thinking is easily converted into workshop outcomes in participatory planning. In foresight a longer term process beyond single engagements or outputs, such as a scenario workshop, and a competitive new paradigm may be required to change conventional paradigms (Nygrén, 2019). In transnational research environments, such as in this study, contextualization is necessary because scenario narratives carry also assumptions. When using the future, for instance in a backcasting mode that aims for a specific target, there should actually be room for polyphonic voices. Openness to peripheral options enables a consideration of overlooked alternatives and keeps the future open (Sardar, 1993).

At present, the interface of long-term planning with a systematic and systemic foresight approach is lacking in many countries. The formulation of long-term strategies or public policy directions omits foresight methodologies and tools because their use may not be well-known or understood. When in place, a lack of public investment in high-quality foresight methods and processes and a simultaneous failure to integrate their results into mainstream policy-making hampers their effective use (Boston, 2017). At times, prevailing hierarchies have prevented the opening up of foresight efforts to social and cultural needs, and broader requirements (Proskuryakova, 2017). However, foresight could be introduced, enriched, or even adapted into multiple levels of energy research and decision-making, for instance, by amending existing or establishing novel institutional arrangements. In some economies, renewable energy foresight may have to pay attention to risks and uncertainties associated with economic or political instability.

Conventional sustainability transition frameworks have often struggled in acknowledging power structures, values, agency, and local, contextual issues. In this exercise a hybrid theoretical and methodological approach was used, with a combination of foresight methods and tools, to position the analysis into a present—futures nexus. This opened up a dialogue to test and map multiple possibilities for transformative change in an approachable space. In essence, deliberative foresight advocates taking into consideration power dynamics and elements of criticality but builds on them in a constructive manner. Ideally, when transition frameworks and deliberative foresight are combined, the creation of futures images is enabled so that different audiences can even create novel visions and pathways.

15.7 Conclusion

This chapter has studied how energy transitions, as a type of sustainability transitions, take place in the context of the knowledge and power dynamics related to technological and social change. In a foresight process, transformative scenarios that focused on the sociocultural aspects of energy were tested. This created a space for constructive and critical discussions to explore the dynamics of a renewable energy-based transition. Drawing on the results of a Futures Clinique, argumentative and evidence-based proposals on Chile's energy futures were generated. The foresight process recognizes the determined work in Chile in adopting nonconventional renewable energy technologies and opens up emerging directions and uncertainties that can further shape the transition, making Chile advance in the field as a whole. From a reflexive position the analysis illustrates how knowledge creation, expectations, and institutional arrangements inextricably relate to technology adoption and adaptation.

The actual research speaks of cultural change and inequality. Beyond a coevolution of technology and markets, the results imagine how increasing decentralized energy resources could be supported with regionalized political power, thereby hastening Chile's energy transition. Improved education and openness contribute to more educated societies and aspirations to accumulate and create knowledge. In knowledge production, two directions emerge: a desired scenario where Chile owns its renewable energy transition as a knowledge and technology innovator and another one where it is unable to generate novel business ideas or retain their intellectual property rights. Moving beyond traditional research and development with collaborative arrangements as well as the build-up of future-oriented capabilities may be worthwhile strategies in efforts to promote transformative change.

With regards to the research agenda of transition studies, this chapter has brought forth the potential of futures research and foresight approaches. As a theoretical contribution, it has aimed to explain their benefits and value when exploring energy transitions, as sustainability transitions. Identifying and understanding pioneers is an example of an anticipatory approach, including a concrete element of agency—of making futures. When exploring energy futures or formulating long-term public policy goals, there is a specific value in thinking of alternatives. The reason is that novel, transformative ideas may be ignored if co-creative foresight processes are absent. Methodologically, the scenario-testing through a Futures Clinique exercise illustrates how new perspectives can be discussed in a transnational context. Cultural, institutional, and social dynamics are in this regard of particular importance.

Future horizons are moving constantly. In the coming years, it may be anticipated that the transition debates evolve from the analyses of energy production to address also transport, heating/cooling, and food and water systems in order to explore the emergence of multiple nonfossil fuel-based alternatives. A systematic use of foresight methodologies, tools, research designs, and practices enables aiming far and

can prepare even for radical changes. Chile already has an evolving interface of foresight and energy transition analysis for long-term planning, whereas many countries are still lacking such arrangements.

Thinking about the future enables an escape from the shackles of the past. Moreover, rather than imagining a single outcome, multiple futures can be probed for. If foresight enables the identification of opportunities in the present, systematic futures approaches can be used to unleash dormant forces—for shaping preferred futures.

Acknowledgments

This research was funded through the public financing of Tekes, the Finnish funding agency for innovation (now: Business Finland), for the "Neo-Carbon Energy" project under the number 40101/14. We would also like to thank El Consejo Chileno de Prospectiva y Estrategia (CChPE) and Paula Castro for the support, and Carlos Benavides, Carlos Finat, Claudio Huepe Minoletti, and Guillermo Jiménez for their comments that contributed to this manuscript.

References

Aghahosseini, A., Bogdanov, D., Barbosa, L.S.N.S., Breyer, C., 2019. Analysing the feasibility of powering the Americas with renewable energy and inter-regional grid interconnections by 2030. Renew. Sustain. Energy Rev. 105, 187–205. Available from: https://doi.org/10.1016/j.rser.2019.01.046.

Agora Energiewende, 2015. Current and future cost of photovoltaics – long-term scenarios for market development system prices and LCOE of utility-scale PV systems. In: Study Prepared by Fraunhofer Institute for Solar Energy Systems. <https://www.ise.fraunhofer.de/content/dam/ise/de/documents/publications/studies/AgoraEnergiewende_Current_and_Future_Cost_of_PV_Feb2015_web.pdf> (accessed 06.03.19.).

Ahlqvist, T., Rhisiart, M., 2015. Emerging pathways for critical futures research: changing contexts and impacts of social theory. Futures 71, 91–104. Available from: https://doi.org/10.1016/j.futures.2015.07.012.

Alvial-Palavicino, C., Opazo-Bunster, J., 2018. Looking back to go forward? The interplay between long-term futures and political expectations in sustainability transitions in Chile. Futures 104, 61–74. Available from: https://doi.org/10.1016/j.futures.2018.07.005.

Amara, R., 1981. The futures field, searching for definitions and boundaries. Futurist XV (1), 25–29.

Amer, M., Daim, T.U., Jetter, A., 2013. A review of scenario planning. Futures 46, 23–40. Available from: https://doi.org/10.1016/j.futures.2012.10.003.

Balcom Raleigh, N., Heinonen, S., 2018. Entangling and elevating creativity and criticality in participatory futuring engagements. World Futures Rev. Available from: https://doi.org/10.1177/1946756718807014.

Barbosa, L.S.N.S., Farfan, J., Bogdanov, D., Breyer, C., 2016. Hydropower and power-to-gas storage options: the Brazilian energy system case. Energy Procedia 99, 89–107. Available from: https://doi.org/10.1016/j.egypro.2016.10.101.

Bauwens, M., 2007. The Peer to Peer Manifesto: The Emergence of P2P Civilization and Political Economy. P2P Foundation. <http://www.masternewmedia.org/news/2007/11/03/the_peer_to_peer_manifesto.htm>.

Bell, W., 1997. Foundations of Futures Studies: Human Science for a New Era, volume 1: History, Purposes, and Knowledge. Transaction Publishers, New Brunswick, NJ.

Benedikter, R., Siepmann, K., 2015. Chile in Transition: Prospects and Challenges for Latin America's Forerunner of Development, first ed. Springer, New York.

Bengston, D.N., 2016. The futures wheel: a method for exploring the implications of social–ecological change. Soc. Nat. Resourc. 29 (3), 374–379. Available from: https://doi.org/10.1080/08941920.2015.1054980.

Benkler, Y., 2017. Peer production, the commons, and the future of the firm. Strat. Org. 15 (2), 264–274.

Bezerra, B., Mocarquer, S., Barroso, L., Rudnick, H., 2012. Expansion pressure: energy challenges in Brazil and Chile. IEEE Power Energy Mag. 10, 48–58.

Boston, J., 2017. Governing for the Future: Designing Democratic Institutions for a Better Tomorrow. Emerald Publishing, Bingley.

Breyer, C., Khalili, S., Bogdanov, D., 2019. Solar photovoltaic capacity demand for a sustainable transport sector to fulfil the Paris Agreement by 2050. In: Progress in Photovoltaics: EU PVSEC PAPER. <https://doi.org/10.1002/pip.3114>.

Brown, T.W., Bischof-Niemz, T., Blok, K., Breyer, C., Mathiesen, B.V., 2018. Response to 'Burden of proof: a comprehensive review of the feasibility of 100% renewable-electricity systems'. Renew. Sustain. Energy Rev. 92, 834–847. Available from: https://doi.org/10.1016/j.rser.2018.04.113.

Caiati, G., Quinti, G., D'andrea, L., 2019. Energy transition, anticipation and change: a study on the anticipatory experiences of the low carbon societyAnticipation Science 4 In: Poli, R., Valerio, M. (Eds.), Anticipation, Agency and Complexity. Springer Books, pp. 121–136.

Casanueva, H., 2013. ALC-UE: ¿Construir juntos el futuro? Estudios internacionales (Santiago) 45, 176. Available from: https://doi.org/10.5354/0719-3769.2013.30017.

Chang, R.-D., Zuo, J., Zhao, Z.-Y., Soebarto, V., Zillante, G., Gan, X.-L., 2017. Approaches for transitions towards sustainable development: status quo and challenges. Sustain. Dev. 25 (5), 359–371. Available from: https://doi.org/10.1002/sd.1661.

Cordeiro, J.L. (Ed.), 2012. Latinoamérica 2030: Estudio Delphi y Escenarios. Millennium Project, Washington, DC.

Cordoves-Sánchez, M., Vallejos-Romero, A., 2019. Social construction of risk in non-conventional renewable energy: risk perception as a function of ecosystem services in La Araucanía, Chile. Ecol. Econ. 159, 261–270. Available from: https://doi.org/10.1016/j.ecolecon.2019.01.031.

Dahle, K., 2007. When do transformative initiatives really transform? A typology of different paths for transition to a sustainable society. Futures 39, 487–504.

Dator, J., 2009. Alternative futures at the Manoa School. J. Futures Stud. 14 (2), 1–18. <http://www.jfs.tku.edu.tw/14-2/A01>.

EFP, 2019. European Foresight Platform: What Is Foresight? <http://www.foresight-platform.eu/community/forlearn/what-is-foresight/> (accessed 26.02.19.).

Escenarios Energéticos, 2018. Futuro de la Energía en Chile: Factores de Cambio y Tendencias. Escenarios Energéticos. <http://futuroenergia.escenariosenergeticos.cl/wp-content/uploads/2018/09/futuro.pdf> (accessed 08.03.19.).

EU, 2014. European Union Research in Foresight. Publications Office of the European Union, Luxembourg. <https://ec.europa.eu/research/social-sciences/pdf/project_synopses/research_in_foresight.pdf>.

Fattah, H.M., 2002. P2P: How Peer-to-Peer Technology is Revolutionizing the Way We Do Business. Dearborn Trade Publishing, a Kaplan Professional Company, Chicago, IL.

Freeman, C., 1992. A green techno-economic paradigm for the world economy. In: Freeman, C. (Ed.), The Economics of Hope: Essays on Technical Change, Economic Growth, and the Environment. Pinter Publishers, London, pp. 190–211.

Gallardo, L., Barraza, F., Ceballos, A., Galleguillos, M., Huneeus, N., Lambert, F., et al., 2018. Evolution of air quality in Santiago: the role of mobility and lessons from the science-policy interface. Elem. Sci. Anth. 6 (1), 38. Available from: https://doi.org/10.1525/elementa.293.

Geels, F.W., 2002. Technological transitions as evolutionary reconfiguration processes: a multi-level perspective and a case study, Res. Policy, 31. pp. 1257–1274.

Geels, F.W., 2004. From sectoral systems of innovation to socio-technical systems: Insights about dynamics and change from sociology and institutional theory, Res. Policy, 33. pp. 897–920.

Geels, F.W., 2010. Ontologies, socio-technical transitions (to sustainability), and the multi-level perspective. Res. Policy 39 (4), 495–510. Available from: https://doi.org/10.1016/j.respol.2010.01.022.

Geels, F.W., 2011. The multi-level perspective on sustainability transitions: responses to seven criticisms. Environ. Innov. Societal Transit. 1, 24–40.

Geels, F.W., Schot, J., 2007. Typology of sociotechnical transition pathways. Res. Policy 36 (3), 399–417. Available from: https://doi.org/10.1016/j.respol.2007.01.003.

Gibbs, G.R., 2007. Analyzing Qualitative Data. The SAGE Qualitative Research Kit, London and Thousand Oaks.

Gils, H., Simon, S., Soria, R., 2017. 100% renewable energy supply for Brazil - the role of sector coupling and regional development. Energies 10 (11), 1859.

Glenn, J.C., 2009. Futures wheel [CD-ROM] In: Glenn, J.C., Gordon, T.J. (Eds.), Futures Research Methodology—Version 3.0. The Millennium Project, Washington, DC.

Gobierno De Chile, 2015. Energía 2050: Política Energética de Chile. Ministerio de Energía, Santiago de Chile, <www.sec.cl/sitioweb/baners/LIBRO-ENERGIA-2050.pdf> (accessed 06.03.19.).

Greenberg, M.R., 2014. Energy policy and research: the underappreciation of trust. Energy Res. Soc. Sci. 1, 152–160. Available from: https://doi.org/10.1016/j.erss.2014.02.004.

Haas, J., Palma-Behnke, R., Valencia, F., Araya, P., Díaz-Ferrán, G., Telsnig, T., et al., 2018. Sunset or sunrise? Understanding the barriers and options for the massive deployment of solar technologies in Chile. Energy Policy 112, 399–414. Available from: https://doi.org/10.1016/j.enpol.2017.10.001.

Hansen, K., Breyer, C., Lund, H., 2019. Status and perspectives on 100% renewable energy systems. Energy 175, 471–480. Available from: https://doi.org/10.1016/j.energy.2019.03.092.

Heinonen, S., 2017. Pioneer analysis and international cultural changes. Application of positrend and negatrend analysis in the identification of cultural changeActa Futura Fennica nro 10 In: Heinonen, S., Kuusi, O., Salminen, H. (Eds.), How Do We Explore Futures? Finnish Society for Futures Studies, Helsinki, pp. 241–251.

Heinonen, S., Hiltunen, E., 2012. Creative foresight space and the futures window: using visual weak signals to enhance anticipation and innovation. Futures 44, 248–256. Available from: https://doi.org/10.1016/j.futures.2011.10.007.

Heinonen, S., Karjalainen, J., 2019. Pioneer analysis as a futures research method for analysing transformationsAnticipation Science 4 In: Poli, R., Valerio, M. (Eds.), Anticipation, Agency and Complexity, 2019. Springer Books, <https://doi.org/10.1007/978-3-030-03623-2>.

Heinonen, S., Ruotsalainen, J., 2013. Futures Clinique – method for promoting futures learning and provoking radical futures. Eur. J. Futures Res. 15 (7). Available from: https://doi.org/10.1007/s40309-013-0007-4.

Heinonen, S., Karjalainen, J., Ruotsalainen, J., Steinmuller, K., 2017a. Surprise as the new normal – implications for energy security. Eur. J. Futures Res. 5 (12), 1–13. <http://rdcu.be/A19U> (accessed 06.03.19.).

Heinonen, S., Vähäkari, N., Karjalainen, J., 2017b. Neo-Carbon Energy World – What Opportunities for Chile? Neo-Carbon Energy Futures Clinique IV. FFRC eBOOK 3/2017, Finland Futures Research Centre, University of Turku. <https://www.utu.fi/fi/yksikot/ffrc/julkaisut/e-tutu/Documents/eBook_3-2017.pdf> (accessed 06.03.19.).

Heinonen, S., Ruotsalainen, J., Karjalainen, J., 2017c. Transformational Energy Futures 2050. Neo-Carbon Energy Societal Scenarios. FFRC eBOOK10/2017, Finland Futures Research Centre, University of Turku. <http://www.utu.fi/fi/yksikot/ffrc/julkaisut/e-tutu/Documents/eBook_10-2017.pdf>.

Herreras Martínez, S., Koberle, A., Rochedo, P., Schaeffer, R., Lucena, A., Szklo, A., et al., 2015. Possible energy futures for Brazil and Latin America in conservative and stringent mitigation pathways up to 2050. Technol. Forecast. Soc. Change 98, 186–210. Available from: https://doi.org/10.1016/j.techfore.2015.05.006.

Hölscher, K., Wittmayer, J.M., Avelino, F., Giezen, M., 2019. Opening up the transition arena: an analysis of (dis)empowerment of civil society actors in transition management in cities. Technol. Forecast. Soc. Change 145, 176–185. Available from: https://doi.org/10.1016/j.techfore.2017.05.004.

IEA. Electricity generation by fuel: Chile 1990–2016 and Key stats for Chile, 1990–2016. In: Statistics Data Browser. International Energy Agency. <https://www.iea.org/countries/Chile/>.

Inayatullah, S., 2009. Questioning scenarios. J. Futures Stud. 13 (3), 75–80. <http://www.jfs.tku.edu.tw/13-3/S01.pdf>.

Jasanoff, S., 2018. Just transitions: a humble approach to global energy futures. Energy Res. Soc. Sci. 35, 11–14. Available from: https://doi.org/10.1016/j.erss.2017.11.025.

Jungk, R., Müllert, N., 1987. Future Workshops: How to Create Desirable Futures, first ed. Institute for Social Inventions, London.

Karjalainen, J., Heinonen, S., 2018. Using deliberative foresight to envision a neo-carbon energy innovation ecosystem – a case study of Kenya. Afr. J. Sci., Technol. Innov. Dev. 10 (5), 625–641. Available from: https://doi.org/10.1080/20421338.2017.1366133.

Kelly, K., 2016. The Inevitable. Understanding the 12 Technological Forces That Will Shape Our Future. Viking Press, New York.

Kiessling, T., Salas, S., Mutafoglu, K., Thiel, M., 2017. Who cares about dirty beaches? Evaluating environmental awareness and action on coastal litter in Chile. Ocean Coastal Manage. 137, 82–95. Available from: https://doi.org/10.1016/j.ocecoaman.2016.11.029.

Köhler, J., Geels, F., Kern, F., Markard, J., Onsongo, E., Wieczorek, A., et al., 2019. An agenda for sustainability transitions research: state of the art and future directions. Environ. Innov. Societal Transit. 31, 1–32.

Kuosa, T., 2011. Evolution of futures studies. Futures 43 (3), 327–336. Available from: https://doi.org/10.1016/j.futures.2010.04.001.

Loorbach, D., Frantzeskaki, N., Avelino, F., 2017. Sustainability transitions research: transforming science and practice for societal change. Annu. Rev. Environ. Resourc. 42, 599–626. Available from: https://doi.org/10.1146/annurev-environ-102014-021340.

Lund, H., 2014. Renewable Energy Systems: A Smart Energy Systems Approach to the Choice and Modeling of 100% Renewable Solutions, second ed. Academic Press, Cambridge, MA.

Manuschevich, D., 2016. Neoliberalization of forestry discourses in Chile. Forest Policy Econ. 69, 21–30. Available from: https://doi.org/10.1016/j.forpol.2016.03.006.

Markard, J., Raven, R., Truffer, B., 2012. Sustainability transitions: an emerging field and its prospects. Res. Policy 41 (2012), 955–967. Available from: https://doi.org/10.1016/j.respol.2012.02.013.

Masini, E., 2006. Rethinking futures studies. Futures 38, 1158–1168. Available from: https://doi.org/10.1016/j.futures.2006.02.004.

Medina Vasquez, J., 2006. The legacy of Eleanora Masini: reflections from Latin America. Futures 38, 1187–1195. Available from: https://doi.org/10.1016/j.futures.2006.02.007.

Miller, C.A., Iles, A., Jones, C.F., 2013. The social dimensions of energy transitions. Sci. Cult. 22 (2), 135–148. Available from: https://doi.org/10.1080/09505431.2013.786989.

Futures literacy: transforming the future. In: Miller, R., Miller, R. (Eds.), Transforming the Future: Anticipation in the 21st Century. UNESCO and Routledge, London and New York.

Miller, R., Poli, R., Rossel, P., 2014. The Discipline of Anticipation: Exploring Key Issues. UNESCO, The Rockefeller Foundation.

Milojević, I., Inayatullah, S., 2015. Narrative foresight. Futures 73, 151–162. Available from: https://doi.org/10.1016/j.futures.2015.08.007.

Mojica, F.J., 2010. The future of the future: strategic foresight in Latin America. Technol. Forecast. Soc. Change 77 (9), 1559–1565. Available from: https://doi.org/10.1016/j.techfore.2010.07.008.

Murphy, J., 2006. Building trust in economic space. Prog. Hum. Geogr. 30 (4), 427–450.

Nasirov, S., Silva, C., Agostini, C., 2015. Investors' perspectives on barriers to the deployment of renewable energy sources in Chile. Energies 8, 3794–3814. Available from: https://doi.org/10.3390/en8053794.

Neo-Carbon Energy, 2017. Emission-Free Future Now Available. Neo-Carbon Energy Project: Impact and Results. VTT Technical Research Centre of Finland Ltd, Lappeenranta University of Technology and University of Turku, Finland Futures Research Centre, <https://urly.fi/WDs> (accessed 03.03.19.).

Neo-Carbon Energy, 2018. Internet of Energy: Simulation of a Global 100% Renewable Electricity System. Lappeenranta University of Technology, <www.neocarbonenergy.fi/internetofenergy/> (accessed 05.11.18.).

Nygrén, N.A., 2019. Scenario workshops as a tool for participatory planning in a case of lake management. Futures 107, 29–44. Available from: https://doi.org/10.1016/j.futures.2018.10.004.

Patterson, J., Schulz, K., Vervoot, J., Van Der Hel, S., Widerberg, O., Adler, C., et al., 2017. Exploring the governance and politics of transformations towards sustainability. Environ. Innov. Societal Transl. 24, 1–16. Available from: https://doi.org/10.1016/j.eist.2016.09.001.

Pesch, U., 2015. Tracing discursive space: agency and change in sustainability transitions. Technol. Forecast. Soc. Change 90 (B), 379–388. Available from: https://doi.org/10.1016/j.techfore.2014.05.009.

Poli, R., 2017. Introduction to Anticipation Studies. Anticipation Science 1. Springer, <https://doi.org/10.1007/978-3-319-63023-6>.

Anticipation Science 4 Poli, R., Valerio, M. (Eds.), 2019. Anticipation, Agency and Complexity. Springer Books, Switzerland.

Pollitt, M., 2004. Electricity reform in Chile lessons for developing countries. In: Working Paper 04-016. Center for Energy and Environmental Policy Research. MIT Sloan Institute. <http://ceepr.mit.edu/files/papers/2004-016.pdf>.

Proskuryakova, L., 2017. Energy technology foresight in emerging economies. Technol. Forecast. Soc. Change 119, 205–210. Available from: https://doi.org/10.1016/j.techfore.2016.05.024.

Rauschmayer, F., Bauler, T., Schäpke, N., 2015. Towards a thick understanding of sustainability transitions –

linking transition management, capabilities and social practices. Ecol. Econ. 109, 211–221. Available from: https://doi.org/10.1016/j.ecolecon.2014.11.018.

Rhisiart, M., Miller, R., Brooks, S., 2015. Learning to use the future: developing foresight capabilities through scenario processes. Technol. Forecast. Soc. Change 101, 124–133. Available from: https://doi.org/10.1016/j.techfore.2014.10.015.

Rifkin, J., 2014. The Zero Marginal Cost Society. The Internet of Things, the Collaborative Commons, and the Eclipse of Capitalism. Palgrave MacMillan, New York.

Rotmans, J., Loorbach, D., 2009. Complexity and transition management. J. Ind. Ecol. 13 (2), 184–196. Available from: https://doi.org/10.1111/j.1530-9290.2009.00116.x.

Ruotsalainen, J., Karjalainen, J., Child, M., Heinonen, S., 2017. Culture, values, lifestyles, and power in energy futures: a critical peer-to-peer vision for renewable energy. Energy Res. Soc. Sci. 34, 231–239. Available from: https://doi.org/10.1016/j.erss.2017.08.001.

Sanchez-Alfaro, P., Sielfeld, G., Van Campen, B., Dobson, P., Fuentes, V., Reed, A., et al., 2015. Geothermal barriers, policies and economics in Chile – lessons for the Andes. Renew. Sustain. Energy Rev. 51, 1390–1401. Available from: https://doi.org/10.1016/j.rser.2015.07.001.

Santana, C., 2014. Energías Renovables en Chile, El Potencial eólico, solar e hidroeléctrico de Arica a Chiloé. Ministerio de Energía and Deutsche Gesellschaft für Internationale Zusammenarbeit (GIZ) GmbH, Santiago de Chile, Chile, <http://www.minenergia.cl/archivos_bajar/Estudios/Potencial_ER_en_Chile_AC.pdf>.

Sardar, Z., 1993. Colonising the future: the 'other' dimension of futures studies. Futures 25 (2), 179–187. Available from: https://doi.org/10.1016/0016-3287(93)90163-N.

Schoemaker, P., 2019. Attention and foresight in organisations. Futures Foresight Sci. 1 (1). Available from: https://doi.org/10.1002/ffo2.5.

Schot, J., Kanger, L., 2018. Deep transitions: emergence, acceleration, stabilization and directionality. Res. Policy 47 (6), 1045–1059. Available from: https://doi.org/10.1016/j.respol.2018.03.009.

Schueftan, A., Sommerhoff, J., González, A.D., 2016. Firewood demand and energy policy in south-central Chile. Energy Sustain. Dev. 33, 26–35. Available from: https://doi.org/10.1016/j.esd.2016.04.004.

Smith, A., Stirling, A., Berkhout, F., 2005. The governance of sustainable sociotechnical transitions. Res. Policy 34, 1491–1510. Available from: https://doi.org/10.1016/j.respol.2005.07.005.

Son, H., 2015. The history of Western futures studies: an exploration of the intellectual traditions and three-phase periodization. Futures 66, 120–137. Available from: https://doi.org/10.1016/j.futures.2014.12.013.

Sovacool, B.K., 2016. How long will it take? Conceptualizing the temporal dynamics of energy transitions. Energy Res. Soc. Sci. 13, 202–215. Available from: https://doi.org/10.1016/j.erss.2015.12.020.

Taleb, N.N., 2007. The Black Swan: The Impact of the Highly Improbable, first ed. Random House, New York.

Tapio, P., Heinonen, S., 2018. Focused futures from Finland. World Futures Rev. 10 (2), 111–135. Available from: https://doi.org/10.1177/1946756718754847.

Van Der Heijden, K., 2005. Scenarios: The Art of Strategic Conversation. John Wiley & Sons, Chichester.

Van Lange, P., 2015. Generalized trust: four lessons from genetics and culture. Curr. Dir. Psychol. Sci. 24 (1), 71–76. Available from: https://doi.org/10.1177/0963721414552473.

Van Notten, P.W.F., Rotmans, J., Van Asselt, M.B.A., Rothman, D.S., 2003. An updated scenario typology. Futures 35 (5), 423–443. Available from: https://doi.org/10.1016/S0016-3287(02)00090-3.

Vargas Payera, S., 2018. Understanding social acceptance of geothermal energy: case study for Araucanía region, Chile. Geothermics 72, 138–144. Available from: https://doi.org/10.1016/j.geothermics.2017.10.014.

WEF, 2018. Frameworks for the Future of Electricity: Leading the Transformation Through Multistakeholder Cooperation. World Economic Forum, Cology/Geneva, <http://www3.weforum.org/docs/Frameworks_Future_Electricity_2018.pdf>.

Williams, S., Doyon, A., 2019. Justice in energy transitions. Environ. Innov. Societal Transit. 31, 144–153. Available from: https://doi.org/10.1016/j.eist.2018.12.001.

Wittmayer, J.M., Avelino, F., Van Steenbergen, F., Loorbach, D., 2017. Actor roles in transition: insights from sociological perspectives. Environ. Innov. Societal Transit. 24, 45–56. Available from: https://doi.org/10.1016/j.eist.2016.10.003.

Further reading

Rifkin, J., 2011. The Third Industrial Revolution: How Lateral Power is Transforming Energy, the Economy, and the World. Palgrave MacMillan, New York.

Integrating social and justice dimensions to energy transitions: the case of Mexico

Carlos Tornel

Department of Sustainability Studies, Iberoamericana University, Mexico City, Mexico

16.1 Introduction

Energy systems are among the biggest human enterprises. The large-scale investments and physical infrastructure required to sustain energy systems worldwide are now a multimillion-dollar endeavor that manifests itself in different cultural, economic, and political contexts (Bakke, 2016). Despite a relative lack of engagement with the role of energy systems, in terms of social and spatial considerations during most part of the 20th century, a new-found interest in the development and transition toward low-carbon energy systems has now become a major global concern in the 21st century (Newell and Mulvaney, 2013).

With the advent of climate change, calls to globally accelerate a total decarbonization of the energy sector are now a constant in international negotiations aimed at addressing environmental, social, and economic problems. This is not surprising, considering that the energy sector plays a key role in the production of global warming, contributing with roughly 35% of global greenhouse gas (GHG) emissions [Intergovernmental Panel for Climate Change (IPCC), 2015]. A global transformation of the energy system is now considered a necessary condition to address the climate crisis, along with a series of other social, political, and economic concerns such as granting access to energy services and reducing world poverty (World Energy Council, 2017). For example, the Sustainable Development Goals (SDGs) include energy (SDG 7) as a means to "ensure access to affordable, reliable, sustainable and modern energy for all" (United Nations, 2015), which, based on other considerations, can be a link to reduce poverty, achieve prosperity, and mitigate the ills of climate change (Miller et al., 2015b).

These tasks represent a major transformation of the global energy system, which, among other things, would imply a rapid deployment of renewable energy technologies, and a substitution of fossil fuel and their

The Regulation and Policy of Latin American Energy Transitions
DOI: https://doi.org/10.1016/B978-0-12-819521-5.00016-4

codependent and coproduced social and physical systems (Unruh, 2002). However, this issue presents two main difficulties:

- A rapid substitution of fossil fuels and their related infrastructures and coproduced social routines and actions would face significant resistances (economic and political) and institutional challenges to overcome (Newell and Mulvaney, 2013).
- The rapid substitution for renewable energy would also need to provide access to electricity services, through low-carbon electricity generation technologies to at least 1 billion people globally who do not have access to energy services and increase and maintain reliable services to approximately 2−3 billion people who lack access to reliable and affordable clean energy (Poor People's Energy Outlook, 2018).

According to the International Energy Agency, the current trend of GHG emissions associated with energy intensity shows that, not only are energy systems failing to decarbonize at the necessary rate to meet the 1.5°C trajectory presented by the Intergovernmental Panel on Climate Change and addressing the temperature limits adopted in the Paris Agreement [United Nations Framework Convention on Climate Change (UNFCCC), 2015; Intergovernmental Panel for Climate Change (IPCC), 2018], but also they are increasing fossil fuel demands overall.[1] Despite the clear and necessary transformation of the energy system, the pathway through which the global energy transition is being designed and implemented is ambiguous at best (Jasanoff, 2018). Geopolitical and geological constraints have been at the center of a global transformation of the energy systems, where issues over energy security and independence have now shifted efforts to accelerate a global transition toward unconventional fossil fuel resources on the one hand, and toward a renewable energy policy on the other (Calvet, 2015). This apparent paradoxical condition is brought about, among other things, by understanding how the energy transition is being designed, imagined, and who is framing and producing the "solutions" to address climate change and maintain the energy transition process.

One of the major advancements of the social sciences in relation to energy research in the last two decades has been the notion of coproduction, which basically shows how the production of knowledge and living in the world are inexorably linked (Tidwell and Tidwell, 2018). Coproduction is understood as the process that constitutes the physical reality with the often invisible role of knowledge, expertise, technical practices, and material objects in shaping, sustaining, subverting, or transforming relations of authority (Jasanoff, 2005). In other words the notion of the coproduction shows that energy systems are in fact sociotechnical systems where particular configurations of the energy systems are as much a matter of the artifacts, infrastructures, technologies, and physical representations of the system as they are the result of political, economic, and socially managed routines, decisions, and designs that shape and determine the way the system evolves, changes, and adapts (Miller et al., 2013). This realization is fundamental in comprehending that power relations and specific configurations of the future through particular socio-technical configurations shape the design, development, deployment, and distribution of the energy systems.

---

[1] "Global energy consumption in 2018 increased at nearly twice the average rate of growth since 2010, driven by a robust global economy and higher heating and cooling needs in some parts of the world (. . .) Energy-related CO_2 emissions rose 1.7% to a historic high of 33.1 Gt CO_2. While emissions from all fossil fuels increased, the power sector accounted for nearly two-thirds of emissions growth." (International Energy Agency, 2019).

Hence, a more thorough understanding of how these transitions unfold and how they are intertwined with international, national, and local politics is necessary. At the same time a broader approach to the understanding of energy problems must be integrated into the design and development of this transition processes. In other words, there is no "silver bullet" or "one-size-fits-all" solution to decarbonize the energy sector and still address the economic and social considerations that constitute the energy transition process. A just energy transition would then require major reconfiguration of social, technological, political, and economic issues surrounding energy systems. The concept of the Energy Trilemma (World Energy Council, 2017) presents an interesting approach to address this issue, which includes a triple challenge of obtaining safe, affordable, and environmentally sustainable energy. This approach argues addressing energy transitions must necessarily address these three components to guarantee a real system change.

Similar to this Energy Trilemma, an important part of academic literature has recently ventured into the notion of energy justice (Miller et al., 2013; Jenkins et al., 2016, 2018; Jenkins, 2018), which has emerged as a cross-cutting issue over the distributional, recognition, and representational issues of energy use, access and production (Jenkins et al., 2018). These analyses aim to pinpoint how issues of justice in other long-standing fields, such as climate and environmental justice, have been unable to articulate adequate solutions to problems of distribution, representation, and recognition, where the notion of energy justice could deliver a more coherent, less confrontational approach to address issues related to justice with a more clear-cut definition of the problem and its solutions (Jenkins, 2018).

Issues over energy justice prevail in the way that energy systems are designed, built, and imagined. Or in other words, the lack of adequate design can be a major obstacle for energy transitions that could perpetuate existing inequalities and produce forms of inequality (Miller et al., 2015a). The pressing issue of climate change makes a straightforward case that energy systems need to be decarbonized; however, there is little evidence that renewable energy systems will have the capabilities of sustaining energy futures with increasing social demands of energy, consumption, and population growth, without facing real challenges to meet future demand (Gras, 2017).

In Mexico the framing of the energy transition process has emerged as a reaction to international negotiations and market prices, which have shifted the discourse and economic costs of renewable energy technologies. In 2013 the Mexican government reformed the Mexican Constitution allowing private sector participation in the electricity generation, leaving a state-owned monopoly for the transmission, distribution, and power supply (Presidencia de la República, 2013). This new configuration of the electricity sector included the adoption of long-term energy auctions (LTEA) sponsored by the Federal Commission of Electricity (CFE) to obtain electricity at the lowest possible cost.[2] This was also accompanied by the adoption of the Energy Transition Law (2015) (ETL) that included national mandatory targets to generate at least 35% of electricity with "clean energy" sources and the commitment of ambitious goals to reduce GHG emissions through a Nationally Determined Contribution (NDC) in 2015, before the signing and ratification of the Paris Agreement [Secretaría de Medio Ambiente y Recursos Naturales (SEMARNAT), 2015].

---

[2] Secretaría de Energía. Regulación del Mercado Eléctrico Mayorista (MEM) [Regulation for the Wholesale Power Market, 2014]. Available from: <https://www.cenace.gob.mx/paginas/publicas/Transparencia/REC_MEM.aspx>.

Yet, despite a considerable increase in renewable energy generation as a result of these changes, little-to-no attention has been paid to the social and spatial impacts generated by the inclusion of mostly large-scale, privately owned, and internationally financed wind and solar farms, which have produced new forms of spatial and social inequalities, justified under the national pursuit of climate change and energy transition goals (Avila-Calero, 2017). The acceleration in renewable energy integration has also overlooked important considerations of distributional, procedural, and recognition-based approaches necessary to address the just transformation of the energy system under a framework of justice: currently more than 11 million households (36%) still experience some form of energy poverty (Garcia-Ochoa and Graizbord, 2016), despite an almost universal interconnection, and most of the distributional costs of energy generation are being unevenly distributed (Baker, 2018).

This chapter's main contribution is to offer clear insights as to how to increase renewable energy integration by incorporating the notion of socio-energy systems into the energy transition process in Mexico, which places considerable emphasis in the interaction between the technical, spatial, and social dimensions of the energy transition (Bridge et al., 2013). The chapter focuses on a review of academic literature, and the legal and regulatory framework in Mexico to show how energy futures have been framed in remarkably narrow configurations. This chapter presents the argument in the following way:

- Section 16.2 presents a review of the socio-technical configurations of the energy sector and a framework to understand how energy futures are coproduced through Sheila Jasanoff's notion of socio-technical imaginaries (STIs).
- Section 16.3 gives a broad overview of the energy transition process in Mexico and its impact in the regulatory framework. This section will put particular emphasis on how the energy system has been designed through specific regulatory and policy instruments that work as *dispositifs*—that is, particular laws, policies, and regulatory instruments that seek to maintain and control specific power relations—that aim to accelerate renewable energy integration and the energy transition in Mexico, with little-to-none attention paid to socio-technical and socio-spatial configurations of such a transition.
- Section 16.4 argues that, despite a broad and overreaching energy sector reform, policy instruments and actions have had a remarkable narrow framing, which in turn is manifested in how energy transitions are framed and addressed through specific STIs. This section reviews the main outcomes of the prospectives developed by the Mexican government to address the future of the electricity sector and compares them to other peer-reviewed papers that also make specific projections for the future of the energy system in Mexico.
- Section 16.5 concludes this chapter by offering a brief outline of some of the main challenges facing the sector in the development of an energy transition process and with a series of recommendations for further research that seek to incorporate the social and spatial dimensions in the future design of the energy systems, particularly in developing countries and for Latin America.

16.2 Socio-technical transitions

The notion of socio-technical configurations of energy systems has shown that energy systems must be understood in much broader considerations that go beyond the choices of fuels we make as a society (Miller et al., 2013).

However, the ways energy transitions are usually framed are presented as a mere substitution of technology without considering the spatial and social configurations that are built around the use of such technologies. In other words, technologies, as well as the society that use them, change overtime and are configured differently over space. Yet, the narrow framing of actual energy transitions shows that energy injustices still prevail in the way that energy systems are designed, built, and imagined. The lack of adequate design can be a major obstacle for energy transitions, which could perpetuate existing inequalities and produce new ones (Miller and Richter, 2014). Some of the main challenges of framing the energy transition as mere substitution of fossil fuels are as follows:

- First, the fact that the energy transition is seen as a linear and progressive arrow, in which more progress, or in this case, more deployment, innovation, and technology will be enough to satisfy future demand and reduce GHG emissions (Jasanoff, 2018).
- Second, little attention is paid to the disparities within societies and hence to the local political and social contexts in which renewable energy projects are developed and undertaken. This issue is normally addressed through a series of techno-fixes that present substituting technologies as long-term sustainable solutions (Huesemann and Huesemann, 2011).
- Finally, Jasanoff (2018) argues that environmental policy remains largely a reactive enterprise, where "there seems to be no principles in place for how to think about our common energy future(s), especially if the objective is not merely to effect a transition, but to do so with attention to social justice."

These issues have become remarkably important with the global deployment of renewable energy technologies, particularly wind and solar, which are now developed through large-scale interventions and internationally financed mechanisms (Labussière et al., 2018). A global demand to address climate change and implement an energy transition has followed the same reactive principles, universalizing solutions, producing remarkably narrow policy framings, and presenting technological substitutions, improvements, and innovations as "one-size-fits-all" or "silver bullets" to address the problem globally (Bellamy-Foster et al., 2010).

These universalized solutions and technological optimisms have now led some scholars to point out that renewable energies are increasingly becoming fossilized (Raman, 2013; Evrard, 2013), or in other words, a new source of extractivism and exploitation (Boyer and Howe, 2016; Avila-Calero, 2017). This condition, they argue, has to do with how the deployment of renewable technologies has systematically overlooked the socio-political—commercial—material networks that underpin the production of many renewable energy technologies, which are now starting to resemble the fossil fuel regime they were intended to supersede. At the same time the lack of recognition of the new spatial reconfigurations that are brought about by renewable energy technologies and projects must also be taken into account. In other words, this means that energy transitions not only imply a change in technology but a reconfiguration of how society experiences space and time. The geographical characteristics of energy transitions must then be understood as spatial reconfigurations that change the way society experiences and relates with technology and infrastructure.

This "spatial turn" in energy research has effectively demonstrated that space and place play a role in changing and rearranging landscapes and producing uneven patterns of development between spaces (i.e., the city and the rural communities) (Labussière et al., 2018) that change and shift the form, structure, and function of the energy sector (Bridge, 2018). Understanding spaces as socially produced, changes its nature, since it is not merely as an empty platform but a factor in the ways that new spatialities from renewable energy infrastructure and value-chains change the dynamics of power between places. Local, national, and transnational flows of technology, funding, and ideology are entangled with issues of political discourses, framings, and power that are ultimately manifest in the use of land and the territory (i.e., clearing people, wildlife, or forests form to meet national energy targets).[3]

The remarkable narrow configuration of energy transitions policies is then translated into specific concerns such as the reduction of prices and overall cost of electricity generation, the possibility of maintaining and delivering energy services in the cheapest manner, and more recently the reduction of GHGs associated with that generation (Miller et al., 2013). These configurations lack a comprehensive analysis of both the socio-technical and socio-spatial components that make up and coproduce energy systems.

The notion of STIs is a useful concept to understand how particular configurations of the energy sector originate and how they are embedded, resisted, and extended. For Sheila Jasanoff, STIs are defined as "collectively held, institutionally stabilized, and publicly performed visions of desirable futures, animated by shared understandings of forms of social life and social order attainable through, and

supportive of, advances in science and technology" (Jasanoff and Kim, 2015, pp. 21). Yet STIs are not presented in well-defined boxes, groups, and coalitions; they are instead contested in the sphere of public debate, and hence dominant or hegemonic STIs are the result of power relations and political structures. Indeed, as Tidwell and Tidwell argue, STIs must not be understood as perspectives of society as a whole but must be understood as broadly held understandings of what is socially desirable, and not necessarily manifestations of individual or specific groups understanding science and technology (Jasanoff and Kim, 2015, pp. 104). The analysis of STIs can help understand how specific social norms, values, laws, and other dispositives can be used to perpetuate particular (re)configurations of the politics and social organizations around technology and, in this case, energy systems.

Hence, energy systems must be understood as hybrids: coproduced through the physical manifestations of energy technologies and the social relations that produce, maintain, and dictate the way that these energy systems are shaped, and the way they influence social and political life. In other words, energy systems can only change if and when people make specific choices, while these technologies reshape social values, relationships, practices, institutions, business models, forms of work, and ways of knowing and living (Miller et al., 2013, pp. 136). The same argument can be made for particular energy futures, where

[3] Labussière et al. (2018); for example, Yanetti et al. (2017) show how the spatial development of large-scale solar projects in India has now become a major contestation for local villagers and farmers, where the demand of land and the framing of the space as "wastelands" (Baka, 2013) has now given way to new forms of land enclosures and new forms of accumulation by dispossession in order to meet global and national climate change and energy transition targets. This is normally framed as an issue of spatial justice, where large-scale renewables are increasingly becoming new competitors for the use of space, increasingly expanding commodity frontiers (Yanetti et al., 2016). Similarly, Baka (2013) shows how biodiesel plantations in India respond to urban demands for liquid fuel but are, at the same time, also acts of energy dispossession as they require the clearance of trees that provide fuelwood for households and industry (Baka, 2013 quoted in Bridge, 2018).

social, political, and economic relations are deeply embedded into the choices that people make.

16.3 Framing the energy transition in Mexico: a brief review of the development of the regulatory framework

Mexico has maintained a complex relationship with energy resources since the first half of the 20th century. In the case of the electricity sector, which grew through a series of sporadic investment by different private companies in order to address specific local problems, no national plan for electrification was set in place, and hence most of the investment lacked a coherent narrative (Jano-Ito and Crawford-Brown, 2016). However, starting in 1960 the electricity sector was progressively nationalized, following the long standoff between the Mexican Government and international oil companies that started in 1938. The adoption of the first Electricity Industry Law in 1939 by the Congress culminated in an overall nationalization of the electricity sector under the CFE and the Company *Luz y Fuerza del Centro* (LyFC), effectively consolidating a National Electricity System in 1976 (Jano-Ito and Crawford-Brown, 2016).

During this time, no private investment was allowed into the electricity sector and both CFE and LyFC operated under a national mandate to maintain cheap, accessible, and reliable electricity. However, by 1992 the Mexican government—advocating a less interventionist agenda by the state, following a series of economic crisis, the increase deficit of both CFE and LyFC and a series of neoliberal policies implemented after the Washington Consensus (Baker, 2018)—allowed private participation through the adoption of public–private partnerships and contracts with CFE and the LyFC. However, a sweeping reform to allow private entities into the energy sector was rejected by the Congress. This reform allowed CFE and LyFC to enter into contracts with private entities to supply natural gas for electricity generation (Vietor and Sheldahl-Thomason, 2017). It was not until 2008 that another attempt by the Federal Government to reform the energy sector was presented to the Congress; however, it failed to open the sector for foreign and private participation. Nevertheless, two specific laws were adopted, which aimed to promote renewable energy and energy efficiency but lacked any effective economic and financial support for renewable energy development. This was seen as problematic at the time, given the substantial economic costs of renewable energy (Averchenkova and Guzman Luna, 2018).

Following the failed attempt to reform the energy sector, in 2012 the former Mexican President Felipe Calderón (2006–12) signed into law The General Climate Change Law (GCCL), which, among other things, retrieved and mandated Mexico's commitments presented during the COP 16 in Cancún, Mexico, to reduce GHG emissions by 30% in 2020 and by 50% based on 2000 emissions. The Law included an additional aspirational goal for energy transition, which aimed to achieve 35% of electricity generation from clean energies by 2024 (General Climate Change Law, 2012). The GCCL also included a series of planning and policy instruments, such as the National Strategy on Climate Change (ENCC), the Special Program on Climate Change, which articulates mitigation and adaptation goals to be implemented by each federal administration in accordance with that ENCC mid- and long-term goals (Averchenkova and Guzman Luna, 2018).

By 2013 the new Federal Administration (2012–18) reformed the Constitution presenting the most sweeping energy reform to date, opening the electricity sector to private investment and participation in energy generation, leaving the rest of the electricity services—transmission, distribution, and retail—to the now public company, CFE (Miller, 2014). The CFE holds a mandate for universal coverage, which demands that energy be delivered at the lowest possible cost to benefit the user (Jano-Ito and Crawford-Brown, 2016). Under the new regulatory procedures, CFE would buy electricity directly to qualified generators and private entities through LTEA and Public Private Partnerships that would allow CFE to acquire electricity at the lowest possible cost. All these changes were incorporated into the Electricity Industry Law (EIL) that was adopted in 2014 (EIL, 2014).

Two other relevant instruments were included into the energy transition process:

- A mandate to implement a Social Impact Assessment in the energy sector was incorporated in the EIL and the Hydrocarbon Laws and their respective Bylaws (Secretaría de Energía, 2018a,b).
- EIL also recognized the right of the indigenous communities to be previously consulted through a process of Free, Prior and Informed Consent (FPIC) with the objective of producing a consent before the development of infrastructure projects related to both the electricity and hydrocarbons sectors (EIL, 2014).

Public participation was recognized as an important inclusion into the legislative framework. However, criticisms have arisen from this recognition. First, there is a limited recognition of how the "public" and "groups" can participate in the process is highly limited; second, these mechanisms for participation are inserted into highly technocratic and knowledge-based process that renders much of other forms of knowledge and organization procedures inherently excluded. This approach also narrows down the possible imaginaries to be carried out and orchestrated by government agencies. This approach hence produces a sort of passive subject that is limited by design to a series of deliberative actions that offer no expansion of democratic principles such as disagreements and the possibility of refusing the development of projects (Chilvers and Longhurst, 2016).

In 2015 Mexico adopted the ETL, which transformed the aspirational goal established in the GCCL to generate at least 35% of electricity form clean energy sources, reducing GHG emissions from the electricity sector and increasing (solar) distributed generation (DG). In that same year the Mexican government presented an INDC before the UNFCCC, which would later be considered as the target to meet the emission pathways identified in the Paris Agreement (2015). This contribution included an unconditional commitment to reduce 22% of emissions by 2030, with a 36% conditional commitment by the same year [Secretaría de Medio Ambiente y Recursos Naturales (SEMARNAT), 2015]. In 2018 these reforms were incorporated into the legal framework of the GCCL with a new recognition of the Paris Agreement and the temperature limits established in Article 2, which aims to limit the temperature rise by 2°C with continued efforts to limit at 1.5°C by the end of the present century (Averchenkova and Guzman Luna, 2018).

The combinations of laws approved after the energy reform and public commitment expressed by Mexico to address climate change effectively increased renewable energy integration and, through the three LTEA, reached electricity generation prices of 43, 33, and 20.57 USD/MWh, respectively, allocating 56 large-scale renewable energy projects to make up for approximately 3% of additional generation needed to meet the 2024 national target (Secretaría de Energía, 2018a). The ETL also mandated a series of bylaws to increase DG, which included compensation schemes such as Net Metering and Net Billing (Diario Oficial de la Federación, 2019). Table 16.1 shows the composition of the regulatory framework supporting the energy transition in Mexico.

TABLE 16.1 Regulatory framework for the energy transition in Mexico.

	Constitutional Energy Reform: enables the participation of private entities in the generation of electricity		
	EIL	**ETL**	**GCCL**
Instruments	*ELT A*: Auctions are mandated by the wholesale electricity market to obtain power, energy at the lowest possible costs. Three auctions took place, reducing costs for wind and solar energy up to 20.57 USD/MWh in 2018. The ETL A also included a CEC market to address obligations to reduce emissions in the sector to meet the 2024 clean energy goal	*Benefits of distributed generation*: The law mandates SENER to develop a study of the potential cobenefits in public health, CO_2 mitigation, and other cobenefits	*National policy framework*: The Law establishes an interministerial commission on climate change, a council with civil society and academia and the policy instruments: the National Climate Change Strategy, with long-term planning, a Spatial Program for Climate Change, with obligations for each sector, and a national climate change fund to support mitigation and adaptation actions
	PRODESEN: The Program for the Development of the National Electricity System is mandated in the EIL as a way to assess the rentability and quality of the sector, as well as the evolution and modernization of the national electric grid. The program makes a 15-year forecast with the integration and retirement of power plants	*Net Billing and Net Metering dispositions*: The law also recognizes the possibility of having compensation schemes for installations smaller than 500kW. The cost provided is calculated using the marginal generation cost of each region	*National Climate Change Goals*: Mandates the reduction of 30% and 50% of GHG emissions by 2020 and 2050 respectively, based on emissions of the year 2000. No sectorial targets and or obligations where established
	SIA: Incorporated as a way to address the social impacts of the electricity industry's infrastructure in local populations. The developer of each project is obligated to identify affected local communities and to present compensation measures	*Goals to generate 35% from renewables by 2024*: The Law includes a mandatory goal of achieving at least 35% generation by 2024 through the use of clean energy. There is no mandatory goal for distributed generation	*Recognition of the Paris Agreement*: The Law recognized the 2°C and 1.5°C temperature limits as an objective, and advocates for further ambition to meet this emission pathways. It also mandates the NDC goals by 2030 to reduce 22% of emission unconditionally and 36% conditionally, based on the emissions of the year 2013. The unconditional goals contain sectorial obligations and costs
	Indigenous Consultations: Adopted by Mexico in 1990. Established the principle of auto-determination for indigenous communities and the right to be consulted. If indigenous populations are recognized as affected by the project, they must grant a previously informed consent to the project and its possible impacts		

CEC, Clean Energy Certificate; *EIL*, Electricity Industry Law; *ETL*, Energy Transition Law; *GCCL*, General Climate Change Law; *GHG*, greenhouse gas; *LTEA*, long-term energy auctions; *NDC*, Nationally Determined Contribution; *SENER*, Ministry of Energy; *SIA*, social impact assessment.
Based on: Energy Transition Law, 2015. Available from: < http://dof.gob.mx/nota_detalle.php?codigo = 5421295&fecha = 24/12/2015 > ; Electricity Industry Law, 2014. Available from: < http://www.diputados.gob.mx/LeyesBiblio/pdf/LIElec_110814.pdf > ; General Climate Change Law, 2012. Available from: < http://www.diputados.gob.mx/LeyesBiblio/pdf/LGCC_130718.pdf > ; Secretaría de Energía, 2018a. Programa de Desarrollo del Sector Eléctrico Nacional 2018-2032. Available from: <https://www.gob.mx/cms/uploads/attachment/file/331770/PRODESEN-2018-2032-definitiva.pdf> (in Spanish); Secretaría de Energía, 2018b. Reporte de Energías Limpias en México. Available from: <https://www.gob.mx/sener/documentos/informe-sobre-la-participacion-de-las-energias-renovables-en-la-generacion-de-electricidad-en-mexico-al-30-de-junio>.

TABLE 16.2 Generation by technology for 2029–32 based on Mexico's PRODESENs.

Technology	PRODESEN 2029 (2015–29)	PRODESEN 2030 (2016–30)	PRODESEN 2031 (2017–31)	PRODESEN 2032 (2018–32)	PRODESEN 2033 (2019–33)
Combined cycle	44.05	58.08	44.63	50.95	50.41
Conventional thermoelectric	0.55	0.00	0.14	1.56	4.20
Coal-based power	0.23	0.22	6.32	5.97	5.88
Turbogas	0.69	0.21	0.38	0.09	0.48
Internal combustion	0.23	0.01	1.08	0.27	0.23
Fluidized bed	0.00	0.94	1.58	1.49	0.62
Hydropower	9.23	11.25	8.51	8.37	8.30
Wind	20.09	10.68	14.80	12.84	12.90
Geothermal	2.72	2.81	3.32	2.49	1.81
Solar PV	1.63	2.86	2.93	4.13	8.61
Thermosolar	0.00	0.01	0.01	0.01	0
Nuclear	6.52	8.78	8.46	8.00	2.22
Bioenergy	1.63	0.06	2.77	1.24	0.51
Efficient Cogeneration	12.49	4.09	5.07	1.61	3.81
Imports	0.00	0	0	0.98	0
Total (%)	**100**	**100**	**100**	**100**	**100**

Based on: Secretaría de Energía, 2015. Programa de Desarrollo del Sector Eléctrico Nacional 2015–2029. Available from: <http://base.energia.gob.mx/prodesen/PRODESEN2015/PRODESEN_2015-2029.pdf> (in Spanish); Secretaría de Energía, 2016. Programa de Desarrollo del Sector Eléctrico Nacional 2016–2030. Available from: <http://base.energia.gob.mx/prodesen/PRODESEN2016/PRODESEN-2016-2030.pdf> (in Spanish); Secretaría de Energía, 2017. Programa de Desarrollo del Sector Eléctrico Nacional 2017–2031. Available from: <http://base.energia.gob.mx/prodesen/PRODESEN2017/PRODESEN-2017-2031.pdf> (in Spanish); Secretaría de Energía, 2018a. Programa de Desarrollo del Sector Eléctrico Nacional 2018–2032. Available from: <https://www.gob.mx/cms/uploads/attachment/file/331770/PRODESEN-2018-2032-definitiva.pdf> (in Spanish); Secretaría de Energía, 2018b. Reporte de Energías Limpias en México. Available from: <https://www.gob.mx/sener/documentos/informe-sobre-la-participacion-de-las-energias-renovables-en-la-generacion-de-electricidad-en-mexico-al-30-de-junio>; Secretaría de Energía, 2019. Programa de Desarrollo del Sector Eléctrico Nacional 2019–2033. Available from: <https://www.gob.mx/sener/documentos/prodesen-2019-2033> (in Spanish) (Secretaría de Energía, 2015, 2016, 2017).

Finally, based on a mandate by EIL, the Ministry of Energy (SENER) is required to develop a series of projections on how the National Electric System (SEN) would be constituted in the future. The outcomes of these projections are included into the Program for the Development of the National Electricity System (PRODESEN), which makes a 15-year forecast into the future of the energy systems in Mexico. Table 16.2 shows the composition of the electricity generation for 2029, 2030, 2031, 2032, and 2033 based on the four programs that have been published by the Federal Administration, and the first Program published by the incumbent federal administration of Andres Manuel López Obrador (2018–24). The constitution of the energy system is projected as highly centralized, where almost 50% of total generation will be supplied through combined cycle technologies. Although wind plays an important role in the 2029 forecast, it systematically decreases to make way for coal, nuclear energy, and other centralized sources. Geothermal, solar Photovoltaic (PV), thermosolar, and bioenergy make up less than 7% of total generation, despite a clear availability of these resources (Alemán-Nava et al., 2014). Despite a clear availability of solar energy, the amount considered by each of the PRODESENs does not go beyond 4%. Finally, the analysis shows that all PRODESENs does not include a specific category for DG and relies mainly in highly centralized systems. Notably, the last PRODESEN (2019–33) presents some interesting changes, mainly by substantially increasing state-led generation through solar power and efficient cogeneration, while reducing nuclear power, and eliminating new additions that were consistent until 2030 (Secretaría de Energía, 2019).

16.4 Narrow configurations: an assessment of energy futures in Mexico

Hence it can be argued that an energy transition in Mexico is already underway; however, the actual future of that transition is uncertain. The adoption of a wholesale electricity market, which includes the LTEA and the creation of a Clean Energy Certificate market, has made important contributions to address the mandated goals to generate at least 35% of electricity through clean sources by 2024. Similarly, the adoption of regulation to promote DG has responded to the fact that the cost of technology for solar photovoltaic panels and wind turbines have declined rapidly, which has been the primary factor shaping the energy transition in the country (Hancevic et al., 2017).

Nevertheless, the inclusion of 56 large-scale energy projects through the LTEA has systematically overlooked the social impacts implicit in the development of such large-scale development. Based on Mexico's adoption of the indigenous consultation rights after the 169 ILO Convention, the EIL incorporated the right as a fundamental issue to address FPIC. However, allocation of the auctioned projects where granted without such prior consultation (Del Pozo and Gutiérrez, 2019) that propelled a series of contestations, criticism, and conflicts from civil society and indigenous groups at the local level.[4] Most of these criticisms were developed over the lack of adequate means of access to information, lack of consultation and community participation and involvement in the design, development, and operation of the projects and the lack of adequate compensation measures and the potential environmental impacts (Júarez-Hernández and León, 2014).

Issues over the territory in these spaces show that the energy transition process has systematically considered spaces as a "platform and not a product" (Bridge, 2018). This characterization has avoided the fact that space and place play an important role in the coproduction of the energy systems and its infrastructure and physical manifestations. Framing most of the adoption of renewables as benign and desirable normally obscures the spatial injustices that occurred in their development and hence the production of uneven power relations that derive from them. Geographers have been the key in showing that energy projects, infrastructure, and resources, all occupy spaces. Understanding geographical location

---

[4] For a deeper examination on the impact of large-scale wind projects in Oaxaca, Mexico and the role of the social dimensions of energy transitions, see Chapter 17, The role of social resistance in shaping energy transition policy in Mexico: the case of wind power in Oaxaca.

as more than fixed coordinates, but as the locations of physical infrastructure relative to other spaces and places, is a key part of the issues that reconfigure the power dynamics in the territory and alter and produce new configurations of energy landscapes (Bridge, 2018).

When analyzing the energy reforms' own goals—that is, to reduce electricity costs to consumers and reduce GHG emissions in the wake of climate change—few benefits can be counted. Mexico achieved most of the energy-related emission reductions through the substitution of coal and fuel oil power plants (PRODESEN, 2018). The regulatory landscape has favored this substitution by exempting natural gas from the carbon tax and labeling efficient cogeneration as a source of clean energy in the EIL. This approach has warranted several criticisms, mostly by environmental groups that recognize this as a unilateral decision to push for a steady integration of natural gas in the future.[5]

SENER's latest PRODESEN (2019—33), which presents the first prospective of the new federal government (2019—24), maintains more or less the same production of electricity but makes special emphasis in a state-led development of energy generation projects (Secretaría de Energía, 2019). While the adoption of solar PV can be seen as a positive turn, the adoption of new cogeneration power plants and maintaining coal-based generation until 2033 reaffirms the techno-managerial, cost-effectiveness approach toward energy generation in the future, which questions the role of climate change into future perspectives and the social and economic impacts it entails.

On the one hand, it is important to consider how the future of energy transition process has been framed by the regulatory framework and the energy prospective. The analysis of each PRODESEN shows there are no mechanisms in place to reduce energy consumption and estimates an increase of almost 3% of electricity generation yearly, which will be met by a highly centralized and fossil fuel—dependent electricity system. The last PRODESEN includes an estimate for emission reductions for the electricity sector, complying with Mexico's 22% emission reduction commitment on the unconditional NDC presented before the UNFCCC. However, this goal is far from placing Mexico in an adequate path to meet its commitments established in the GCCL and would follow a trajectory where temperatures would increase from 2°C to 3°C by 2100 (Climate Action Tracker).

On the other hand an important number of households in Mexico are currently unable to satisfy their energy needs. According to Garcia-Ochoa and Graizbord (2016), 36% of Mexican households are not able to meet their energy needs that include lighting, entertainment, water heating, cooking, efficient refrigeration, and thermal comfort. This is although electricity tariffs for 96% households in Mexico are heavily subsidized by the Federal Government. In 2015 the Government reported a subsidy that amounted to almost USD 6 billion. IEAs own estimate is that the subsidy, if let to grow at current rate and with a demand of almost 3% increase a year, would amount to a cumulative subsidy of around USD 90 billion by 2035 (International Energy Agency, 2016). Departing from this position, both civil society organizations and academics have put forward plans and studies over how the energy subsidy could be better focalized and designed, in order to reduce energy poverty [Iniciativa Climática de México (ICM), 2017; Hancevic et al., 2017; Tornel, 2018], address increasing demands of the electricity in the residential sector, and contribute to the energy transition and climate change mitigation targets. Yet, no

---

[5] Climate Transparency Initiative, 2018. Brown to Green Report. Mexico Country Profile. Available from: <https:// www.climate-transparency.org/g20-climate-performance/g20report2018> (accessed 12.01.19.).

official plan to address this issue has been officially adopted by the Government or CFE.

Considering the basic principles set forth by the Energy Trilemma and energy justice, it can be argued that the distribution of costs and benefits has been allocated unevenly within the Mexican society. There is little recognition of other, often local, considerations, particularly in the reconfiguration of space and the social dimension of electricity systems.[6] As mentioned above, energy justice is not only about the distributional aspects of costs and benefits that in the case of Mexico have been systematically unevenly distributed among large-scale development of renewable energy projects, enclosing communal territories and producing uneven patterns of development among regions and social groups (Avila-Calero, 2017), but it has to do with issues of procedure, which in this case is represented by the lack of consideration of inputs by consumers and local communities, and recognition of the social, environmental, and economic impacts technologies and the process of transition itself (Jenkins et al., 2016). These conditions are the foundation for the continuation and subsistence of more unequal energy futures.

The future of the electricity sector is entangled with a series of complex power relations. As this chapter has argued, issues over the role energy ownership and energy sovereignty were a recurring issue for debates in the Mexican Congress, and these issues are still deeply entangled with the Mexican national identity and ideas of progress and modernization (Melgar, 2013). However, the declining availability of fossil fuel reserves since 2013 has systematically changed the discourse toward finding and producing other forms to supply electricity and address the increasing energy needs of the population.[7]

In this sense, almost all of the reviewed papers that address futures of the electricity system in Mexico either directly or indirectly argue that there is a large potential of renewable energy resources available in the country with little support mechanisms in place to promote more integration. Most of the analysis focuses on increasing renewable energy integration and participation, with the development of technologies and industries around an increasing energy demand.[8] However, other accounts of the transition present a more comprehensive approach of the energy transition process, particularly those concerned with issues over energy justice, the socio-technical dimension of the energy systems, and local involvements and participation in the development of the energy systems (Pérez-Denicia et al., 2017; Jano-Ito and Crawford-Brown, 2016; Rennkamp et al., 2017). Finally, another set of academic papers and activists have effectively analyzed the socio-spatial consequences and the uneven geographical distribution of costs and benefits that have resulted from this particular energy transition model.[9] However, a more coherent narrative is warranted in the

---

[6] A similar analysis of the "double burden" of energy access and climate change mitigation can be found in Chapter 6, The challenges and contradictions of Peru's *Proyecto Masivo de Energía Solar*. Similarly, Chapter 18, Energy transition in isolated communities of the Brazilian Amazon, addresses the important issue that granting energy access does not equate the end of energy poverty.

[7] Oswald (2017) and Vidal-Amaro et al. (2015). For other examples, particularly the case of Chile's long-term energy futures, see Chapter 15, Foresight for Chile's energy transition—unleashing societal transformations.

[8] Only a few of the discussions brought by these papers fully integrate the socio-technical and socio-spatial characteristics of the evolution of the National Electricity System into their analysis such as Melgar (2013), Sandoval-García (2013), Alemán-Nava et al. (2014), Guevara et al. (2016), Vidal-Amaro et al. (2017), Oswald (2017), Elizondo et al. (2017).

[9] See: Del Pozo and Gutierrez (2018), Baker (2018), Avila-Calero (2017), Oceransky (2008), Dunlap (2017), Boyer and Howe (2016), Nahmad et al. (2014).

case of a just energy transition, one that is able to bring the socio-spatial dimension of the energy transition to the design and governance of energy systems.

Likewise, a deeper engagement with the way that energy transitions are imagined and understood is still an issue largely unexplored in countries such as Mexico. On the one hand the actual social–technical configurations and the social and geographical impacts of the transition process have already been present in an important number of academic literature and activist/civil society criticism and policy recommendations. However, a discussion between these two narratives seems to be mutually exclusive in the current state of affairs of the electricity system. What this chapter has shown is that energy systems must be understood as democratic spaces, with a clear definition over how costs and burdens are distributed among different actors and groups, making particular emphasis in issues over procedure and recognition—that is, how energy systems, energy projects, and the energy transition itself are designed and what consequences have arisen from such a process (Miller and Richter, 2014).

While this review of the academic literature concerning energy futures in Mexico was not exhaustive, it helps one to show that the configurations of the energy futures are normally framed in very narrow and technical terms, where little consideration is given to energy justice, particularly to issues over the spatial (re)configurations that occur as a direct and indirect consequence of the substitution of one energy source to another, and the process by which societies tend to adapt, use and behave around that new energy source. Parting form the framework of STIs, it is possible to argue that the energy transition has been framed almost exclusively in the need to substitute

polluting technologies with cheaper alternatives. But these substitutions have not warranted any changes in social behaviors, energy demands, and for that matter, the reconfiguration of the electricity system in the future. Hence it is safe to say that the factors that have driven the energy transition in Mexico thus far have been based purely on economic reasons, where GHG emission reductions and renewable energy integration have been obtained only as marginal cobenefits associated with the low generation costs of large-scale renewable energy projects. Nevertheless, prospectives for the integration of additional capacity from renewable technologies in the future of electricity sector are substantially limited.[10]

Expectations of the future of the Mexican energy transition seem to perpetuate energy injustices, by a lack of recognition of the uneven consumption patterns of electricity, a lack of consideration of the spatial inequalities generated by the adoption of large-scale energy projects and the lack of recognition of those affected by these projects (Boyer and Howe, 2016). The energy transition in Mexico has tended to focus on individual technological developments and projects without considering the systemic implications of such developments. A just energy transition would necessarily need to be constructed parting form the notion of socio-technical systems, where issues over how the energy transition are designed, imagined, and produced, must be understood as a process of coproduction and coconstitution between the technical and social dimensions that make up the energy system.

Similarly, rather than looking at communities and local populations as obstacles, the energy transition process has demonstrated elsewhere that active engagement with a wider

---

[10]   For an analysis of thse issues, particularly for techno-economic framings in energy transitions for the case of Brazil, see Chapter 14, The role and impacts of policies in hampering the biogas transition in Brazil.

range of social actors can produce more democratic and legitimate energy systems that account for a wider array of political and economic factors in technological change. Future research in energy transitions must consider these issues in order to address a more just energy system in the future. The notion of STIs can be a major first step to change the way we think and allow ourselves to imagine alternative energy futures.

16.5 Conclusion

This chapter has argued that the energy transition in Mexico has been framed into a narrow configuration of what could be called the "techno-economics paradigm," based on technological substitutions and the reduction of energy generation costs. Issues over political, social, spatial, and justice dimensions have systematically been placed outside the spectrum of design, ownership, and imaginaries shaping the energy transition. This narrow framing of the energy system has then maintained the inequalities of the current energy system in Mexico, where energy consumption remains highly unequal and the distribution of costs and benefits have now produced new spatial inequalities associated with the development of energy infrastructures.

Socializing, politicizing, and democratizing energy policy by recognizing the role of people and social groups in shaping the design, scales, and forms of ownerships and participation in the energy systems are then crucial to guarantee a justified energy transition. This chapter has shown that energy systems are coproduced through socially mediated issues and physical infrastructure, and hence the design and objectives of the energy sector are embedded into the configurations of the energy futures. Likewise, socio-technical imaginers are present in shaping and mediating the role that technology will have in the future.

The analysis of Mexico's PRODESENs and a series of peer-reviewed publications have shown that the energy futures are presented in the same narrow technical and economic framework, where issues over the growth in energy demand are not addressed and instead are framed as issues over supply and costs.

Imagining a future of energy systems must be understood as a social and democratic issue. Although groups with more power are usually the ones shaping the energy transition process, this chapter has shown that disregarding the social and spatial dimensions that coconstitute present and future energy systems will necessarily yield uneven patters of distribution of costs and benefits, lack of access to enough energy services and resources and at the same time will be unable to recognize and develop procedures for participation and representation. Mexico, as well as the rest of developing countries in Latin America, is currently in the process of imagining and developing future energy systems; doing so must then acknowledge the role of the social and spatial issues that coproduce the energy systems to develop a more comprehensive approach that is based on a redistribution to guarantee access and equity to clean and affordable energy services. This will most likely be one of the main challenges of the near future for countries such as Mexico.

References

Alemán-Nava, G., et al., 2014. Renewable energy research progress in Mexico: a review. Renew. Sustain. Energy Rev. 32, 140−153.

Averchenkova, A., Guzman Luna, S.L., 2018. Mexico's General Law on Climate Change: Key Achievements and Challenges Ahead. Grantham Research Institute on Climate Change and the Environment and Centre for Climate Change Economics and Policy, London School of Economics and Political Science.

Avila-Calero, S., 2017. Contesting energy transitions: wind power and conflicts in the Isthmus of Tehuantepec. J. Polit. Ecol. 24, 992−1012.

Baka, J., 2013. Governmentality, land acquisition and social inequality in South India. Dev. Change 44 (2), 409–428.

Baker, S., 2018. Emerging challenges in global energy transitions: a view from the frontlines. In: Salter, R., González, C.G., Warner, E.A.K. (Eds.), Energy Justice. US and International Perspectives. Edward Elgar Publishing.

Bakke, G., 2016. The Grid: The Fraying Wires Between Americans and Our Energy Future. Bloomsbury Publishing, New York.

Bellamy-Foster, J., Clark, B., York, R., 2010. The Ecological Rift: Capitalism's War on the Earth. Monthly Review Press, New York.

Boyer, C., Howe, D., 2016. Aeolian extractivism and community wind in Southern Mexico. Public Cult. 28 (2), 215–235.

Bridge, G., 2018. The Map is not the territory: a sympathetic critique of energy research's spatial turn. Energy Res. Soc. Sci. 36, 11–20.

Bridge, G., Bouzarovski, S., Bradshaw, M., Eyre, N., 2013. Geographies of energy transition: space, place and the low-carbon economy. Energy Policy 53, 331–340.

Calvet, K., 2015. From 'Energy geography' to 'Energy geographies': perspectives on a fertile academic borderland. Prog. Hum. Geogr. 40, 1–21.

Chilvers, J., Longhurst, N., 2016. Participation in transition (s): reconceiving public engagements in energy transitions as co-produced, emergent and diverse. J. Environ. Policy Plann. 18 (5), 585–607.

Climate Action Tracker. Country Summary: Mexico. Available from: <https://climateactiontracker.org/countries/mexico/> (accessed 12.04.19.).

Del Pozo, E., Gutiérrez, R., 2019. De la consulta a la libre determinación de los pueblos: Informe sobre la implementación del derecho a la consulta y al consentimiento previo, libre e informado en México. IIJ, UNAM; FUNDAR, DPLF.

Diario Oficial de la Federación, 2019. Disposiciones administrativas de carácter general de generación distribuida y generación limpia distribuida. Available from: <http://www.dof.gob.mx/nota_detalle.php?codigo = 5474790&fecha = 07/03/2017> (accessed 12.01.19.).

Dunlap, A., 2017. Counterinsurgency for wind energy: the Bíi Hioxo wind park in Juchitán, Mexico. J. Peasant Stud. 0 (0), 1–23.

Electricity Industry Law, 2014 [Ley de la Industria Eléctica]. Available from: <http://www.diputados.gob.mx/LeyesBiblio/pdf/LIElec_110814.pdf>.

Elizondo, A., Pérez-Cirera, V., Strapsson, A., Fernández, J. C., Cruz-Cano, D., 2017. Mexico's low carbon futures: an integrated assessment for energy planning and climate change mitigation by 2050. Futures 93, 14–26.

Energy Transition Law, 2015. [Ley de Transición Energética]. Available from: <http://dof.gob.mx/nota_detalle.php?codigo = 5421295&fecha = 24/12/2015>.

Evrard, A., 2013. Contre vents et marées. Politiques des énergies renouvelables en Europe. Presses Académiques de Sciences Po, Paris.

Garcia-Ochoa, R., Graizbord, B., 2016. "Caracterización espacial de la pobreza energética en México. Un análisis a escala subnacional" Economía, Sociedad y Territorio, vol. XVI, núm. 51, mayo-agosto. El Colegio Mexiquense, A.C., pp. 289–337.

General Climate Change Law, 2012. Available from: <http://www.diputados.gob.mx/LeyesBiblio/pdf/LGCC_130718.pdf> (accessed 18.01.19.).

Gras, A., 2017. The deadlock of the thermo-industrial civilization: the (impossible?) energy transition in the Anthropocene. In: Garcia, E., Martinez-Iglesias, M., Kirby, P. (Eds.), Transitioning to a Post-Carbon Society. International Political Economy Series. Palgrave Macmillan, London.

Guevara, Z., Sousa, T., Domingos, T., 2016. Insights on energy transitions in Mexico from the analysis of useful energy 1971–2009. Energies 9, 488.

Hancevic, P.I.; Nuñez, H.M.; Rosellón Diaz, J.d.D.E., 2017. Distributed photovoltaic power generation: Possibilities, benefits, and challenges for a widespread application in the Mexican residential sector. In: DIW Discussion Papers, 2017. No. 1663, Deutsches Institut für Wirtschaftsforschung (DIW), Berlin.

Huesemann, M., Huesemann, J., 2011. Techno-Fix: Why Technology Won't Save Us or the Environment. New Society Publishers, Canada.

Iniciativa Climática de México (ICM), 2017. Análisis de Costo Beneficio del Programa Bono Solar Fase 1. Available from: <http://www.iniciativaclimatica.org/wp-content/uploads/2018/04/Bono-Solar-ICM.pdf> (accessed 11.01.19.).

Intergovernmental Panel for Climate Change (IPCC), 2015. Fifth Assessment Report. Available from: <https://www.ipcc.ch/assessment-report/ar5/>.

Intergovernmental Panel for Climate Change (IPCC), 2018. Special Report on 1.5°C. Available from: <https://www.ipcc.ch/sr15/>.

International Energy Agency, 2016. México Energy Outlook. Available from: <https://www.iea.org/publications/freepublications/publication/MexicoEnergyOutlook.pdf> (accessed 12.11.18.).

International Energy Agency, 2019. Global Energy & CO_2 Status Report: The Lates Trends in Energy and Emissions in 2018. Available from: <https://www.iea.org/geco/>.

Jano-Ito, M.A., Crawford-Brown, D., 2016. Socio-technical analysis of the electricity sector of Mexico: its historical evolution and implications for a transition towards low-carbon development. Renew. Sustain. Energy Rev. 55, 567–590.

Jasanoff, S., 2005. States of Knowledge. The Co-production of Science and the Social Order International Library of Sociology, 15. Routledge, New York.

Jasanoff, S., 2018. Just energy transitions: a humble approach to global energy futures. Energy Res. Soc. Sci. 35, 11–14.

Jasanoff, S., Kim, S.-H., 2015. Dreamscapes of Modernity: Socio-Technical Imaginaries and the Fabrication of Power. The University of Chicago Press Books.

Jenkins, K., 2018. Setting energy justice apart from the crowd: lessons from environmental and climate justice. Energy Res. Soc. Sci. 39 (2018), 117–121.

Jenkins, K., McCauleya, D., Heffronb, R., Stephan, H., Rehner, R., 2016. Energy justice: a conceptual review. Energy Res. Soc. Sci. 11, 174–182.

Jenkins, K., Sovacool, B.K., McCauley, D., 2018. Humanizing socio-technical transitions through energy justice: An ethical framework for global transformative change. Energy Policy 117, 66–74.

Júarez-Hernández, S., León, G., 2014. Energía eólica en el istmo de Tehuantepec: desarrollo, actores y oposición social. Revista Problemas del Desarrollo 178 (45), 157.

Labussière, O., Banos, V., Fontaine, A., Verdiel, E., Nadaï, A., 2018. The spatialities of energy transition processes. In: Labussière, O., Nadaï, A. (Eds.), Energy Transitions, Energy, Climate and the Environment. Palgrave, Macmillan.

Melgar, L., 2013. Energy transitions: a path toward sustainable development for Mexico. Latin Am. Policy 1, 98–113.

Miller, C., Iles, A., Jones, C.F., 2013. The social dimensions of energy transitions. Sci. Cult. 22 (2), 135–148.

Miller, C., Richter, J., 2014. Social planning for energy transitions. In: Current Sustainable/Renewable Energy Reports, Urban Planning. vol. 1, no. 3, pp. 77–84.

Miller, C., Moore, N., Altamirano-Allende, C., Irshad, N., Biswas, S., 2015. The social value of mid scale energy in Africa: redefining value and redesigning energy to reduce poverty. Energy Res. Soc. Sci. 19, 67–69.

Miller, C.A., Richter, J., O'Leary, J., 2015b. Socio-energy systems design: a policy for energy transitions. Energy Res. Soc. Sci. 6, 29–40.

Miller, J.L.C., 2014. Reforma energética, ¿era realmente necesaria? Economía Informa núm. 385, 3–45.

Nahmad, S., Nahón, A., Langlé, R., 2014. La visión de los actores sociales frente a los proyectos eólicos del Istmo de Tehuantepec. CIESAS, Mexico.

Newell, P., Mulvaney, D., 2013. The political economy of the 'just transition'. Geogr. J. 179 (2), 132–140.

Oceransky, S., 2008. Wind conflicts in the isthmus of Tehuantepec: the role of ownership and decision-making models in indigenous resistance to wind projects in Southern Mexico. Commoner, Winter (13), 9.

Oswald, Ú., 2017. Seguridad, disponibilidad y sustentabilidad energética en Mexico. Energy Security, Availability, and Sustainability in Mexico. In: Revista Mexicana de Ciencias Políticas y Sociales Universidad Nacional Autónoma de México Nueva Época. Año LXII, Núm. 230, pp. 155–196.

Pérez-Denicia, E., Fernández-Luqueño, F., Vilariño-Ayala, D., Montaño-Zetina, L.M., Maldonado-López, L.A., 2017. Renewable energy sources for electricity generation in Mexico: a review. Renew. Sustain. Energy Rev. 78, 597–613.

Poor People's Energy Outlook 2018. Achieving Inclusive Energy Access at Scale. Available from: <https://policy.practicalaction.org/resources/publications/item/ppeo-2018-summary-report-achieving-inclusive-energy-access-at-scale>.

Presidencia de la República, 2013. Reforma Energética. In: Gobierno de la República. Available from: <https://www.gob.mx/cms/uploads/attachment/file/10233 Explicacion_ampliada_de_la_Reforma_Energetica1.pdf>.

Raman, S., 2013. Fossilizing renewable energies. Sci. Cult. 22 (2), 172–180.

Rennkamp, B., Haunss, S., Wongsa, K., Ortega, A., Casamadrid, E., 2017. Competing coalition: the politics of renewable energy and fossil fuels in Mexico, South Africa and Thailand. Energy Res. Soc. Sci. 34, 214–232.

Sandoval-García, ER., junio 2013. Proyección sobre energía eléctrica en México mediante la Identidad de Kaya. Economía Informa núm. 380 mayo.

Secretaría de Energía, 2015. Programa de Desarrollo del Sector Eléctrico Nacional 2015–2029. Available from: <http://base.energia.gob.mx/prodesen/PRODESEN2015/PRODESEN_2015-2029.pdf> (in Spanish).

Secretaría de Energía, 2016. Programa de Desarrollo del Sector Eléctrico Nacional 2016–2030. Available from: <http://base.energia.gob.mx/prodesen/PRODESEN2016/PRODESEN-2016-2030.pdf> (in Spanish).

Secretaría de Energía, 2017. Programa de Desarrollo del Sector Eléctrico Nacional 2017–2031. Available from: <http://base.energia.gob.mx/prodesen/PRODESEN2017/PRODESEN-2017-2031.pdf> (in Spanish).

Secretaría de Energía, 2018a. Programa de Desarrollo del Sector Eléctrico Nacional 2018–2032. Available from: <https://www.gob.mx/cms/uploads/attachment/file/331770/PRODESEN-2018-2032-definitiva.pdf> (in Spanish).

Secretaría de Energía, 2018b. Reporte de Energías Limpias en México. Available from: <https://www.gob.mx/sener/documentos/informe-sobre-la-participacion-de-las-energias-renovables-en-la-generacion-de-electricidad-en-mexico-al-30-de-junio>.

Secretaría de Energía, 2019. Programa de Desarrollo del Sector Eléctrico Nacional 2019−2033. Available from: <https://www.gob.mx/sener/documentos/prodesen-2019-2033> (in Spanish).

Secretaría de Medio Ambiente y Recursos Naturales (SEMARNAT), 2015. Nationally Determined Contribution NDC. Available from: Disponible en: <https://www.gob.mx/cms/uploads/attachment/file/162973/2015_indc_ing.pdf>.

Tidwell, J.H., Tidwell, A.S.D., 2018. Energy ideals, visions, narratives, and rhetoric: examining socio-technical imaginaries theory and methodology in energy research. Energy Res. Soc. Sci. 39, 103−107.

Tornel, C., 2018. Energy poverty in Mexico: a Solution in the Solar Bonus Program. In: Presentation at the Eradicating Poverty through Energy Innovation Conference. Arizona State University. February 12−14, 2018.

United Nations, 2015. Sustainable Development Goals. Available from: <https://www.un.org/sustainabledevelopment/sustainable-development-goals/> (accessed 09.02.19.).

United Nations Framework Convention on Climate Change (UNFCCC), 2015. The Paris Agreement. Available from: <https://unfccc.int/process-and-meetings/the-paris-agreement/the-paris-agreement>.

Unruh, G.C., 2002. Escaping carbon lock-in. Energy Policy 30, 317−325.

Vidal-Amaro, J.J., Østergaard, P.A., Sheinbaum-Prado, C., 2015. Optimal energy mix for transitioning from fossil fuels to renewable energy sources—the case of the Mexican electricity system. Appl. Energy 150, 80−96.

Vietor, R.H.K., Sheldahl-Thomason, H., 2017. Mexico's energy reform. Harvard Bus. School 717 (27), 1−32.

World Energy Council, 2017. World Energy Trilemma. Changing Dynamics − Using Distributed Energy Resources to Meet the Trilemma Challenge. Available from: <https://www.worldenergy.org/wp-content/uploads/2017/11/World-Energy-Trilemma-2017_Full-report_WEB.pdf> (accessed 08.02.18.).

Yanetti, K., Day, R., Golubchikov, 2016. Spatial Justice and the land politics of renewables: dispossessing vulnerable communities through solar energy mega-projects. Geoforum 76, 90−99.

Further reading

Castan-Broto, V., 2016. Innovation territories and energy transitions: energy, water and modernity in Spain, 1939−1975. J. Environ. Policy Plann. 18 (5), 712−729.

Castree, N., 2017. Anthropocene and planetary boundaries. In: Richardson, D., Castree, N., Goodchild, M.F., Kobayashi, A., Liu, W., Marston, R.A. (Eds.), The International Encyclopedia of Geography: People, the Earth, Environment, and Technology. John Wiley & Sons Ltd, Chichester, United Kingdom, pp. 161−174.

Conde, M., Walter, M., 2014. Commodity Frontiers. In: D'Alisa, G., Demaria, F., Kallis, G. (Eds.), Degrowth: A Vocabulary for a New Era. Routledge Books.

Diario Oficial de la Federación (DOF), 2018. ACUERDO por el que se emiten las Disposiciones Administrativas de Carácter General sobre la Evaluación de Impacto Social en el Sector Energético. 1 de junio, disponible en: <http://www.dof.gob.mx/nota_detalle.php?codigo = 5524885&fecha = 01/06/2018>.

Grunstein-Dickter, M., 2016. Against the wind: regulation, social crisis and institutional change in the Wind Corridor in the Isthmus. Economía, Sociedad y Territorio 16 (51), 485−517.

Hydrocarbons Law, 2014. Available from: <http://www.diputados.gob.mx/LeyesBiblio/pdf/LHidro_151116.pdf>.

International Renewable Energy Agency, 2018. International Renewable Power Generation Costs in 2017. Abu Dhabi. Available from: <https://www.irena.org/-/media/Files/IRENA/Agency/Publication/2018/Jan/IRENA_2017_Power_Costs_2018.pdf>.

International Renewable Energy Agency. Renewable Energy in Latin America 2015. An Overview of Policies. Available from: <https://www.irena.org/-/media/Files/IRENA/Agency/Publication/2015/IRENA_RE_Latin_America_Policies/IRENA_RE_Latin_America_Policies_2015.pdf>.

Kemp, R., Loorbach, D., Rotmans, J., 2007. Transition management as a model for managing processes of co-evolution towards sustainable development. Int. J. Sustain. Dev. World Ecol. 14 (1), 78−91.

Kuzemko, C., Lockwood, M., Mitchell, C., Hoggett, R., 2016. Governing for sustainability energy systems: politics, contexts and contingency. Energy Res. Soc. Sci. 12, 96−105.

Labussière O. and Nadaï, A. (2018). New energy resoruces in the making. in: Energy transitions: A Socio-Technical Inquiry. Olivier Labussiere, Alain Nadaï (Eds.) Palgrave, 2018, Energy, Climate and the Environment series, 978-3-319-77024-6. ⟨halshs-01962685⟩.

Lazzard, 2018. Lazard's Levelized Cost of Energy Analysis—Version 12.0. Available from: <https://www.lazard.com/media/450784/lazards-levelized-cost-of-energy-version-120-vfinal.pdf>.

Miller, C., Wybron, C., 2018. Co-production in global sustainability: histories and theories. Environ. Sci. Policy. Available from: < https://www.sciencedirect.com/science/article/pii/S1462901117306366 >.

Mulvaney, D., 2013. Opening the black box of solar energy technologies: exploring tensions between innovation and environmental justice. Sci. Cult. 22 (2), 230–237.

Myers, R., Larson, A.M., Ravikumar, A., Kowler, L.F., Yang, A., Trench, T., 2018. Messiness of forest governance: how technical approaches suppress politics in REDD + and conversation projects. Global Environ. Change 50, 314–324.

SENER, 2014a. Regulación del Mercado Eléctrico Mayorista (MEM) [Regulation for the Wholesale Energy Market]. Available from: <https://www.cenace.gob.mx/paginas/publicas/Transparencia/REC_MEM.aspx>.

SENER, 2014b. Explicación Ampliada de la Reforma Energética. Available from: <https://www.gob.mx/sener/documentos/explicacion-ampliada-de-la-reforma-energetica>.

Swyngedouw, E., 2015. Liquid Power: Contested Hydro-Modernities in Twentieth-Century Spain. MIT Press, Cambridge, Massachusetts.

The role of social resistance in shaping energy transition policy in Mexico: the case of wind power in Oaxaca

Adolfo Mejía-Montero[1], *Lourdes Alonso-Serna*[2] *and Carlo Altamirano-Allende*[3]

[1]University of Edinburgh, Edinburgh, United Kingdom [2]University of Manchester, Manchester, United Kingdom [3]Arizona State University, Tempe, AZ, United States

17.1 Introduction

Over the last decade, renewable energy has had a dynamic development in Latin America. Investments in renewables in the region have grown 11-fold, and Brazil, Mexico, and Chile are among the top 10 largest energy markets in the world (IRENA, 2016). National energy policies play a decisive role in enabling the deployment of renewable energy systems, and the potentials for increased and sustained economic growth derived from these systems are dependent on the policy instrument and their mechanisms for implementation at play. Without a doubt, Latin America is part of a global effort to transition to a carbon-free world. Nevertheless, clean energy transitions are much more than shifts of technologies and resources; contemporary developments in the energy sector point to a global transformation with historical geographical, technical,

economic, and social implications. Thus attention to the social dimensions of energy transitions is critical to guarantee socially robust policies in the energy sector.

Mexico has set an ambitious goal to produce 35% of its electricity with clean sources of energy by the year 2024. This goal is underpinned by an active leadership within international negotiations and the commitment to adopt mitigation strategies to address some of the most pressing effects of climate change. At the national level, the government has also sought to regulate the renewable energy sector by creating attractive market conditions for private investments. Besides hydropower, wind energy was the first technology to be deployed at a commercial scale. Among the different regions in the country suitable to produce electricity via wind, the region known as the Isthmus of Tehuantepec, in the Southern state of Oaxaca, is the one with the largest

The Regulation and Policy of Latin American Energy Transitions
DOI: https://doi.org/10.1016/B978-0-12-819521-5.00017-6

developments of wind energy plants in Mexico (Fig. 17.1). This region is also home to a strong and diverse population that includes indigenous communities, and the deployment of wind farms has faced a series of conflicts between the government, utilities, energy developers, and different sectors of the local population. Several scholars have given extensive attention to these episodes of conflict with a focus on analyzing the social impacts of wind farm deployment as a social justice issue (Dunlap, 2017a,b; Howe and Boyer, 2015; Howe et al., 2015; Huesca Pérez et al., 2016; Juárez Hernández and León, 2014); however, little attention has been paid to the interlinkages between these social movements and the formulation of the policy instruments that are currently guiding the transformation of the energy sector in the country.

The chapter analyzes the importance of the social dimension in Mexico's energy transition. It calls attention to the role of grassroots movements in inducing some changes in the regulatory framework of renewables. It seeks to understand the ways in which top-down public

policies to boost renewables are modified by bottom-up social movements. It is considered that the understanding of these relations can contribute to developing a more comprehensive public policy that effectively incorporates the social dimension of energy transitions. Furthermore, this practical approach can build bridges between the academic sector and policymakers to contribute to the understanding and implementing of just energy transitions for present and future generations in Latin America.

The chapter is organized as follows:

- The second section explores Mexico's active role within international negotiations related to climate change, which eventually led to the definition of an ambitious goal of decarbonization aiming to produce 35% of the country's electricity through clean energy sources by 2024 and 50% by 2050.
- The third section describes the techno-economic rationale behind the development of wind energy and the top-down development of the first policy framework for the industry. The main concern of the

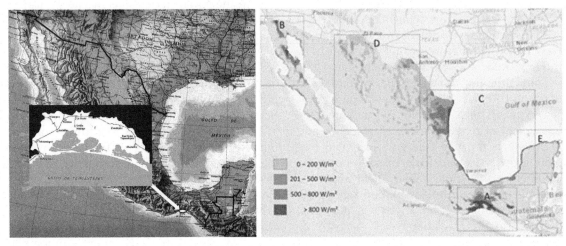

FIGURE 17.1   On the left: Location of the Oaxaca's Isthmus region. On the right: Geographic distribution of the existing wind power resources in Mexico. Source: *From (Left) Jaramillo, O.A., Borja, M.A., 2004. Wind speed analysis in La Ventosa, Mexico: a bimodal probability distribution case. Renew. Energy 29, 1613–1630 (Jaramillo and Borja, 2004) and (right) Alemán-Nava, G.S., Casiano-Flores, V.H., Cárdenas-Chávez, D.L., Díaz Chavez, R., Scarlat, N., Mahlknecht, J., et al., 2014. Renewable energy research progress in Mexico: a review. Renew. Sustain. Energy Rev. 32, 140–152 (Alemán-Nava et al., 2014).*

policy was to give certainty to the international wind power developers that were interested in investing in wind energy.

- The fourth section centers around the implementation of utility-scale projects in the indigenous region of the Isthmus of Tehuantepec and the emergence of movements of resistance to particular wind energy projects. Such emergence, as argued in this chapter, represents the necessity to reconsider the techno-economical framework of wind power development.
- The fifth section looks at the changes in regulation and the introduction of key elements that relate to the social dimension of energy within the existing legislation.
- Finally, the chapter concludes with a reflection on the potential links between the top-down evolution of policy frameworks related to wind power projects in the country and the bottom-up processes of social resistance against such projects.

17.2 The influence of "Our Common Future" and the fight against climate change in Mexico

The 1988 Brundtland report marks a groundbreaking moment in international cooperation, as for the first time, climate change was framed as a global environmental problem that needed global engagement. The report highlighted the human origin of the accelerated depletion of the world's natural resources, and the global impacts of 200 years of industrialization, particularly from Western societies. The report placed humans at the center stage of environmental degradation and gave birth to the notion of the Anthropocene, as an era in which humans radically transformed the living conditions of the biosphere (Clémençon, 2012). In 1992 at the International Summit of Rio de Janeiro, leaders of 100 countries adopted the United Nations Convention on Climate Change and pledged to protect the environment for the present and future generations. This convention acknowledges, "the global nature of climate change calls for the widest possible cooperation [...] in accordance with their common but differentiated responsibilities and respective capabilities and their social and economic conditions" (United Nations, 1992).

In 1988 immediately after the publication of "Our Common Future," the Mexican government passed the General Law for Ecologic Equilibrium and Environmental Protection (Cámara de Diputados del H. Congreso de la Unión, 2015), whose main objective was to foster the recently coined concept of sustainable development from the Brundtland report. In 1992 it created the Ministry for Social Development to provide a better institutional framework for social and environmental policies; the National Institute of Ecology and the Federal Attorney Office for Environmental Protection. Furthermore, the Ministry of the Environment, Fisheries, and Natural Resources was created in 1994 (SEMARNAT, 2013). These institutions aimed to create mechanisms to ensure the country's sustainable development and the protection of the natural resources and ecosystems.

The institutional changes adopted in the 1990s exemplify the way in which national policies and institutions are able to adapt to international environmental concerns. However, it was not until 1997 that the causes of climate change were addressed at the international arena and measures to reduce greenhouse gas (GHG) emissions were put in place via the Kyoto Protocol.[1] In this juncture, Mexico, along with other Latin American

---

[1] The historic protocol implemented a legally binding scheme for developed countries and big polluters to reduce the amount of GHG emissions. However, countries with a low carbon footprint also signed and ratified the protocol and set their own voluntary goals (UNFCCC, 2008).

countries, signed the protocol in 1998, and in 2012, it became one of the first Latin American countries to enact a General Law on Climate Change. Under this law, Mexico pursued a goal to produce 35% of the country's clean electricity by 2024 (Cámara de Diputados del H. Congreso de la Unión, 2012).

The Kyoto protocol made of energy decarbonization the fundamental strategy to address climate change. Industrialized countries with heavy energy consumption could focus on energy efficiency to account for emission reductions. However, developing economies such as Mexico still had growing energy needs, giving it a protagonist role to clean energy transition in the fight against climate change. Since then, countries have made a strong focus in renewable energy technologies such as solar, wind, and hydro as the best solutions for an energy transition.

17.3 Laying the path for wind power in Mexico: a techno-economic phase

Renewable energy technologies are currently seen as a silver bullet against climate change and a means for society to satisfy its energy consumption with a more limited greenhouse emission. However, renewable energy technology was not initially developed to tackle climate change. For instance, wind power technological development initiated before the Brundtland report. It was intended as a local alternative to diminish the dependency on fossil fuels (Pasqualetti et al., 2004). The development of this technology was related to concerns for energy security, rather than environmental sustainability.

As a result of the oil crisis in the 1970s, Denmark developed large-scale wind farms in a nationwide effort to incentivize wind power and as a response to civil society rejection of nuclear power (Albizu et al., 2018, pp. 1–44). Those actions produced the landmark success

in wind energy as hundreds of community-owned wind turbines rose in the country. Currently, wind provides around 40% of the country's electricity consumption and its manufacturing industry plays a key role in the Danish economy accounting for 6.7% of total exports of goods (Danish Wind Industry Association, 2018). This early wind energy adoption called the attention of the United States, and in 1986 the first wind power farm with a capacity of 1200 MW was built in California (AWEA, 2019). This first wind rush in the 1980s proved to be technical and economically feasible, and it also diminished dependency on imports of fossil fuels. In the 1990s a group of successful wind power utilities was already collecting their first economic successes.

In Mexico the first wind farm, known as La Venta I, was built in the Isthmus of Tehuantepec in 1992. It was a project of five turbines of 225 kW each (The Wind Power, 2018). Despite the small installed capacity, La Venta I played a key role in assessing the technical feasibility of wind power in the region as well as its feasibility as an alternative source of energy. Early developments of wind energy are strongly linked to a set of neoliberal policies that started in the 1980s and culminated in the government of Carlos Salinas de Gortari (1988–94) with the signing and implementation of the North American Free Trade Agreement in 1994. As part of this neoliberal restructuring, the regulatory framework of the electric system went through profound changes. The Law of Public Service of Electric Power was modified to allow private companies to produce and sell electricity under the schemes of independent power producers and self-supply (Vargas, 2017).

Wind power companies with their *know-how*, capital, and experience were the main beneficiaries of this legal reform. In this context the government of the state of Oaxaca organized six international colloquiums to promote wind energy in the Isthmus of Tehuantepec from 2001 to 2006.[2] The federal government saw this as an opportunity to accomplish the international commitments to address climate change and, at the same time, receive the economic benefits from a developing wind power industry, while private utilities would get legal certainty and exclusive access to a region with a high-quality wind for electricity production. In those years the federal government delineated the contracts and tariffs for transmission and distribution for renewable energy companies and provided tax incentives to promote the private investments for more than 6000 MW of renewable energy capacity in the region of the Isthmus of Tehuantepec.[3]

In 2006 both the government and wind energy companies agreed to the terms of a *temporada abierta* (open-season mechanism) to finance the infrastructure of transmission that was needed to evacuate the electricity produced in the Isthmus. According to experts, the lack of transmission infrastructure was one of the main technical barriers for wind power in the region.[4] These public—private partnerships laid the foundations for an explosion of large-scale developments of wind energy in Oaxaca. La Venta II became the first large-scale wind farm in the area (83.3 MW). This first phase took off with a narrowly defined regulatory framework that addressed feasibility issues of wind power potential in only technical and financial

terms. In 2008 the government passed the Law for the Use of Renewable Energy and the Financing of Energy Transition, and in 2015, 15 years from the first colloquium in Oaxaca, the Law of Energy Transition was enacted. Overall, it is possible to see the techno-economic rationale that underpins wind energy in Mexico.[5]

17.4 The local deployment of wind power projects: conflict brewed from an invisible social dimension

As mentioned in the previous section, the region of the Isthmus of Tehuantepec, in the state of Oaxaca, has excelled in the production of wind energy. This is the first region in Mexico that holds a high-scale wind energy development, and it is also Mexico's spearhead in its transition to renewables. The Isthmus is a geographical and cultural region in Oaxaca where strong winds are generated by the difference in atmospheric pressure between the Gulf of Mexico and the Pacific Ocean that causes a wind flow through the Isthmus of Tehuantepec. According to the "Wind Energy Resource Atlas of Oaxaca," the coastal plain of the Isthmus of Tehuantepec holds the highest (class 7) wind resource potential that extends over a land area of about 1200 km^2 that can potentially support about 6 GW of installed capacity.[6]

[2] Oaxaca Economy Ministry. Invitación para el Coloquio 2010. Available from: <https://www.amdee.org/Coloquio_Oaxaca/invitacion.pdf> (accessed 05.04.19.).

[3] AMDEE. Visión del Trabajo Previo, Retos y Oportunidades. Available from: <https://www.amdee.org/Proyectos/AMDEE-Vision-general-121109prev-1.pdf> (accessed 05.04.19.).

[4] Interview February 16, 2018.

[5] Chapter 16, Integrating social and justice dimensions to energy transitions: the case of Mexico, gives a comprehensive analysis of the legal and regulatory framework to enable the energy transition.

[6] NREL. Atlas de Recursos Eólicos del Estado de Oaxaca. Available from: <https://www.nrel.gov/docs/fy04osti/35575.pdf> (accessed 05.04.19.).

Currently, as of March 2019, there is a network of 24 wind farms in the Isthmus, with 1401 turbines producing over 2756 MW of electricity in an area of approximately 25,000 ha. These projects belong to the first phase of the wind developments in the region; however, plans to install at least 13 new wind farms to reach the 6000 MW of installed capacity that the region can hold are currently underway. To achieve this goal the government announced in 2018 the construction of 1221 km of another transmission line from the Isthmus of Tehuantepec to the Central Mexico to evacuate 3000 MW, but the project has already faced opposition from sectors of the local community.

Wind farms in the Isthmus of Tehuantepec are located in five municipalities: Asunción Ixtaltepec, El Espinal, Juchitán, Santo Domingo Ingenio, and Unión Hidalgo. A large percentage of the population in these municipalities identify itself as indigenous, particularly in Juchitán, where around 60% of the population speaks Zapotec as a first language. As described in previous sections, the development of wind energy was predominantly approached from a techno-economic/top-down perspective, focusing mainly in permits to produce electricity or planning for siting infrastructure in windy areas. This meant that the relationship with the local population had a top-down approach, leaving vast sectors of the population excluded from the debates and decisions of wind energy in the area.

The government of Oaxaca and utilities approached landholders to arrange the terms to lease land to deploy wind farms, selectively. The outreach focused on a dialog with key representatives of local organizations and not with the entire population. Thus *ejido*[7] commissioners and representatives of small-holders organizations attended the before-mentioned colloquiums in Huatulco[8] to negotiate characteristics of investments. They, in turn, would be in charge of informing landlords on the technical characteristics of the wind farms, as well as on the economic, environmental, and social advantages of the industry. However, according to landlords, commissioners, and other representatives often gave incomplete information. Furthermore, some landlords argue that some representatives and commissioners used the information provided in those meetings for their advantage, as they bought land in the areas where they knew wind farms would be deployed.[9] Thus the relationship with landlords was limited to the lease of land. Some companies visited the region in the 1990s to secure land for the future deployment of windmills. Landholders in areas such as La Venta, La Ventosa, Juchitán, and Unión Hidalgo signed leasing contracts with utilities, but they did not have the opportunity to negotiate the contracts. Furthermore, landlords complained that they did not receive any type of support from the state or federal governments to negotiate higher rents and better benefits for them.

The allocation of the first contracts can be seen as a sign of a land rush. The land was secured years ahead the state conducted the regulation of the wind energy sector. Therefore it was secured before utilities had the permits to produce electricity and before the transmission lines were built. Most of the companies that secured land transferred the contracts to

---

[7] Ejido is a form of land tenure created in Mexico in the early 20th century to grant land to peasants who were dispossessed in the 19th century. Although peasants had individual plots of land, they did not have land titles and thus land could not be sold or mortgaged. Each ejido had three governance bodies: the general assembly, the commissioner, and the vigilance council. A constitutional reform in 1992 allowed individual land titles in ejidos.

[8] Huatulco is a touristic beach resort in the Pacific Coast of Oaxaca.

[9] Interview January 28, 2018; interview January 29, 2018; interview February 17, 2018.

other utilities that could get the permits and the financial credits to build wind farms. The transfer of contracts is a common practice in the region of which landlords complain because the original companies—locally known as *coyotes*[10]—created higher expectations on the economic revenues they could get from utilities, while at the same time offered minimal payments in the contracts they signed, paving the road for a future break of trust.

The first wind farm in La Venta, owned by the state utility, Federal Electricity Commission, fixed the baseline payments to which other utilities had to adjust.[11] One of the *coyote* companies in La Venta, a company called Timber and Seeds from the Lagoon, adopted CFE's baseline payments. According to landlords in La Venta, the company approached them individually, and some of them signed the contracts. However, with a desire to negotiate higher rents, a group of landlords organized collectively and formed the Solidarity group of La Venta.

In La Venta, land tenure is *ejido*, and members of Grupo Solidario wanted the contract to be discussed and negotiated by the whole assembly of the ejido members known as ejidatarios. This idea gained support from the local population; however, the company refused any collective negotiation. As the pressure from the *ejidatarios* increased, the company sought the support of the government to stop collective negotiations. The solution was to oust the ejido commissioner, a strong supporter of collective negotiation. So with the commissioner gone, ejidatarios of *Grupo Solidario La Venta* were isolated, and the company simply denied to sign any contract with them and divided the group by advancing some payments to ejidatarios that had already signed the contract.[12]

After this first episode of conflict, some members of *Grupo Solidario La Venta* openly criticized how wind energy in the region was being developed. Other sectors of the local population in Juchitán ignited a local debate on the social aspects of wind energy. In 2006 a group of teachers and peasants formed an organization of teachers and grassroots movements.[13] The group organized regular meetings to discuss the social, economic, and political problems in Juchitán. Members were concerned about the environmental and social impacts of wind farm deployment and a group of peasants joined the group asking for advice to terminate the contracts they had with a company. According to members of the *Colectivo*, they sought for external legal advice and pushed the company to terminate the contracts.[14]

This first organization disappeared but very soon evolved into the APIIDTT (Assembly of Indigenous Peoples of the Isthmus in Defense of Land and Territory).[15] To this day, APIIDTT is the most important organization in the region with a very strong critical stance toward wind farm deployments. This organization has some coordinating functions with the social movements against wind energy in

---

[10] The term *coyote* is used in Mexico to refer to a person who, in exchange for money, carries out some paperwork in a public office. Coyote is also a person who crosses people from Mexico to the United States. In all cases, coyote has a negative meaning, as the duties performed are not legal.

[11] Interview February 16, 2018.

[12] Interview March 6, 2018.

[13] The organization was called Colectivo Magisterial y Popular 14 de junio.

[14] Interview February 8, 2018.

[15] Asamblea de Pueblos Indígenas del Istmo en Defensa de la Tierra y el Territorio.

the region. Furthermore, APIIDTT has been able to articulate a coherent agenda of social aspects of wind energy, highlighting the importance of including the indigenous communities in the social and environmental impacts of a massive deployment of wind farms. The organization underscores that the Mexican government has to enforce the rights of indigenous communities in the development of wind energy. This organization also highlights the unequal distribution of economic profits between utilities and landholders.

Along with APIIDTT, there are other social movements with a critical stance toward wind energy in Juchitán, Unión Hidalgo, and San Dionisio del Mar. In these municipalities, there have been episodes of opposition to the deployment of wind farms. In Juchitán, for example, the APPJ, or Popular Assembly of the Peoples of Juchitán opposed the construction of a wind farm owned by Gas Natural Fenosa. Members of this organization were particularly worried about potential damages to the lagoon, in which a sector of the local population finds a regular source of food and income. APPJ organized in 2013 when the wind farm was under construction, and despite the opposition from different sectors of the local population, the project was completed.

In Unión Hidalgo a group of landholders organized as the *comuneros*[16] of Unión Hidalgo to terminate some contracts with the company *Desarrollos Eólicos Mexicanos*, a subsidiary of Renovalia Energy. This company began the construction of the wind farm in 2011, ignoring the concerns of a group of landlords regarding a large number of trees cut down to install the wind pads and other infrastructure. They were also asking for higher compensation for the damages to their properties; however, the company did not agree to negotiate with the

group, who then decided to seek for legal advice to cancel the contracts. Despite this opposition, the wind farm started its operation in 2012. The *comuneros* of Unión Hidalgo have been fighting in the Agrarian Courts to cancel the contracts and to seek recognition of communal land tenure rights.

In 2012 a large sector of the local population in San Dionisio del Mar opposed the construction of the Mareña Renovables wind farm. This would be the largest wind farm in Latin America, with 132 windmills and a total capacity of 396 MW (REVE, 2015). When the wind farm construction began, a large proportion of the population claimed to have no information on the project and demanded the company and the government not to go any further until the population approved the construction. Municipal authorities conducted some informative meetings, but the information did not satisfy the local population. One of the main concerns was that the wind farm would be located in a sandbar by the lagoon, potentially affecting fisheries, the primary source of food in the region.

The population also complained of the lack of accountability from the local government and the executive board of communal land that granted permits to build the project without informing the population. Although the major attempted to inform the population, he could not satisfy their main demand, which at that stage was to call off the project. After this, the population openly condemned the major and overthrown it from Town Hall, restraining access to the area near the sandbar to stop any attempt from the company to start the construction. Despite all the attempts of negotiation among the government, the group of investors, and the Asamblea de San Dionisio, the conflict turned into a deadlock, with the local population demanding the wind farm to

---

[16] Comuneros are peasants who own land collectively under a form of tenure known as "comunidad." This land tenure is similar to ejido; it has its own governance bodies. As ejido, since 1992, comuneros can get individual land titles over plots of land.

be called off and the company and the authorities unwilling to address this demand. The conflict escalated to some episodes of violence with the police, drawing the attention of international human rights organizations. After 3 years the company decided to cancel the project and moved inland.

In all cases, the diverse sectors of the population, which identify as indigenous—Zapotecs or Ikojts—were not informed of the projects in the early stages. To those sectors that were informed, such as landlords, the information given was also limited, technical, and, therefore, not easy to understand. Overall, the general demand across all cases of conflict is to uphold the right to free, prior, and informed consent by conducting indigenous consultations according to the 169 International Labor Organization (ILO) Convention. The San Dionisio del Mar case discussed here is the pioneer case in which a large-scale renewable energy project was halted mainly due to the organized social resistance. The social movement grew to such an extent that it gathered national and international attention due to the violations of the ILO Convention 169 and the negative response from the government offices toward the growing unrest.

These episodes of resistance to wind farm deployment are all diverse and include a large list of claims of injustice; however, social movements agree that wind power is not in itself the problem, but "the way in which these projects are being forced and introduced in the region" without previous consultation and a fair distribution of benefits and impacts.[17] These claims, however, have little resonance among the general population in the Isthmus of Tehuantepec, as people actively engaging in social movements are a minority within most communities. Thus the sector of the population who actively participates in actions against wind farm deployment faces limitations in term of resources and capabilities against the backdrop of the large budgets of transnational companies for legal advice and lobby activities with government decision makers.[18] In counterpart the social movement's lack of material and economic resources restrains actions as simple as traveling to Mexico City to campaign and meet civil servants.[19] They also rely strongly in their members' will to volunteer, and in the pro bono advisory skills and legal expertise of civil organizations. For this reason the Mareña Renovables case became emblematic for members of resistance groups, as an example of what would happen if large sectors of the local population engaged actively against wind power projects.

The social movements with a critical stance toward wind energy in the Isthmus are part of a wider network against mega infrastructure projects, such as mining, hydropower, and transport (Silva et al., 2018). This alliance of indigenous groups labels wind energy and the other projects as "Proyectos de Muerte" (deathly projects), in stark contrast with international organizations, government, and the business sector, which frame renewables as climate change mitigation strategies, and the survival of humanity. Therefore under the current scenario, it seems impossible to reconcile the views of these two groups. This is not of minor importance for the Mexican renewable transition, as this labeling could produce an ex ante opposition from communities against renewable energy in other parts of the country. Under this context the Isthmus' network of resistance against wind power projects is able to provide support and share its experience and strategies of resistance to other

[17] Interview November 5, 2017; interview December 12, 2017.

[18] Interview December 8, 2017.

[19] Interview November 27, 2017.

communities in Mexico where there are new developments of utility scale solar or wind projects. This is the case of Mayan communities in Yucatan that have been supported by Zapotec social movements and academics which engaged with this first stage of wind power projects (Zarate-Toledo and Fraga, 2016) (Table 17.1).

TABLE 17.1   Wind power projects in the Tehuantepec Isthmus at 2016.

Name	Location	Capacity (MW)	Operational since
La Venta I	La Venta	1.125	1994
La Venta II	La Venta	83	2006
Parques Ecologicos de Mexico (La Ventosa I and II)	La Ventosa	80	2008
Eurus, Juchitan de Zaragoza Oaxaca	La Venta	251	2009
Instituto de Investigaciones Electricas	La Ventosa	0	2010
Bii Nee Stipa I	El Espinal	26.35	2010
Electrica del Valle de Mexico (La Mata- La Ventosa)	La Ventosa/Asunción Ixtaltepec	68	2010
Energías Renovables Venta III, La Venta III	Santo Domingo Ingenio	103	2011
Energias Ambientales de Oaxaca, Oaxaca I	Santo Domingo Ingenio	102	2012
CE Oaxaca Dos, Oaxaca II	Santo Domingo Ingenio	102	2012
CE Oaxaca Tres, Oaxaca III	Santo Domingo Ingenio	102	2012
CE Oaxaca Cuatro, Oaxaca IV	Santo Domingo Ingenio	102	2012
Fuerza Eólica Del Istmo I and II	La Ventosa	80	2012
Stipa Nayaa	Juchitan	74	2012
DEMEX 1	Union Hidalgo	90	2012
Bii Nee Stipa II (Eólica El Retiro)	La Ventosa	74	2012
Bii Nee Stipa III (Zopiloapan)	LaVentosa	70	2012
Parques Ecológicos de México (La Ventosa III)	La Ventosa	22	2013
Eoliatec del Istmo	Juchitan	164	2013
Eoliatec del Pacifico	Santo Domingo	160	2014
Bii Nee Stipa IV (Eolica Dos Arbolitos)	Juchitan	70	2014
Fuerza y Energia Bii Hioxo	Juchitan	234	2014
DEMEX 2	Union Hidalgo	138	2014
PE Ingenio	Santo Domingo Ingenio	50	2015
Energías Renovables La Mata, La Mata (Sureste I fase II)	Asunción Ixtaltepec	102	2015
	Total	*2348*	

DEMEX, Desarrollos Eólicos Mexicanos.

17.5 Developing mechanisms to include the social dimensions of energy transitions

In 2013 a constitutional reform paved the way for an aggressive set of policies aimed at advancing the deployment of clean and renewable energy systems through various public—private partnership schemes with very ambitious objectives. These objectives, measured for the most part in terms of a gradual reduction in GHG emissions, have vast implications in the controversies and social resistance based on an articulation of indigenous autonomy and the rights of communities living in and around these energy systems. The speed with which these legal changes took place was only matched by the magnitude and transcendence of what was proposed, as within just 1 year, the Mexican congress discussed and changed key articles of the constitution (articles 25, 26, and 28), including a package of secondary laws.[20] The legislation that was enacted, coupled with the growing attention to social resistance to renewable energy projects, illustrates the unresolved tension between the carbon emission reduction goals, the exploitation of Mexico's vast renewable energy potential, and the notion of free, prior, and informed consent by communities directly affected by those developments.

The ILO Convention 169 (ILO, 1989), ratified by the Mexican State in 1990, was the first international legally binding instrument to recognize indigenous peoples as a collective subject of rights. The signature of this instrument contributed to recognizing the highly diverse ethnic composition of Mexico. This acknowledgment was later introduced in the Mexican constitution, and it endowed a degree of

autonomy to indigenous communities. On the other hand the United Nations Declaration on the Rights of Indigenous Peoples of 2007 is an instrument for the states to guarantee the rights of indigenous peoples, in particular, the right to own their lands, territories, and resources. Although this declaration is not legally binding, it obligates states to create mechanisms to prevent any form of dispossession of land, territory, and resources of indigenous communities. Both the UN Declaration and ILO Convention 169 have taken a predominant place in both the claims from resistance groups against wind power projects and the later analysis around the applicability of free, prior, and informed consent in infrastructure and resource extraction projects in indigenous territories. The regulatory framework derived from the energy reforms in regards to the consultation and social assessment processes are defined in two documents—The Law on the Electricity Industry and the Regulation of the Electricity Industry Law. Article 119 of the Electric Industry Law states,

> In order to take into account the interests and rights of the communities and indigenous peoples in which projects of the electricity industry sector are developed, the Ministry of Energy shall carry out the necessary consultation procedures and any other activity necessary for their safeguarding, in coordination with the Ministry of the Interior and the corresponding agencies.

In an effort to reflect on the social dynamics around energy developments and to incorporate the social dimension within the final version of the new energy legislation, the federal government and congress conveyed national dialogs with experts across a multiplicity of sectors and expertise, which allowed for the

---

[20] Article 25 refers to the participation of private companies in strategic sectors to incentivize competition, economic growth, and development in the country. Article 26 refers to mechanisms of civil consultation, planning, and evaluation for development programs and projects. Article 28 refers to monopolies and competition, opening the generation and consumption to private companies and reserving state ownership for power transmission and distribution.

design of a regulatory proposal on the social impacts and land occupation for electrification projects.

As some of the social controversies around wind energy projects already existed at the time of negotiations in 2013, representatives from NGOs, unions, academics, and energy experts provided input, and as a consequence, the General Direction of Social Impact and Surface Occupation was created in 2014 under the Ministry of Energy to ensure that companies incorporate principles of sustainability and human rights into the planning of energy projects. New regulation, produced after the energy reforms, obliged companies to conduct an Social Impact Assessment[21] related to the potential impacts that an energy project might have in the population in the surrounding areas of the project and to propose measures to prevent or, in case of inevitability, to mitigate negative impacts to the local population.

Even though this requirement for a Social Impact Assessment was published in 2014, the specific guidelines that the social impact assessment ought to follow were not defined but until June of 2018, when the Ministry of Energy published them after 3 years of public comment process on the first draft of the document (Cámara de Diputados del H. Congreso de la Unión, 2018). The purpose of the social impact assessment is to ensure that projects have a continued beneficial and productive relationship between impacted communities and the electric industry, in this case the wind energy developers. The lack of clarity on the normative applications of the new standards for social assessment practices has not contributed to reducing social resistance. Indigenous groups such as APIIDTT, which were not invited to the negotiations of the energy reform, keep gaining notoriety in the public perception of wind energy projects as their social demands are unsuccessfully met with an increasing number of energy projects in their territories.

Correspondingly, the Regulation of the Electrical Industry Law stipulates in articles 86−92 that the permits for projects related to the generation and distribution of electricity will be granted once the social impact assessment is submitted for review to the Ministry of Energy. When the projects locate in indigenous territory, the ministry will be the sole responsible for consultation procedures with the local population in accordance to article 119 of the Electric Industry Law and in cooperation with other federal, state, and local agencies. According to article 91 of the referred Law, the consultations are guided by the principles of good faith, cultural relevance, transparency, and compromise. These principles match the highest international standards on this matter. The consultation is divided into six phases: delineation of the consultation protocol, prior agreements, informative phase, deliberative phase, consultative phase, and monitoring and following up on agreements.

In 2014 and 2015 took place the first consultation in the Isthmus of Tehuantepec. The wind farm of Eólica del Sur, comprising the follow-up project of Mareña Renovables, relocated to the municipalities of El Espinal and Juchitán. Various public offices from the federal, state, and municipal branches were involved in the development of the Protocol. The process in Juchitán extended for 9 months, from October 2014 to July 2015, and it was highly contested. The social movements in the area, such as APIIDTT and APPJ, participated in the processes and highlighted anomalies during the different phases of the consultation. First, they argued that the fact that the consultation was taking place after the companies got the permits to produce electricity and the leasing agreements with landholders, violated the very essence of the free, prior, and informed

---

[21] Evaluación de Impacto Social.

consent. The strongest critiques from these movements, as well as some NGOs and groups of scholars, maintain that the consultation aimed to legitimize projects with tick-box simulative processes.[22]

Furthermore, the fact that the DGISOS and other government offices elaborated the protocol and set the format of the consultation, constrained the role of indigenous communities in actively engaging in the consultation. According to the international standards and other national protocols, the indigenous consultation is an exercise of autonomy in which the indigenous communities design and conduct the whole process with the minimum interference from government offices (Instituto Estatal Electoral y de Participación Ciudadana de Oaxaca, 2014). The active role of the federal government throughout the process was a permanent issue of contention with the social movements, and it was one of the main reasons for them to question the legality of the consultation. After 8 months of a complex process, a sector of the population in Juchitán accepted the construction of the wind farm; however, other sectors of the local population and members of APPIIDTT and APPJ called for a *writ to amparo*.[23] However, in the end, even if the *writ to amparo* stalled the project for some months, the wind farm was built.

On the other hand, the Mexican government has also addressed the demands from landlords that rent land to utilities. Landlords complaints relate to compensations for damages in their properties and on the amount of rents that utilities pay. The Ministry of Energy published the *Action Protocol on Shared Social Benefits of Energy Projects* (Mexico Energy Ministry, 2018). This document provides a minimal set of rules for the distribution of revenues from energy projects. This minimal set of rules intends to create the conditions for a horizontal and open dialog among social groups, dwellers, and communities that conform to it. This document also aims to widen the social acceptance of energy projects by providing a minimal set of economic benefits to the wider population. However, the appeal to social movements that oppose wind energy might be limited as this instrument, among others such as the indigenous consultations, are labeled as mechanisms to give legitimacy to energy projects (Dunlap, 2017a).

In parallel to these top-down mechanisms, there have been attempts to develop pioneer community wind projects in the region. However, a lack of financial and planning support mechanisms combined with a regulatory framework exclusively tailored for utility-scale private wind projects has kept community-owned schemes out of picture in Mexico (Oceransky, 2010). However, there are other examples of countries in Latin America, which are aiming to develop comprehensive policy mechanisms in order to include indigenous communities cosmovision in the energy transition (Ministerio de Energía de Chile, 2017) or to facilitate the partnership of private renewable companies and local indigenous communities to develop jointly owned energy projects.[24]

The creation of the General Direction of Social Impacts and Land Occupation in 2014 within the Ministry of Energy was a step forward in the policy-making formulation efforts to account for ongoing controversies in

---

[22] Interview November 21, 2017.

[23] Writ to amparo is a legal procedure in the Mexican Law that seeks to protect individuals or groups against the violation of human rights.

[24] Chapter 19, An inclusive and participative model for energy transition in Latin America: the case of Chilean *Generación Comunitaria*, presents a case of public policy to favor communities.

indigenous territories; however, the complex dynamics involved in the implementation of public consultation mechanisms leaves that office with the conflicting position of both determining the need for a consultation to be made but also to push an attractive ecosystem for investments as part of the Ministry of Energy. The legal framework that followed the energy reform of 2013 seeks to prevent nuances related to the social impacts derived from energy projects; however, the promotion of new investments in the sector reinforces the techno-economic perspective. The historic reforms in the energy sector of 2013 deal mostly with two issues: energy security and a sustainable transition toward a carbon-free economy. As it has been shown, social resistance stems not from a disconnection with these two issues but with a broader problem of governance of the energy systems.

17.6 Conclusion

The first experience with wind power has made of the Tehuantepec Isthmus a laboratory for the Mexican energy transition where the rationale followed by policymakers did not incorporate the social dimension. The latest reforms in the energy framework aim to include the social dimensions to the energy transition. However, opposition to renewable energy projects in Mexico, and the resistance movements that characterized this opposition, are only one of many ingredients that contributed to the inclusion of instruments within the energy policy ecosystem that account to the social dimension of energy. Policy processes are influenced and shaped by a multiplicity of actors, interests, values, and contexts and Mexico accounts for a long history of resistance movements against energy infrastructure and extractive projects.

However, this chapter shows a clear before-and-after regarding the existence of a social dimension in the Mexican policy-making for energy transition, suggesting that the influence of claims coming from bottom-up resistance movements toward wind power can be found printed in new policy frameworks and mechanisms developed after the Mexican energy reform.

There is still a gap between current policy and claims from resistance groups, manifested through conflicts around wind power in the Isthmus. However, it remains to be seen the long-term consequences of the incorporation of a nuanced understanding of the social dynamics around the newly built energy systems in Mexico. As the rules of engagement continue to change and adapt to the complexities of the relation between social organizations, investors, and government actors.

It is even possible that the episodes of conflict brewed from the wind power rush in the Isthmus of Tehuantepec has produced an irretrievable opposition to wind energy from the social movements referred in the chapter. This would mean that opposition groups would disqualify in advance the new instruments introduced in the law such as the social impacts assessments and the indigenous consultations.

However, interests from transnational renewable companies and national government targets for climate change mitigation are strongly aligned. Even if it has been proved that social resistance movements have had some influence in modifying energy policy frameworks, it seems unlikely that they can put a brake to the accelerated Mexican renewable transition. Therefore there is a big risk that in its impulse to reach the targets of renewable energy capacity in the country, the essential issues related to due process, human rights, transparency, and accountability are overlooked a second time.

In this scenario, stakeholders looking for a just and fair transition process must support the development and implementation of comprehensive policies that conciliate local population interests with global goals to address climate change and to incorporate a social dimension in the energy transition.

Therefore this chapter finalizes with a call for empirical, interdisciplinary, and multiactor research aiming to deepen in understanding between top-down policy-making and bottom-up resistance processes in energy transitions in Latin America. Academics and policymakers must cooperate toward more just renewable energy systems that truly enforce due process through transparent and participative mechanisms. This would allow for wider population to be properly informed and actively involved in the development of clean energy projects, avoiding the threat that current conflicts comprise for both local populations and energy transition.

References

Albizu, L.G., Pagani, D., Brink, T., 2018. Denmark. In: Identifying Success Factors for Wind Power. World Wind Energy Association, Germany, pp. 1–44.

Alemán-Nava, G.S., Casiano-Flores, V.H., Cárdenas-Chávez, D.L., Díaz Chavez, R., Scarlat, N., Mahlknecht, J., et al., 2014. Renewable energy research progress in Mexico: a review. Renew. Sustain. Energy Rev. 32, 140–152.

AWEA, 2019. History of wind: 1980's—turbine timeline. Available from: <https://www.awea.org/wind-101/history-of-wind/1980s> (accessed 04.04.19.).

Cámara de Diputados del H. Congreso de la Unión, 2012. Ley General de Cambio Climático. Cámara de Diputados del H. Congreso de la Unión, México.

Cámara de Diputados del H. Congreso de la Unión, 2015. Ley General Del Equilibrio Ecologico y la Proteccion al Ambiente. Cámara de Diputados del H. Congreso de la Unión, México.

Cámara de Diputados del H. Congreso de la Unión, 2018. Acuerdo por el que se emiten las Disposiciones Administrativas de Carácter General sobre la Evaluación de Impacto Social en el Sector Energético. Cámara de Diputados del H. Congreso de la Unión, México.

Clémençon, R., 2012. Welcome to the Anthropocene: Rio + 20 and the meaning sustainable development. J. Environ. Dev. 21, 311–338.

Danish Wind Industry Association, 2018. Employment, export and revenue. Available from: <https://en.windpower.org/wind-in-denmark/statistics/employment-export-and-revenue> (accessed 04.04.19).

Dunlap, A., 2017b. 'The town is surrounded': from climate concerns to life under wind turbines in La Ventosa, Mexico. Hum. Geogr. 10, 16–36.

Dunlap, A., 2017a. Counterinsurgency for wind energy: the Bíi Hioxo wind park in Juchitán, Mexico. J. Peasant. Stud. 0, 1–23.

Howe, C., Boyer, D., Barrera, E., 2015. Wind at the margins of the state: autonomy and renewable energy development in Southern Mexico. Contested Powers. The Politics of Energy and Development in Latin America. Zed Books, London, pp. 92–115.

Howe, C., Boyer, D., 2015. Aeolian politics. J. Soc. Theory 16, 31–48.

Huesca Pérez, M.E., Sheinbaum Pardo, C., Köppel, J., 2016. Social implications of siting wind energy in a disadvantaged region – the case of the Isthmus of Tehuantepec, Mexico. Renew. Sustain. Energy Rev. 58, 952–965.

ILO, 1989. C169—Indigenous and Tribal Peoples Convention, 1989 (núm. 169). Available from: <https://www.ilo.org/dyn/normlex/es/f?p = NORMLEXPUB:12100:0::NO::P12100_INSTRUMENT_ID:312314> (accessed 15.03.19.).

Instituto Estatal Electoral y de Participación Ciudadana de Oaxaca, 2014. Guía de actuación para juzgadores en materia de Derecho Electoral Indígena. México. Available from: <http://www.ieepco.org.mx/archivos/documentos/2016/Gu%C3%ADa%20de%20actuaci%C3%B3n%20para%20juzgadores%20en%20materia%20de%20Derecho%20Electoral%20Ind%C3%ADgena.pdf> (accessed 19.05.19.).

IRENA, 2016. Renewable Energy Market Analysis: Latin America. IRENA, Abu Dhabi. Available from: <https://www.irena.org/-/> (accessed 10.04.19.).

Jaramillo, O.A., Borja, M.A., 2004. Wind speed analysis in La Ventosa, Mexico: a bimodal probability distribution case. Renew. Energy 29, 1613–1630.

Juárez Hernández, S., León, G., 2014. Energía eólica en el Istmo de Tehuantepec: desarrollo, actores y oposición social. Probl. Desarro. 178, 139–162.

Mexico Energy Ministry, 2018. Protocolo de actuacion sobre beneficios sociales compartidos de proyectos energeticos. Mexico Energy Ministry, Mexico.

Ministerio de Energía de Chile, 2017. Capitulo Indígena. In: Política Energética 2050. Santiago de Chile, p. 94.

Oceransky, S., 2010. Renewable energy as a common resource. In: Sparking a Worldwide Energy Revolution. p. 670.

Pasqualetti, M., Righter, B., Gipe, P., 2004. History of wind energy use. Encycl. Energy 6, 419–433.

REVE, 2015. Anuncia SENER construcción de parques eólico en Oaxaca. Available from: <https://www.evwind.com/tags/marena-renovables/> (accessed 15.03.19.).

SEMARNAT, 2013. Antecedentes: Secretaría de Medio Ambiente y Recursos Naturales. Available from: <http://www.semarnat.gob.mx/conocenos/antecedentes> (accessed 04.04.19.).

Silva, L., Munro, P.G., De Lourdes, M., Zurita, M., 2018. Proyectos de Muerte: Energy justice conflicts on Mexico's unconventional gas frontier. Extr. Ind. Soc. 5, 481–489.

The Wind Power, 2018. La Venta II (Mexico)—Parques eólicos. Available from: <https://www.thewindpower.net/windfarm_es_4093_la-venta-ii.php> (accessed 04.04.19.).

UNFCCC, 2008. Kyoto protocol reference manual. Available from: <https://unfccc.int/resource/docs/publications/08_unfccc_kp_ref_manual.pdf> (accessed 04.04.19.).

United Nations, 1992. United Nations Framework Convention on Climate Change. Available from: <https://unfccc.int/sites/default/files/conveng.pdf> (accessed 14.05.19.).

Vargas, A.M., 2017. Estructura de la industria eléctrica mexicana: El Modelo de Comprador Único. Economía. Teoría y Práctica 46, 71–95.

Zarate-Toledo, E., Fraga, J., 2016. La política eólica mexicana: Controversias sociales y ambientales debido a su implantación territorial: Estudios de caso en Oaxaca y Yucatán. Trace. Travaux et Recherches dans les Amériques du Centre 69, 65–95.

Further reading

AMDEE, 2012. Visión del Trabajo Previo, Retos y Oportunidades. Available from: <https://www.amdee.org/Proyectos/AMDEE-Vision-general-121109prev-1.pdf> (accessed 04.04.19.).

NREL, 2004. Atlas de Recursos Eólicos del Estado de Oaxaca. Available from: <https://www.nrel.gov/docs/fy04osti/35575.pdf> (accessed 04.04.19.).

Oaxaca Economy Ministry, 2010. Invitación para el Coloquio 2010. Available from: <https://www.amdee.org/Coloquio_Oaxaca/invitacion.pdf> (accessed 04.04.19.).

Energy transition in isolated communities of the Brazilian Amazon

Antonella Mazzone

King's College London, London, United Kingdom

18.1 Introduction

Modern energy services are considered one of the most fundamental requirements for economic and social development.[1] Energy provides clean water, sanitation, health care, education, enhance economic opportunities, and reduce hunger and, potentially, may address some gender disparities.[2] Despite its recognized importance for human development, there are currently 2.7 billion people relying on traditional biomass for subsistence and 1.3 billion people lack access to modern energy services,[3] of which the vast majority live in rural areas of the southern hemisphere. International Energy Agency (IEA) (2016) defines energy poverty as the lack of modern energy services (i.e., electricity and diesel) and the reliance on what is available in nature (i.e., biomass and dung). The absence of reliable and sustainable energy services, defined as energy poverty, creates a relentless dependency on traditional biomass sources for heating and cooking, such as firewood, dung, and organic residue. The most common way to "end" energy poverty is via "energy access," which is universally understood as household's uses of modern energy services. Initially, access was conceptualized and measured via grid expansion and universal rural and urban electrification. However, this concept has often been approached as a binary (have/have not) connection to electric energy, leading to the development of more complex

[1] Goldemberg, J., Johansson, T.B., Reddy, A.K.N., Williams, R.H., 2004. A global clean cooking fuel initiative. Energy Sustain. Dev. 8, 5–12.

[2] IEA, 2016. World Energy Outlook 2016. International Energy Agency, Paris, France. Available from: <http://www.iea.org/publications/freepublications/publication/WEB_WorldEnergyOutlook2015ExecutiveSummaryEnglishFinal.pdf> (accessed May 2017).

[3] IEA, 2016. World Energy Outlook 2016. International Energy Agency, Paris, France. Available from: <http://www.iea.org/publications/freepublications/publication/WEB_WorldEnergyOutlook2015ExecutiveSummaryEnglishFinal.pdf> (accessed May 2017).

The Regulation and Policy of Latin American Energy Transitions
DOI: https://doi.org/10.1016/B978-0-12-819521-5.00018-8

methodologies to conceptualize and measure energy access (e.g., multitier framework).[4]

In the case of Brazil, energy access is commonly measured via access to energy services (mostly electricity) via grid expansion or off-grid deployment. In 2015 the National Institute of Geography and Statistics [Ministerio de Minas e Energia (MME), 2015] calculated that 99.7% of the Brazilian residences had access to electric energy. Compared to other developing nations in the world,[5] Brazil has almost reached its universal target. So, why is Brazil so important in the understanding of energy transition?

First, it is important to establish a definition for energy transition. Energy transition is defined by Smil (2017) as "the change in the composition (structure) of primary energy supply, the gradual shift from a specific pattern of energy provision to a new state of an energy system." A shifting structure of energy supply implies that the transformation of primary energy sources[6] is central to sophisticated production and human lifestyles. This definition of energy transition has been increasingly associated to a low-carbon and renewable energy production, which is the main focus of this chapter.

Despite the most recent statistics,[7] in Brazil, there is a persistent number of households who still do not have access to modern energy services (155,000 rural households) (Sánchez et al., 2015). The persistent reliance of rural isolated households on biomass and solid fuels situates them in the so-called category of the "energy poor." To ensure access to modern energy services, the Brazilian program Light for All in 2003 was designed to supply the remaining areas without electricity by 2014 but was successively delayed until 2018 with the Decree 8387/2014 [Ministerio de Minas e Energia (MME), 2015] after a series of alterations on an annual basis. The program has been successively extended to 2022 by Decree 9357 of 2018 because of persistent geographical and economic issues in achieving the "very last mile." According to the latest directives of the program, rural villages should receive electricity mainly via low-carbon energy decentralized solutions. Low-carbon decentralized solutions are believed to improve living conditions and increase the use of energy for productive purposes and support the implementation of basic services (i.e., health, education, sanitation, and sewage). Transitioning to renewable technologies is assumed to substitute "traditional" fuels in rural areas, indirectly fostering the assumption that rural people will adopt new technologies automatically (e.g., energy ladder theory). Renewable technologies are meant to bring about not only economic development to rural impoverished areas but also social goals, such as gender equality via women's "empowerment," inclusion, health, education, mobility and community empowerment, amongst other benefits. By 2030, decentralized renewable solutions are expected to provide 70% of rural connections (World Bank and International Energy Agency, 2015). New sophisticated forms of renewable energy, particularly solar, are now occupying the scene as a catalyst for rural energy transition.

However, how likely is that transition to renewable systems guarantees the end of "energy poverty" to ensure people's needs are met? This chapter explores the effectiveness of

---

[4] https://www.seforall.org/sites/default/files/MTFpresentation_SE4ALL_April5.PDF

[5] Ninety-five percent of the people without access to modern energy services are either in sub-Saharan Africa or developing Asia (85% in rural areas) (SE4All, 2015) https://seforall.org/sites/default/files/1/2013/09/EnergyAccess.pdf.

[6] Primary source of energy occurs naturally (both for fossil fuels and renewables).

[7] https://trackingsdg7.esmap.org/country/brazil

low-carbon decentralized energy systems in isolated communities of the Brazilian Amazon from an end-user perspective. Given the key role of energy transition in providing sustainable, reliable, and affordable energy access in rural areas, this chapter focuses on the pilot project called *Miniusinas Solares* or solar PV mini-grid. The program was launched in 2011 in 12 remote villages of the Brazilian Amazon, with an initial investment equivalent to US\$ 1.6 million (Governo do Brazil, 2011). *Miniusinas Solares* is now considered a benchmark for future installations in the North Brazil.

18.2 Context

The Brazilian energy domestic supply is characterized by the large contribution of renewable sources (43.5%), sugarcane biomass (17.5%), and hydraulic energy (12.6%).[8] If taken alone, the electricity system relies mostly on hydraulic energy (71.5% in 2016). Given the high reliance on hydraulic energy stocked in large dams, the National Grid has one of the largest power transmission lines in the world.[9] The Brazilian Electricity Regulator defines the National Grid as in "permanent expansion" as it allows the integration of new large hydropower plants into the system and ameliorates

the exchange of electricity between regions. In 2010 the length of National Grid covered an area of 141,388 km (third worldwide), with a prediction to increase to 185,484 by 2024.[10] Despite its planned expansion, the National Grid cannot reach remote areas in Amazon forest because of geographical, economic, and demographic factors.[11]

18.2.1 Amazon region

The Amazon forest acts as one of the essential sources of heat to the global atmosphere through its intense evapotranspiration and release of latent heat of condensation in the middle and upper troposphere in tropical convective clouds, contributing to the generation and maintenance of the atmospheric circulation on regional and global scales.[12] The Amazon tropical evergreen forest has an important role for the regulation of the world's climate and atmospheric composition as an important component of the global carbon dioxide.[13] The Amazon rainforest is home to one-third of the world's species, one-fourth of the world's freshwater, one-fifth of the world's forests, a stock of 48 billion tons of carbon dioxide in its trees, and home to over 200 indigenous and traditional communities.[14]

---

[8] EPE, 2016. Sistemas Isolados—Energia solar para suprimento de sistemas isolados do Amazonas. Available from: <http://www2.aneel.gov.br/aplicacoes/editais_geracao/documentos/EPE-DEE-PT-084-2016-r1%20(Grupo%20B%20-%20Vers%C3%A3o%20Edital).pdf> (accessed June 2017).

[9] Kable (2017).

[10] ONS, 2019. <http://ons.org.br/paginas/sobre-o-sin/o-sistema-em-numeros>.

[11] ANEEL, 2012. Resolução Normativa n° 482 de 17 de Abril de 2012. Available from: Aneel 1. <http://www2.aneel.gov.br/arquivos/PDF/Resolu%C3%A7%C3%A3o%20Normativa%20482,%20de%202012%20-%20bip-junho-2012.pdf> (accessed 18.08.15.).

[12] Fearnside, P.M., 2005. Brazil's Samuel Dam: lessons for hydroelectric development policy and the environment in Amazonia. Environ. Manage. 35 (1), 1–19.

[13] Castanho et al. (2016).

[14] Fearnside, P.M. (2005).

18.2.2 Climate and seasonality

The Brazilian Amazon forest covers for 4.1 million km² the states of Acre, Amazonas, Roraima, Rondônia, Amapá, Tocantins, Pará, and Maranhão, and is called Amazonia Legal. The tropical equatorial climate of the region is hot and humid with average temperatures of 25°C changing along the day with different thermic amplitudes. The population of the Brazilian Amazon reached 9694,728 habitants (5.1% of the Brazilian population) and a demographic density of 2.51 habitants/km² [National Water Agency (ANA), 2016]. The case studies chosen for this research are in the state of Amazonas, which is also Brazil's fourth poorest state and home of 7.1% of the extreme poor of the country.[15]

The state of Amazonas is distributed over 1,559,148,890 km². Its dense tropical forest expands across 98% of its territory.[16] The isolation is due to the dense forest and by the absence of major roads and highways. The Amazon Basin is composed of hundreds of tributaries used as waterways for transportation, fishing, and traveling. The most important rivers are Amazonas (the world's biggest), Rio Negro, Solimões, Tapajós, and Madeira.

18.2.3 Isolated System

The Isolated System has its energy demand met by means of midsize thermo-power plants supplying energy to the consumers of the northern states (Amazonas, Pará, Rondônia, Roraima, Amapá, and Mato Grosso as well as the island of Fernando de Noronha). The systems are characterized by the use of nonrenewable thermic generation and for the

logistic impediments to attend all the consumers living in this area (GTON Eletrobras, 2016). There are 146 oil and diesel-fired thermo-power plants only in the state of Amazonas. Despite the efforts to expand the National Grid (SIN) to the isolated regions, there are still important geographical and economical impediments to its realization. Since 2013, the Brazilian Government planned the construction of the transmission line Manaus–Macapá, which was interpreted as the starting point of the integration of two cities in the Isolated System with the National Grid [Ministerio de Minas e Energia (MME), 2015]. The transmission line was made operational in 2015.[17] Yet communities located at over 300 m from the nearest Isolated System (midsize thermo-power plant) are not supplied with electricity. Local politicians, in the municipalities where the villages are located, generally donate a small diesel generator (working for some hours daily). The donation is often part of the plan aiming at supporting education in isolated areas. The diesel generator is often used initially to supply the local school and then also the households through a rudimental mini-grid. These micro systems deliver precarious (Van Els et al., 2012) electricity for a period of time that varies between 3 and 10 h/day. This has a surprising factor for Gómez and Silveira (2012) who argue that local resources would have been a better solution for this region. However, the only off-grid projects carried out in this region failed as it lacked operational sustainability and were not addressed properly.

Moreover, the fact that these small diesel engines are distributed by local municipalities rather than national programs for energy

---

[15]   See World Bank and International Energy Agency (2015).

[16]   Governo do Amazonas (2017). Available from: http://www.amazonas.am.gov.br/o-amazonas/.

[17]   ANEEL. Aneel Resolução 687. Aneel 25. Available from: <http://www.aneel.gov.br/documents/656827/14866914/M%C3%B3dulo1_Revisao_9/1b78da82-6503-4965-abc1-a2266eb5f4d7> (accessed October 2017).

supply may give fertile opportunities for clientelist behaviors. According to Gómez and Silveira (2012), off-grid solutions in the Amazon need to consider the variety of resources available locally. These two main options regard the supply of electricity to the entire community or the single household supply. The main source of electricity supply available in the Amazon is the traditional diesel engine, commonly used only at night times, and since 2011 12 communities experimented the PV mini-grid-system and the individual solar panel. For Sánchez et al. (2015), diesel systems have a low initial investment cost but a higher operation and maintenance and environmental risks.[18] Moreover, according to the authors, the price of diesel in isolated areas can be very high due to transportation costs. In some regions of the state of Amazonas the cost of transporting diesel to isolated communities of São Gabriel da Cachoeira is four times more expensive than the cost of diesel itself. In those communities without any kind of modern energy services, wealthier people have their own micro-diesel generator, using this energy for private consumption or to sell it to their neighbors.[19] This indicates that the absence of public services exacerbates local inequalities.

18.2.4 Energy transition in the Amazon

Energy access in the Brazilian Amazon has been tackled since the 1990s with public policies implemented at the state level, using treasury resources and having the programs implemented by concessionaries controlled by state government.[20] Operating under a different name in each state, the Rural Poverty Alleviation Program, established in 1993, was sponsored by the World Bank. Grants were allocated to local association to finance projects in the Northeast of Brazil in which communities were making their own decision with their own development terms. These projects included grid-connected rural electrification projects and off-grid solar systems in addition to several other rural-development projects.

Arguably, the sustainable use of local resources, such as biomass, seeds, wood, solar, and hydric power rather than diesel, would have been a better solution for this region (Di Lascio, 2009). However, the only off-grid projects carried out in this region failed as it lacked operational sustainability and were not addressed properly. For example, in 1996, a project called "Vila Campinas," which received the financial support of USAID in the municipality of *Manacapuru* (Cartaxo, 2000) in the state of Amazonas, was one of the first hybrid PV/diesel system installed in the area. In spite of the great expectations that this system could be reproduceable in other parts of the Amazon, economic constraints together with the increasing demand of energy for productive activities by locals, shifted the system from hybrid PV into full diesel. Other cases, such as *Comunidade de Joanes*, which was the first hybrid solar PV-wind system installed in 1994 is currently inoperative because of maintenance and operational issues.[21]

18.3 Methods

This section draws upon Winther's (2008) book "The impact of electricity: Development, desires and dilemmas" in which the author

[18] Hoffman, L., Schmidt, A., 2010. Life Cycle Assessment, Public Policy. European Environment Agency.

[19] Hoffman, L., Schmidt, A., 2010. Life Cycle Assessment, Public Policy. European Environment Agency.

[20] ANEEL (2005).

[21] BARBOSA (2006).

argues for an "anthropology of energy" with electricity being the particular object of research. As Winther (2008, p. 4) argues in the book, "in anthropology, relatively little has been done to understand people's use of energy, with electricity being a particular case in point." A growing literature on anthropology of energy is now emerging with cases from sub-Saharan Africa, India, and South Africa.[22] There are currently no studies approaching lived energy experiences in the Amazon. Ethnographic techniques such as participant observation and field notes can bring about a deeper understanding of the impacts of energy access from the individual perspective and might be the key to understand energy transition in rural areas. This research recognizes the role of individual level analysis to understand the uses of energy precisely to explain how energy can be a causal factor for decision-making, aspirations, desires, and energy practices, which ultimately lead to understanding energy transition and technological acceptance.

18.3.1 Case study and data collection

Primary data were collected between July and October 2015 in two rural villages located in two macro-regions:

- Case 1: The village named "Interconnected" is connected to a thermo-power plant (3840 kW), which supplied a larger town in the vicinity. The village "Interconnected" had 23 consumer units (20 households, 1 church, 1 community center, and 1 cooperative) and electricity provision for 24 h/day.
- Case 2: The second village, named "Solar" because supplied by a solar PV mini-grid installed in 2012, had 24 consumer units.

The PV mini-grid in "Solar" constitutes six solar panels of 135 Wp each and the batteries of 140 A. The system supplied 16 households and four buildings (a school, a health center, a church, and a community center) through a mini-grid of 735 m for 24 h/day.

One of the criteria used to select the villages was to work on an even and manageable number of households. The idea was to facilitate the comparison between villages. The choice to study villages with the same demography, equal distance to urban and periurban areas, and different energy access is precisely to isolate the variable "energy" from other socioeconomic factors determining a change in income and gender relations. Interviews were carried with the members of the households. The interviews aimed at getting a personal view on energy use, energy collection, cooling habits, household energy management, income, income-generating activities, management of the monetary resources, spousal relations, mobility, ownership, cash-transfer program, among others.

The access to different types of energy demonstrates how local people change their livelihood strategies in respect to energy availability, consumption, and technologies (renewable and nonrenewable). One of the conditions that established trust between the informants and the researcher is the guarantee that their identities would be anonymized.

Observations on different phases of household and productive activities as well as manioc production and pottery manufacturing were collected; in particular, how women manufacture their clothes, and what they cook and how. Passive participation was used as methodology as observer and not participant (Spradley, 1980) who was talking and taking notes, pictures and recordings moments. These participatory moments were fundamental not only to create a

---

[22] see Matinga (2010) and Winther (2012).

TABLE 18.1 Energy technology used for each economic activity (Interconnected only).

Products	Technology used	Energy service (direct, indirect)
Fruit pulp	Grinder, freezer	Electricity, transportation
Guaraná production	Grinder, refrigerator	Electricity, transportation
Snacks for children	Blender, oven, and refrigerator	LPG, electricity
Bread, cakes, etc.	Oven and refrigerator	LPG, electricity, transportation
Acai	Grinder, refrigerator and freezer	Electricity, transportation
Brazil nuts (roasted)	Oven	LPG or Firewood
Wood	Chain saw	Transport
Manioc	Grinder, oven	Electricity, transportation
Tambaqui and pirarucu	Refrigerator, freezer	Electricity, transportation
Other nuts and fruits	Grinder, refrigerator, and freezer	Electricity, transportation

Author's own fieldwork, 2015.

deeper bond and reinforce trust with the research participants but also to understand better their energy uses and aspirations.

18.4 Results

Energy use in rural areas includes illumination, cooking, cooling/heating, mechanical energy for agriculture, transportation, home and commercial appliances. Therefore central to rural development in literature is access to electricity, as well as modern energy for irrigation and intensification of land use (Barnes et al., 1994), changes in farming practices (Barnes et al., 1994) and job creation (Akella et al., 2008), and a change in the volume of production for existing and new firms. Energy transition can only be successful in the long term if it satisfies the needs of local people, which include not only the energy for residential needs (lighting, cooking, heating, and cooling) but also accommodates small informal activities. A system interconnected to a grid is notoriously more flexible in terms of satisfying energy for productive activities. In these areas, access to technology and energy is linked to livelihood diversification (Mazzone, 2019).

Table 18.1 shows how the technology found in the Interconnected system helped the diversification of productive activities, which was completely absent in the Solar village.

Table 18.1 suggests that a higher installed capacity could lead to increased productivity levels and diversification of livelihood activities with modern technology. While the Solar village fits into the current definition of energy transition (e.g., access to sustainable and reliable modern energy services), it is evident that the system cannot support more than a TV, three light bulbs, a TV—DVD combo, and an efficient refrigerator. While providing electricity for 24 h/day, the Solar village cannot explore energy for productive activities.

In-depth interview with Dom Pedro (33 years old) in Solar village shows how local people tried to diversify their livelihood strategies in the primary sector (farming) but without success:

I have saved enough money to buy the machinery needed to start an egg farm. The machine would heat the eggs to the right temperature until they are hatched, and I would sell chicken too to the neighbouring villages. I think this is a great idea, but we do not have enough electricity to support this system.

Cooling energy (refrigeration) also plays an important role in reshaping the product value for perishable products. In the Interconnected village, a reliable provision of electricity has allowed families to conserve their food and commercial products such as fresh fruits and fruit pulps. For example, Renato (26 years old, farmer) reported how the price of cupuaçu (*Theobroma grandiflorum*) has changed since the family purchased a freezer. Before the arrival of electricity the fruit would rot within days, so he had to sell the fruit at a lower price, while now, with better refrigeration, he sells the products at a double price:

> Five years ago, when we didn't have electricity, I use to sell the cupuacu at R$2 per kilogram. That's because it rots so easily. Now we freeze it and last all year long. And I sell it for 6 R$/kg. Electricity was a blessing here.

Cooling energy is also fundamental in the negotiation between local fishermen and traders. In the Amazon, hot temperatures and high humidity levels decompose food products easily. Small fishermen and large boats have increasingly exploited Amazonian basins in the last 30 years. Local fishermen own small boats with an engine between 7 and 30 hp. The power of the engine also determines expenses for the diesel fuel, as energy consumption is proportional to the capacity (tons) and speed (horsepower) of the boat. The large distances and the lack of a refrigeration system in the Solar village push local fishermen to

1. buy ice to preserve the catch or
2. depend on commercial boats passing nearby.

Both options are disadvantageous for the fishermen. In option (1) fishermen transport their catch to the nearest municipality to commercialize their products. However, the cost of diesel or gasoline for transportation is high, hence their deducted earnings. Moreover, to keep the fish fresh during transportation, they buy ice cubes acquired in the nearest towns for the equivalent of US $ 4.97 per bag of 5 L. In the other case, larger commercial boats (*regatões*) pass by the villages regularly to buy fish and other produce. If the village has a steady source of electricity, the fish can be stored appropriately. Fish can be frozen and sold to *regatões*, who often buy fish and other products from small villages at lower prices. Passing commercial boats in isolated areas appear to be an excellent opportunity for small fishermen and small rural producers who cannot afford to sell their products in the nearest municipality. However, interviews show that the owners of *regatões* tend to buy local products from the villagers without negotiating the price.

The importance of refrigeration for food conservation and hygiene has already been documented (Kirubi et al., 2009). However, this study sheds new light on a different need for refrigeration in isolated and marginalized communities. Refrigeration constitutes a new bargaining tool against the exploitation of small fishermen by larger buyers. A refrigeration system would offer small fisherman the opportunity to increase their bargaining power, because of the ability to conserve their catch.

It is clear that the concept of "energy transition" as shifted to more sophisticated and technologically advanced energy systems does not equate with current needs of local people, which appear to be more satisfied in the Interconnected village (on-grid connection). This does not suggest a reliance to polluting energy systems, rather this paper suggests a more complete analysis of energy transition in which planning is complemented with sociocultural analysis.

18.4.1 Energy to support migrations and demographic growth

Geographical location and economic constraints often limit people's mobility and curb opportunities for interactions with others; interregional migration is not uncommon, offering a chance to understand other realities and exchange ideas. Religious missions, NGOs, and electrification programs are some of the reasons for local people's migration and offer an opportunity to exchange ideas and lifestyles. These missionaries often spend up to 5 years in each remote village, which creates opportunities for them to train and divulge new business ideas.

However, while interregional migration represented an opportunity for exchanging ideas and spur new business, it also represented a way to worsen local tension, especially when the resources are limited. For example, in the solar PV mini-grid, people from other communities moved in to Solar village because they were informed about a electrification project. However, because the system does not allow energy demand growth, these villagers were left out of the provision of electricity. The coexistence in the same community of electrified and nonelectrified households was the main reason for local tension in Solar village. This finding goes against the common belief that rural electrification reduces social inequalities (Da Silveira Bezerra et al., 2017).

18.5 Conclusion

This chapter analyzed the effectiveness of decentralized renewable solutions in rural isolated villages of the Brazilian Amazon. Findings show that the villages with an overall higher consumption of energy (including electricity, LPG, firewood, and diesel/gasoline) also correspond with a larger impact on livelihood diversification.

In the primary sector (farming) the extension of working hours at night increased the production of certain products, such as manioc flour and bread. A reliable and stable supply of electricity encourages the acquisition of technology and facilitates and accelerates food processing and farming. In fact, the only village found to use technology for productive purposes (e.g., irrigation pumps, chain saw, grinders, and trucks) was the Interconnected village. The Solar village, due to its limited installed capacity of 16.2 kWp, does not meet the demand for certain productive uses of energy in the primary sector.

This chapter suggests, based on evidence found in four villages in the Brazilian Amazon, that "energy transition" conceptualized globally as the adoption of more sophisticated and renewable energy sources is still challenging in the Brazilian Amazon. This is because, in rural isolated villages, renewable systems (solar PV mini-grid in this case) do not support income-creation activities or migrations and demographic growth. Moreover, higher levels of electricity allow residences to level geographically based inequalities by increasing the bargaining power of local fishermen.

Current understanding of energy transition in Brazil is therefore more likely to be related to the definition of "energy access" as linked to a binary "have" and "have not" connection to the grid. However, studies in Europe show that even if energy access is achieved, this does not necessarily mean energy consumption. Different variables influence households' energy uses, for example, reliability, affordability, sustainability, and resilience. In other words, households may have received access to modern energy services, but they might not use energy services because of affordability issues, or because the quality of energy is poor (e.g., intermittent and insufficient for basic needs) to mention a few. The inability to pay for energy services creates the household's dependence on solid fuels for heating and cooking purposes, which is a well-known health hazard. Energy access, therefore, does not equate the end of energy poverty, and so energy transition.

This chapter urges Brazil policy-makers to define the dimensions of the "lack" of modern energy services, which can be linked to the family's financial constraints to afford energy services, or whether it is the result of geographical isolation (Herrero and Bouzarovski, 2014) or a "complex combination of factors, including lack of physical availability of certain energy types and high costs associated with using energy" (Pachauri and Spreng, 2011) Understanding the way in which energy poverty is understood in Brazil may unpack how the government place importance on energy for human well-being and how it will ensure that no one is left behind.

To ensure reliable, sustainable, and affordable energy transition in rural isolated areas in the long term, current Brazilian energy policies should design energy transitions that consider a more accurate projection of energy demand based on local people's sociocultural profile. By looking at isolated communities in the Amazon, it is clear that energy-transition paradigms, in order to be successful in the long term, need to be designed around the needs of the final consumer.

References

Akella, A.K., Saini, R.P., Sharma, M.P., 2008. Social, economical and environmental impacts of renewable energy systems. Renew. Energy 34 (2), 390–396. Available from: https://doi.org/10.1016/j.renene.2008.05.002.

Barnes, D.F., Openshaw, K., Smith, K.R., van der Plas, R., 1994. What Makes People Cook with Improved Biomass Stoves, Tech. Rep. 242. World Bank. Available from: <http://ehs.sph.berkeley.edu/krsmith/publications/94_barnes_1.pdf>.

Cartaxo, E., 2000. Análise técnica e econômica de um sistema de distribuição de energia elétrica híbrido solardiesel: um estudo de caso. <http://www.fem.unicamp.br/~jannuzzi/documents/vilacps2-AM.pdf>.

Da Silveira Bezerra, P.B., Callegari, C.L., Ribas, A., Lucena, A.F.P., Portugal-Pereira, J., Koberle, A., et al., 2017. The power of light: socio-economic and environmental implications of a rural electrification program in Brazil. Environ. Res. Lett. 12. Available from: https://doi.org/10.1088/1748-9326/aa7bdd.

Di Lascio, M.A., 2009. Energia e desenvolvimento sustentável para a Amazônia rural brasileira: eletrificação de comunidades isoladas/Marco Alfredo Di Lascio, Eduardo José Fagundes Barreto; com a colaboração, Daniel Pioch, Écio Rodrigues. Ministério de Minas e Energia, Brasília. ISBN 978-85-62491-00-9.

Gómez, M.F., Silveira, S., 2012. Delivering off-grid electricity systems in the Brazilian Amazon. Energy Sustain. Dev. 16, 155–167. Available from: https://doi.org/10.1016/j.esd.2012.01.007.

Governo do Brazil, 2011. Eletrobras inaugura na Amazônia quatro miniusinas a luz solar em sistema pré-pago. Available from: <http://www.brasil.gov.br/noticias/infraestrutura/2011/07/eletrobras-inaugura-na-amazonia-quatro-miniusinas-a-luz-solar-em-sistema-pre-pago> (accessed November 2018).

GTON Eletrobras, 2016. Programa Mensal de Operação Dos Sistemas Isolados. Available from: <http://eletrobras.com/pt/AreasdeAtuacao/12%20Relat%C3%B3rio%20PMO%20Dezembro2016.pdf> (accessed March 2018).

Herrero, S.T., Bouzarovski, S., 2014. Energy Transitions and Regional Inequalities in Energy Poverty Trends: Exploring the EU Energy Divide. SSRN. Available from: https://doi.org/10.2139/ssrn.2537067.

International Energy Agency (IEA), 2016. <https://www.iea.org/media/weowebsite/energymodel/Poverty_Methodology.pdf>.

Kirubi, C., Jacobson, A., Kammen, D.M., Mills, A., 2009. Community-based electric micro-grids can contribute to rural development: evidence from Kenya. World Dev. 37, 1208–1221. Available from: https://doi.org/10.1016/j.worlddev.2008.11.005.

Matinga, M.N., 2010. We Grow Up With It: An Ethnographic Study of the Experiences, Perceptions and Responses to the Health Impacts of Energy Acquisition and Use in Rural South Africa (Ph.D. diss.). University of Twente, The Netherlands.

Mazzone, A., 2019. Decentralised energy systems and sustainable livelihoods, what are the links? Evidence from two isolated villages of the Brazilian Amazon. Energy Build. 186 (March), 138–146. Available from: https://doi.org/10.1016/j.enbuild.2019.01.027.

Ministerio de Minas e Energia (MME), 2015. <https://www.mme.gov.br/luzparatodos/downloads/manual_sistemas_isolados_16_11_15.pdf> (accessed October 2017).

National Water Agency (ANA), 2016. Relatório de Gestão do Exercício 2016. <http://www3.ana.gov.br/portal/ANA/todos-os-documentos-do-portal/documentos-aud/relatorio-de-gestao-do-exercicio-2013-2016/relatorio-de-gestao-do-exercicio-2016.pdf/view> (accessed October 2017).

Pachauri, S., Spreng, D., 2011. Measuring and monitoring energy poverty. Energy Policy 39 (12), 7497−7504. Available from: https://doi.org/10.1016/j.enpol.2011.07.008.

Sánchez, A.S., Torres, E.A., Kalid, R.A., 2015. Renewable energy generation for the rural electrification of isolated communities in the Amazon Region. Renew. Sustain. Energy Rev. Available from: https://doi.org/10.1016/j.rser.2015.04.075.

Smil, V., 2017. Energy Transitions: Global and National Perspectives. *Praeger*.

Spradley, J.P., 1980. Ethnography for what? In: Participant Observation. pp. 13−25.

Van Els, R.H., De Souza Vianna, J.N., Brasil, A.C.P., 2012. The Brazilian experience of rural electrification in the Amazon with decentralized generation - the need to change the paradigm from electrification to development. Renew. Sustain. Energy Rev. Available from: https://doi.org/10.1016/j.rser.2011.11.031.

Winther, T., 2008. The impact of electricity: development, desires and dilemmas. In: The Impact of Electricity: Development, Desires and Dilemmas. <https://doi.org/10.1111/j.1548-1433.2009.01161_25.x>.

Winther, T., 2012. Electricity theft as a relational issue: A comparative look at Zanzibar, Tanzania, and the Sunderban Islands, India. Energy for Sustainable Development 16, 111−119. Available from: https://doi.org/10.1016/j.esd.2011.11.002.

World Bank, International Energy Agency, 2015. Sustainable Energy for All 2015: Progress Toward Sustainable Energy. World Bank, Washington, DC, <https://openknowledge.worldbank.org/handle/10986/22148> © World Bank; International Energy Agency License: CC BY 3.0 IGO.

Further reading

Agencia Nacional de Energia Eletrica- Relatorio Aneel 2005. http://www.aneel.gov.br/documents/656835/14876457/2006_Relatorio2005/fee7cccf-dc50-0ee2-900e-a41e664afb56.

Attigah, B., Mayer-Tasch, L., 2013. The impact of electricity access on economic development: a literature review. In: Productive Use of Energy - PRODUSE. p. 26.

Bhattacharyya, S., Palit, D., Sarangi, G.K., 2015. Towards Scaling Up of Electricity Access: Summary and Policy Recommendations from OASYS South Asia Project. The Energy and Resources Institute, New Delhi.

Castanho, A.D.A., Galbraith, D., Zhang, K., Coe, M.T., Costa, M.H., Moorcroft, P., 2016. Changing Amazon biomass and the role of atmospheric CO_2 concentration, climate, and land use. Global Biogeochem. Cycles 30, 18−39. Available from: https://doi.org/10.1002/2015GB005135.

Coelho, S.T., Goldemberg, J., 2013. Energy access: lessons learned in Brazil and perspectives for replication in other developing countries. Energy Policy 61 (October), 1088−1096. Available from: https://doi.org/10.1016/j.enpol.2013.05.062.

Cook, P., 2011. Infrastructure, rural electrification and development. Energy Sustain. Dev. 15, 304−313. Available from: https://doi.org/10.1016/j.esd.2011.07.008.

Covre, S., Marques F., Mattos, E., 2008. Oferta de trabalho e transferências: Evidências do efeito das condições impostas pelo programa Bolsa-Família. <http://www.anpec.org.br/encontro2008/artigos/200807141223420-.pdf>.

DFID, 1999. Sustainable Livelihoods Guidance Sheets. Introduction: Overview. p. 10. <https://doi.org/10.1002/smj>.

Eletrobras Amazonas Energia, 2012. Solução de energia limpa para o interior do Amazonas está em fonte fotovoltaica. Available from: <http://www.eletrobrasamazonas.com/cms/index.php/solucao-de-energia-limpa-para-o-interior-do-amazonas-esta-em-fonte-fotovoltaica/> (accessed November 2017).

Eletrobras Amazonas Energia, 2018. Programa Luz para Todos. Available from: <http://www.eletrobrasamazonas.com/cms/index.php/institucional/programas-e-projetos/programa-luz-para-todos/> (accessed November 2018).

Ellis, F., 1998. Household strategies and rural livelihood diversification. J. Dev. Stud. 35, 1−38. Available from: https://doi.org/10.1080/00220389808422553.

Fearnside, P.M., 2005. Deforestation in Brazilian Amazonia: history, rates, and consequences. Conserv. Biol. 19, 680−688.

Fearnside, P.M., 2015. Environment: Deforestation soars in the Amazon. Nature. Available from: https://doi.org/10.1038/521423b.

Fernández-Baldor, Á., Boni, A., Lillo, P., Hueso, A., 2014. Are technological projects reducing social inequalities and improving people's well-being? A capability approach analysis of renewable energy-based electrification projects in Cajamarca, Peru. J. Hum. Dev.

Capabilities 15, 13–27. Available from: https://doi.org/10.1080/19452829.2013.837035.

GNESD, 2013. Country report (Brazil). Energy poverty in developing countries' urban poor communities: assessments and recommendations. In: Urban and Peri-urban Energy Access III. Report Prepared for the Global Network on Energy for Sustainable Development (GNESD) by CENBIO/USP, Centro Clima/COPPE/UFRJ and POLICOM/POLI/UPE. Roskilde, Denmark.

Goldemberg, J., 2006. Energia e Desenvolvimento. Estud. Avançados 12 (33), 7–15. Available from: https://doi.org/10.1590/s0103-40141998000200002.

Gómez, M.F., Silveira, S., 2015. The last mile in the Brazilian Amazon - a potential pathway for universal electricity access. Energy Policy 82, 23–37. Available from: https://doi.org/10.1016/j.enpol.2015.02.018.

IBGE. Synopsis of the 2010 demographic census. <https://censo2010.ibge.gov.br/sinopse/index.php?dados = P13& uf = 00>.

Johansson, T.B., Goldemberg, J., 2002. A policy agenda to promote energy for sustainable development. Energy Sustain. Dev. 6 (4), 67–69.

Kooijman-van Dijk, A.L., 2012. The role of energy in creating opportunities for income generation in the Indian Himalayas. Energy Policy 41, 529–536. Available from: https://doi.org/10.1016/j.enpol.2011.11.013.

Ministerio de Desenvolvimento agrario, 2018. Agricultura Familiar e do Desenvolvimento Agrário. <http://www.mda.gov.br/sitemda/secretaria/safseaf/apresenta%C3%A7%C3%A3o> (accessed November 2018).

Schaeffer, R., Szklo, A., Frossard Pereira De Lucena, A., Soria, R., Chavez-Rodriguez, M., 2013. The vulnerable Amazon: the impact of climate change on the untapped potential of hydropower systems. IEEE Power Energy Mag. 11, 22–31. Available from: https://doi.org/10.1109/MPE.2013.2245584.

Seplancti (Secretaria do Estado do Amazonas), 2017. <http://www.seplancti.am.gov.br/wp-content/uploads/2017/03/bc_de_janeiro_2017.pdf> (accessed October 2017).

Todaro, M., Smith, S.C., 2011. Economic Development, 11th ed. Addison-Wesley. Pearson Education Limited.

Trace, S., 2016. Measuring access for different needs. In: Guruswamy, L. (Ed.), International Energy and Poverty. Routledge Studies in Energy Poverty, p. 178.

UNDP, 2005. Smoke in the kitchen: health impacts of indoor air pollution in developing countries. In: Seminar Proceedings. <http://www.undp.org/content/dam/aplaws/publication/en/publications/environment-energy/www-ee-library/sustainable-energy/smoke-in-the-kitchen-health-impacts-of-indoor-air-pollution/Smoke%20in%20Kitchen_2005.pdf> (accessed June 2017).

UNDP, IPEA, FJP, 2013. Human Development Report in Brazil. Brasilia. ISBN: 978-85-7811-171-7. Available from: <http://www.atlasbrasil.org.br/2013/data/rawData/publicacao_atlas_municipal_pt.pdf> (accessed November 2018).

UNIDO, 2010. Energy for a Sustainable Future 2010, Summary Report and Recommendations. New York. <https://www.unido.org/sites/default/files/2010-05/AGECCsummaryreport_0.pdf> (accessed November 2017).

Van Dijk, A., 2012. The role of energy in creating opportunities for income generation in the Indian Himalayas. Energy Policy 41, 529–536. Available from: https://doi.org/10.1016/j.enpol.2011.11.013.

An inclusive and participative model for energy transition in Latin America: the case of Chilean *Generación Comunitaria*

Francisco Merino[1]*, Adolfo Mejía Montero*[2] *and Cecilia Dastres*[3]

[1]Coordinator of Generación Comunitaria Partnership Framework, Participation and Community Engagement Division, Ministry of Energy, Chile [2]University of Edinburgh, Edinburgh, United Kingdom [3]Head of Content, Methodology and Partnerships Unit, Participation and Community Engagement Division, Ministry of Energy, Chile

19.1 Introduction

Since the late 1980s the natural resources sector has assumed greater importance for the global economic and social development, and Latin America is not an exception. Nevertheless, the extractive and energy sectors have impacted resource-rich countries in Latin America, both positively and negatively, in economic and social ways. Positively, they have generated higher revenues for GDP and increased employability. Negatively, they have resulted in high risks for communities located in the areas affected by these specific projects. The last is particularly acute in developing countries, such as Peru,[1] Mexico,[2]

[1] Several social-environmental mappings have illustrated conflicts in Peru, Mexico, and Chile about energy conflicts. For example, in Peru the Ombudsman Office demonstrated that at least 35 social-environmental conflicts have relation with energy projects. Defensoría del Pueblo, Adjuntía para la Prevención de Conflictos Sociales y la Gobernabilidad, 2016. Reporte Mensual de Conflictos Sociales No. 154. Available from: <https://sinia.minam.gob.pe/documentos/reporte-mensual-conflictos-sociales-ndeg-154-diciembre-2016> (accessed 01.07.19.).

[2] In the same way, in Mexico as reported by the Instituto de Investigaciones en Ecosistemas y Sustentabilidad de la Universidad Nacional Autónoma de México around 35 social-environmental conflicts have relation with energy projects. Observatorio Socioambiental de la Unión de Científicos Comprometidos con la Sociedad y del Centro de Ciencias de la Complejidad de la Universidad Nacional Autónoma de México (OSA-UNAM), 2019. Available from: <http://osa.fisica.unam.mx/> (accessed 01.07.19.).

331

and Chile.[3] Therefore these countries should consider the financial and regulatory frameworks to address social conflicts through investment programs and mechanisms in the areas of influence. Furthermore, some communities live closely with energy projects without realizing the concrete benefits coming from them.

Socially, large projects may directly affect the area of influence through the requirement of resettlement, alienate communities from their land, or affect sociocultural groups, the traditions of which are not addressed by the government or the project. High social risks associated with developing projects in these countries may also affect the investment, since interruptions can lead to project downtime, legal injunctions can lead to expensive delays and larger compensation packages resulting in cost overruns. Such interruptions may influence a project's ability to repay debt in a timely manner.[4]

Energy projects are exposed during all processes to social conflicts, fact of which leads to the adoption of remedial steps to both protect their own investments and ensure that their business is not impacting negatively on those affected by violent conflict or the wider objectives of their own institutions. For this reason the due diligence should be focused to study the mitigation of the social risk under longer time frames and complex legal arrangements.[5]

Social risks have a huge influence on large-scale natural resource projects, since they can incur a significant cost for companies, communities, and government.[6] The social risk is sensitive, since it does not just require a revenue stream or a product but rather a whole way of being and coexisting. The area of influence may be affected for several reasons, for example, employment, tradition, environmental, cultural, indigenous and nonindigenous tradition and health.

Hence, local communities have demanded flexible mechanisms or instruments for developing partnership inside the area of influence, according to the social life of each local community. Following the same line, mechanisms used by companies for the project, in particular, the social responsibility or sustainability department in both private and public sector, cannot embed methodologies of participation or social development focusing solely on market necessities. As an example, methodologies considering variables, such as "modern," "educated," or "responsible", could result in developing issues rather than becoming part of the solution. A partnership framework should always be thought over fundamental topics, such as governance, human rights, and local development.

The energy sector in Chile has experienced great transformations in the last years. For instance, the energy matrix has been evolving according to existent energy demand and infrastructure installed over the last 40 years. Such transformation implies a transition from a fossil fuel—based energy system to one based on an enormous potential for Non-Conventional Renewable Energy, supported by a novel set of

[3] In contrast, in Chile the Instituto Nacional de Derechos Humanos stated that 24 social-environmental conflicts have relation with energy projects. Instituto Nacional de Derecho Humano (INDH), 2019. Available from: <https://mapaconflictos.indh.cl/#/> (accessed 01.07.19.).

[4] Crossin, C., Banfield, J., 2006. Conflict and Project Finance: Exploring Options for Better Management of Conflict Risk. International Alert.

[5] Crossin, C., Banfield, J. Conflict and Project Finance.

[6] Davis, R., Franks, D., 2011. The Cost of Conflict With Local Communities in the Extractive Industry. SRMining 2011, Santiago, de Chile.

policy frameworks. Today, the Chilean challenge is to achieve an engagement level among stakeholders so that a social, technical, and political approval is obtained, following the Sustainable Development Goals (SDG), Organization for Economic Cooperation and Development (OECD) Guidelines, International Finance Corporate (IFC) Standards, Inter-American Development Bank Guideline, and Business and Human Rights principles.

The proposal of the Chilean Ministry of Energy is to implement a partnership framework between private energy companies or nongovernmental organizations (NGOs) and indigenous and nonindigenous communities to mitigate risks in local communities. Such framework aims to improve the social and environmental performance in the natural resource industry, through participation and dialogue between local communities and industry. This topic is addressed by the ministry in several policies and documents.[7]

The objective is to achieve a consensus mechanism, work from the core of communities, to mitigate local community risks. However, the discussion now lays over which institution must lead the participation and dialogue in a partnership framework. In some cases, community engagement and planning is promoted by the public sector toward the private sector like, for example, in the Scottish and the Chilean Government. In contrast, some authors highlight that the same industry should encourage this participation to create trust inside the core of relations.[8]

Both the private sector and local communities may benefit if the principles are adopted and complied without resulting in negative impacts

in the area of influence. As a consequence, several international institutions that foster social risk standards have been supported by a worldwide requirement for national and international investment using due diligence. Therefore such assessment and analysis will more likely become a way to detect the area of influence and to determine if the private sector is in compliance with their burden and actions.

The partnership framework has shown that energy projects can become key instruments to promote both energy transition and a relationship between all the actors involved in the development of projects. Such frameworks encourage the company—community—government linkage, facilitating initiatives to foster community local development. Hence, it is role of the Ministry of Energy to promote appropriate orientations and incentives to foster those indigenous and nonindigenous communities. In this regard, indigenous and nonindigenous communities will be included as active contributors for renewable energy initiatives aligned with the principles of energy justice.

Energy justice aims to serve as an interdisciplinary analytic tool to engage in energy systems as socio-technic constructions (Jenkins et al., 2018), beyond a traditional techno-economic rationale. The scholarship has experienced an accelerating development in the last years, since it was first mentioned in academic literature in 2010 (Guruswamy, 2010).

Over the last years the engagement of energy justice literature with empiric research based in Latin America remains limited to papers exploring energy democracy in southern Mexico[9] and energy affordability in

---

[7] Compromiso de Diálogo: Guía de Estándares de Participación para el Desarrollo de Proyectos de Energía, Capítulo Indígena de Energía, Política de Desarrollo Local Sostenible y Asociativo, Ruta Energética 2018—22, and Política Energética Nacional 2050.

[8] Bebbington, A.J., 2014. Socio-environmental conflict: an opportunity for mining companies. J. Cleaner Prod. 84, 34.

[9] Baker, S.H., 2016. Mexican energy reform, climate change, and energy justice in indigenous communities. Nat. Resour. J. 56.

Chile.[10] In order to help contributing to fulfill this concept gap, this chapter makes use of energy justice's vast conceptual toolkit to engage with the Chilean community partnership framework *Generación Comunitaria*. Such exercise aims a dual objective by contributing toward widening a necessary contribution of academia to policy making in Latin America while answering to the call for nonwestern study cases in energy justice growing literature.

Under this context, this chapter is divided into four main subsections focusing in the *Generación Comunitaria* partnership framework. Section 19.2 provides the reader with a contextual understanding of the international and national landscape from which the development of *Generación Comunitaria* is understood and justified. Section 19.3 elaborates on the rationale, the development process, and the aims of *Generación Comunitaria*. It also provides some information regarding the current state of the program along with the main challenges to overcome regarding supporting communities in local governance, economic and technical capabilities, and access to finance schemes. Section 19.4 consists on a brief assessment of *Generación Comunitaria* through the energy justice applied principles framework. Such analysis reflects upon how *Generación Comunitaria* follows principles of availability, affordability, due process, transparency and accountability, sustainability, responsibility, and intragenerational equity. Finally, Section 19.5 draws some conclusions from the chapter, its significance to the Latin American region and its potential future in Chile.

19.2 Context

Conceptually, a community can be defined as a group of people living in the same place, usually sharing a common cultural or ethnic identity. Hence, a place where a particular body of people lives.[11]

In 1960 the UN recognized the importance of communities into improving the standard of living of lower income groups living in congested urban areas through their mechanisms. Furthermore, the UN acknowledged the significant role of governments in planning, financing, and executing programs for communities' facilities.[12] Such ideas made certain for different notions of development to be bounded to the concept of community over this period. The last is reflected by Zachariah and Sooryamoorthy (as cited by Mathbor), who highlights that the notion of development to promote sustainable economic growth should consider three objectives, through UN resolution, regarding communities: fair, equitable distribution and the creation of opportunities for everyone.[13]

Under this context the 1992 Earth Summit has been an agent of social change influenced by the Brundtland Commission in 1983. Over these last 25 years the community concept has been instrumental to develop social policy around the world based in two factors: government support/NGOs and companies/media. First, the Earth Summit in Rio de Janeiro was the most important conference in the last 50 years with regard to the environment and social development. The purpose of the meeting was to discuss the next generation of policy related to global environment and the

---

[10] Alvial-Palavicino, C., Ureta, S., 2017. Economizing justice: turning equity claims into lower energy tariffs in Chile. Energy Policy 105, 642–647.

[11] The Oxford English Dictionary, 2013. OED Online. In: OU PRESS (Ed.), Oxford English Dictionary.

[12] UNGA, 1960. Low-cost housing and related community facilities. In: A/RES/1508. Available from: <https://www.refworld.org/docid/3b00f06748.html> (accessed 01.07.19.).

[13] Mathbor, G.M., 2008. Effective Community Participation in Coastal Development. Monmouth University.

relationship between economics, science, and the environment.[14] During the summit, every member signed a convention agreeing to new and additional financial resources in order to meet the goals of the convention: the promotion of technology transfer to developing countries; and an institutional mechanism to enable the international community to manage the common threat of climate change over the long term, working with the Intergovernmental Panel on Climate Change (UNGA, 1992).

According to Kemp and Owen,[15] since the beginning of the 1990s, the natural resources sector has gradually focused their efforts to develop new tools, such as community relations and sustainable development, inside their main instruments.

A high level of social risk may be associated with a conflict situation, which involves a complex relationship of breakdown and violence between the companies and local communities. Hence, the social risk may address a cost known as the cost of productivity, which can delay the design, construction, assembly, and start-up of the project. The greatest cost is the opportunity cost that arises from the inability to pursue future projects and opportunities related to damage in community relation within the expansion or sale of the company.[16]

Hence, the partnership framework of *Generación Comunitaria* between the private sector, public sector, and communities seeks for stakeholders to integrate their own partnership mechanism to evaluate compliance standards, ensuring viability and avoiding expensive costs to their initiatives.

These community relations and sustainable development mechanisms should be ingrained into business's core competence.[17]

On one hand the Ministry of Energy proposes a partnership framework based on stakeholders mapping methodologies, evaluated at each stage independently, aiming to reduce communities expectations and avoid the information asymmetry. In fact, stakeholder mapping will be useful as a guide to work at the first stage of the initiatives, by providing an alternative approach to work with local communities. This can result in achieving a positive impact through the distribution of employment opportunities, which translates in socioeconomic security for local population, relationships with political actors, improvement of environmental impact assessment through participatory planning and inclusion of local knowledge and social investments like, for example, the ones represented by *Generación Comunitaria*.

On the other hand, international standards in the energy sector may be the milestone for a new policy and discussion in relation to democracy, development, and sustainability.[18,19]

Social and environmental issues have started to be managed by the use of guidelines,

---

[14]   Meakin, S., 1992. The Rio Earth Summit: summary of the united nations conference on environment and development. In: BP-317E. Brazil.

[15]   Kemp, D., Owen, J.R., 2013. Community relations and mining: core to business but not "core business". Resour. Policy 38, 523–531.

[16]   Davis, R., Franks, D., 2011. The Cost of Conflict With Local Communities in the Extractive Industry. SRMining 2011. Santiago, de Chile.

[17]   Crossin, C., Banfield, J., 2006. Conflict and Project Finance: Exploring Options for Better Management of Conflict Risk. International Alert (Kemp and Owen, Community relations and mining).

[18]   Bebbington, A.J., 2014. Socio-environmental conflict: an opportunity for mining companies. J. Cleaner Prod. 84, 34.

[19]   Bebbington, A.J., Franks, D.M., Davis, R., Ali, S.H., Kemp, D., Scurrah, M., 2014. Conflict translates environmental and social risk into business costs. Proc. Natl. Acad. Sci. USA 111, 7576–7581.

which aimed to create a mix between business and local communities or host societies. One of these standards was the environmental impact assessment. In the first stage, these standards sought to fill some existing gaps in these relations by enabling an accurate understanding of the potential social conflicts. From the perspective of the stakeholders, capacity is taken here as the capacity to understand the prospective costs and the consequences of these risks. Furthermore, stakeholders fostered the ability to appreciate the spectrum of influence implicit in a project for such conflict, directly, indirectly, and to varying degrees.

During the last two decades the international NGOs and the private sector have been working to foster a debate regarding business and local communities. The international forum has been focused on sharing experiences and solutions to deal with both social and environmental issues through the adoption of a multistakeholder process, for instance, the World Bank, IFC Performance Standards, the UN Guiding Principle on Business and Human Right, ISO 26000, OECD Guideline for Multinational Enterprises, the Principles for Responsible Investment, UNEPFI Statement of Commitment, and the Equator Principles. However, the financial sector has not actively participated in the process.[20]

For example, the World Bank developed new standards for large infrastructure projects after facing a number of public outcries from local communities in the area of influence. Therefore it fostered several sets of sustainability policies, which must be complied with before funding was approved. The World Bank standards were followed by several international financial institutions, including the Asian Development Bank, the Inter-American Development Bank, and the African Development Bank. However, the World Bank started to create and use its own standard and policies.

Finally, at the national level, the government approach should be to foster the relationship between the stakeholder and community in a large infrastructure project through a legal framework, institution, and agencies. These will help to address and reduce environmental and social disputes. Under this context, governments must assume leadership to provide and build the necessary capacity to assess and monitor jointed work with communities (Udall and Udall, 2015). Thus a public policy should incorporate legal remediation strategies, monitoring or enforcement standards so that the energy project and stakeholders may avoid negative impacts on the national interest of the society, for example, local communities or taxpayers. The lack of experience or expertise to support each reality requires the government to be an intermediator to lease between all stakeholders, aiming to build and create standards for on-ground implementation.

19.3 *Generación Comunitaria* as a partnership framework

The Chilean Ministry of Energy promoted a series of partnership mechanisms through several policies and documents[21] to address social conflicts and promote the Guiding Principles on Business and Human Rights, SDG, and the Principles of ILO Convention No. 169. These mechanisms state that all projects developed in the country should foster a partnership framework that contributes to local development and improves social and

[20] Crossin, C., Banfield, J., 2006. Conflict and project finance: exploring options for better management of conflict risk. International Alert.

[21] *Compromiso de Diálogo: Guía de Estándares de Participación para el Desarrollo de Proyectos de Energía, Capítulo Indígena de Energía, Política de Desarrollo Local Sostenible y Asociativo, Ruta Energética 2018-2022 y la Política Energética Nacional 2050.*

environmental performance of the projects. Therefore it ensures that energy development favors the paradigm of local development defined by the communities, in a manner consistent with the national and regional strategy. Such view promotes the implementation of energy projects designed by small producers and communities interested in harnessing native energy resources available in their territory.

Several indigenous and nonindigenous communities have been demanding better ways of addressing their local development as well as their environmental and cultural commitment. Under this line, clean energy has been a way to achieve both goals. Currently, at least 10 indigenous and nonindigenous communities have been promoting their energy initiatives, either through the management of their own resources or through a partnership framework with a technical and economic partner, such as private energy companies or even NGOs. Both ways are viable and contribute to their goals.

This mechanism was driven by the request of communities themselves, as a demand to make use of their territories to promote local development. This was taken up by several open workshops throughout the country involving hearing of different actors, from public sector and civil society, including academia, NGOs, environmental groups, neighborhood associations, unions, companies, and representatives of indigenous communities and peoples. The guideline principles of these workshops were participation and dialogue. Under this context, participatory processes considered an international consensus around the role that companies have in the defense of human rights and in the promotion of sustainable development. Nowadays, partnership frameworks became central actors of development because of their contribution to growth, employment, and innovation and to provide access to goods and services that improve the quality of life. Given the growing influence of business activity, the principle of company's responsibility to respect human rights and to subject their conduct to due diligence is based on the international scenario.

The stakeholders discussed under the conviction that indigenous and nonindigenous communities are fundamental agents for achieving their own sustainable development. Such aim must be always developed under a multidimensional approach to fulfill the citizens' energy needs and concerns based on international standards.

This processes incorporated the human rights approach in the actions carried out by the Ministry of Energy, implying change within the scope of orientation, not only in the actions but also in the criteria considered to address each scenario. Under this way the Ministry of Energy has promoted an institutionalized and regulated process, ensuring that all relevant actors, organizations, and communities are informed, trained, and strengthened in energy matters of interest, as well as to strengthen their capacities to generate opportunities for local energy development.

On a second phase the proposals emanating over the workshops were analyzed and compared with national and international experiences. The most important comparative experiences and researches were found in the United Kingdom (UK Government. 2014), Scotland (Scottish Government, 2015b), British Columbia (Canada),[22] Australia,[23]

---

[22] British Columbia Government, 2016. First Nations Clean Energy Business Fund. Available from: <https://www2.gov.bc.ca/assets/gov/environment/natural-resource-stewardship/consulting-with-first-nations/agreements/fncebf_-_guidelines_-_2016-09-01.pdf> (accessed 01.07.19.).

[23] Australian Government, 2013. National Community Energy Strategy. Available from: <https://arena.gov.au/assets/2017/02/ERP_Guidelines.pdf> (accessed 01.07.19.).

New Zealand,[24] the United States (US Government, 2013), Denmark,[25] and others. The first practical experiences were developed around 20 years ago. However, most of these policies, strategies, and statements related to community energy were elaborated in the last 8 years.

The *Generación Comunitaria* is part of the short-, medium-, and long-term policies of the Ministry of Energy, such as National Energy Policy 2050, Indigenous Energy Chapter, and Energy Route 2018–22. The objective of the *Generación Comunitaria* is to promote participation in the ownership of energy projects by indigenous and nonindigenous organizations, prior contribution of these, in the association with energy companies and NGOs (Ministerio de Energía, 2018), in order to promote clean energies, along with local community development and promoting agents of climate change.

Currently, *Generación Comunitaria* initiatives are in the pilot stage with some indigenous and nonindigenous communities in several locations in Chile, such as Antofagasta, Bío-Bío, Araucanía, Los Ríos, and Los Lagos. These initiatives will inject a total of nearly 150 MW into the electricity system (with a total private investment of 350 million USD).

However, the introduction into the financial ecosystem of the importance of associativity for the financing of energy infrastructures promoted by communities has not yet been managed. As a result, the initiative still does not have the financial means from national and international banks and investors, nor have the Chilean Government scaled the number of initiatives. Therefore it is important to improve managerial capabilities for *Generación Comunitaria*, in order to solve financial access to energy infrastructure and to promote the local development of each community. In contrast, indigenous and nonindigenous communities depend heavily on subsidies granted by the state for their local development, without reaching economic autonomy for decision-making. Therefore an improvement in financial access will allow for scaling up of new productive initiatives and strengthening of governance capabilities. This will allow communities to generate their own resources, access financing, and even leverage it with state resources, managing to reduce public spending or redirect it to other needs. In conclusion the communities will strengthen their model of governance and local development.

The *Generación Comunitaria* policy seeks to

1. support social organization to achieve the inclusion of indigenous and nonindigenous communities in energy initiatives;
2. provide technical and economic support for the realization of projects, promoting public management innovation by the communities and reducing dependence on subsidies; and
3. support access to financing for community energy infrastructure, fostering a competitive and fair society based on local development.

19.3.1 Local governances support

The Ministry of Energy will support communities in the formation and strengthening of local governances to generate understanding and agreement, regarding the development of energy projects. Therefore the willingness of the communities will be supported by education and training provided by the academy, guilds, law firms, NGOs and consultants, and

---

[24]   New Zealand Government, 2011. National Policy Statement for Renewable Electricity Generation. Available from: <https://www.mfe.govt.nz/sites/default/files/nps-reg-2011.pdf> (accessed 01.07.19.).

[25]   Danish Government, 2015. The Danish Energy Model. Available from: <https://ens.dk/sites/ens.dk/files/Globalcooperation/the_danish_energy_model.pdf>.

the Ministry of Energy This will be a solid construction of an organizational model that will give social validation to the different decisions that will have to be made in the next years.

19.3.2 Technical and economic support

The Ministry of Energy must support the technical and economic viability of the potential renewable energy projects to be developed by interested communities. This type of projects may arise from different processes or actors, such as social organizations, indigenous communities, energy companies, and NGOs. The ministry will provide skills to the communities to empower them and offer access to technical and economic information to make decisions regarding the development of energy projects by their internal units and external support from private companies, as well as consultants, NGOs, and universities by means of voluntary agreements.

In addition, the initiatives will be supported by other public services through the energy policy denominated + Energía. This gives free and open access information and knowledge about how to develop energy projects.

19.3.3 Financial access support

The Ministry of Energy has been working to promote that national and international banking and investors fund these initiatives, supported by one of the four energy guilds in Chile, universities, NGOs, and cooperative associations (such as savings and credit cooperatives).

In the short term the Ministry of Energy will seek to coordinate and articulate instances of private, national, and international financing. This will be done so that communities can carry out all necessary technical studies related to energy potential, technical and economic design of the projects, as well as their construction. In the medium and long term, it will promote conditions to investments in *Generación Comunitaria* and access to project development both in preinvestment and construction phases.

19.4 An energy justice perspective

Energy justice has developed a series of conceptual frameworks to engage with some of the most important moral questions embedded in energy systems.[26] These conceptual frameworks include existing tenets from environmental justice literature,[27] applied principles to use as guidelines for decision-making (Sovacool and Dworkin, 2015) and whole systems frameworks involving energy life cycle stages.[28]

19.4.1 Availability

Availability has been described in energy justice literature as the ability of the population to access "sufficient energy resources of high quality (suitable to meet their end uses)" (Sovacool et al., 2017). The idea of "sufficient energy resources" is normally linked to the generation and distribution infrastructure used to provide energy services for the final

[26] Sovacool, B.K., Dworkin, M.H., 2014. Introduction. In: Global Energy Justice: Problems, Principles, and Practices. pp. 1–391.

[27] McCauley, D., Heffron, R.J., Stephan, H., Jenkins, K., 2013. Advancing energy justice: the triumvirate of tenets. In: International Energy Law Review.

[28] Heffron, R.J., McCauley, D., 2014. Achieving sustainable supply chains through energy justice. Appl. Energy 123, 435–437.

consumer. In the case of electric power and rural and indigenous communities in developing countries, most of the time this means first to bring electricity to general population. In contexts like, for example, the ones existing in many African or South East Asian countries, grid access is still a major challenge for national governments (IEA, 2017). However, Latin American countries such as Chile provide access to electricity to around 99.8% of the population (Ministerio de Energía de Chile, 2017), representing an interesting case on how to frame the principle of availability.

Availability of basic electric services for isolated communities (which in its majority are indigenous communities) is still an important axis for the energy strategy in the country. Even if the country has provided general access to electricity, 0.2% of the Chilean population remains off-grid. Under this context, one of the aims of the *Generacion Comunitaria* mechanism is to improve energy access through facilitating the implementation of distributed renewable energy technologies in off-grid communities (Ministerio de Energía, 2018).

On the other hand the principle of availability also "includes the amount of investment needed to keep the system functioning" (Sovacool and Dworkin, 2015). Under Chilean policies aiming to include rural and indigenous communities as active actors in the energy systems, this aim is being pursued at two different levels:

- At a "specific energy project level" by making available for communities the necessary financial mechanisms, capabilities, and skills become energy project developers. The availability of these mechanisms also seeks to give communities the opportunity to partner with a private developer, which can study, develop, and implement energy generation projects within the community territory (Ministerio de Energía de Chile, 2017). The energy

produced by the community could then be sold to the network as a regular energy company.
- At a wider federal level the Chilean Government's participatory mechanisms provided historically isolated groups with access to become active actors in the decision-making process that shapes the energy reality of their community and country.

Such approach from the Chilean Government exceeds the traditional conception of "availability." Their approach not only aims to provide the technical infrastructure for electricity access in Chile but also aims to recognize communities' discourses and visions in energy policy-making processes.

19.4.2 Affordability

The principle of affordability has been traditionally linked to the concept of energy poverty, object of a long academic debate in recent years. For energy justice literature and many scholars studying energy poverty, energy poverty means that "all people, including the poor, should pay no more than 10% of their income for energy services" (Sovacool et al., 2017). Such definition of affordability limits the role of society to passive consumer, by looking at low tariffs as the unique way in which realization of justice regarding affordability can be achieved. However, this chapter intends to draw the attention to communities assuming an active role as potential energy producers. Such approach calls for reframing affordability out from an energy poverty/consumer logic to a point of view where society is an energy producer. Under this approach the lack of support schemes and financial mechanisms comprise a barrier to make affordable for communities to be included and assume their active role in the renewable energy transition.

The development of these strategies and mechanisms to include civil society is very relevant in the Latin American context. Most renewable energy projects are being pushed by large private companies participating competitively in tendering processes that "requires significant expertise as well as access to large reserves of capital, which smaller actors do not have to the same degree as large specialist" (REN21, 2017). Under this context, availability is strongly interlinked with affordability, as support and finance schemes are necessary for communities to afford and develop renewable energy generation systems. Such approach to provide financial assistance and mechanisms to incentive the development of community energy projects has been already explored by some countries, such as Scotland (The Scottish Government, 2014).

The *Generación Comunitaria* scheme aims to develop the necessary conditions for making affordable for communities to become actors in the energy reality in Chile. Going back to the traditional meaning of affordability, the inclusion of communities in the development of energy projects could mean more affordable energy tariffs by making energy projects more competitive through the reduction of social risks. Moreover, a deeper realization of recognition and procedural justice within the Chilean energy system could be achieved by incorporating traditionally isolated and poor sectors of civil society to decision-making processes shaping the energy landscape in the country.

19.4.3 Due process

This principle states that "countries should respect due process and human rights in their production and use of energy" (Sovacool et al., 2017) and that "communities must be involved in deciding about projects that will affect them" (Sovacool and Dworkin, 2015). Under

this context the conceptual foundations of the *Generación Comunitaria* based on free, prior, and informed consent aim to follow international standards regarding indigenous people's rights described in the ILO Convention No. 169 (Dunbar-Ortiz et al., 2015).

These conceptual foundations were the base for an intense, long process to effectively engage with different communities in the country. This was made with the objective to build an inclusive energy sector containing the different perspectives and aspirations of social actors, aiming to give communities the right to decide about their own development while contributing to climate change mitigation.

Due process is a key element to lay strong foundations to diminish social risk in energy projects. Such social risk can become a real treat not only to the energy projects, but more importantly due to the social impacts and violent conflicts, which could be generated because of the break of trust between a community and project developers (Huesca-Pérez et al., 2016).

For the case of the Chilean *Generación Comunitaria*, due process could play even a stronger role as it seeks not only to obtain mere "acceptance" of the project but also to get communities to involve actively in the energy project. However, such objective involves the need to overcome barriers related to disparity of skills, knowledge, and cosmovision between community members and energy project specialists (Ministerio de Energía de Chile, 2017).

Under this context, *Generación Comunitaria* aims to incorporate legal remedies, monitoring and enforcement standards, so that the energy project and stakeholders may avoid negatively affecting the national interest of the society. Such mechanisms were created through years of intercultural dialogue with indigenous and nonindigenous community representatives (Ministerio de Energía de Chile, 2017).

An effective due process in the development of energy projects aims to reinforce existing capabilities, knowledge, and skills in local population. By reinforcing such capabilities, communities will be able to engage actively in the development of energy projects. This new set of skills, knowledge, and community cohesion has been also perceived as positive lateral effect of community energy projects (Forman, 2017).

Under this context the methodologies deployed by Chilean Government seeking to include the indigenous and nonindigenous vision into the energy transition could be used as a model for incentivizing due process in Latin American energy transitions.

19.4.4 Transparency and accountability

Universal "access to high quality information about energy and the environment and fair, transparent, and accountable forms of energy decision-making" (Sovacool et al., 2017) is a key element in the *Generación Comunitaria*, as it heavily determines the extent in which the community can participate.

A lack of early information could lead to failure to engage the community in an effective way as people are not fully aware of the positive and negative impacts that energy initiatives could bring. Transparency is important in this context in order to manage expectations around energy initiatives, avoiding a future break of trust due to unexpected negative impacts or misunderstood about the benefits which could lead in the long term to a break of trust and potential conflicts.

Experiences collected by the Chilean Government while engaging with communities around energy projects' potential issues also highlighted the asymmetry of information between communities and project developers as one of the most relevant issues. Such asymmetries represent a challenge, especially when

the project involves indigenous communities, which need information to be delivered in a culturally adequate way. Moreover, it will be a major task for *Generación Comunitaria* to overcome preconceptions that many communities could have regarding incoming energy projects due to negative past experiences (Ministerio de Energía de Chile, 2017). This process will involve clear and transparent information regarding the existing difference between traditional and renewable energy projects, their technical characteristics and potential impacts and benefits, in a language that is accessible and clear for local population.

Under this context an early process to map relevant community stakeholders, which can help to effectively engage the community in a clear and transparent informative process is considered key. This needs to be followed by a process in which community representatives are provided with the necessary skills and knowledge to function as a bridge for the community to reduce the expectations for communities and avoid the information asymmetry and future breaks of trust and conflicts.

19.4.5 Intragenerational equity

One of the main objectives for the Chilean Government *Generación Comunitaria* is to foster communities' local development and break the intragenerational gap through the implementation of energy projects. Under this context, intragenerational equity through the eyes of energy justice is framed as the access to fairly access energy services (Sovacool et al., 2017).

However, the *Generación Comunitaria* scheme also seems to follow the intersectionality principle "acknowledging how the realization of energy justice is linked to other forms of justice, e.g. socio-economic, political and environmental" (Sovacool et al., 2017). *Generación Comunitaria* aims to bring equity not only through providing equal access to energy services but also through the fair distribution of the overall positive and negative impacts of energy projects in the economic, technical, and social spheres among local indigenous and nonindigenous communities.

Under this context, academic literature have highlighted related community schemes as an alternative to restore distributive justice in renewable energy projects (The British Academy, 2016). This is because in most cases traditionally isolated local populations bare the impacts of projects to mitigate climate change for the sake of global population (Cowell et al., 2012). Under this context, unfair distribution of benefits many times cause a clear gap between wider social acceptance for renewable energy projects and social opposition at the local level (Fast, 2013).

If local communities are supported and assisted to fairly internalize the socioeconomic, environmental, and political benefits coming from renewable energy projects, then the social acceptance of these projects and the social risk is more likely to decrease. The internalization of benefits coming from community projects could be translated to community benefits used to foster regional development and productive projects (Oceransky, 2010, p. 670), clean energy for self-supply and strengthening local governance through a stronger sense of local identity and the development of managerial skills as an outcome of the active engagement of local population in the project.

19.4.6 Sustainability and responsibility

Regarding the principle of sustainability and responsibility toward natural resources protection, it is possible to say that *Generación Comunitaria* follows these principles in theory as it comprises a tool for the Chilean renewable energy transition. This means that the mechanism inherently seeks to protect energy and natural resources at the same time it looks to minimize external treats (Sovacool et al., 2017).

However, these principles are achieved not only through the generation of clean energy by community projects. This mechanism also aims to effectively include local knowledge into an energy project planning process through local stakeholders' participation in environmental impact assessment. Such participative planning strategies for energy projects have been the object of a considerable amount of academic work. It is most of the time considered a best practice aiming to avoid impacts to the local environment, by including mitigation and adaptation strategies nourished by local knowledge.

19.5 Conclusion

The Chilean Energy Ministry has incorporated into the short-, medium- and long-term policy, strategy, and statements a series of principles and actions that are promoted by international organizations regarding to stakeholder engagement in the energy sector. Such is the case of the *Generación Comunitaria* mechanism, with which the ministry seeks to reduce barriers and achieve a fair, safe, economic, and sustainable energy, based on consent and willingness of the communities.

Generación Comunitaria is developed as a mechanism to bond the goal of both the indigenous and nonindigenous communities, to develop energy initiatives using their own natural resources of their land; and the goal at a national level for a clean energy transition. Under this context, *Generación Comunitaria* will promote energy justice and the community itself and motivate the communities to be agents of change and achieve local development. Such development will be based in strengthening the sustainable growth of the community, based on the principles of justice, equality, and the creation of opportunities for all.

In conclusion, *Generación Comunitaria's* aim is to offer communities the opportunity to reach an autonomous development through the use of their own energy resources, as long as the initiatives are implemented according to the principles and actions of national and international regulations, policies, standards, and guides.

There is still to be done within the Ministry of Energy in order to achieve these objectives. In the short term the Ministry of Energy should convene stakeholders on the energy and finance sector to coordinate and articulate a national and international instance to promote the financing access for communities to energy infrastructure. In the medium and long term, the ministry must seek to scale the number of initiatives. Therefore financial inclusion is the key to fund small, medium, and large energy infrastructure in the communities, in order to strengthen its role as an agent of change and for its local development. Goals of the Ministry of Energy for the next years should include (from pilot phase to 2021) the access to financial mechanisms to fund the initiatives and strengthen the community governance.

If the Chilean partnership framework *Generación Comunitaria* succeeds, it could become a banner in the Latin American renewable transitions, which would incentivize other Latin American countries to develop a new generation of energy policy in compliance of energy justice guidelines. Such new energy policy frameworks could then contribute toward redressing both historic and socioeconomic inequalities present in all the countries of our region and future global concerns such as climate change.

References

Cowell, R., Bristow, G., Munday, M., 2012. Wind Energy and Justice for Disadvantaged Communities. Joseph Rowntree Foundation.

Dunbar-Ortiz, R., Sambo-Dorough, D., Alfredsson, G., Swepston, L., Wille, P., 2015. Indigenous Peoples' Rights in International Law: Emergence and Application, vol. 1. Gáldu & IWGIA. Available from: https://doi.org/10.1017/CBO9781107415324.004.

Fast, S., 2013. Social acceptance of renewable energy: trends, concepts, and geographies. Geogr. Compass 7 (12), 853–866. Available from: https://doi.org/10.1111/gec3.12086.

Forman, A., 2017. Energy justice at the end of the wire: enacting community energy and equity in Wales. Energy Policy 107, 649–657. Available from: https://doi.org/10.1016/j.enpol.2017.05.006.

Guruswamy, L., 2010. Energy Justice and Sustainable Development. pp. 1–416.

Huesca-Pérez, M.E., Sheinbaum-Pardo, C., Köppel, J., 2016. Social implications of siting wind energy in a disadvantaged region – the case of the Isthmus of Tehuantepec, Mexico. Renew. Sustain. Energy Rev. 58, 952–965. Available from: https://doi.org/10.1016/j.rser.2015.12.310.

IEA, 2017. Energy Access Outlook 2017: From Poverty to Prosperity, vol. 94. OECD—The International Energy Agency. Available from: https://doi.org/10.1787/9789264285569-en.

Jenkins, K., Sovacool, B.K., Mccauley, D., 2018. Humanizing sociotechnical transitions through energy justice: an ethical framework for global transformative change. Energy Policy 117, 66–74. Available from: https://doi.org/10.1016/j.enpol.2018.02.036.

Ministerio de Energía, 2018. Ruta Energética 2018–2022. Retrieved from: <https://www.cne.cl/wp-content/uploads/2018/05/rutaenergetica2018-2022.pdf> (accessed 01.07.19.).

Ministerio de Energía de Chile, 2017. Capitulo Indígena. Política Energética 2050. Ministerio de Energía de Chile, Santiago, de Chile.

Oceransky, S., 2010. Renewable energy as a common resource. Sparking a Worldwide Energy Revolution.

REN21, 2017. Renewable Energy Tenders and Community [Em]power[ment]: Latin America and Caribbean. REN21. Retrieved from: <http://www.ren21.net/wp-content/uploads/2017/09/LAC-Report.pdf>.

Scottish Government, 2015b. Community Energy Policy Statement. Available from: <https://www.webarchive.org.uk/wayback/archive/20170401055145/http://www.gov.scot/Topics/Business-Industry/Energy/CEPS2015>.

Sovacool, B.K., Dworkin, M.H., 2015. Energy justice: conceptual insights and practical applications. Appl. Energy 142, 435–444. Available from: https://doi.org/10.1016/j.apenergy.2015.01.002.

Sovacool, B.K., Burke, M., Baker, L., Kotikalapudi, C.K., Wlokas, H., 2017. New frontiers and conceptual frameworks for energy justice. Energy Policy 105, 677–691. Available from: https://doi.org/10.1016/j.enpol.2017.03.005.

The British Academy, 2016. Cultures of Community Energy International – Case Studies.

The Scottish Government, 2014. Community Renewable Energy Toolkit.

Udall, M.K., Udall, S.L., 2015. Udall Foundation [Online]. Available from: <https://http://www.udall.gov/AboutUs/AboutUs.aspx> (accessed 07.07.15.).

UK Government, 2014. Community Energy Strategy. Available from: <https://assets.publishing.service.gov.uk/government/uploads/system/uploads/attachment_data/file/275163/20140126Community_Energy_Strategy.pdf>.

UNGA, 1992. Report of the United Nations conference on environment and development - Rio declaration on environmental and development. In: A/CONF.151/26, vol. I. UNGA, Brazil.

US Government, 2013. Community Energy Strategic Planning. Available from: <https://www.energy.gov/sites/prod/files/2014/05/f15/cesp_guide.pdf> (accessed 01.07.19.).

Further reading

Scottish Government, 2015a. Community Empowerment (Scotland) Act 2015.

UNGA, 2003. The Human Rights Based Approach to Development Cooperation: Towards a Common Understanding Among UN Agencies.

UNGA, 2015. Transforming Our World: The 2030 Agenda for Sustainable Development. Available from: <https://sustainabledevelopment.un.org/content/documents/21252030%20Agenda%20for%20Sustainable%20Development%20web.pdf>.

Index